S0-AWV-374

Topics in Applied Physics Volume 28

Topics in Applied Physics Founded by Helmut K. V. Lotsch

Vol. 1 **Dye Lasers** 2nd Edition Editor: F. P. Schäfer

Vol. 2 **Laser Spectroscopy** of Atoms and Molecules Editor: H. Walther

Vol. 3 **Numerical and Asymptotic Techniques in Electromagnetics** Editor: R. Mittra

Vol. 4 **Interactions on Metal Surfaces** Editor: R. Gomer

Vol. 5 **Mössbauer Spectroscopy** Editor: U. Gonser

Vol. 6 **Picture Processing and Digital Filtering** Editor: T. S. Huang

Vol. 7 **Integrated Optics** Editor: T. Tamir

Vol. 8 **Light Scattering in Solids** Editor: M. Cardona

Vol. 9 **Laser Speckle** and Related Phenomena Editor: J. C. Dainty

Vol. 10 **Transient Electromagnetic Fields** Editor: L. B. Felsen

Vol. 11 **Digital Picture Analysis** Editor: A. Rosenfeld

Vol. 12 **Turbulence** 2nd Edition Editor: P. Bradshaw

Vol. 13 **High-Resolution Laser Spectroscopy** Editor: K. Shimoda

Vol. 14 **Laser Monitoring of the Atmosphere** Editor: E. D. Hinkley

Vol. 15 **Radiationless Processes** in Molecules and Condensed Phases Editor: F. K. Fong

Vol. 16 **Nonlinear Infrared Generation** Editor: Y.-R. Shen

Vol. 17 **Electroluminescence** Editor: J. I. Pankove

Vol. 18 **Ultrashort Light Pulses.** Picosecond Techniques and Applications
Editor: S. L. Shapiro

Vol. 19 **Optical and Infrared Detectors** Editor: R. J. Keyes

Vol. 20 **Holographic Recording Materials** Editor: H. M. Smith

Vol. 21 **Solid Electrolytes** Editor: S. Geller

Vol. 22 **X-Ray Optics.** Applications to Solids Editor: H.-J. Queisser

Vol. 23 **Optical Data Processing.** Applications Editor: D. Casasent

Vol. 24 **Acoustic Surface Waves** Editor: A. A. Oliner

Vol. 25 **Laser Beam Propagation in the Atmosphere** Editor: J. W. Strohbehn

Vol. 26 **Photoemission in Solids I.** General Principles Editors: M. Cardona and L. Ley

Vol. 27 **Photoemission in Solids II.** Case Studies Editors: L. Ley and M. Cardona

Vol. 28 **Hydrogen in Metals I.** Basic Properties Editors: G. Alefeld and J. Völkl

Vol. 29 **Hydrogen in Metals II.** Application-Oriented Properties
Editors: G. Alefeld and J. Völkl

Vol. 30 **Excimer Lasers** Editor: C. K. Rhodes

Hydrogen in Metals I

Basic Properties

Edited by G. Alefeld and J. Völkl

With Contributions by
G. Alefeld R. M. Cotts K. W. Kehr H. Kronmüller
H. Peisl A. Seeger K. Sköld T. Springer
A. C. Switendick J. Völkl F. E. Wagner H. Wagner
W. E. Wallace G. Wortmann

With 178 Figures

Springer-Verlag Berlin Heidelberg New York 1978

Professor Dr. *Georg Alefeld*
Dr. *Johann Völkl*
Physik-Department der Technischen Universität München,
D-8046 Garching, Fed. Rep. of Germany

ISBN 3-540-08705-2 Springer-Verlag Berlin Heidelberg New York
ISBN 0-387-08705-2 Springer-Verlag New York Heidelberg Berlin

Library of Congress Cataloging in Publication Data. Main entry under title: Hydrogen in metals. (Topics in applied physics; v. 28–29). Includes bibliographical references and indexes. Contents: v. 1. Basic properties. v. 2. Application-oriented properties. 1. Metals-Hydrogen content. I. Alefeld, G., 1933—, II. Völkl, J., 1936—, TH690.H97 669'.94 78-4487.

Monophoto typesetting, offset printing and bookbinding: Brühlsche Universitätsdruckerei, Lahn-Gießen
2153/3130-543210

Preface

Progress in solid-state sciences results from many effects: the preparation of new materials, the development of new experimental methods, or increase in technological interest. All three aspects are valid for research on hydrogen in metals.

Although these systems are not completely new, advanced preparation techniques yielding well-defined samples not only made the results on diffusion, solubility, and phase transitions for example, more and more reproducible, but also stimulated the application of new methods like neutron scattering and Mössbauer effect. The technological interests range from old problems like hydrogen embrittlement to applications of metal-hydrogen systems in fission and fusion reactors, fuel cells, as energy storage systems, etc. The concept for future energy transport and supply known as "hydrogen economy" also caused appreciable increase in research on interaction of hydrogen with metals.

Due to this broad interest, this group of materials presently is studied by many scientists affiliated with various disciplines such as physics, chemistry, physical chemistry, metallurgy and engineering. Therefore, research results are scattered throughout many different journals or conference reports.

The books *Hydrogen in Metals I. Basic Properties* and *Hydrogen in Metals II. Application–Oriented Properties* contain reviews written by experts in the field. Those topics which have experienced the greatest progress over the recent years have been selected. Areas for which comprehensive review papers exist have been omitted. Although the editors have coordinated arrangement and content of the various contributions, they have purposely not eliminated all diverging points of view which necessarily exist in a rapidly expanding area of research.

The great number of subjects made it necessary to divide the contributions into two volumes, the first one devoted more to basic properties, the second one more to application-oriented properties. It is evident that ambiguities in this subdivision were unavoidable.

The editors are grateful to the authors for being very cooperative, for fast response to changes, and for being on time. The editors are also grateful to the publishers and their staff, and Dr. H. Lotsch for encouragement.

Munich, March 1978

G. Alefeld · J. Völkl

Contents

Contributors

Alefeld, Georg
Physik-Department der Technischen Universität München,
D-8046 Garching, Fed. Rep. of Germany

Cotts, Robert M.
Laboratory of Atomic and Solid State Physics, Cornell University,
Clark Hall, Ithaca, NY 14853, USA

Kehr, Klaus W.
Institut für Festkörperforschung, Kernforschungsanlage Jülich,
D-5170 Jülich 1, Fed. Rep. of Germany

Kronmüller, Helmut
Max-Planck-Institut für Metallforschung, Institut für Physik,
D-7000 Stuttgart 80, Fed. Rep. of Germany

Peisl, Hans
Sektion Physik der Ludwig-Maximilians-Universität München,
D-8000 München 22, Fed. Rep. of Germany

Seeger, Alfred
Max-Planck-Institut für Metallforschung, Institut für Physik,
D-7000 Stuttgart 80, Fed. Rep. of Germany

Sköld, Kurt
Solid State Science Division, Argonne National Laboratory,
Argonne, ILL 60439, USA and
AB Atomenergi, Fack, S-61101 Nyköping, Sweden

Springer, Tasso
Institut Laue-Langevin, Centre de Tri, F-38042 Grenoble, Cedex, France, and
Institut für Festkörperforschung, Kernforschungsanlage Jülich,
D-5170 Jülich 1, Fed. Rep. of Germany

Switendick, Alfred C.
Division 5151, Solid State Physics, Sandia Laboratories,
Albuquerque, NM 87185, USA

Völkl, Johann
Physik-Department der Technischen Universität München,
D-8046 Garching, Fed. Rep. of Germany

Wagner, Friedrich E.
Physik-Department der Technischen Universität München,
D-8046 Garching, Fed. Rep. of Germany

Wagner, Herbert
Sektion Physik der Ludwig-Maximilians-Universität München,
D-8000 München 22, Fed. Rep. of Germany

Wallace, William E.
University of Pittsburgh, Department of Chemistry, Chemistry Building,
Pittsburgh, PA 15260, USA

Wortmann, Gerhard
Physik-Department der Technischen Universität München,
D-8046 Garching, Fed. Rep. of Germany

1. Introduction

G. Alefeld and J. Völkl

In 1866, *Thomas Graham*, one of the founders of modern chemistry in Great Britain, discovered the ability of the metal palladium (Pd) to absorb large amounts of hydrogen [1.1]. *Graham* also observed that hydrogen can permeate Pd-membranes at an appreciable rate, and thus Pd-membranes can be used either to extract H from a gas stream or to purify hydrogen [1.1, 2]. Further basic experiments on the changes of physical properties of Pd due to absorption of H, such as lattice expansion, electrical resistivity, or magnetic susceptibility [1.3], as well date back to *Graham* who also introduced the concept of metal-hydrogen alloys without definite stoichiometry, in contrast to stoichiometric nonmetallic hydrides [1.2]. A brief review of *Graham*'s life and work was given by *Flanagan* [1.4].

Since *Graham*, many outstanding chemists, physicists, physical chemists, metallurgists, etc., have devoted much of their scientific work to the metal-hydrogen systems, and presently an abundance of meetings on this subject can be noted. What is so special about these systems and what are the areas of interest?

Research in hydrogen in metals has attracted attention for reasons motivated both from a basic as well as an applied point of view. It is evident that there is an overlap between both approaches. The possible application of metal-hydrogen systems will be discussed in [1.5]. In this volume we shall, therefore, confine ourselves to topics of fundamental nature, e.g., the basic interactions leading to phase transitions of H in metals, the change in electronic properties, and the hydrogen mobility. Metal-hydrogen systems, being interesting in themselves, turned out to be model substances or prototypes for certain physical properties:

Nonstoichiometric compounds have attracted great interest over the last decade. The hydrogen-palladium system was the first nonstoichiometric system for which equilibrium measurements existed. *Sieverts* [1.6] found that the concentration in solution is proportional to $p^{1/2}$ (p = hydrogen pressure) at small concentrations, an observation which can be considered as direct evidence for the dissociation of the H molecule in solution. *Lacher*'s theory [1.7] of the nonideal solubility isotherms for H in Pd was the first successful statistical mechanical treatment of a nonstoichiometric compound. *Lacher*'s work on H–Pd anticipates the theoretical concept of a three-dimensional lattice gas [1.8] which plays an important role in the understanding of phase transitions of second order.

The discussion of *Mott* and *Jones* [1.9] of the atomistic processes occurring in the solution of H in Pd was the first application of band theory of metals to a chemical process. Hydrogen as a quasimetal, a concept introduced by *Graham*, is still a fruitful working hypothesis.

Metal-hydrogen systems are furthermore a prototype for fast solid-state diffusion in spite of the fact that, due to the metallic character, they cannot be used as solid electrolytes [1.10]. For metal-hydrogen systems, the largest isotope effects and the smallest activation energies have been found. Below room temperature the H-diffusion coefficients in several transition metals or alloys are the largest known for long-range diffusion in equilibrium solid-state systems (see Chap. 12).

A quantitative understanding of the $\alpha-\alpha'$ phase transition of H in metals has, in spite of *Lacher*'s pioneering work, been achieved only more recently. From the point of view of phase transitions, these systems serve today simultaneously as a prototype for three physical aspects, which certainly are related to each other: 1) The critical point for the lattice-gas lattice-liquid transition of H in metals like Nb, Pd, Pd Ag alloys, etc., is one of the rare cases (besides superconductivity) in which Landau's mean-field theory for phase transitions of second order [1.11] apparently holds exactly (see Chap. 2 by *H. Wagner*). It has been shown [1.12] that, close to the critical point, van der Waals' equation describes the state of the hydrogen-lattice gas better than that of any free gas. 2) The system Nb–H serves today as a prototype for what is known in metallurgy as "spinodal decomposition" [1.13, 14]. Four "spinodals" have been found experimentally [1.15], and even more exist according to theory [1.16] (see Chap. 2). 3) Metal-hydrogen systems are the prototype for systems in which the elastic interaction is responsible for the phase transition. The deformation of the lattice due to the hydrogen atom on the interstitial site, leading to long-range strain fields, i.e., also to macroscopic volume and lattice-parameter changes, is the cause for the attractive interaction leading to the gas-liquid-like condensation. With the knowledge of the strain field (more exactly the dipole-moment tensor) a quantitative prediction of the transition temperature is possible, a situation which is very rare for metallurgical systems. The long range of the elastic interaction leads to nonlocal thermodynamic potentials and thus to the unusual phenomena of a sample-shape dependent and stability curve and a sample-shape dependent macroscopic diffusion coefficient.

The whole field of the theory of elastic interaction with the corresponding mode theory as well as the spinodal decomposition and the shape dependence of physical properties is presented in Chapter 2. Chapter 3 by *Peisl* deals with the methods for measurements and the results for the strain field around H atoms dissolved in a lattice. The knowledge of the strain field is important for the quantitative understanding of the phase transitions of H in metals but also for questions like hydrogen embrittlement. The strength of interaction with dislocations or other lattice defects, the attraction of hydrogen to a crack tip, can quantitatively only be understood if the strain field is known. Chapter 4 by

Springer presents an additional aspect of interaction and changes of interaction, namely the changes of the phonon-dispersion curves and the appearance of local modes or optical modes due to hydrogen in metals. In Chapter 5 the changes of electronic properties due to solution of hydrogen are presented by *Switendick*. Chapter 6 by *F. E. Wagner* and *Wortmann* summarizes the large amount of recent data on changes of electronic properties due to hydrogen in metals as investigated by the Mössbauer effect. In this chapter a short description of the recently discovered effect of motional narrowing of the Mössbauer line of Ta due to hydrogen diffusion [1.17] is also given. In Chapter 7, *Wallace* presents the magnetic properties of metal-hydrogen alloys. In Chapters 8–13 the emphasis is laid on presentation and discussion of hydrogen-diffusion data, which does not mean that in Chapters 9, 10, and 11, which deal with the results obtained be certain experimental techniques, the additional information specific for the corresponding techniques is suppressed. In Chapter 8, *Kehr* summarizes the state of theoretical knowledge for the rate theory of H diffusion. In Chapter 9, *Cotts* summarizes the information which has been collected by nuclear magnetic resonance techniques. In Chapter 10, *Sköld* describes the method of quasielastic neutron scattering and presents results. Chapter 11 by *Kronmüller* is devoted to the magnetic disaccommodation technique. In Chapter 12 by *Völkl* and *Alefeld* the results of H diffusion in metals are collected and presented as a function of temperature, isotope, structure, etc. In Chapter 13, *Seeger* compares the proton diffusion with the very recent results for muon diffusion. Such a comparison of the diffusion properties of a positive particle with small mass, yet not as light as the positron, with those of the proton may yield clues which allow to identify the diffusion mechanisms.

In the second volume [1.5], contributions are presented which deal with application-oriented properties. It is evident that a decision whether a subject is more basic or closer to application is ambiguous. For some areas like electrotransport (Chap. by *Wipf*), it is expected that applications will result, whereas hydrogen storage (Chap. by *Wiswall*) is certainly very close to applications. The second volume mainly provides that information required if applications are intended. *Schober* and *Wenzl* treat the present knowledge about phase diagrams for H in V, Nb, and Ta. *Wicke* and *Brodowsky* describe that system which received the most attention in the past, i.e., the Pd–H system. *Baranowski* presents the highpressure solubility work, and *Stritzker* and *Wühl* the latest results on superconductivity. Finally, *Wert* discusses the interaction of hydrogen interstitials with other defects.

The thermodynamics of hydrogen solution is dealt with both by *Wicke* and *Brodowsky*, and by *Baranowski* in [1.5]. Since a comprehensive summary on these questions exists [1.18] and a new one is in preparation by *Flanagan* and *Oates* [1.19], we have omitted a chapter dealing explicitly with this subject.

The neutron-scattering work on site position of the hydrogen intersitial and on hydride structures has been summarized by *Somenkov* [1.20]. Since then, not many new facts have been added. Some of the results are presented in the contribution by *Schober* and *Wenzl*.

References

1.1 T.Graham: Phil. Trans. Roy. Soc. (London) **156**, 399 (1866)
1.2 T.Graham: Proc. Roy. Soc. (London) **16**, 422 (1868); Phil. Mag. **36**, 63 (1868); C. R. Acad. Sci. **66**, 1014 (1868); Ann. Chim. Phys. (Paris) **14**, 315 (1868)
1.3 T.Graham: Proc. Roy. Soc. (London) **17**, 212, 500 (1869); C.R. Acad. Sci. **68**, 101, 1511 (1869); Ann. Chim. Phys. (Paris) **16**, 188 (1869); Ann. Chem. Pharm. **150**, 353; **152**, 168 (1869)
1.4 T.B.Flanagan: Technical Bulletin Engelhard Industries, Inc. **VII**, 9 (1966)
1.5 G.Alefeld, J.Völkl (eds.): *Hydrogen in Metals II, Application-Oriented Properties*. Topics in Applied Physics, Vol. 29 (Springer, Berlin, Heidelberg, New York 1978) in preparation
1.6 A.Sieverts: Z. Physik. Chem. **88**, 451 (1914)
1.7 J.R.Lacher: Proc. Roy. Soc. (London) A **161**, 525 (1937)
1.8 T.D.Lee, C.N.Yang: Phys. Rev. **87**, 410 (1952)
1.9 N.F.Mott, H.Jones: *The Theory of Metals and Alloys* (Oxford University Press, Oxford 1936)
1.10 S.Geller (ed.): *Solid Electrolytes*. Topics in Applied Physics, Vol. 21 (Springer, Berlin, Heidelberg, New York 1977)
1.11 L.Landau: Zh. Eksper. Teor. I. Fiz. **7**, 627 (1937)
1.12 H.Buck, G.Alefeld: Phys. Stat. Sol. (b) **49**, 317 (1972)
1.13 J.W.Cahn: Acta Met. **9**, 795 (1961); **10**, 907 (1962)
1.14 M.A.Krivoglaz: Soviet Phys.-Solid State **5**, 2526 (1964)
1.15 H.Conrad, G.Bauer, G.Alefeld, T.Springer, W.Schmatz: Z. Physik **266**, 239 (1974)
H.C.Bauer, J.Tretkowski, U.Freudenberg, J.Völkl, G.Alefeld: Z. Physik B **28**, 255 (1977)
J.Tretkowski, J.Völkl, G.Alefeld: Z. Physik B **28**, 259 (1977)
1.16 H.Wagner, H.Horner: Advan. Phys. **23**, 587 (1974)
H.Horner, H.Wagner: J. Phys. C: Solid State Phys. **7**, 3305 (1974)
R.Bausch, H.Horner, H.Wagner: J. Phys. C: Solid State Phys. **8**, 2559 (1975)
1.17 A.Heidemann, G.Kaindl, D.Salomon, H.Wipf, G.Wortmann: Phys. Rev. Lett. **36**, 213 (1976)
1.18 T.B.Flanagan, W.A.Oates: Ber. Bunsenges. Phys. Chem. **76**, 706 (1972)
1.19 T.B.Flanagan, W.A.Oates: To be published
1.20 V.A.Somenkov: Ber. Bunsenges. Phys. Chem. **76**, 733 (1972)

2. Elastic Interaction and Phase Transition in Coherent Metal-Hydrogen Alloys

H. Wagner

With 13 Figures

2.1 Background

In some metal-hydrogen alloys, phase transitions occur at moderate hydrogen concentrations which differ from the structural transition observed at higher concentrations. Examples are the $(\alpha-\beta)$- and the $(\alpha-\alpha')$-transition in PdH_x and NbH_x, respectively. In these transitions the crystal retains its cubic structure and the host lattice apparently plays a passive role. The form of the phase diagram (see [Ref. 2.1, Chap. 2, 3]) resembles that of a normal gas-liquid transition. This seems to support the picture, promoted by *Alefeld* [2.2], that hydrogen in transition metals at low to moderate concentrations behaves like a one-component fluid interpenetrating the metal lattice. Palladium-hydrogen alloys have indeed been treated as a textbook example of a lattice gas [2.3].

In this chapter, we summarize recent theoretical work on the nature of these transitions. As a byproduct we shall see whether the above liquid-gas analogy survives a closer examination.

Atomic hydrogen is the simplest solute in a metal one can think of. Despite this fact, a quantitative electronic theory of metal-hydrogen alloys is practically nonexistent. Consequently, the form of the direct interaction between the screened protons and the forces between the latter and the metal ions are not known in any detail. Some knowledge concerning these basic interaction mechanisms is, of course, indispensable if one tries to compute thermodynamic quantities from statistical mechanics. Therefore, we are forced to proceed in a phenomenological manner.

We shall be mainly concerned with the thermodynamic properties of metal-hydrogen alloys in the vicinity of the critical points associated with the above transitions. These critical properties are expected to be insensitive to the detailed form of the direct electronic interaction between the protons. Therefore, we simply assume this part of the interaction to be short ranged and mainly repulsive. Further, there are experimental indications [2.4, 5] that the phase transitions in question obey Landau's classical theory. This hints at a long-ranged attractive interaction between protons as the driving force of these transitions. A candidate is the elastic interaction; an interstitial proton deforms the host lattice and creates a long-ranged strain field which is felt by other protons. These strain fields have been investigated by x-ray and neutron-scattering methods (Chap. 3, 4, 10). In coherent crystals the resulting elastic interaction is weak, but purely attractive, and of macroscopic range. The meaning of coherency will be discussed in detail later.

The elastic interaction of lattice defects has been studied extensively in the past [2.6–11]. Its possible relevance for phase transitions in metal-hydrogen alloys has also been noted by a number of authors (see, for instance, [2.12]). The elastic interaction has been advocated most explicitly by *Alefeld* [2.2, 13] as the dominant attractive force between protons in metals.

In thermodynamic applications, the elastic interaction is often treated in a crude and sometimes obscure manner which does not do justice to its delicate nature. Therefore, it is worthwhile to examine its properties in some detail. This leads us to the concept of elastic modes. The latter may be considered as the normal modes of a macroscopic strain field caused by internal stress. The expansion of the elastic interaction in terms of elastic modes turns out to be rather convenient for the discussion of the free energy, the phase transition, for the equilibrium fluctuations in the hydrogen density and for the description of nonequilibrium problems as, for instance, anelastic relaxation and spinodal decomposition. In the final section the theoretical predictions are compared with recent experimental results.

2.2 Elastic Interaction

Although the elastic interaction of protons in a metal lattice is easier to treat analytically in the continuum approximation, there are conceptual advantages in first discussing a simple lattice model*.

2.2.1 Lattice Model

Let us consider a crystal with N_L metal ions, loaded with N protons. The protons induce displacements $\boldsymbol{u}^m$, $m=1 \ldots N_L$ in the metal lattice. To specify these displacements we need a reference configuration which we take to be the equilibrium sites $\{\boldsymbol{R}^m\}=R$ of the pure crystal together with the ideal interstitial sites $\{\boldsymbol{S}^a|a=1\ldots N_H\}=S$ of the protons. The latter are defined to be the rest positions of the protons if the metal ions are fixed at the ideal lattice positions R. The distribution of the N protons over the N_H available sites S will be specified by a set $\tau=\{\tau_a|a=1\ldots N_H\}$ of occupation numbers $\tau_a=1$ (occupied) or 0 (empty), $\sum_a \tau_a=N$. For the purpose of this section we need not be concerned with the direct electronic interaction between the screened protons. Therefore, we focus attention on the lattice potential energy. The difference of the latter between the strained and the reference configuration will be denoted by $\mathcal{U}$. Our

* For an introduction into the theory of point defects in metals see the recent monograph by *Leibfried* and *Breuer* [2.65]

first model assumption is that $\mathscr{U}$ is given by the harmonic approximation

$$\mathscr{U}=\tfrac{1}{2}u_m\cdot\Phi_{mn}\cdot u_n \tag{2.1}$$

in matrix notation.

$\Phi_{mn}(=\Phi^{mn}_{\mu\nu};\mu,\nu=x,y,z)$ denote the force constants of the pure metal lattice. The displacements $\boldsymbol{u}^m(\tau)$ caused by the protons are, of course, not known in advance. They are related to forces $\boldsymbol{K}^m$ which balance the lattice forces in the strained configuration,

$$K_m(\tau)=\Phi_{mn}\cdot u_n(\tau), \tag{2.2}$$

with

$$\sum_m \boldsymbol{K}^m(\tau)=0, \quad \sum_m \boldsymbol{K}^m(\tau)\times\boldsymbol{R}^m=0, \tag{2.3}$$

as follows from translational and rotational invariance. The $\boldsymbol{K}^m$ may obviously be interpreted as (fictitious) forces which have to be introduced in the pure lattice R to produce, in harmonic approximation, the same displacements as those due to the real hydrogen-lattice forces in the loaded crystal.

By (2.2) we parameterize the unknown displacements by equally unknown forces and nothing seems to be gained. However, (2.2) allows us to relate the forces $\boldsymbol{K}^m$ to the measurable volume change ΔV of the metal lattice caused by the protons. Following the arguments given in [2.14], one finds

$$\frac{\Delta V}{V}=\frac{1}{3}\kappa_{\mathrm{T}}\frac{N}{V}\sum_\alpha \Pi_{\alpha\alpha}(\tau), \tag{2.4}$$

where

$$\Pi_{\alpha\beta}(\tau)=\frac{1}{N}\sum_m K^m_\alpha(\tau)R^m_\beta=\Pi_{\beta\alpha}(\tau) \tag{2.5}$$

defines the (force-) dipole tensor associated with the forces $\boldsymbol{K}^m(\tau)$, and $\kappa_{\mathrm{T}}=3/(C_{11}+2C_{12})$ denotes the compressibility of the pure host lattice with elastic constants C_{ij}. The relation (2.4) was originally derived by *Eshelby* [2.15] for the elastic continuum. The experimental implications of (2.4) are discussed by *Peisl* in Chapter 3.

In order to proceed, we have to make an assumption about the dependence of the forces $\boldsymbol{K}^m(\tau)$ on the proton configuration τ. Provided the real forces between metal ions and screened protons are short ranged, it seems plausible to assume that $\boldsymbol{K}^m(\tau)$ depends only on the presence of the protons in the vicinity of the metal ion "m". In contrast to $\boldsymbol{K}^m(\tau)$, the displacement $\boldsymbol{u}^m(\tau)$ will in general contain contributions from distant protons. However, one has to take into account that the protons not only execute direct forces on the metal ions but also cause a local change in the force constants of the host lattice. The effect of this renormalization of the force constants propagates through the lattice and influences the force $\boldsymbol{K}^n(\tau)$ even at large separations $|\boldsymbol{R}^m - \boldsymbol{R}^n|$. In other words, $\boldsymbol{K}^m(\tau)$ represents a many-body force which, in principle, does not allow a simple superposition.

We are therefore lead to the following Ansatz:

$$K_\mu^m(\tau) = \sum_a \chi_\mu^{ma} \tau_a + \sum_{ab}{}' \chi_\mu^{mab} \tau_a \tau_b + \dots, \tag{2.6}$$

with phenomenological constants $\chi_\mu^{ma_1 \dots a_n}$ ($\sum' : a = b$ excluded). Since the first sum in (2.6) also contributes if there is only a single proton in the lattice, we may assume that χ_μ^{ma} is short ranged in $|\boldsymbol{R}^m - \boldsymbol{S}^a|$. The second sum in (2.6) arises from the aforementioned change in the force constants and distant protons will enter.

The latter term also contributes via the dipole tensor to the nonlinear variation of $\Delta V/V$ with hydrogen concentration. On the other hand, one finds experimentally for NbH_x in the single-phase region that $\Delta V/V$ does behave linearly with N/N_H up to concentrations of 100 at% (Chap. 3). We take this as an indication that the combined effect of the nonlinear terms in (2.6) is negligible in the $\boldsymbol{K}^m(\tau)$ and hence in $\Pi_{\alpha\beta}(\tau)$. It should be kept in mind, however, that (2.4) is derived from (2.2) which is valid only in the harmonic approximation. It is conceivable, for instance, that some cancellation occurs in $\Delta V/V$ with anharmonic terms not included in (2.2).

In the following, we shall omit the nonlinear terms in the expansion (2.6). Going back to (2.1), we now express the excess potential energy in terms of the forces $\boldsymbol{K}^m(\tau)$ by means of the Legendre transformation

$$\mathcal{U} \to \mathcal{H} = \mathcal{U} - K_m(\tau) \cdot u_m(\tau). \tag{2.7}$$

With (2.2) we find

$$\begin{aligned} \mathcal{H} &= -\tfrac{1}{2} K_m(\tau) \cdot u_m(\tau) \\ &= -\tfrac{1}{2} K_m(\tau) \cdot G_{mn} \cdot K_n(\tau), \end{aligned} \tag{2.8}$$

where we introduced the static Green function $G_{mn} = \Phi_{mn}^{-1}$ (matrix inverse of Φ_{mn}) of the pure lattice. $\mathcal{H}$ represents the harmonic elastic energy stored in the crystal which is strained by the protons in the configuration τ. Obviously, it may be regarded as a configurational energy of the dissolved hydrogen and it describes an effective interaction of the protons. With the general expression (2.6) for $\boldsymbol{K}^m(\tau)$ inserted into $\mathcal{H}$ we have a many-body interaction. If we neglect the renormalization of the force constants by omitting the nonlinear terms in (2.6), we arrive at a sum of pairwise interactions,

$$\mathcal{H} = -\tfrac{1}{2} \sum_{ab} \tau_a W_{ab} \tau_b, \tag{2.9}$$

where

$$W_{ab} = \sum_{mn} \chi_{\mu}^{ma} G_{\mu\nu}^{mn} \chi_{\nu}^{nb} . \tag{2.10}$$

From lattice theory, one knows that the static lattice Green function G_{mn} is long ranged in the distance $R_{mn} = |\boldsymbol{R}^m - \boldsymbol{R}^n|$ and depends in a delicate way on the boundary conditions prescribed at the surface of the crystal. From (2.10) it is clear that these features are shared by W_{ab}.

There are two important cases in which W_{ab} can be treated more explicitly:

1) Crystal with periodic boundary conditions. In this case, the lattice Green function is denoted by $\tilde{G}_{mn}$ and can be expressed in terms of phonon modes,

$$\tilde{G}_{\mu\nu}^{mn} = \sum_{\boldsymbol{q}s}{}' \frac{e_{\mu}(\boldsymbol{q}, s) e_{\nu}(\boldsymbol{q}, s)}{N_{\mathrm{L}} M_{\mathrm{L}} \omega^2(\boldsymbol{q}, s)} \exp(\mathrm{i}\boldsymbol{q} \cdot \boldsymbol{R}^{mn}) . \tag{2.11}$$

M_{L} is the mass of a metal ion, $\omega(\boldsymbol{q}, s)$ and $\boldsymbol{e}(\boldsymbol{q}, s)$ are the phonon frequencies and polarization vectors of the pure host lattice. The sum over wave vectors $\boldsymbol{q}$ is restricted to the first Brillouin zone $\left(\sum_{\boldsymbol{q}}{}' : \boldsymbol{q} = 0 \text{ excluded}\right)$. Provided the force constants Φ_{mn} are known, e.g., from a *Born-von Karman* fit to measured phonon frequencies, and an assumption is made concerning χ_{μ}^{ma}, then $\tilde{W}_{ab}$ for periodic boundary conditions can be computed numerically. This has been done for Nb [2.16], and the results will be used later.

2) Crystal with free surface and homogeneous proton distribution. Here, the occupation numbers τ_a are replaced by their average value $\bar{\tau}_a = N/N_{\mathrm{H}}$. Taking into account that a site can only be singly occupied, $\tau_a^2 = \tau_a$, one finds

$$\begin{aligned} \bar{\mathscr{H}} &= -\frac{1}{2} \sum_a W_{aa} (\bar{\tau}_a - \bar{\tau}_a^2) - \frac{1}{2} \sum_{ab} W_{ab} \bar{\tau}_a \bar{\tau}_b \\ &= -\frac{1}{2} \left[\frac{N}{N_{\mathrm{H}}} - \left(\frac{N}{N_{\mathrm{H}}} \right)^2 \right] \sum_a W_{aa} - \frac{1}{2} \left(\frac{N}{N_{\mathrm{H}}} \right)^2 \sum_{ab} W_{ab} . \end{aligned} \tag{2.12}$$

For a crystal with a free surface, $\sum_{ab} W_{ab}$ can be calculated exactly with the result [2.14]

$$\sum_{ab} W_{ab} = \frac{N_{\mathrm{H}}^2}{V} P_{\alpha\beta} S_{\alpha\beta\mu\nu} P_{\mu\nu} . \tag{2.13}$$

$S_{\alpha\beta\mu\nu}$ are the elastic compliances and $P_{\mu\nu}$ is the dipole tensor

$$P_{\mu\nu} = \sum_m \chi_{\mu}^{ma} (R_{\nu}^m - S_{\nu}^a) . \tag{2.14}$$

With an isotropic dipole tensor, $P_{\mu\nu}=P\delta_{\mu\nu}$, the right-hand side of (2.13) reduces to $\kappa_T(N_H P)^2/V$.

In the self-energy term $\sum_a W_{aa}$, only the Green function G_{mn} at small R_{mn} on the order of a few lattice distances enters. Thus, $\sum_a W_{aa}$ will be insensitive to boundary conditions for large crystals. Specifically, we shall have $\sum_a (W_{aa}-\tilde{W}_{aa}) \sim V^{2/3}$ and therefore the self-energy term can be computed from (2.10) and (2.11). An estimate for an isotropic crystal with $P_{\mu\nu}=P\delta_{\mu\nu}$ yields [2.14]

$$\sum_a W_{aa} \approx \frac{N_L N_H}{VC_{11}} P^2 . \tag{2.15}$$

In the general case of a finite crystal with a free surface and an arbitrary proton configuration, the elastic energy $\mathscr{H}$, because of its macroscopic range, depends on the shape of the crystal. This fact will be explicitly demonstrated later. In the lattice model the shape dependence arises in a rather indirect and analytically intractable way through the difference between the force constants Φ_{mn} on the surface and those in the bulk. We therefore proceed by going over to a continuum description, where boundary conditions are easier to formulate and less difficult to deal with than in lattice theory.

Let us make a few final remarks on the lattice model.

1) We have been talking about protons in metals, but the nature of the interstitial ions never entered explicitly. Therefore, the present discussion of the elastic interaction may be applied to any kind of point defect.

2) The forces $\boldsymbol{K}^m$ can be introduced more generally than was done here by simply defining them through (2.2), without referring to the harmonic approximation for the defect-lattice potential. In this way, one arrives at the well-known *Kanzaki* forces [2.17, 18]. One has to realize, however, that the Kanzaki forces not only depend on the defect configuration but also on the displacements through the anharmonicity of the defect-lattice potential. It is possible to calculate the displacements induced by a single defect and the elastic interaction of an isolated pair of defects, but this method becomes troublesome for higher concentration of defects.

3) We have suppressed any dependence of the dipole tensor on the type of interstitial sites. Although hydrogen in NbH_x occupies tetrahedral sites with three distinguishable types (x-, y-, z-sites), one finds [2.19] that the protons act like pure dilation centers, as they strictly do on the octahedral sites in Pd. Therefore, for simplicity, we shall assume from now on that there is only one type of site present.

2.2.2 Continuum Model

We are mainly concerned with hydrogen concentrations in the range of the critical concentration, about 30% [H/Met]. The average proton spacing is then on the order of lattice distances. Therefore, one might object from the beginning to the application of continuum elasticity theory to describe the strain-induced elastic interaction between such closely spaced defects. An isolated defect in the elastic continuum is represented by a singular body force and it is indeed well known that the resulting strain field obtained from elasticity theory correctly describes the lattice distortion only far away from the defect. Here, however, we are primarily interested in a macroscopic description of a proton "fluid" with a smoothly varying density which causes nonsingular internal stresses. For this purpose, the continuum approach should be legitimate. Accordingly, we consider lattice displacements $\boldsymbol{u}^m$ which can be interpolated by a continuous displacement field $\boldsymbol{u}(\boldsymbol{x})$, varying slowly within the range of lattice forces, such that $\boldsymbol{u}^m = \boldsymbol{u}(\boldsymbol{x} = \boldsymbol{R}^m)$. Neglecting surface effects in the force constants Φ_{mn}, the continuum version of $\mathscr{U}$ may be written as

$$\mathscr{U} = \tfrac{1}{2} \int \varepsilon_{\alpha\beta}(\boldsymbol{x}) C_{\alpha\beta\mu\nu} \varepsilon_{\mu\nu}(\boldsymbol{x}) dV, \tag{2.16}$$

where

$$\varepsilon_{\alpha\beta}(\boldsymbol{x}) = \tfrac{1}{2}[\partial_\alpha u_\beta(\boldsymbol{x}) + \partial_\beta u_\alpha(\boldsymbol{x})] \tag{2.17}$$

denotes the strain tensor. $C_{\alpha\beta\mu\nu}$ are the elastic constants of the pure metal. The forces which act upon the strained metal are deduced from the variational principle

$$\delta_u \mathscr{U} = \int_V \boldsymbol{K} \cdot \delta \boldsymbol{u} \, dV + \oint_\Sigma (\boldsymbol{k} + \boldsymbol{k}^e) \cdot \delta \boldsymbol{u} \, df, \tag{2.18}$$

where the forces $\boldsymbol{K}(\boldsymbol{x})$ and $\boldsymbol{k}(\boldsymbol{x})$ correspond to the lattice forces $\boldsymbol{K}^m$ and will be specified below. We have also included an additional external force $\boldsymbol{k}^e$ applied to the surface Σ to study both the effect of boundary conditions and the anelastic response.

Equation (2.18) with $\mathscr{U}$ from (2.16) leads to

$$C_{\alpha\beta\mu\nu} \partial_\beta \varepsilon_{\mu\nu}(\boldsymbol{x}) = -K_\alpha(\boldsymbol{x}), \quad \boldsymbol{x} \text{ in } V \tag{2.19}$$

and

$$C_{\alpha\beta\mu\nu} n_\beta(\boldsymbol{x}) \varepsilon_{\mu\nu}(\boldsymbol{x}) = k_\alpha(\boldsymbol{x}) + k_\alpha^e(\boldsymbol{q}), \quad \boldsymbol{x} \text{ on } \Sigma, \tag{2.20}$$

where $\boldsymbol{n}(\boldsymbol{x})$ is an outward-directed normal unit vector on Σ. Equations (2.19) and (2.20) determine the forces in terms of the displacements. As in the lattice model we change variables by transforming from displacements to forces via

$$\mathscr{U} \to \mathscr{H} = \mathscr{U} - \int \boldsymbol{K} \cdot \boldsymbol{u} \, dV - \oint (\boldsymbol{k} + \boldsymbol{k}^e) \cdot \boldsymbol{u} \, df, \tag{2.21}$$

with the result

$$\mathscr{H} = -\tfrac{1}{2} \int \boldsymbol{K} \cdot \boldsymbol{u} \, dV - \tfrac{1}{2} \oint (\boldsymbol{k} + \boldsymbol{k}^e) \cdot \boldsymbol{u} \, df. \tag{2.22}$$

In order to specify the functional form of $\boldsymbol{K}$ and $\boldsymbol{k}$ in terms of the hydrogen density ϱ, we compare $\mathscr{H}$ in (2.22) for $\boldsymbol{k}^{\mathrm{e}}=0$ with its lattice form in the first line of (2.8), which reads explicitly

$$\mathscr{H}_{\text{lattice}}=-\tfrac{1}{2}\sum_{ma}u_{\mu}^{m}\chi_{\mu}^{ma}\tau_{a}, \tag{2.23}$$

where we have used (2.6) with the nonlinear terms omitted. In (2.23) we substitute $\tau_a\to\tau(\boldsymbol{S}^a)$, $\boldsymbol{u}^m\to\boldsymbol{u}(\boldsymbol{R}^m)$ and expand

$$u_\alpha(\boldsymbol{R}^m)=u_\alpha(\boldsymbol{S}^a)+(R_\beta^m-S_\beta^a)\partial_\beta u_\alpha(\boldsymbol{S}^a)+\dots\,. \tag{2.24}$$

After insertion of (2.24) into (2.23), the contribution of $\boldsymbol{u}(\boldsymbol{S}^a)$ vanishes because of translational invariance [cf. (2.3)]. Thus we find, for $\boldsymbol{k}^{\mathrm{e}}=0$,

$$\mathscr{H}\approx-\tfrac{1}{2}\int\varepsilon_{\alpha\beta}(\boldsymbol{x})P_{\alpha\beta}\varrho(\boldsymbol{x})\,dV \tag{2.25}$$

after replacing the sum on $\boldsymbol{S}^a$ by an integral. $\varrho(\boldsymbol{x})=\tau(\boldsymbol{x})N_{\mathrm{H}}/V$ denotes the proton density and $P_{\alpha\beta}$ is given by (2.14). We have kept only the second term on the right-hand side of (2.24). Higher order derivatives of $\boldsymbol{u}(\boldsymbol{x})$ could easily be included and would lead to higher order multipole tensors of χ_μ^{ma}. After a partial integration in (2.25), the comparison with (2.22) yields in the "dipole approximation"

$$K_\alpha(\boldsymbol{x})=-P_{\alpha\beta}\partial_\beta\varrho(\boldsymbol{x}),\quad \boldsymbol{x}\ \text{in}\ V, \tag{2.26}$$

and

$$k_\alpha(\boldsymbol{x})=n_\beta(\boldsymbol{x})P_{\alpha\beta}\varrho(\boldsymbol{x}),\quad \boldsymbol{x}\ \text{on}\ \Sigma. \tag{2.27}$$

With the stress tensor $\sigma_{\alpha\beta}(\boldsymbol{x})$ defined by

$$\sigma_{\alpha\beta}=C_{\alpha\beta\mu\nu}\varepsilon_{\mu\nu}-P_{\alpha\beta}\varrho(\boldsymbol{x}), \tag{2.28}$$

(2.19) and (2.20) read

$$\partial_\beta\sigma_{\alpha\beta}(\boldsymbol{x})=0,\quad \boldsymbol{x}\ \text{in}\ V, \tag{2.29}$$

$$n_\beta(\boldsymbol{x})\sigma_{\alpha\beta}(\boldsymbol{x})=k_\alpha^{\mathrm{e}},\quad \boldsymbol{x}\ \text{on}\ \Sigma. \tag{2.30}$$

These equations constitute a boundary value problem for determining $\boldsymbol{u}(\boldsymbol{x})$, and therefore also the elastic energy $\mathscr{H}$, as functional of $\varrho(\boldsymbol{x})$ and $\boldsymbol{k}^{\mathrm{e}}(\boldsymbol{x})$.

In passing we note that an alternative form of the elastic energy can be found in the literature, e.g., [2.7, 20]. For simplicity, let us take $\boldsymbol{k}^{\mathrm{e}}=0$. One writes $\sigma_{\alpha\beta} = C_{\alpha\beta\mu\nu}\hat{\varepsilon}_{\mu\nu}$, where

$$\hat{\varepsilon}_{\alpha\beta} = \varepsilon_{\alpha\beta} - S_{\alpha\beta\mu\nu} P_{\mu\nu} \varrho(\boldsymbol{x}) . \tag{2.31}$$

An elastic energy $\hat{\mathscr{H}}$ is then defined by

$$\hat{\mathscr{H}} = \tfrac{1}{2} \int \sigma_{\alpha\beta} \hat{\varepsilon}_{\alpha\beta} dV . \tag{2.32}$$

It is easily shown that this $\hat{\mathscr{H}}$ is related to our $\mathscr{H}$ (for $\boldsymbol{k}^e = 0$) by

$$\hat{\mathscr{H}} = \mathscr{H} + \tfrac{1}{2} P_{\alpha\beta} S_{\alpha\beta\mu\nu} P_{\mu\nu} \int \varrho^2(\boldsymbol{x}) dV . \tag{2.33}$$

We now follow *Bausch* et al. [2.21] and separate the internal stress $\sigma^i(\boldsymbol{x})$ induced by $\varrho(\boldsymbol{x})$, from the stress $\sigma^e(\boldsymbol{x})$ induced by $\boldsymbol{k}^e$. We put $\boldsymbol{u}(\varrho, \boldsymbol{k}^e) = \boldsymbol{u}^i(\varrho) + \boldsymbol{u}^e(\boldsymbol{k}^e)$ and correspondingly

$$\sigma^e_{\alpha\beta} = C_{\alpha\beta\mu\nu} \varepsilon^e_{\mu\nu} , \tag{2.34}$$

$$\partial_\beta \sigma^e_{\alpha\beta}(\boldsymbol{x}) = 0, \quad \boldsymbol{x} \text{ in } V ; \quad n_\beta(\boldsymbol{x}) \sigma^e_{\alpha\beta}(\boldsymbol{x}) = k^e_\alpha(\boldsymbol{x}), \quad \boldsymbol{x} \text{ on } \Sigma , \tag{2.35}$$

and

$$\sigma^i_{\alpha\beta} = C_{\alpha\beta\mu\nu} \varepsilon^i_{\mu\nu} - P_{\alpha\beta} \varrho , \tag{2.36}$$

$$\partial_\beta \sigma^i_{\alpha\beta}(\boldsymbol{x}) = 0, \quad \boldsymbol{x} \text{ in } V, \tag{2.37}$$

$$n_\beta(\boldsymbol{x}) \sigma^i_{\alpha\beta}(\boldsymbol{x}) = 0, \quad \boldsymbol{x} \text{ on } \Sigma . \tag{2.38}$$

Then $\sigma = \sigma^i + \sigma^e$, and $\mathscr{H}$ as given by (2.22) with the forces (2.26) and (2.27) may be written in the form

$$\begin{aligned} \mathscr{H} &= -\tfrac{1}{2} \int \varepsilon^i_{\alpha\beta} P_{\alpha\beta} \varrho \, dV - \int \varepsilon^e_{\alpha\beta} P_{\alpha\beta} \varrho \, dV - \tfrac{1}{2} \oint \boldsymbol{u}^e \cdot \boldsymbol{k}^e \, df \\ &= \mathscr{H}^i(\varrho) + \mathscr{H}^{ie}(\varrho, \boldsymbol{k}^e) + \mathscr{H}^e(\boldsymbol{k}^e) . \end{aligned} \tag{2.39}$$

The first term, $\mathscr{H}^i(\varrho)$, describes the internal elastic energy of the protons. Its analysis is the topic of the next sections. The second term, $\mathscr{H}^{ie}(\varrho, \boldsymbol{k}^e)$, contains the interaction of the protons with the externally induced strain field. It produces the anelastic response (Gorsky effect) and provides a tool to study long-ranged correlations in the local proton density. Finally, $\mathscr{H}^e(\boldsymbol{k}^e)$ represents the external elastic energy due to $\boldsymbol{k}^e$ and is unimportant in the present context.

2.2.3 Incoherent and Coherent States

To compute $\mathscr{H}^i(\varrho)$ for a given distribution $\varrho(\boldsymbol{x})$ of protons, we have to solve the boundary value problem for $\boldsymbol{u}(\boldsymbol{x})$ defined by (2.36–38) and the geometrical equation (2.17). Here and in the next sections, we drop the label "i" (or take $\boldsymbol{k}^e = 0$).

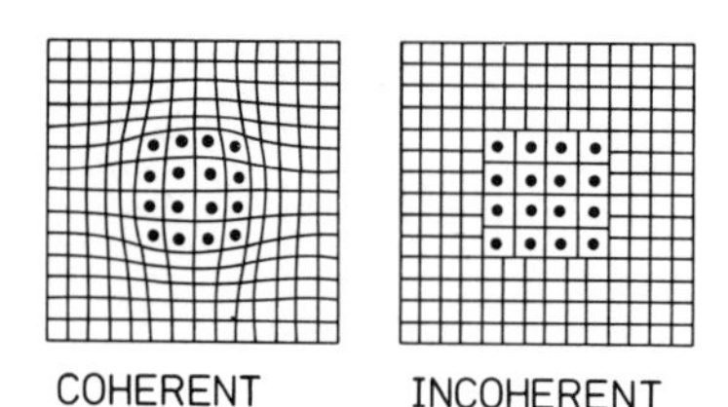

Fig. 2.1. Illustration of the coherent and incoherent state. Dots mark subvolumes containing interstitials

The boundary value problem completely specifies $\boldsymbol{u}(\boldsymbol{x})$. On the other hand, if one prefers to work with strains instead of displacements, one has to take into account the compatibility relations (see, e.g., [2.22]),

$$\partial_\mu \partial_\nu \varepsilon_{\alpha\beta} + \partial_\alpha \partial_\beta \varepsilon_{\mu\nu} - \partial_\beta \partial_\nu \varepsilon_{\alpha\mu} - \partial_\alpha \partial_\mu \varepsilon_{\beta\nu} = 0 . \tag{2.40}$$

For a simply connected body, these relations guarantee that $\varepsilon_{\alpha\beta}$ is a proper strain tensor associated with a continuously differentiable displacement field. Although the relations (2.40) (which can be reduced to three independent ones) are automatically obeyed if $\varepsilon_{\alpha\beta}$ is given by (2.17), they may be violated if an Ansatz or an approximation is made in terms of $\varepsilon_{\alpha\beta}$ (or $\sigma_{\alpha\beta}$) instead of u_α. Thus, for instance, the $\hat{\varepsilon}_{\alpha\beta}$ defined by (2.31) does not in general fulfill (2.40) for arbitrary $\varrho(\boldsymbol{x})$.

A second important example is the following: the equilibrium and boundary conditions (2.37–38), are trivially satisfied by setting $\sigma_{\alpha\beta}=0$. From (2.36) we then obtain immediately

$$\varepsilon_{\alpha\beta}(\boldsymbol{x}) = S_{\alpha\beta\mu\nu} P_{\mu\nu} \varrho(\boldsymbol{x}) , \tag{2.41}$$

and therefore,

$$\mathscr{H}(\varrho) = -\tfrac{1}{2} P_{\alpha\beta} S_{\alpha\beta\mu\nu} P_{\mu\nu} \int \varrho^2(\boldsymbol{x}) dV . \tag{2.42}$$

However, the above $\varepsilon_{\alpha\beta}(\boldsymbol{x})$ violates the compatibility relations, except in those cases where $\varrho(\boldsymbol{x})$ is either a constant or a linear function of $\boldsymbol{x}$, i.e., $\partial_\alpha \partial_\beta \varrho = 0$. Thus, in general there is no unique continuous displacement field associated with the strain tensor (2.41). Although this stress-free strain is not a proper solution of our boundary value problem, it allows for a physical interpretation. In a stress-free strained crystal, the internal stresses arising originally from an inhomogeneous distribution of protons have been removed either by plastic deformation or by tearing the crystal into pieces which are then able to relax individually.

Of course, there is no mechanism built into our model which accounts for the creation of dislocations to achieve the plastic deformation. The situation here is reminiscent of the original *van der Waals* theory of condensation in which the phase equilibrium was put in by hand via the Maxwell construction.

Following usual practice, we shall say that a stress-free strained crystal with $\partial_\alpha \partial_\beta \varrho \neq 0$ is "incoherent". The opposite case is then the "coherent" crystal with compatible strains obtained by honestly solving the boundary value problem. The stresses associated with the compatible strains are called "coherency stresses". Figure 2.1 is a schematic picture of the coherent and the incoherent states.

The concept of (in-)coherency is well known in metallurgy. It enters crucially in *Kröner*'s continuum theory of dislocation [2.23] and plays a role in *Cahn*'s theory of spinodal decomposition [2.20, 24].

As we shall see, the thermodynamic properties of coherent metal-hydrogen alloys differ in a remarkable way from those obtained for the incoherent state.

2.2.4 Elastic Interaction in the Coherent State

Consider two isolated droplets of hydrogen, $\varrho_1(\boldsymbol{x})$, $\varrho_2(\boldsymbol{x})$, confined to regions V_1 and V_2, with V_1 centred at the origin and V_2 at $\boldsymbol{r}$; the sizes $l_{1,2}$ of the droplets are assumed to be much smaller than their separation r.

The elastic interaction energy of the droplets is

$$\mathscr{H}_{12} = -\int_{V_2} \varrho_2(\boldsymbol{x}) P_{\alpha\beta} \varepsilon^{(1)}_{\alpha\beta}(\boldsymbol{x}) dV. \tag{2.43}$$

The displacement field $\boldsymbol{u}^{(1)}(\boldsymbol{x})$ induced by the first droplet can be written as

$$u^{(1)}_\alpha(\boldsymbol{x}) = \int_{V_1} G_{\alpha\beta}(\boldsymbol{x}, \boldsymbol{x}') P_{\beta\gamma} \partial'_\gamma \varrho_1(\boldsymbol{x}') dV', \tag{2.44}$$

where we formally solved (2.37) with the help of the elastic Green function

$$C_{\alpha\beta\mu\nu} \partial_\beta \partial_\nu G_{\mu\gamma}(\boldsymbol{x}, \boldsymbol{x}') = \delta_{\alpha\gamma} \delta(\boldsymbol{x} - \boldsymbol{x}'). \tag{2.45}$$

Infinite Continuum

If the host material is an ∞-continuum, then $G_{\alpha\beta}(\boldsymbol{x}, \boldsymbol{x}') = \overset{\infty}{G}_{\alpha\beta}(\boldsymbol{x} - \boldsymbol{x}')$ with the boundary condition $\overset{\infty}{G}_{\alpha\beta}(\boldsymbol{x}) \to 0$ for $x \to \infty$, replacing (2.38). For $x \gg l_1$, we find asymptotically from (2.44)

$$\overset{\infty}{u}{}^{(1)}_\alpha(\boldsymbol{x}) \approx \partial_\gamma \overset{\infty}{G}_{\alpha\beta}(\boldsymbol{x}) P_{\beta\gamma} N_1, \tag{2.46}$$

where N_1 is the total number of protons in V_1. [If $\varrho_1(\boldsymbol{x}) = \delta(\boldsymbol{x})$ ("single proton"), then (2.46) formally holds for all $\boldsymbol{x} \neq 0$.] In an anisotropic continuum the Green function is of the form (see, e.g., [2.18, 25]).

$$\overset{\infty}{G}_{\alpha\beta}(\boldsymbol{x}) = \frac{1}{x} g_{\alpha\beta}(\hat{\boldsymbol{x}}), \qquad \hat{\boldsymbol{x}} = \boldsymbol{x} \cdot x^{-1}. \tag{2.47}$$

Hence $\overset{\infty}{\boldsymbol{u}}{}^{(1)}(x) \sim x^{-2}$ for $x \gg l_1$, and

$$\overset{\infty}{\mathscr{H}}_{12}(\boldsymbol{r}) \approx \frac{N_1 N_2}{r^3} P_{\alpha\beta} f_{\alpha\beta\mu\nu}(\hat{\boldsymbol{r}}) P_{\mu\nu}. \tag{2.48}$$

This r^{-3}-behavior of the elastic dipole-dipole interaction is a well-known result. For an arbitrary anisotropic lattice the angular factor $f_{\alpha\beta\mu\nu}(\hat{\boldsymbol{r}})$ cannot be computed in closed form. It may be shown, however, that its average over the unit sphere vanishes. Therefore, $\overset{\infty}{\mathscr{H}}_{12}(\boldsymbol{r})$ has attractive and repulsive parts, depending on the direction of $\boldsymbol{r}$.

For anisotropic cubic materials the elastic displacement field of point defects has been calculated numerically [2.26–28]. Approximations for weak anisotropy are studied, for example, in [2.29].

In the special case of an isotropic dipole tensor, $P_{\alpha\beta} = P\delta_{\alpha\beta}$, and an isotropic material [where $\overset{\infty}{G}_{\alpha\beta}(\boldsymbol{x})$ is known exactly], one finds from (2.46)

$$\overset{\infty}{u}{}^{(1)}_{\alpha}(\boldsymbol{x}) \approx \frac{PN_1}{4\pi C_{11}} \frac{x_\alpha}{x^3}. \tag{2.49}$$

The local dilatation $\overset{\infty}{\varepsilon}{}^{(1)}_{\alpha\alpha}(\boldsymbol{x})$, arising from the asymptotic contribution, vanishes identically for $\boldsymbol{x} \neq 0$. Hence, the r^{-3}-interaction is absent in this case [2.30].

The right-hand side of (2.46) constitutes the leading term in a multipole expansion of the integral in (2.44), and corrections to (2.46) involving higher order moments of the density distribution $\varrho_1(\boldsymbol{x})$ could easily be written down. However, this would not yield a systematic asymptotic expansion of the displacement field, since terms competing with higher order multipole contributions arise from other sources. For instance, one should then include, in the forces $\boldsymbol{K}(\boldsymbol{x})$ and $\boldsymbol{k}(\boldsymbol{x})$, higher order moments of χ^{ma}_{μ}. Furthermore, lattice correction to the elastic Green function would have to be taken into account also.

Finite Continuum

We now consider the two droplets to be in a sphere with radius R, centered at the origin. We also assume elastic isotropy and $P_{\alpha\beta} = P\delta_{\alpha\beta}$. The displacement field $\boldsymbol{u}^{(1)}(\boldsymbol{x})$ induced by the droplet at the origin may be written as [2.31]

$$\boldsymbol{u}^{(1)}(\boldsymbol{x}) = \overset{\infty}{\boldsymbol{u}}{}^{(1)}(\boldsymbol{x}) + \boldsymbol{v}^{(1)}(\boldsymbol{x}), \tag{2.50}$$

where $\overset{\infty}{\boldsymbol{u}}{}^{(1)}(\boldsymbol{x})$ is the displacement field of the ∞-continuum and $\boldsymbol{v}^{(1)}$, the "image"-displacement field, is a solution to the homogeneous ($\varrho_1 = 0$) equilibrium condition (2.37), which is required to satisfy the boundary condition (2.38)

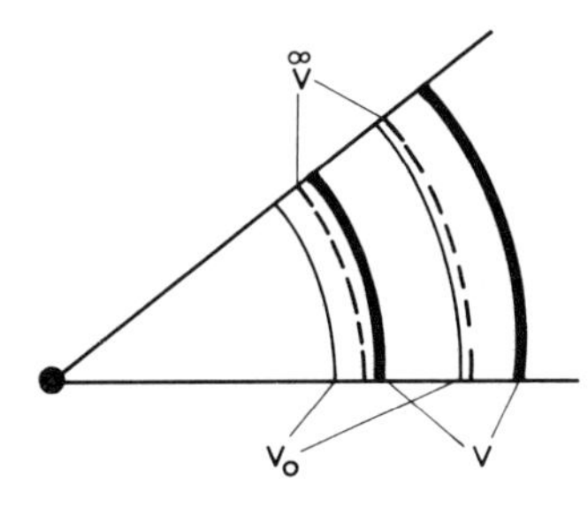

Fig. 2.2. Volume changes in an isotropic material caused by a dilatation center. v_0: original volume element in the pure material, $v_0 \to \overset{\infty}{v}$: volume change in the infinite material with a dilatation center (dot); $|v_0| = |\overset{\infty}{v}|$. $\overset{\infty}{v} \to v$: volume change after removal of the material outside of a sphere; $|v| > |\overset{\infty}{v}|$

for $\boldsymbol{u}^{(1)}$ at the surface of the sphere. Since $l_1 \ll r, R$, we can employ (2.46) for $\overset{\infty}{\boldsymbol{u}}{}^{(1)}$. As for the image field, we set $\boldsymbol{v}^{(1)}(\boldsymbol{x}) = a\boldsymbol{x}$, which is easily checked to be a solution of (2.37) with $\varrho = 0$. The local dilatation entering the interaction energy (2.43) is then given by $\varepsilon_{\alpha\alpha}^{(1)}(\boldsymbol{x}) = 3a$, since the contribution of $\overset{\infty}{\boldsymbol{u}}{}^{(1)}$ vanishes identically for $\boldsymbol{x} \neq 0$. Inserting (2.50) into the boundary condition (2.38) we find[1]

$$a = \frac{PN_1}{3V}\left(\kappa_{\mathrm{T}} - \frac{1}{C_{11}}\right), \tag{2.51}$$

and finally $\mathscr{H}_{12}(\boldsymbol{r}) \approx \mathscr{H}_{12}^{\mathrm{as}}(\boldsymbol{r})$ for $r \gg l_{1,2}$, with

$$\begin{aligned}\mathscr{H}_{12}^{\mathrm{as}}(\boldsymbol{r}) &= -3aN_2P \\ &= -\frac{N_1N_2}{V}P^2\left(\kappa_{\mathrm{T}} - \frac{1}{C_{11}}\right).\end{aligned} \tag{2.52}$$

This simple expression displays some remarkable features of the asymptotic part of $\mathscr{H}_{12}(\boldsymbol{r})$.

1) $\mathscr{H}_{12}^{\mathrm{as}}$ is entirely due to the induced image displacement (see Fig. 2.2),
2) $\mathscr{H}_{12}^{\mathrm{as}}$ is independent of $\boldsymbol{r}$,
3) $\mathscr{H}_{12}^{\mathrm{as}}$ represents an attractive interaction, since normally $\kappa_{\mathrm{T}} > 1/C_{11}$.

Although these properties of $\mathscr{H}_{12}^{\mathrm{as}}$ reflect to some extent the special proton configuration of our example, they hint at the general nature of the elastic interaction in coherent materials. Thus, for instance, there will be an $\boldsymbol{r}$-dependence in $\mathscr{H}_{12}^{\mathrm{as}}$ if both droplets are located off-center within the sphere, but this variation of $\mathscr{H}_{12}^{\mathrm{as}}$ with the location of the droplets will be found to occur on the macroscopic scale R (=radius of the sphere). Furthermore, property 3), namely that the elastic interaction via the free surface through image displacements is attractive, holds quite generally. The importance of the boundary condition may be demonstrated in our example by considering a clamped ($\boldsymbol{u}^{(1)} = 0$ for $x = R$) instead of a free surface. In this case $\mathscr{H}_{12}^{\mathrm{as}} = P^2N_1N_2/(VC_{11})$, which is repulsive.

---

[1] The constant "a" may also be determined by using (2.4) for the volume change.

2.2.5 Elastic Modes

We now turn to a description of the internal elastic energy in a finite material of arbitrary shape with an arbitrary distribution $\varrho(\boldsymbol{x})$ of protons. In the equilibrium condition (2.37) and in the boundary condition (2.38), both the strain $\varepsilon_{\alpha\beta}$ and the hydrogen distribution $\varrho(\boldsymbol{x})$ enter linearly. Consequently, the internal strain is linearly related to $\varrho(\boldsymbol{x})$,

$$\varepsilon_{\alpha\beta}(\boldsymbol{x}) = \int W_{\alpha\beta\mu\nu}(\boldsymbol{x}, \boldsymbol{x}') P_{\mu\nu} \varrho(\boldsymbol{x}') dV' . \tag{2.53}$$

With this expression we obtain for the internal elastic energy

$$\mathscr{H} = -\tfrac{1}{2} \iint \varrho(\boldsymbol{x}) W(\boldsymbol{x}, \boldsymbol{x}') \varrho(\boldsymbol{x}') dV \, dV' , \tag{2.54}$$

where $W = P_{\alpha\beta} W_{\alpha\beta\mu\nu} P_{\mu\nu}$. Equation (2.54) is the continuum version of (2.9). The problem is now to compute $W(\boldsymbol{x}, \boldsymbol{x}')$.

For the incoherent state one finds immediately from (2.42),

$$W^{\text{inc}}(\boldsymbol{x}, \boldsymbol{x}') = P_{\alpha\beta} S_{\alpha\beta\mu\nu} P_{\mu\nu} \delta(\boldsymbol{x} - \boldsymbol{x}') . \tag{2.55}$$

We recall here that (2.55) is a proper solution to the boundary value problem only if $\partial_\alpha \partial_\beta \varrho(\boldsymbol{x}) = 0$ for all α, β. To proceed with the coherent crystal, we introduce the concept of elastic modes [2.14, 21]. The $W(\boldsymbol{x}, \boldsymbol{x}')$ is a real, symmetrical kernel. Therefore, it can be diagonalized in terms of its eigenfunctions $\psi_{\mathrm{L}}(\boldsymbol{x})$ and (real) eigenvalues E_{L},

$$\int W(\boldsymbol{x}, \boldsymbol{x}') \psi_{\mathrm{L}}(\boldsymbol{x}') dV' = E_{\mathrm{L}} \psi_{\mathrm{L}}(\boldsymbol{x}) . \tag{2.56}$$

The eigenfunctions $\psi_{\mathrm{L}}(\boldsymbol{x})$ are taken to be orthonormal and may (but need not) be chosen to be real. They are assumed to constitute a complete basis. Thus

$$\int \psi_{\mathrm{L}}(\boldsymbol{x}) \psi_{\mathrm{L}'}(\boldsymbol{x}) dV = \delta_{\mathrm{LL}'} ; \qquad \sum_{\mathrm{L}} \psi_{\mathrm{L}}(\boldsymbol{x}) \psi_{\mathrm{L}}(\boldsymbol{x}') = \delta(\boldsymbol{x} - \boldsymbol{x}') \tag{2.57}$$

and we have

$$W(\boldsymbol{x}, \boldsymbol{x}') = \sum_{\mathrm{L}} E_{\mathrm{L}} \psi_{\mathrm{L}}(\boldsymbol{x}) \psi_{\mathrm{L}}(\boldsymbol{x}') . \tag{2.58}$$

With the expansion

$$\varrho(\boldsymbol{x}) = \sum_{\mathrm{L}} \varrho_{\mathrm{L}} \psi_{\mathrm{L}}(\boldsymbol{x}) , \tag{2.59}$$

the internal elastic energy may now be written as

$$\mathscr{H}(\varrho) = -\tfrac{1}{2} \sum_{\mathrm{L}} E_{\mathrm{L}} \varrho_{\mathrm{L}}^2 , \tag{2.60}$$

which is convenient form for thermodynamic applications.

We still have to reformulate the eigenvalue problem (2.56) in such a way that it is tractable for practical calculations. This is achieved as follows [2.21].

Each mode $\psi_L(\boldsymbol{x})$ represents a special density distribution and induces its own strain field $\varepsilon^L_{\alpha\beta}(\boldsymbol{x})$ via (2.53) namely

$$P_{\alpha\beta}\varepsilon^L_{\alpha\beta}(\boldsymbol{x}) = E_L \psi_L(\boldsymbol{x}) . \tag{2.61}$$

The coherency stress (2.36) associated with this single mode can be written as

$$\sigma^L_{\alpha\beta}(\boldsymbol{x}) = C^L_{\alpha\beta\mu\nu}\varepsilon^L_{\mu\nu}(\boldsymbol{x}) , \tag{2.62}$$

with

$$C^L_{\alpha\beta\mu\nu} = C_{\alpha\beta\mu\nu} - \frac{1}{E_L} P_{\alpha\beta} P_{\mu\nu} . \tag{2.63}$$

After insertion of $\sigma^L_{\alpha\beta}$ into the equilibrium and boundary conditions (2.37) and (2.38), we arrive at a (generalized) eigenvalue problem for determining the elastic energies E_L and the eigendisplacements $\boldsymbol{u}_L(\boldsymbol{x})$. The eigenfunctions $\psi_L(\boldsymbol{x})$ are then obtained from (2.61). The advantage of this formulation is that the calculation of the internal stresses and strains induced by $\varrho(\boldsymbol{x})$ has been transformed to a normal elastic problem[2] with the elastic constants (2.63).

In general, the spectrum of elastic modes in a material of compact[3] shape consists of two parts:

1) Macroscopic modes $\{\psi_j(\boldsymbol{x}), E_j\}$ which depend on the boundary conditions, i.e., one the shape of the material. For compact shapes, their energy levels are macroscopically discrete.

2) Bulk modes $\{\psi_q(\boldsymbol{x}), E_q\}$ which are insensitive to boundary conditions. Bulk modes are related to long wavelength phonons.

We now consider a few examples.

Isotropic Sphere with Dilatation Centers

In this case, the eigenvalue problem for the free surface can be solved exactly [2.14]. The eigenfunctions of the macroscopic modes are [$j=(l,m)$, R = radius of the sphere]

$$\psi_{lm}(\boldsymbol{x}) = \left(\frac{2l+3}{R^3}\right)^{1/2} \left(\frac{x}{R}\right)^l Y_{lm}(\hat{\boldsymbol{x}}) , \tag{2.64}$$

---

[2] Obviously, this method can also be applied to thermoelastic problems where internal stresses and strains are created by inhomogeneous heating of the material.

[3] By "compact" we mean that all linear dimensions of the body are of the same order of magnitude.

where Y_{lm}, $l \geqq 0$, $|m| \leqq l$, denote spherical harmonics. The corresponding eigenvalues are

$$E_l = P^2 \frac{2l^2 + 4l + 3}{(2l^2 + 4l + 3)C_{12} + 2(l^2 + l + 1)C_{44}}. \tag{2.65}$$

The bulk modes are completely degenerate, with an energy $E_q = P^2/C_{11}$. The eigenfunctions $\psi_q(\boldsymbol{x})$ of the bulk modes are only specified insofar as they have to be normalizable and orthogonal to the $\psi_{lm}(\boldsymbol{x})$. Otherwise, they can be constructed arbitrarily such that they form a complete basis together with the macroscopic modes (2.64).

With the help of the stability conditions $C_{44} > 0$, $C_{11} + 2C_{12} > 0$, $C_{11}(= C_{12} + 2C_{44}) > 0$, it is easy to show that

$$E_0 = P^2 \kappa_T = E_1 > E_2 > \ldots > \frac{P^2}{C_{12} + C_{44}} > E_q = \frac{P^2}{C_{11}}. \tag{2.66}$$

The level spacing is independent of the volume. The degeneracy of the modes $l = 0$ and $l = 1$ reflects the fact that these modes, being constant or linear in $\boldsymbol{x}$, do not create coherency stresses.

The interaction kernel $W(\boldsymbol{x}, \boldsymbol{x}')$ is given by

$$W(\boldsymbol{x}, \boldsymbol{x}') = M(\boldsymbol{x}, \boldsymbol{x}') + \frac{P^2}{C_{11}} \delta(\boldsymbol{x} - \boldsymbol{x}'), \tag{2.67}$$

where

$$M(\boldsymbol{x}, \boldsymbol{x}') = \sum_{lm} \left(E_l - \frac{P^2}{C_{11}} \right) \psi_{lm}(\boldsymbol{x}) \psi^*_{lm}(\boldsymbol{x}'). \tag{2.68}$$

The dominant attractive terms in $M(\boldsymbol{x}, \boldsymbol{x}')$ are those for small values of l. These terms describe a macroscopically long-ranged interaction varying on the scale R. In the special case $\varrho(\boldsymbol{x}) = \delta(\boldsymbol{x}) + \delta(\boldsymbol{x} - \boldsymbol{r})$, the interaction part of $\mathscr{H}(\varrho)$ is equal to $-W(0, \boldsymbol{r})$, and one recovers the result (2.52) with $N_1 = N_2 = 1$.

For large l, the eigenfunctions $\psi_{lm}(\boldsymbol{x}) \sim (x/R)^l$ are localized at the surface of the sphere and decay rapidly as x decreases. Clearly, any rapid variation of $\psi_{lm}(\boldsymbol{x})$ within one lattice distance is physically meaningless. To eliminate unphysical modes one has to introduce a Debye-type cutoff $l_{\max}$. A reasonable choice is $l_{\max} = R/d$, where d is a length on the order of the lattice parameter. Then $(x/R)^{l_{\max}} \approx \exp[(R - x)/d]$, and the number of macroscopic modes is $\sim V^{2/3}$.

Wegner [2.32] has shown that the spectrum of elastic modes in an isotropic material with dilatation centers and a free surface of arbitrary shape is bounded by $P^2\kappa_T > E_j > P^2/(C_{11}-C_{44})$. He also calculated the spectrum of macroscopic modes of a spherical shell. The macroscopic modes in an elastic sphere with a weakly anisotropic dipole tensor have recently been studied by *May* [2.33]. The main effect of anisotropic defects is to lift the m-degeneracy of the eigenvalues, as expected.

For comparison, we also quote the result for an isotropic sphere with a clamped surface $\boldsymbol{u}(R)=0$

$$\hat{E}_l = \frac{P^2}{C_{11}}\left\{1 - \frac{(l+1)C_{44}}{[lC_{12}+(3l+1)C_{44}]}\right\}. \tag{2.69}$$

Thus $P^2/C_{11} > \hat{E}_l \geqq 0$.

Thin Plates (Foils) and Wires with Dilatation Centers

Consider a rectangular plate in the (x, y)-plane with thickness h much less than the linear dimensions l_x, l_y in the x- and y-direction. On the top and bottom face, (2.38) requires $\sigma^{\mathrm{L}}_{\alpha z}=0$. For a thin plate we may then employ the plane-stress approximation whereby one assumes, first, that $\sigma^{\mathrm{L}}_{\alpha z}=0$ holds also in the interior of the plate and second, that σ^{L}_{xx}, σ^{L}_{yy} and σ^{L}_{xy} are independent of z. Although the plane-stress approximation violates the compatibility relations for a three-dimensional body, this inconsistency is known [2.34] to be insignificant for $h/l_{x,y} \ll 1$. Furthermore, with reference to *De St. Venant*'s principle [2.34], we replace the correct boundary conditions on the corner faces by periodic boundary conditions in the x- and y-direction.

With these simplications one finds [2.35] for a cubic material (cubic axes $\| \hat{\boldsymbol{x}}, \hat{\boldsymbol{y}}, \hat{\boldsymbol{z}}$)

$$E_q = P^2 \frac{2(C_{11}-C_{12})-Q(\boldsymbol{q})}{C_{11}^2-C_{12}^2-C_{11}Q(\boldsymbol{q})}, \tag{2.70}$$

with

$$Q(\boldsymbol{q}) = \frac{(C_{11}-C_{12})q_x^2 q_y^2 \zeta}{2\zeta q_x^2 q_y^2 + (\boldsymbol{q}^2)^2}, \tag{2.71}$$

where $\zeta=(C_{11}-C_{12}-2C_{44})/C_{44}$. The eigenfunctions are of the form $\psi_q(\boldsymbol{x}) = g(z)\exp(\mathrm{i}q_x x + \mathrm{i}q_y y)$, with $g(z)$ linear in z. Clearly, these modes are of a mixed nature as far as the distinction between macroscopic and bulk modes is concerned. The obvious reason for this feature is the large difference in the linear dimensions.

The above spectrum contains the bending modes studied in [2.21] as special cases ($q_x=0$, $q_y=q$ or $q_x=q_y=q/\sqrt{2}$). In addition to the modes (2.70), one still has the stress-free modes with energy $P^2\kappa_T$.

A long, thin wire (diameter d, length l; $d \ll l$) resembles a one-dimensional system. In one dimension there are no coherency stresses and the macroscopic modes are obtained from $\sigma^{\mathrm{L}}_{\alpha\beta}=0$. We therefore conclude that the macroscopic modes in a long, thin wire are nearly degenerate ($E_j \approx P_{\alpha\beta} S_{\alpha\beta\mu\nu} P_{\mu\nu}$), and the kernel $W(\boldsymbol{x}, \boldsymbol{x}')$ is given by (2.55), with corrections vanishing as $d/l \to 0$.

Bulk Modes

If the boundary condition (2.38) for $\sigma^{\mathrm{L}}_{\alpha\beta}(\boldsymbol{x})$ is replaced by periodic boundary conditions for $\boldsymbol{u}_{\mathrm{L}}(\boldsymbol{x})$, then (2.37) can be solved in terms of plane waves, $\exp(\mathrm{i}\boldsymbol{q} \cdot \boldsymbol{x})$. For the eigenvalues $E_q = E(\hat{\boldsymbol{q}})$, $\hat{\boldsymbol{q}} = \boldsymbol{q}/q$, one finds [2.7, 21]

$$E(\hat{\boldsymbol{q}}) = P^2 \left\{ C_{12} + C_{44} \left[1 + \frac{\Omega(\zeta, \hat{\boldsymbol{q}})}{\partial \Omega(\zeta, \hat{\boldsymbol{q}})/\partial \zeta} \right] \right\}^{-1}, \tag{2.72}$$

with ζ given below (2.71) and

$$\Omega(\zeta, \hat{\boldsymbol{q}}) = 1 + \zeta + \zeta^2 (\hat{q}_x^2 \hat{q}_y^2 + \hat{q}_x^2 \hat{q}_z^2 + \hat{q}_y^2 \hat{q}_z^2) + \zeta^3 \hat{q}_x^2 \hat{q}_y^2 \hat{q}_z^2 . \tag{2.73}$$

The eigenvalues $E(\hat{\boldsymbol{q}})$ are related to phonon frequencies $\omega(\boldsymbol{q}, s)$ in the long wavelength limit. In the symmetry directions one has

$$\begin{aligned} E(111) &= 3P^2/(C_{11} + 2C_{12} + 4C_{44}), \\ E(110) &= 2P^2/(C_{11} + C_{12} + 2C_{44}), \\ E(100) &= P^2/C_{11} . \end{aligned} \tag{2.74}$$

2.3 Thermodynamics and Phase Transition

2.3.1 Free Energy

The total configurational Hamiltonian of the protons is now taken to be

$$\mathscr{H}_{\mathrm{tot}} = \tfrac{1}{2} \sum_{ab} (V_{ab} - W_{ab}) \tau_a \tau_b . \tag{2.75}$$

We have included the direct electronic interaction V_{ab} of protons in a rigid lattice. For the temperature region of interest we may assume that classical statistics is applicable. Accordingly, the canonical partition functions reads

$$Z = \sum_{\{\tau\}}{}' \exp(-\beta \mathscr{H}_{\mathrm{tot}}) = \exp[-\beta F(T, V, N)], \tag{2.76}$$

from which we obtain the (configurational) canonical free energy $F(T, V, N)$. In (2.76), $\beta = 1/k_B T$ and the sum over configurations is constrained by $\sum_a \tau_a = N$. The contribution to $\mathscr{H}_{tot}$ from the elastic modes will be split into two parts,

$$W_{ab} = \overset{\infty}{W}_{ab} + \Delta W_{ab}. \tag{2.77}$$

$\overset{\infty}{W}_{ab}$ denotes the elastic interaction in an infinite crystal and is obtained from $\tilde{W}_{ab}$ given by (2.10) after inserting the Green function (2.11) and replacing the sum on $\boldsymbol{q}$ by an integral. The correction ΔW_{ab} embodies the macroscopic modes due to the free surface of the crystal. $\overset{\infty}{W}_{ab}$ is further separated into a short-ranged part,

$$\overset{\infty}{W}{}^s_{ab} = \begin{cases} \overset{\infty}{W}_{ab}, & |\boldsymbol{S}^a - \boldsymbol{S}^b| < R_0 \\ 0, & |\boldsymbol{S}^a - \boldsymbol{S}^b| > R_0 \end{cases}, \tag{2.78}$$

with some suitably chosen cutoff R_0, and into a long-ranged part $\overset{\infty}{W}{}^l_{ab} = \overset{\infty}{W}_{ab} - \overset{\infty}{W}{}^s_{ab}$. The latter contains the r^{-3}-interaction discussed previously. The contribution from $\overset{\infty}{W}{}^s_{ab}$ to $\mathscr{H}_{tot}$ will be combined with the electronic interaction into a "reference" Hamiltonian

$$\mathscr{H}_0 = \tfrac{1}{2} \sum_{ab} (V_{ab} - \overset{\infty}{W}{}^s_{ab}) \tau_a \tau_b. \tag{2.79}$$

$\overset{\infty}{W}_{ab}$ can be calculated from lattice dynamics, but we are lacking detailed knowledge about the form of V_{ab}, as was stressed at the beginning. Fortunately, this is not too serious a drawback for a study of the qualitative features of phase transitions driven by macroscopic elastic modes. Occasionally we shall replace V_{ab} by a simple hard-core repulsion. For the main part of the following it will be sufficient to assume that the reference system defined by (2.79) does not undergo a phase transition in the critical region of the total system.

It is not the place here to elaborate on the technical steps necessary to cast the partition function into a manageable form. The details may be found in [2.14, 16]. After exploiting the long-range nature of the elastic interaction which tends to suppress fluctuations in the proton density, one arrives at a variational expression for the total free energy,

$$F(T, V, N) = \operatorname*{Min}_{\varrho(\boldsymbol{x})} [\mathscr{F}_0(T, \varrho) - \mathscr{M}(\varrho)], \tag{2.80}$$

where $\mathscr{F}_0$, arising from the reference system, may be written in the form

$$\mathscr{F}_0 = k_B T \int \left\{ f(T, \varrho(\boldsymbol{x})) + \frac{\xi_0^2}{2} [\nabla \varrho(\boldsymbol{x})]^2 \right\} dV, \tag{2.81}$$

and where

$$\mathscr{M} = \tfrac{1}{2} \iint \varrho(\boldsymbol{x}) M(\boldsymbol{x}, \boldsymbol{x}') \varrho(\boldsymbol{x}') dV dV', \tag{2.82}$$

with $M(\boldsymbol{x}, \boldsymbol{x}')$ representing the continuum approximation to $\Delta W_{ab} + \overset{\infty}{W}{}^{l}_{ab} = W_{ab} - \overset{\infty}{W}{}^{s}_{ab}$. For an isotropic spherical crystal, $M(\boldsymbol{x}, \boldsymbol{x}')$ is given by (2.68). The ξ_0 in (2.81) is a measure of the range of forces entering $\mathscr{H}_0$. The minimum in (2.80) has to be calculated subject to the constraint

$$\frac{1}{V} \int \varrho(\boldsymbol{x}) dV = \bar{\varrho} = \frac{N}{V}. \tag{2.83}$$

To arrive at (2.80), one makes a mean-field approximation for evaluating the contribution to Z of the macroscopic modes. Because of the macroscopic range of the interaction of these modes, one is inclined to believe that (2.80) might turn into an exact expression in the thermodynamic limit. If this conjecture, which has some experimental support [2.4, 5], is correct, then metal-hydrogen systems in the coherent state would provide an example where a van der Waals-type theory is strictly valid.

2.3.2 Phase Transition

The form of the equilibrium density distribution $\varrho_{\mathrm{eq}}(\boldsymbol{x})$, which minimizes $\mathscr{F}_0 - \mathscr{M}$ in (2.80), depends on the temperature and on the average density $\bar{\varrho}$. In the case of a normal liquid, one would expect to find a one-phase region where $\varrho_{\mathrm{eq}}(\boldsymbol{x}) = \bar{\varrho}$, and a two-phase coexistence domain at temperatures below a critical point T_c with nonuniform density $\varrho_{\mathrm{eq}}(\boldsymbol{x})$ varying across the interface. In metal-hydrogen alloys there is also a homogeneous phase and a critical point. However, it turns out that the inhomogeneous phase below T_c in the coherent state is of a different nature than the two-phase coexistence in a normal fluid.

Homogeneous Phase

If we set $\varrho(\boldsymbol{x}) = \bar{\varrho}$ then (2.80) simplifies to

$$F/V \to \bar{F}/V = k_B T f(T, \bar{\varrho}) - \tfrac{1}{2} M_0 \bar{\varrho}^2, \tag{2.84}$$

with

$$\begin{aligned} M_0 &= \frac{V}{N_H^2} \sum_{ab} (W_{ab} - \overset{\infty}{W}{}^{s}_{ab}) \\ &= P^2 \kappa_T - \frac{V}{N_H} \sum_{b} \overset{\infty}{W}{}^{s}_{ab}, \end{aligned} \tag{2.85}$$

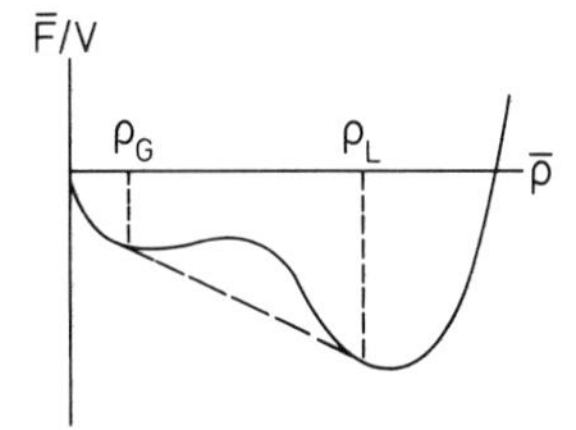

Fig. 2.3. Free energy for uniform hydrogen density, $T<T_c$

where we used the exact result (2.13) with an isotropic $P_{\alpha\beta}$. For isotropic crystals, $M_0=P^2(\kappa_T-1/C_{11})$, but $\sum_b^{\infty} W_{ab}^s$ may also be calculated numerically for anisotropic crystals; see [2.16] for Nb.

It can be shown [2.14] that a constant density indeed yields a minimal free energy for temperatures above the critical point T_c, ϱ_c, given by

$$k_B T \frac{\partial^2}{\partial \bar{\varrho}^2} f(T,\bar{\varrho}) = M_0\,,$$
$$\frac{\partial^3}{\partial \bar{\varrho}^3} f(T,\bar{\varrho}) = 0\,. \tag{2.86}$$

In [2.16] the $f(T,\varrho)$ has been calculated in terms of a low-density expansion

$$f(T,\bar{\varrho}) = \bar{\varrho}\ln\left(\bar{\varrho}\frac{V}{N_H}\right) - \bar{\varrho} + \sum_{n\geqq 2} A_n(T)\bar{\varrho}^n\,, \tag{2.87}$$

for a model where V_{ab} is replaced by a hard core. For NbH_x, the hard-core diameter was chosen to be compatible with the known structure of the β-phase [2.36]. The numerical results for hydrogen in Nb will be compared with experiments in Section 2.6.

Below T_c, $\bar{F}(T,\bar{\varrho})$ is of the form as shown in Fig. 2.3. In the case of a normal liquid one would then draw the double tangent (dashed line) to define the densities $\varrho_G(T)$ and $\varrho_L(T)$ of the gas and the liquid phase coexisting in the interval $\varrho_G(T)<\bar{\varrho}<\varrho_L(T)$. As we shall see, this Maxwell construction is not legitimate as long as the host crystal remains coherent.

Inhomogeneous Phase

In the region $T<T_c$, $\varrho_G(T)<\bar{\varrho}<\varrho_L(T)$ the free energy functional $\mathscr{F}_0-\mathscr{M}$ is minimized by a spatially varying density. To find the correct equilibrium distribution $\varrho_{eq}(\boldsymbol{x};\,T,\bar{\varrho})$, we now have to distinguish between the coherent and the incoherent state. In the following we only consider isotropic crystals with $P_{\alpha\beta}=P\delta_{\alpha\beta}$.

Incoherent State. From (2.55) we find

$$F^{\mathrm{inc}}(T, V, N) = \underset{\varrho(x)}{\mathrm{Min}} \int \left\{ k_B T f(T, \varrho(\boldsymbol{x})) - \frac{1}{2} M_0 \varrho^2(\boldsymbol{x}) + \frac{\xi_0^2}{2} [\nabla \varrho(\boldsymbol{x})]^2 \right\} dV . \tag{2.88}$$

This expression is identical with the free energy in the van der Waals approximation for a normal fluid as analyzed, for instance, by *van Kampen* [2.37]. For an average density $\bar{\varrho} = \lambda \varrho_L(T) + (1-\lambda)\varrho_G(T)$, $0 < \lambda < 1$, one finds a liquid phase of volume $\approx \lambda V$ and a gaseous phase of volume $\approx (1-\lambda) V$. Regions where $\varrho_{eq}(\boldsymbol{x}) = \varrho_G$ are separated from those where $\varrho_{eq}(\boldsymbol{x}) = \varrho_L$ by an interface of width proportional to $\xi_0[\varrho_L(T) - \varrho_G(T)]^{-1}$. The contributions to F^{inc} of the interface regions, where $\varrho_{eq}(\boldsymbol{x})$ varies with $\boldsymbol{x}$, are of the order $\xi_0 V^{2/3}$ and are negligible in the limit of large V. Hence, for $V \to \infty$

$$\frac{1}{V} F^{\mathrm{inc}}(T, V \cdot \bar{\varrho}) = \frac{\lambda}{V} \bar{F}[T, V\varrho_L(T)] + \frac{1-\lambda}{V} \bar{F}[T, V\varrho_G(T)] , \tag{2.89}$$

which is, of course, the free energy as obtained by the Maxwell construction.

In the incoherent state the elastic interaction is short ranged. According to the modern theory of phase transitions and by analogy with a normal fluid, one expects large fluctuations in the local density to occur in the vicinity of the critical point. These fluctuations have been neglected, however, in the steps leading to (2.88). Thus, our treatment of the phase transition in an incoherent crystal reduces to a molecular field-type description with its well-known deficiencies.

Although the incoherent state constitutes the ultimate equilibrium state, it is difficult to measure the incoherent phase diagram in practice. The coherency stresses will not be removed entirely within normal experimental time spans by the creation and motion of dislocation. Rather, residual localized stresses are expected to persist over long time periods. These metastable configurations cause hysteresis effects depending on the history of the sample preparation and on the cooling program after loading with hydrogen. Thus, from a practical point of view, complete incoherency is an idealization, as is the opposite extreme, the coherent state.

Coherent State. To illustrate the effect of coherency on the phase separation, let us consider an isotropic spherical sample with a hydrogen distribution

$$\varrho(\boldsymbol{x}) = \begin{cases} \varrho_L, & 0 \leqq x \leqq r, \\ \varrho_G, & r < x \leqq R . \end{cases} \tag{2.90}$$

In the incoherent case this would be a proper two-phase configuration viewed on a scale where the interface thickness $\sim \xi_0$ can be neglected. The average density is again given by $\bar{\varrho}=\lambda\varrho_L+(1-\lambda)\varrho_G$ with $\lambda=(r/R)^3$. One easily checks that $\mathscr{M}(\varrho)=\mathscr{M}(\bar{\varrho})=-VM_0\bar{\varrho}^2/2$. Therefore, below T_c the two-phase configuration (2.90) does not lead to a reduction of the free energy as compared to $\bar{F}$. The reason for this unusual behaviour lies in the long-ranged nature of the elastic forces in a coherent crystal. If one manipulates $\mathscr{M}(\varrho)$ to extract a gradient term of the form appearing in (2.81), one finds the length corresponding to ξ_0 to be on the order of the radius R of the sphere. This demonstrates that any rapid variation of $\varrho(\boldsymbol{x})$ leads to a contribution $\sim V$ to the elastic energy. Consequently, the Maxwell construction is not legitimate in a coherent crystal. Nevertheless, the hump occurring in $\bar{F}$ for $T<T_c$ can be reduced by minimizing $\mathscr{F}_0-\mathscr{M}$ in terms of density distributions which are linear combinations of elastic modes. Let us put

$$\varrho(\boldsymbol{x})=\varrho_c+\Delta\varrho(\boldsymbol{x})\,; \qquad \frac{1}{V}\int\Delta\varrho(\boldsymbol{x})dV=\bar{\varrho}-\varrho_c\,. \tag{2.91}$$

We expand $\Delta\varrho(\boldsymbol{x})$ according to (2.59),

$$\Delta\varrho(\boldsymbol{x})=\sum_{lm}\varrho_{lm}\psi_{lm}(\boldsymbol{x})+\sum_{q}\varrho_q\psi_q(\boldsymbol{x})\,. \tag{2.92}$$

The expansion coefficients for the macroscopic modes ϱ_{lm} and for the bulk modes ϱ_q will be our variational parameters, with $\varrho_{00}=\sqrt{V}(\bar{\varrho}-\varrho_c)$ fixed.

The subsequent procedure is straightforward but tedious. One first expands $\mathscr{F}_0$ in powers of $\Delta\varrho$. After inserting (2.92) and minimizing, one arrives at a set of coupled nonlinear equations for ϱ_{lm} and ϱ_q. In practice, one has to truncate both expansions of $\mathscr{F}_0$ and $\Delta\varrho$. Because of their discrete spectrum, however, a few macroscopic modes with small values of l dominate for temperatures slightly below T_c. In [2.14] the free energy was minimized both analytically and numerically. In the analytical treatment we only included ϱ_{l0} for $l<2$ with $\mathscr{F}_0$ expanded up to fourth order in $\Delta\varrho$ and the $\partial^n f/\partial\varrho^n$, $n\leqq 4$, left as parameters. In the numerical computations we took into account $l\leqq 5$, $n\leqq 5$ and the leading contribution of suitably constructed bulk modes. The combined effect of the macroscopic modes with $3\leqq l\leqq 5$ and of bulk modes on the free energy turned out to be about 10% of the ($l\leqq 2$)-modes at $(T-T_c)/T_c=-0.1$. The numerical computation was refined in [2.16] by the use of the expansion (2.87) for $f(T,\varrho)$.

2.3.3 Discussion

A quantitative discussion of the phase diagram of hydrogen in Nb is postponed to Section 2.6. Here the qualitative features of the free energy and the phase diagram of an isotropic sphere are summarized schematically in Fig. 2.4. In the upper part of Fig. 2.4 the free energy $F(T,\bar{\varrho})$ is shown as a function of the

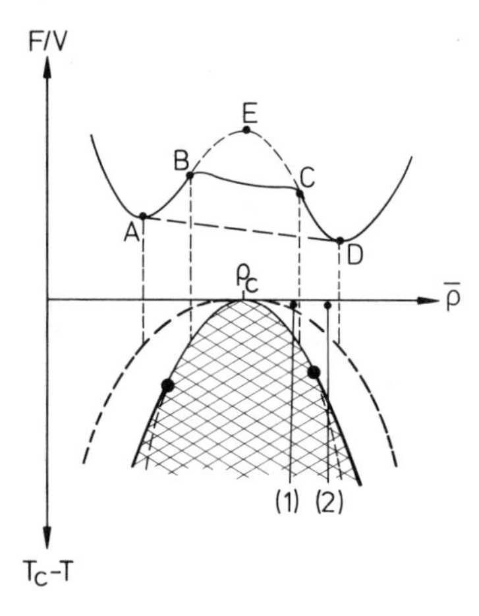

Fig. 2.4. Free energy and phase diagram of a coherent spherical crystal ($T<T_c$; schematically)

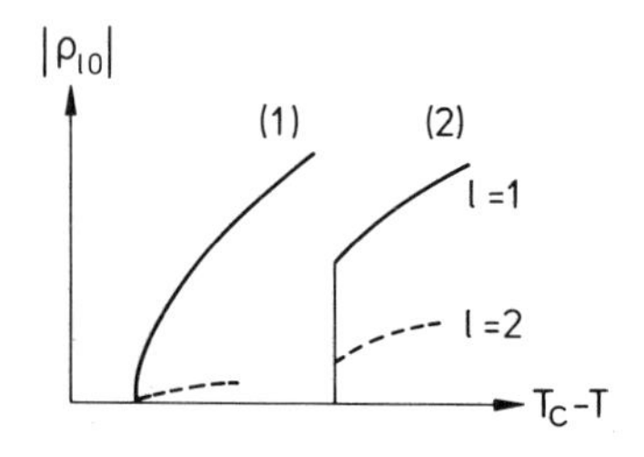

Fig. 2.5. Temperature dependence of the amplitudes of macroscopic modes ($l=1,2$; $m=0$) in the inhomogeneous phase along the paths (1), (2) in Fig. 2.4

average density $\bar{\varrho}$, for some $T<T_c$. The coherent state is described by the curve ABCD. The minimization in terms of elastic modes has indeed reduced the hump of $\bar{F}(T,\bar{\varrho})$ (curve AED) in the density range between B and C, as expected. For temperatures above the tricritical temperatures T_{tr} (see below), the points B and C coincide with the points of inflection on AED.

The free energy of the incoherent state is obtained by drawing the double tangent AD, which gives rise to the coexistence curve (dashed) of the phase diagram in the lower part of Fig. 2.4. In the incoherent case, for densities $\varrho_A(T)<\bar{\varrho}<\varrho_D(T)$ the sample contains a hydrogen-rich ("liquid") phase in coexistence with a hydrogen-poor ("gaseous") phase. The mismatch of the lattice parameters between both phases necessarily implies the presence of dislocations localized in the interface regions.

In the coherent state, the hydrogen density is uniform outside of the cross-hatched area. Inside this area, the density is nonuniform and is given by $\varrho_{eq}(\boldsymbol{x})=\varrho_c+\sum_L \varrho_L(T,\bar{\varrho})\psi_L(\boldsymbol{x})$ [cf. (2.91, 92)] (see Fig. 2.5). If the sample is cooled along the path (1) in Fig. 2.4, the inhomogeneity in $\varrho_{eq}(\boldsymbol{x})$ develops continuously as we enter the cross-hatched area. Along the path (2), however, the amplitudes $|\varrho_L(T)|$ jump discontinuously from zero to a finite value as we cross the heavy line. The latter terminates in a "tricritical" point (heavy dot), where the jump size vanishes[4]. Obviously, the coherent phase transition differs rather drastically from a normal gas-liquid transition. Its unfamiliar features are a consequence of the presence of coherency stresses, which inhibit rapid density variations and thus prevent the appearance of a sharp interface.

The above discussion applies to the phase diagram for a canonical ensemble of spherical crystals. Since the spectrum of elastic modes depends on the shape of the crystal, the canonical free energy will also be shape dependent in that part of the $(\bar{\varrho}, T)$-plane where $\varrho_{eq}(\boldsymbol{x})$ is nonuniform. One may ask how the phase diagram is altered as we deform a sphere continuously into a long thin wire, for instance. Although we cannot offer an explicit calculation, the following behavior suggests itself: In the course of the deformation the tricritical points move along the spinodal towards the critical point. Simultaneously, the lines of first-order transition approach the coexistence curve.

This picture incorporates the fact that for a long, thin wire there is practically no distinction between the coherent and the incoherent phase diagram, since in a nearly one-dimensional crystal the effects of coherency stresses are negligible.

In a normal fluid, the thermal density fluctuations in small subvolumes are described by considering the subvolumes to be members of a grand canonical ensemble with prescribed chemical potential μ. Because of the macroscopic range of the elastic interaction this is no longer true in metal-hydrogen alloys. Instead of the equilibrium condition $\mu = \text{const.}$ within the crystal, we enforce coherency, which is a global requirement for the whole crystal.

It is well known from statistical mechanics that the canonical free energy of any macroscopic piece of matter in thermal equilibrium has to be a concave function of the density ϱ. Therefore, the coherent state with a free energy of the form shown in Fig. 2.4 cannot be a thermal equilibrium state in the strict sense[5]. This is also physically evident, since the free energy $F(T, \bar{\varrho})$ for the region of nonuniform density can be lowered by the nucleation and motion of dislocations. Therefore, the question arises whether coherency below T_c can be achieved experimentally. *Burkhardt* [2.38] has calculated the coherency shear stresses in a Nb-sphere. From a comparison of his results with empirical values of critical yield stresses, he concludes that there might be an accessible temperature interval extending down to 5–10 K below T_c in which the effects of coherency on the phase transition are observable.

As pointed out above, the special features of the elastic energy also enter into the properties of thermal hydrogen density fluctuations, even above T_c where $\varrho_{eq}(\boldsymbol{x})$ is uniform. This topic will be discussed in the next section.

---

[4] In [Ref. 2.14; Eqs. (7.36–40)], this line of first-order transition was erroneously identified with one of the associated spinodals. However, the phase diagram [Fig. 7, Ref. 2.14] has been plotted from the correct expressions.

[5] We recall that the residual hump in the free energy of the coherent state is not due to an inadequate approximation. An analogous minimization procedure as employed here does lead in the case of normal liquid to a concave free energy [2.37].

Lattice gas models with hard-core repulsion and weak, long-ranged attractive forces have also been investigated by *Hall* and *Stell* [2.39]. They obtained phase diagrams resembling the empirical ones for hydrogen in V, Nb, Ta. In their approach the lattice is rigid and the attractive interaction is introduced in an ad hoc fashion.

2.4 Fluctuations and Correlations

2.4.1 Thermal Hydrogen-Density Fluctuations

According to the theory of thermal fluctuations (see, e.g., [2.40]). the Boltzmann factor for a nonuniform hydrogen configuration

$$\tilde{\varrho}(\boldsymbol{x})=\varrho(\boldsymbol{x})-\bar{\varrho} \tag{2.93}$$

is given by

$$w(T,\bar{\varrho};\tilde{\varrho})\sim\exp[-\beta\Delta\mathscr{F}(T,\bar{\varrho};\tilde{\varrho})], \tag{2.94}$$

with

$$\Delta\mathscr{F}(T,\bar{\varrho};\tilde{\varrho})=\mathscr{F}(T,\bar{\varrho}+\tilde{\varrho})-\mathscr{F}(T,\bar{\varrho}), \tag{2.95}$$

where $\mathscr{F}=\mathscr{F}_0-\mathscr{M}$. $\Delta\mathscr{F}$ denotes the minimum work required to alter the density isothermally from its equilibrium value $\bar{\varrho}$ to the configuration $\varrho(\boldsymbol{x})$. In the continuum approximation the quantity $\mathscr{M}$ is related to the (internal) elastic energy $\mathscr{H}$, (2.54), by

$$\begin{aligned}\mathscr{H}(\varrho)+\mathscr{M}(\varrho)&=-\tfrac{1}{2}\left(\frac{V}{N_{\mathrm{H}}}\sum_b^{\infty}W_{ab}^s\right)\int\varrho^2(\boldsymbol{x})dV\\&\equiv\mathscr{W}^s(\varrho).\end{aligned} \tag{2.96}$$

For an isotropic crystal the term in parentheses equals P^2/C_{11}. It will be convenient to work with $\mathscr{H}$ instead of $\mathscr{M}$ and to absorb the term $\mathscr{W}^s(\varrho)$ into the reference free energy by setting $\mathscr{F}_0-\mathscr{W}^s=\hat{\mathscr{F}}_0$ and hence

$$\Delta\mathscr{F}=\Delta\hat{\mathscr{F}}_0+\Delta\mathscr{H}. \tag{2.97}$$

We only consider density fluctuations in the one-phase region. $\Delta\mathscr{F}$ may then be expanded in powers of $\tilde{\varrho}$. Keeping only quadratic terms we have

$$\begin{aligned}\Delta\mathscr{F}=&\tfrac{1}{2}a(T,\bar{\varrho})\int\tilde{\varrho}^2(\boldsymbol{x})dV\\&-\tfrac{1}{2}\int\int\tilde{\varrho}(\boldsymbol{x})W(\boldsymbol{x},\boldsymbol{x}')\tilde{\varrho}(\boldsymbol{x}')dVdV',\end{aligned} \tag{2.98}$$

with $a>0$, assuming that the reference system does not undergo a phase transition in the $(T,\bar{\varrho})$-region of interest here. In the expansion of $\Delta\hat{\mathscr{F}}_0$ we ignored the gradient term because the density fluctuations are governed by the elastic energy. There are no linear terms in (2.98) since the volume integral of $\tilde{\varrho}$ vanishes. With the expansion

$$\tilde{\varrho}(\boldsymbol{x})=\sum_{L\neq 0}\tilde{\varrho}_L\psi_L(\boldsymbol{x}), \tag{2.99}$$

we find [6]

$$\Delta\mathscr{F}=\tfrac{1}{2}\sum_{L\neq 0}[a(T,\bar{\varrho})-E_L]\tilde{\varrho}_L^2. \tag{2.100}$$

Consequently, the Boltzmann factor reduces to a product of decoupled Gaussians and we obtain [2.21]

$$\langle\tilde{\varrho}_L\tilde{\varrho}_{L'}\rangle=\frac{k_BT}{a(T,\bar{\varrho})-E_L}\delta_{LL'}(1-\delta_{L0}), \tag{2.101}$$

where the term in angle brackets on the left-hand side denotes the thermal average. For each mode a "spinodal temperature" $T_L(\bar{\varrho})$ may be defined to be the solution of

$$a(T,\bar{\varrho})=E_L. \tag{2.102}$$

The critical temperature is then given by $T_c=T_1(\varrho_c)=T_0(\varrho_c)$. To leading order in $T-T_L(\bar{\varrho})$, (2.101) takes the Curie-Weiss form

$$\langle\tilde{\varrho}_L^2\rangle=\frac{\lambda_L(\bar{\varrho})}{T-T_L(\bar{\varrho})}\qquad(L\neq 0), \tag{2.103}$$

where $\lambda_L=k_BT_L[\partial a(T_L,\bar{\varrho})/\partial T_L]^{-1}$. With the results of [2.16] for $\mathscr{F}_0$ and $\overset{\infty}{W}{}^s_{ab}$ one finds for Nb that

$$a(T,\bar{\varrho})\approx a_0(\bar{\varrho})+a_1(\bar{\varrho})T, \tag{2.104}$$

which is valid for temperatures extending from T_c up to about 100 K above T_c. This implies that (2.103) also holds approximately in the same temperature interval.

The features of the spectrum of the elastic modes are reflected in the sequence of spinodal temperatures (Fig. 2.6):

Macroscopic Modes $(L=j)$: For crystals with compact shape the sequence of the T_j's is macroscopically discrete. Large values of j correspond to modes

[6] The stress-free modes are denoted by $L=0,1$. Their eigenvalues are $E_0=E_1=P^2\kappa_T$, even for anisotropic crystals of arbitrary shape. Since $\psi_0(\boldsymbol{x})=1/\sqrt{V}$ we have $\tilde{\varrho}_0=0$.

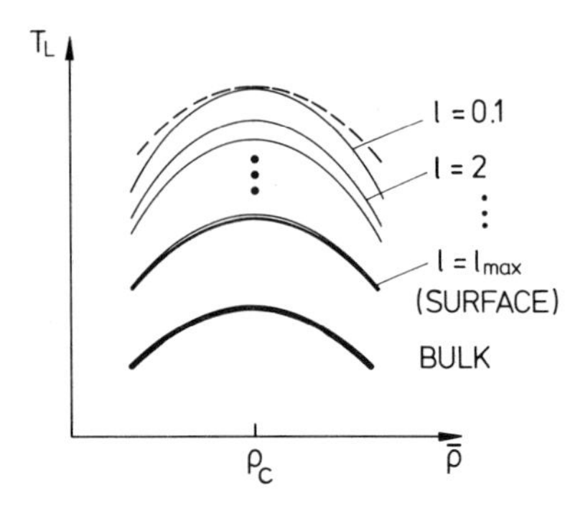

Fig. 2.6. Spinodal temperatures of an isotropic spherical crystal (schematically). Dashed: coexistence curve. In an anisotropic crystal the bulk spinodals are spread out into a band

localized at the surface. As pointed out in the next section, the fluctuations of macroscopic modes can be investigated experimentally by Gorsky relaxation strength measurements.

Bulk Modes: These are labelled by a wave vector $\boldsymbol{q}$ and the eigenvalues are given by (2.72). In anisotropic crystals the corresponding spinodal temperatures depend on the direction of $\boldsymbol{q}$. Since the top of the bulk-mode energy band lies well below $E_1(=E_0)$, (2.102) predicts a substantial suppression of the bulk spinodal temperatures $T_{\boldsymbol{q}}(\varrho_0)$ as compared to T_c.

2.4.2 Fluctuations and Anelastic Relaxation Strength

Let us apply an external force $\boldsymbol{k}^e$ to a loaded crystal at time $t=0$. There will first be an instantaneous elastic strain $\varepsilon^e(\boldsymbol{k}^e)$, which can be calculated from the (2.34) and (2.35). Then, the protons will start to diffuse from compressed to dilatated regions, causing an additional "anelastic" strain $\varepsilon^a(t, \boldsymbol{k}^e)$ (Gorsky effect, see [2.42, 43]). In the final equilibrium or "relaxed" state ($t\to\infty$) the nonuniform density is given by $\langle\varrho(\boldsymbol{x})\rangle_{\boldsymbol{k}^e}=\bar{\varrho}+\langle\tilde{\varrho}(\boldsymbol{x})\rangle_{\boldsymbol{k}^e}$, and the total strain $\varepsilon^r(\boldsymbol{k}^e)$ will be

$$\varepsilon^r(\boldsymbol{k}^e)=\varepsilon^e(\boldsymbol{k}^e)+\varepsilon^a(\boldsymbol{k}^e), \tag{2.105}$$

where

$$\varepsilon^a(\boldsymbol{k}^e)\equiv\varepsilon^a(t=\infty, \boldsymbol{k}^e)=\langle\varepsilon^i\rangle_{\boldsymbol{k}^e}. \tag{2.106}$$

The thermal average of the internal strain is determined by (2.36–38) with $\varrho(\boldsymbol{x})$ replaced[7] by $\langle\tilde{\varrho}(\boldsymbol{x})\rangle$. From (2.53) (where ε denotes the internal strain ε^i) we find

$$\begin{aligned} P_{\alpha\beta}\langle\varepsilon^i_{\alpha\beta}(\boldsymbol{x})\rangle_{\boldsymbol{k}^e} &= P_{\alpha\beta}\varepsilon^a_{\alpha\beta}(\boldsymbol{x};\boldsymbol{k}^e) \\ &= \int W(\boldsymbol{x},\boldsymbol{x}')\langle\tilde{\varrho}(\boldsymbol{x}')\rangle_{\boldsymbol{k}^e} dV'. \end{aligned} \tag{2.107}$$

[7] In the present discussion of the anelastic response, the unstrained reference state is taken to be the state with uniform density $\bar{\varrho}$. Within linear elasticity theory this amounts simply to replace $\varrho(\boldsymbol{x})$ by $\tilde{\varrho}(\boldsymbol{x})$ in the formulae of Section 2.2.2. In this way, one may also take into account the dependence of $C_{\alpha\beta\mu\nu}$ and $P_{\mu\nu}$ on $\bar{\varrho}$.

To compute $\langle\tilde{\varrho}(\boldsymbol{x})\rangle_{\boldsymbol{k}^e}$ we add $\mathscr{H}^{ie}(\tilde{\varrho}, \boldsymbol{k}^e)$ to $\Delta\mathscr{F}$ in the exponent of the Boltzmann factor (2.94). In the linear response approximation we find

$$\langle\tilde{\varrho}(\boldsymbol{x})\rangle_{\boldsymbol{k}^e} = \beta \int \langle\tilde{\varrho}(\boldsymbol{x})\tilde{\varrho}(\boldsymbol{x}')\rangle P_{\alpha\beta}\varepsilon^e_{\alpha\beta}(\boldsymbol{x}'; \boldsymbol{k}^e)\, dV'. \tag{2.108}$$

After expanding (2.107) and (2.108) in terms of elastic modes we arrive at

$$\frac{P\cdot\varepsilon^a_L}{P\cdot\varepsilon^e_L} = \frac{E_L}{k_B T}\langle\tilde{\varrho}^2_L\rangle = \frac{E_L}{a(T,\bar{\varrho}) - E_L}, \tag{2.109}$$

where $P\cdot\varepsilon^{a,e}_L$ abbreviates $P_{\alpha\beta}\int \varepsilon^{a,e}_{\alpha\beta}(\boldsymbol{x})\psi_L(\boldsymbol{x})dV$. The expression (2.109) allows us to obtain information about the spinodal temperatures of macroscopic modes from relaxation strength experiments and thus provides a valuable check of the present theory based on the assumption of a predominant elastic interaction. By bending a foil, for instance, one examines the anelastic response of a single mode. Such bending experiments have been performed by *Tretkowski* et al. [2.44] on Nb foils and wires (see Sect. 2.6).

Equation (2.108) may be also written in the form

$$\langle\tilde{\varrho}(\boldsymbol{x})\rangle_{\boldsymbol{k}^e} = \frac{1}{a(T,\bar{\varrho})} P_{\alpha\beta}\varepsilon^r_{\alpha\beta}(\boldsymbol{x}). \tag{2.110}$$

Inserting (2.110) into the expression (2.28) (cf. footnote 7) we obtain the stress tensor for the relaxed state

$$\sigma^r_{\alpha\beta}(\boldsymbol{x}) = \left[C_{\alpha\beta\mu\nu} - \frac{1}{a(T,\bar{\varrho})} P_{\alpha\beta}P_{\mu\nu}\right]\varepsilon^r_{\mu\nu}(\boldsymbol{x}). \tag{2.111}$$

The brackets define the relaxed elastic constants [2.45] $C^r_{\alpha\beta\mu\nu}$. Hence, the relaxed strain $\varepsilon^r_{\alpha\beta}$ for given external forces is simply obtained from the elastic strain by substituting in $\varepsilon^e_{\alpha\beta}$ the relaxed elastic constants for the "instantaneous" ones, $C_{\alpha\beta\mu\nu}$.

2.5 Nonequilibrium Phenomena

The peculiar nature of the elastic interaction (macroscopic range, shape dependence) is not only reflected in the equilibrium thermodynamics but shows up also in the transport properties of metal-hydrogen alloys in nonequilibrium situations. We demonstrate this fact by two examples, namely the anelastic relaxation above T_c and the spinodal decomposition below T_c.

In both cases one studies the temporal evolution of the hydrogen distribution

$$\varrho(\boldsymbol{x},t)=\bar{\varrho}+\tilde{\varrho}(\boldsymbol{x},t); \qquad \int \tilde{\varrho}(\boldsymbol{x},t)dV=0 \tag{2.112}$$

after either applying an external mechanical force, or quenching the system from above T_c to temperatures below T_c where the uniform hydrogen configuration becomes unstable.

In this section, we closely follow recent work by *Janssen* [2.46] and outline first the arguments leading to a generalized diffusion equation for $\tilde{\varrho}(\boldsymbol{q},t)$. In this approach, which is a modification of the conventional theory [2.20, 24, 47] of spinodal decomposition, one only describes the average time dependence of $\tilde{\varrho}(\boldsymbol{x},t)$ and neglects temporal fluctuations. The main difference to the conventional theory lies in the treatment of macroscopic modes here which are not considered in the former.

2.5.1 Diffusion Equation

The diffusion equation is obtained from the continuity equation

$$\partial_t\tilde{\varrho}+\nabla\cdot\boldsymbol{j}=0 \tag{2.113}$$

with the usual Ansatz for the particle current $\boldsymbol{j}$,

$$\boldsymbol{j}(\boldsymbol{x},t)=-\varrho(\boldsymbol{x},t)B(\varrho)\nabla\frac{\delta\mathscr{F}}{\delta\varrho(\boldsymbol{x},t)}, \tag{2.114}$$

where $\mathscr{F}=\hat{\mathscr{F}}_0+\mathscr{H}$. As for the boundary condition on the surface of the sample we choose

$$\boldsymbol{n}\cdot\boldsymbol{j}=0. \tag{2.115}$$

In general, the kinetic coefficient $B(\varrho)$ is an unknown functional of $\varrho(\boldsymbol{x},t)$. To proceed with a manageable model, we replace the product $\varrho(\boldsymbol{x},t)\cdot B(\varrho)$ in (2.114) by $\bar{\varrho}\cdot B(\bar{\varrho})$, where $B(\bar{\varrho})\equiv\bar{B}$ denotes the hydrogen mobility [8]. Combining (2.113, 114), expanding in terms of elastic modes, and using (2.115), we find

$$\partial_t\tilde{\varrho}_{\mathrm{L}}(t)=-\sum_{\mathrm{L}'}D_{\mathrm{LL}'}\frac{\partial\mathscr{F}}{\partial\tilde{\varrho}_{\mathrm{L}'}(t)}, \tag{2.116}$$

8 For the discussion of the anelastic response above T_c this is presumably not a serious approximation, since, in this case, we shall be content with a linearized version of (2.114). The approximation is harder to justify in the case of the spinodal decomposition, which is an intrinsically nonlinear problem. However, one might argue that near T_c where $|\tilde{\varrho}|$ is small compared to $\bar{\varrho}$, the important nonlinearities are those due to the coupling of modes in $\mathscr{F}$.

where

$$D_{LL'} = \bar{\varrho}\bar{B} \int \nabla\psi_L(\boldsymbol{x}) \cdot \nabla\psi_{L'}(\boldsymbol{x}) dV, \tag{2.117}$$

with $D_{00}=0$. If only bulk modes are taken into account, then (2.116) reduces to Cahn's diffusion equation, provided the gradient term in $\hat{\mathscr{F}}_0$ is kept. With the macroscopic modes included it is advantageous to incorporate the boundary condition (2.115) in the kinetic equation (2.116). This is achieved by making use of the static Green function $\Gamma(\boldsymbol{x}, \boldsymbol{x}')$ of the diffusion problem,

$$\nabla^2 \Gamma(\boldsymbol{x}, \boldsymbol{x}') = \delta(\boldsymbol{x}-\boldsymbol{x}') - \frac{1}{V},$$

$$\boldsymbol{n}(\boldsymbol{x}) \cdot \nabla\Gamma(\boldsymbol{x}, \boldsymbol{x}') = 0, \qquad \boldsymbol{x} \text{ on } \Sigma. \tag{2.118}$$

With the help of (2.118), (2.116) for $L \neq 0$ can be written as

$$\sum_{L' \neq 0} \Gamma_{LL'} \partial_t \tilde{\varrho}_{L'}(t) = -\bar{\varrho}\bar{B} \frac{\partial \mathscr{F}}{\partial \tilde{\varrho}_L(t)}, \tag{2.119}$$

whereby

$$\Gamma_{LL'} = \iint \psi_L(\boldsymbol{x}) \Gamma(\boldsymbol{x}, \boldsymbol{x}') \psi_{L'}(\boldsymbol{x}') dV dV'. \tag{2.120}$$

Going back to (2.116) and (2.117) we note that the coefficients $D_{LL'}$ provide the basic (inverse) relaxation times of the elastic modes. These time scales are determined by the mobility $\bar{B}$, which derives from the individual jump motion of the protons, and by the spatial variation of the elastic modes. In the latter respect, the contribution of the macroscopic modes ($L=j$) and the bulk modes ($L=q$) to $D_{LL'}$ differ qualitatively. From the estimate $\nabla\psi_j \sim \psi_j/l_j$, $\nabla\psi_q \sim \psi_q/l_q$, with typical length scales $l_j \gg l_q$, we have $D_{jj} \sim (l_q/l_j) D_{jq} \sim (l_q/l_j)^2 D_{qq}$ and thus for the time scales $\tau_j \sim 1/\underline{B}D_{jj} \gg \tau_q \sim 1/BD_{qq}$. Hydrogen in metals differs from other interstitials by its uniquely high mobility which is many orders of magnitude higher than, for instance, the mobility of oxygen and nitrogen in metallic solutions. The macroscopic modes of the latter are practically frozen in on usual laboratory time scales and therefore drop out from (2.116). In the case of hydrogen, however, the high mobility reduces the relaxation times drastically, such that the macroscopic modes attain thermal equilibrium with the lattice within accessible times. This is, of course, the reason for including macroscopic modes in the thermodynamics of metal-hydrogen alloys.

Concerning the application to the Gorsky effect and the spinodal decomposition in the vicinity of T_c, we are interested in time scales on the order of the τ_j's for the dominant modes. In this case, since we still have $\tau_j \gg \tau_q$, the bulk modes may be eliminated by assuming that they relax instantaneously towards a constrained equilibrium which is characterized by the $\tilde{\varrho}(t)$. In other words, we set $\partial_t \tilde{\varrho}_q = 0$. In this (adiabatic) approximation the amplitudes $\tilde{\varrho}_q$ are determined by $\partial \mathscr{F} / \partial \tilde{\varrho}_q = 0$ and depend on time only implicitly via their coupling in $\mathscr{F}$ to the macroscopic modes. These arguments allow us to write finally

$$\sum_{j' \neq 0} \Gamma_{jj'} \partial_t \tilde{\varrho}_{j'} = -\bar{\varrho} \bar{B} \frac{\partial \mathscr{F}}{\partial \tilde{\varrho}_j}, \quad (j \neq 0). \tag{2.121}$$

2.5.2 Anelastic Relaxation

If we employ the free energy functional (2.100) and include the interaction $\mathscr{H}^{\mathrm{ie}}$ with an external force, then (2.121) reads

$$\sum_{j' \neq 0} \Gamma_{jj'} \partial_t \tilde{\varrho}_{j'} = -\bar{\varrho} \bar{B} [(a - E_j) \tilde{\varrho}_j - P \cdot \varepsilon_j^{\mathrm{e}}]. \tag{2.122}$$

In this linear approximation, assumed to be valid above T_c, the macroscopic modes decouple from the bulk modes. Provided the external force couples only to a single mode, as in the bending experiments mentioned above, the solution of (2.122) is

$$\tilde{\varrho}_j(t) = \varrho_j(\infty)(1 - \mathrm{e}^{-t/\tau_j}), \tag{2.123}$$

where

$$\tilde{\varrho}_j(\infty) = \frac{P \cdot \varepsilon_j^{\mathrm{e}}}{a - E_j}. \tag{2.124}$$

The relaxation time τ_j is given by

$$\tau_j = \Gamma_{jj} \frac{1}{\bar{D}} \frac{a}{a - E_j}, \tag{2.125}$$

with the diffusion coefficient

$$\bar{D}(\bar{\varrho}, T) = \bar{\varrho} B(\bar{\varrho}) a(T, \bar{\varrho}). \tag{2.126}$$

The geometry of the sample enters through E_j and Γ_{jj}. In the cases of a foil and a wire, the Green function (2.118) and the eigenfunctions $\psi_j(\boldsymbol{x})$ are known [2.46], with the latter in the plane-stress approximation. As a result one finds

$$\Gamma_{jj} = \begin{cases} h^2/10, & \text{foil, thickness } h, \\ 7d^2/96, & \text{wire, diameter } d. \end{cases} \tag{2.127}$$

For Nb-samples we may use the expression (2.104) for $a(T,\bar{\varrho})$. In terms of spinodal temperatures (2.102) we then obtain for the foil

$$\tau_{\mathrm{F}}=\frac{h^2}{10\bar{D}}\left(\frac{\hat{\lambda}_{\mathrm{F}}}{T-T_{\mathrm{F}}}+1\right), \tag{2.128}$$

and for the wire

$$\tau_{\mathrm{W}}=\frac{7d^2}{96\bar{D}}\left(\frac{\hat{\lambda}_{\mathrm{W}}}{T-T_{\mathrm{W}}}+1\right), \tag{2.129}$$

with $\hat{\lambda}_{\mathrm{F,W}}=E_{\mathrm{F,W}}/a_1^{(1)}$. The mode energy E_{F} is given by (2.70) and (2.71) and depends on the orientation of the foil. In the case of the wire we have $E_{\mathrm{W}}=P^2\kappa_{\mathrm{T}}$ and $T_{\mathrm{W}}=T_0(\bar{\varrho})$. The results (2.128) and (2.129) exhibit both the shape dependence and the effect of critical slowing down in the relaxation times. If one introduces an effective diffusion coefficient D_j^* by

$$D_j^*=\frac{1}{\tau_j}\Gamma_{jj}, \tag{2.130}$$

and eliminates a from (2.125), one finds

$$\frac{D_{\mathrm{F}}^*-\bar{D}}{D_{\mathrm{W}}^*-\bar{D}}=\frac{E_{\mathrm{F}}}{E_{\mathrm{W}}}. \tag{2.131}$$

This ratio is specified completely in terms of elastic constants.

As a third example, we consider an isotropic sphere (radius R) where the coefficients $\Gamma_{jj'}=\Gamma_l\delta_{ll'}\delta_{mm'}$ are known exactly [2.46],

$$\Gamma_l=\frac{6l^2+13l+5}{l(2l+1)(2l+3)(2l+5)}R^2, \qquad (l\geqq 1). \tag{2.132}$$

The relaxation time of the $(l=1)$-mode, for instance, is

$$\tau_1=\frac{8\mathrm{R}^2}{35\bar{D}}\,\frac{a}{a-P^2\kappa_{\mathrm{T}}}. \tag{2.133}$$

2.5.3 Coherent Spinodal Decomposition

Consider a crystal at $T_{\mathrm{initial}}>T_{\mathrm{c}}$, with $\bar{\varrho}=\varrho_{\mathrm{c}}$ for simplicity. After quenching to temperatures T slightly below T_{c}, the system will be driven towards a new equilibrium state with a nonuniform density $\varrho_{\mathrm{eq}}(\boldsymbol{x})$, involving only a few modes provided the crystal remains coherent. For an unstable mode with $T_j(\varrho_{\mathrm{c}})>T$, i.e., $\tau_j < 0$, the linearized diffusion equation yields

$$\tilde{\varrho}_j(t) = \tilde{\varrho}_j(0)\mathrm{e}^{t/|\tau_j|}. \tag{2.134}$$

This initial exponential growth will be limited by the nonlinear terms in (2.121) which are omitted in (2.122).

The appearance of the initial value $\tilde{\varrho}_j(0)$ in (2.134) hints at a shortcoming of our deterministic theory, where $\tilde{\varrho}_j(0)$ enters as a free parameter. In a more general approach, spinodal decomposition should be described in terms of a stochastic theory [2.48] which takes into account fluctuations to start the time evolution towards the new equilibrium state.

On the other hand, in the temperature range of interest here, only a few modes are unstable, and fluctuations in these modes are expected to be scarce. In practice, the spinodal decomposition will be initiated at the heterogeneities of the sample.

In order to have a controllable situation one may include an external force which couples to the modes in question such that $\tilde{\varrho}_j(0)$ is given by the relaxed equilibrium value at the initial temperature above T_c. Alternatively one could begin with a coherent equilibrium state at $T_{\text{initial}} \lesssim T_c$ and then study the temporal behavior of the system after the temperature has been changed rapidly by a small amount.

To include the nonlinear mode coupling, we expand $\mathcal{F}$ in powers of $\tilde{\varrho}$ up to quartic terms,

$$\begin{aligned}&\mathcal{F}(T, \varrho_c + \tilde{\varrho}) - \mathcal{F}(T, \varrho_c)\\ &= a(T, \varrho_c) \int \left[\frac{1}{2}\tilde{\varrho}^2(\boldsymbol{x}) + \frac{1}{4\sigma}\tilde{\varrho}^4(\boldsymbol{x})\right] dV,\end{aligned} \tag{2.135}$$

where $\sigma(T) = 6a(T, \varrho_c)\,[k_B T \partial^4 f(T, \varrho_c)/\partial\varrho_c^4]^{-1} \approx \sigma(T_c)$. There are no terms $\sim\tilde{\varrho}^3$ in (2.135) since we have set $\bar{\varrho} = \varrho_c$. The essential feature of the resulting nonlinear diffusion equation may be illustrated by considering the time dependence of a single mode only. In this case (2.121) simplifies to

$$\partial_t \tilde{\varrho}_j(t) = \frac{1}{|\tau_j|}\left\{1 - \left[\frac{\tilde{\varrho}_j(t)}{\tilde{\varrho}_j(\infty)}\right]^2\right\}\tilde{\varrho}_j(t), \tag{2.136}$$

with the solution

$$\tilde{\varrho}_j^2(t) = \frac{\tilde{\varrho}_j^2(0)\exp(2t/|\tau_j|)}{1 + [\tilde{\varrho}_j(0)/\tilde{\varrho}_j(\infty)]^2\,[\exp(2t/|\tau_j|) - 1]}. \tag{2.137}$$

The parameter $\tilde{\varrho}_j(\infty)$ is given by

$$\tilde{\varrho}_j^2(\infty) = \frac{\sigma}{b_j a(T, \varrho_c)}[E_j - a(T, \varrho_c)], \tag{2.138}$$

with

$$
\begin{aligned}
b_j &= V \int \psi_j^{*2}(x) \psi_j^2(x) dV \\
&= \begin{cases} \left.\begin{array}{l} 9/5, \text{ foil} \\ 2\ \ , \text{ wire} \end{array}\right\} \text{ bending modes}, \\ 15/7, \text{ isotropic sphere}, l=1, m=0. \end{cases}
\end{aligned}
\tag{2.139}
$$

Janssen's theory has been applied by *Burkhardt* and *Wöger* [2.49] to study the time dependence of the local lattice parameters during coherent spinodal decomposition. They suggest to map the lattice distortions arising from the macroscopically varying hydrogen density by directing a narrow beam of x-rays at various points on the surface of the sample and measuring shifts in the Bragg maxima. They made predictions about the time dependence of the shifts on the basis of a numerical solution of the diffusion equation for a Nb-sphere. Due to the geometrical factors $\sim R^2$ and the critical slowing down of the relaxation times, the shifts occur on the scale of hours.

In a recent paper, *Kappus* and *Horner* [2.50] investigated the coherent spinodal decomposition in a temperature range between the spinodals of the surface and of the bulk modes. The authors extend Janssen's theory by including fluctuations in an approximate way and offer an explanation of the quasi-periodic precipitation observed [2.51] on the surface of rapidly quenched NbH_x-samples.

In conclusion, we make a few remarks comparing the coherent spinodal decomposition as discussed here with the spinodal phase separation treated in the *Cahn-Hilliard-Langer* theory [2.20, 24, 47, 48]. According to the latter, the initial short wavelength fluctuations grow fastest. In the course of time, the maximum growth rate shifts to longer wavelengths and the shape of the fluctuations changes steadily from a sinusoidal into a droplet form with a sharp interfacial surface.

In the coherent spinodal decomposition the initially unstable modes are of a different nature and the development in the later stages of a well-defined interface is absent. However, the assumption of coherency throughout the decomposition is presumably unrealistic. Rather, the growth of internal stresses in the later stages will be limited by the creation and motion of dislocations and a partially incoherent phase separation will set in quite rapidly. On the other hand, in the vicinity of T_c we expect that coherency is maintained at least in the early stages which might cover time spans on the order of hours. Recent x-ray measurements by *Zabel* [2.52] of local lattice parameters in NbH_x-samples seem to support this picture (see Sect. 2.6).

2.6 Comparison with Experiments

The theory of metal-hydrogen alloys described in the preceeding sections was based on the elastic interaction being the dominant attractive force between dissolved protons. This interaction is conveniently formulated in terms of elastic modes. The macroscopic nature of these modes is reflected in the shape dependence of thermodynamic and transport quantities. In experimental tests of the theory, one naturally tries to focus attention primarily on these rather unusual features.

The hypothesis about the importance of the elastic interaction can be tested to some extent by a comparison of the theoretical with the empirical phase diagram. The most detailed information would be provided by an experimental investigation of the coherent state in the inhomogeneous phase below T_c. Since the coherent state is metastable, such a study would require high resolution measurements on single crystals of different shapes in a narrow temperature range just below T_c where the coherency stresses are still smaller than critical yield stresses. First efforts in this direction have recently been made by *Zabel* [2.52], and some of his results will be briefly discussed in Section 2.6.4.

The main body of experimental data on the phase diagram is concerned with the incoherent state whereby information about shape dependence is lost. Fortunately, the geometry of metal-hydrogen samples also enters in the properties of hydrogen density fluctuations via the discrete spectrum of the macroscopic modes and the associated spinodal temperatures. The density fluctuations can be investigated by Gorsky relaxation and neutron scattering measurements in the stable one-phase region above T_c. These experiments thus provide a crucial test of the predictions of the mode theory.

2.6.1 Phase Diagram

The experimental phase diagram of hydrogen in Nb has been deduced from solubility [2.53–55], x-ray [2.56, 57], and resistivity relaxation [2.58] measurements. Values of T_c are found in the range from 413 K [2.53] to 483 K [2.54]. The results for the critical concentration are within 0.26 [H/Nb] [2.54] and 0.32 [H/Nb] [2.58]. From Gorsky effect measurements one obtains [2.44] $T_c = 450$ K and $c = 0.34$ [H/Nb]. The discrepancies in the reported values of T_c and c_c may partly be due to the difficulty in realizing the ideal incoherent state. One has to be careful to avoid hysteresis effects caused by residual coherency stresses, as emphasized in particular by *Zabel* and *Peisl* [2.57].

The main task in the computation of the $\alpha-\alpha'$ phase diagram near the critical point is to work out a reasonable but still tractable model for the reference system. Here, we shall report on the results of a model calculation described in [2.16]. In this model one replaces the electronic interaction V_{ab} by a hard-core repulsion, which excludes the occupation of up to third neighbor sites in the interstitial lattice. This size of the hard core is compatible with the structure of the β phase [2.36]. The $\overset{\infty}{W}{}^{s}_{ab}$ are computed numerically with lattice force constants obtained from a Born-von Karman fit to neutron scattering results for phonon dispersion curves of pure Nb [2.59]. The reference free energy is then calculated in the form of the low-density expansion, (2.87) including terms up to $n=7$. The coefficients A_n, $n \leqq 4$ are known exactly from a

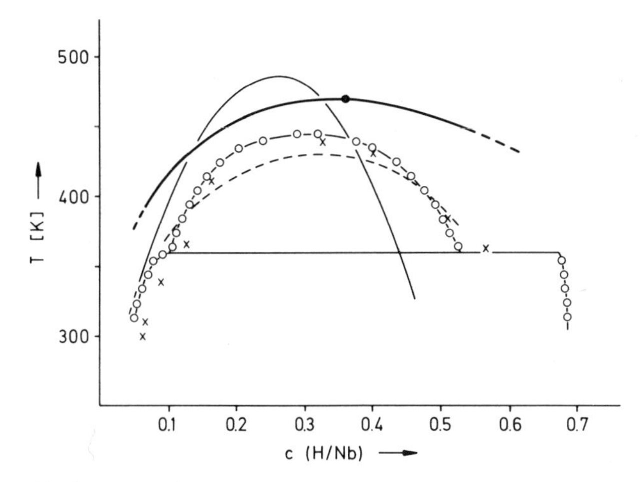

Fig. 2.7. Phase diagram of hydrogen in niobium. ——: *Pryde* and *Titcomb* [2.54], × × : *Walter* and *Chandler* [2.56], ○: *Zabel* and *Peisl* [2.57], – – –: *Schnabel* [2.58]; ——: theory [2.16]

cluster expansion; the values for A_5, A_6, and A_7 are taken from a Monte Carlo calculation. The range R_0 of $\overset{\infty}{W}{}^s_{ab}$ [cf. (2.78)] is chosen to be twice the cubic lattice distance b_L with $b_\mathrm{L} = 3.3$ Å.

The theoretical coexistence curve is shown in Fig. 2.7. For an isotropic dipole tensor with $P = 3.3$ eV [2.60] and a lattice compressibility $\kappa_\mathrm{T} = 5.84 \cdot 10^{-13}\,\mathrm{cm}^2\,\mathrm{dyn}^{-1}$ [2.61], one finds $T_\mathrm{c} = 470$ K and $c_\mathrm{c} = 0.36$ [H/Nb]. Despite the crude approximations, especially concerning V_{ab}, the obtained critical point is within the T- and c-range covered by the experimental results. The strength P of the dipole tensor enters quadratically in M_0 and also, via $\overset{\infty}{W}{}^s_{ab}$, into the reference free energy. More explicitly, the latter depends on P only in the form $Tg(P^2/T)$. From this observation one finds easily that a variation ΔP of P produces a shift in the critical temperature by $\Delta T_\mathrm{c}/T_\mathrm{c} = 2\Delta P/P$, whereas the critical density remains unchanged. For concentrations $c \gtrsim 0.4$ [H/Nb], the low-density expansion of the reference free energy becomes increasingly inaccurate. This is the main reason for the coexistence curve being too flat at higher concentrations.

It should be stressed that there is no freely adjustable parameter involved in the computation of the phase diagram for this model. The value of the cutoff $R_0 = 2b_\mathrm{L}$ is determined such that $\sum_b \overset{\infty}{W}{}^l_{ab}$ is negligible for distances $|S^a - S^b| > R_0$, as required by the continuum approximation.

2.6.2 Gorsky Relaxation Strength and Neutron Scattering Experiments

By bending hydrogen loaded foils and wires one measures the relaxation strength $\Delta = \phi^\mathrm{a}/\phi^\mathrm{e}$, where $\phi^\mathrm{a,e}$ denote the instantaneous elastic and anelastic value of the bending angle, respectively. From the theory of density fluctuations

Table 2.1. Values of R_j and E_j for wire and foil geometry (after [2.21]). $\boldsymbol{M}$: bending torque

Geometry	R_j	E_j
(110) wire	$R_{\mathrm{W}}=\frac{4}{3}\frac{C_{44}(C_{11}-C_{12})}{(C_{11}+2C_{12})(C_{11}-C_{12})+2C_{11}C_{44}}$	$E_{\mathrm{W}}=P^2\kappa_{\mathrm{T}}=P^2\frac{3}{C_{11}+2C_{12}}$
(110) M (100) foil (case 1)	$R_{\mathrm{F}_1}=\frac{(C_{11}-C_{12}+2C_{44})^2}{(C_{11}+C_{12}+2C_{44})(3C_{11}-3C_{12}+2C_{44})}$	$E_{\mathrm{F}_1}=P^2\frac{3C_{11}-3C_{12}+2C_{44}}{C_{11}(C_{11}+C_{12}+2C_{44})-2C_{12}^2}$
(110) (100) M foil (case 2)	$R_{\mathrm{F}_2}=\frac{2C_{44}}{C_{11}+C_{12}+2C_{44}}$	$E_{\mathrm{F}_2}=P^2\frac{2}{C_{11}+C_{12}}$

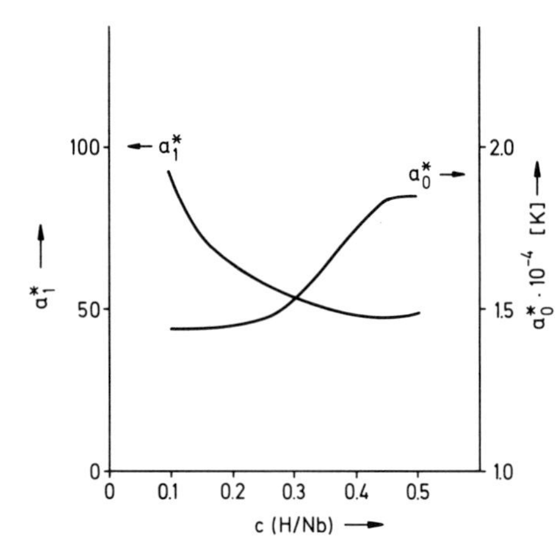

Fig. 2.8. Theoretical concentration dependence of the reduced quantities $a_i^* = a_i N_H / V k_B$, $i=0,1$ [cf. (2.104)]

as outlines in Section 2.4 one finds [2.21]

$$\Delta_j = \frac{R_j E_j}{a(T,\bar{\varrho}) - E_j}. \tag{2.140}$$

This expression derives from the fact that the bending of foils and wires excites, in the plane stress approximation (cf. Sect. 2.2.5), a single mode. R_j is a geometrical factor which depends only on the shape and crystal orientation of the sample and is determined entirely by the elastic constants (see Table 2.1). The quantity $a(T,\bar{\varrho})$ is defined in (2.98) and is inversely proportional to the compressibility of the reference system. In the model described in Section 2.6.1, $a(T,\bar{\varrho})$ is obtained from the low-density expansion (2.87). Although the coefficients $A_n(T)$ depend nonlinearly on T, one finds that a $(T,\bar{\varrho})$ can be well approximated by the linear law (2.104), $a(T,\bar{\varrho}) = a_0(\bar{\varrho}) + a_1(\bar{\varrho})\,T$, within a temperature range extending up to about 100 K above the incoherent spinodal $T_0(\bar{\varrho})$. The concentration dependence of a_0 and a_1 is shown in Fig. 2.8. With the

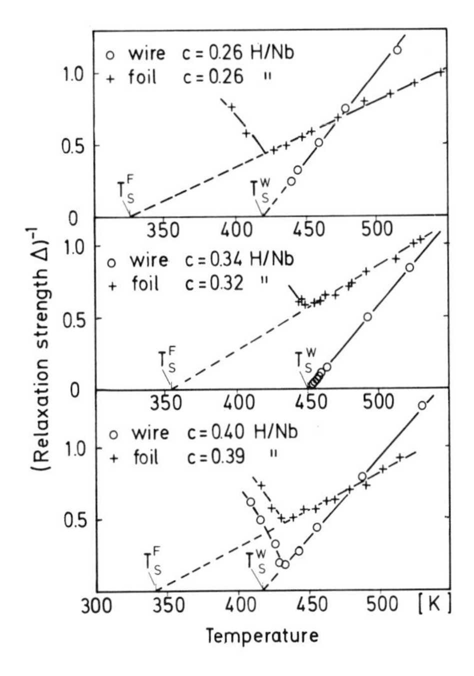

Fig. 2.9. Reciprocal relaxation strength for hydrogen in a niobium foil and a niobium wire at various hydrogen concentrations (after *Tretkowski* et al. [2.44]; see also [2.43])

above approximation for $a(T,\bar{\varrho})$ we have

$$\Delta_j^{-1} = \frac{a_1}{R_j E_j}(T - T_j), \tag{2.141}$$

with the spinodal temperature

$$T_j = \frac{E_j - a_0(\bar{\varrho})}{a_1(\bar{\varrho})}. \tag{2.142}$$

Starting in the one-phase region and lowering the temperature, the Curie-Weiss type relation $\Delta_j^{-1} \sim (T - T_j)$ is expected to hold until the coexistence curve is reached. This behavior is indeed observed in bending experiments [2.44] as shown in Fig. 2.9.

The expressions of R_j and E_j for the geometries of these experiments are given in Table 2.1. The values of T_W, Δ_W^{-1} (wire) and T_F, Δ_F^{-1} (foil) as calculated from (2.141) with elastic constants for pure and for loaded[9] Nb are compared with experimental data in Fig. 2.10. The spinodal temperature of the foil as found by extrapolation is substantially smaller than that of the wire. The corresponding slopes of Δ_j^{-1} vs T also differ between both geometries. These features are at least qualitatively reproduced by the theory.

Clearly, the use of elastic constants for loaded Nb together with values of a_0 and a_1 derived from phonon dispersion curves of pure Nb is not consistent.

[9] The elastic constants of hydrogen-loaded Nb are obtained by linear extrapolation of the low-density data of *Magerl* et al. [2.62].

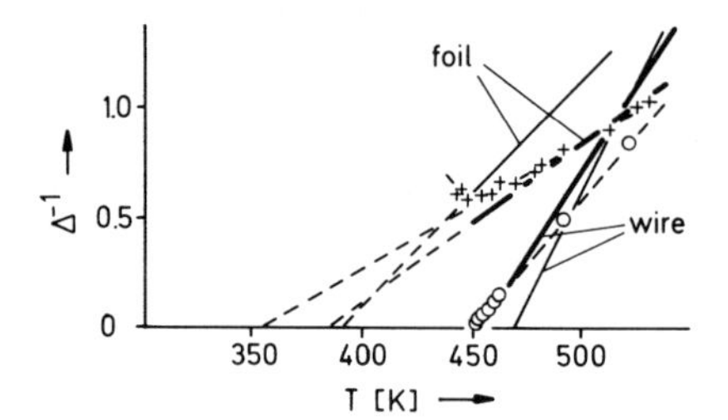

Fig. 2.10. Comparison of experimental and theoretical reciprocal relaxation strengths. +: foil, $c=0.32$ [H/Nb], ○: wire, $c=0.34$ [H/Nb]. ——: theory, with elastic constants of loaded Nb, ——: theory, with elastic constants of pure Nb. The theoretical curves for the foil are computed for the orientation "Case 2" in Table 2.1

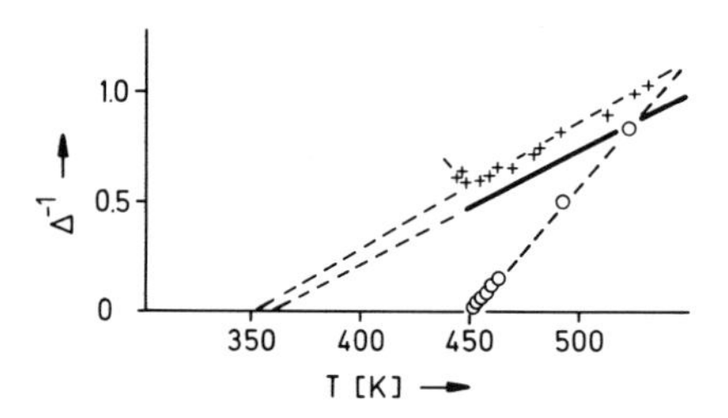

Fig. 2.11. Theoretical reciprocal relaxation strength for the foil ("Case 2" in Table 2.1) as obtained from the data for the wire according to (2.143). Symbols as in Fig. 2.10

This situation may be improved, however, by eliminating $a_0(\bar{\varrho})$ and $a_1(\bar{\varrho})$ from (2.141) and (2.142), and making use of the fact that quantities derived from the reference free energy are independent of the geometry of the sample. In this way we obtain

$$T_{\mathrm{F}}-T_{\mathrm{W}}=\frac{E_{\mathrm{F}}-E_{\mathrm{W}}}{R_{\mathrm{W}}E_{\mathrm{W}}}\left(\frac{\partial\Delta_{\mathrm{W}}^{-1}}{\partial T}\right)^{-1}, \tag{2.143}$$

and

$$\frac{\partial\Delta_{\mathrm{W}}^{-1}/\partial T}{\partial\Delta_{\mathrm{F}}^{-1}/\partial T}=\frac{E_{\mathrm{W}}R_{\mathrm{W}}}{E_{\mathrm{F}}R_{\mathrm{F}}}, \tag{2.144}$$

where the subscripts F and W refer to the bending modes for the foil and wire geometry, respectively. Thus, the spinodal temperature T_{F} and the slope $\partial\Delta_{\mathrm{F}}^{-1}/\partial T$ for the foil can be computed from the corresponding experimental quantities of the wire. In (2.143) and (2.144) we now may legitimately insert the elastic constants of NbH_x. The result is shown in Fig. 2.11.

Information about the spinodal temperatures of bulk modes is obtained from small-angle neutron scattering experiments on NbD_x in the limit of vanishing momentum transfer. The first experiment at low hydrogen concentration was made by *Conrad* et al. [2.41]. Measurements at higher concentration have been performed by *Münzing* et al. [2.63]. The data are given in Table 2.2. The calculation of the spinodal temperatures T_{111}, T_{110}, and T_{100} with the model parameters a_0 and a_1 from Fig. 2.8 yields values (see Table 2.2) which correctly reproduce the observed sequence of these spinodal temperatures, but the numerical agreement with the experimental data is less satisfactory than in the case of the macroscopic modes.

Table 2.2. Spinodal temperatures of bulk modes. Exp.: a) *Münzing* et al. [2.63] b) *Conrad* et al. [2.41]. Theor.: l: with elastic constants of loaded Nb, p: with elastic constants of pure Nb

c	T_{111} [K]			T_{110} [K]			T_{100} [K]		
[H/Nb]	Experimental	Theoretical		Experimental	Theoretical		Experimental	Theoretical	
		l	p		l	p		l	p
0.1	138±15 a) 83±16 b)	193	206	130±11 a) 39±16 b)	182	190	104±10 a)	152	149
0.2	184±15 a)	258	296	170±15 a)	248	272	122±17 a)	223	213
0.3	251±24 a)	266	329	241±15 a)	261	301	208±19 a)	250	233

Table 2.3. Ratios of spinodal temperature of bulk modes, from (2.145). References and symbols as in Table 2.2. Values of T_W taken from *Bauer* et al. [2.64]

c [H/Nb]	$\frac{T_W - T_{110}}{T_W - T_{111}}$			$\frac{T_W - T_{100}}{T_W - T_{111}}$		
	Experimental	Theoretical		Experimental	Theoretical	
		l	p		l	p
0.1	$1.01^{+0.46}_{-0.29}$ a) $1.31^{+0.32}_{-0.23}$ b)	1.12	1.20	$1.39^{+0.56}_{-0.31}$ a)	1.46	1.71
0.2	$1.07^{+0.18}_{-0.15}$ a)	1.07	1.20	$1.31^{+0.22}_{-0.17}$ a)	1.24	1.71
0.3	$1.06^{+0.21}_{-0.19}$ a)	1.03	1.20	$1.23^{+0.28}_{-0.23}$ a)	1.09	1.71

After inserting (2.103) into (2.101) and eliminating a_0 and a_1, we find for any triple L, L', L'' of elastic modes that

$$\frac{T_L - T_{L'}}{T_L - T_{L''}} = \frac{E_L - E_{L'}}{E_L - E_{L''}}, \tag{2.145}$$

whereby the right-hand side is given solely in terms of elastic constants. In Table 2.3 the calculated ratios are compared with the measured ones. The energies of the bulk modes are taken from (2.74).

The theoretical values for the bulk spinodal temperatures are very crude estimates only. The main uncertainty arises through the linear linear extrapolation of the density dependence of elastic constants from a few percent [H/Nb] [2.62] up to 30% [H/Nb]. Furthermore, the discrepancy between the data of *Conrad* et al. [2.41] and those of *Münzing* et al. [2.63] has not yet been resolved.

2.6.3 Diffusion Coefficient

From the relaxation time τ_j of the anelastic strain in a bending experiment, one obtains the effective (macroscopic) diffusion coefficient D_j^* according to (2.130), which may be written as

$$D_j^* = \bar{\varrho}\bar{B}(T,\bar{\varrho})\,[a(T,\bar{\varrho}) - E_j]$$
$$= \bar{\varrho}\bar{B}(T,\bar{\varrho})\,a_1(\bar{\varrho})\,[T - T_j(\bar{\varrho})]\,. \quad (2.146)$$

D_j^* varies with the geometry of the sample through its dependence on E_j and T_j. The factor $T - T_j$ describes the effect of critical slowing down.

In the discussion of diffusion experiments one is usually interested in the so-called tracer diffusion coefficient $D(\bar{\varrho}, T)$ which is defined in terms of the mobility $\bar{B}$ via Einstein's relation,

$$D = k_B T \bar{B}(T,\bar{\varrho})\,. \quad (2.147)$$

We then have

$$D = \frac{k_B}{\bar{\varrho}a_1(\bar{\varrho})} \cdot \frac{T}{T - T_j(\bar{\varrho})} D_j^*\,. \quad (2.148)$$

In the atomistic theory of diffusion (Chap. 8), the mobility is derived from the jump motion of a single proton. Accordingly, one expects $\bar{B}$ and thus D to be independent of the sample geometry. In particular, D should no longer show any critical slowing down effects.

These predictions have recently been tested in detail by *Bauer* et al. [2.64] and *Tretkowski* et al. [2.44]. The procedure is to calculate D with the help of (2.148) from the measured values of D_j^* as obtained from both foil and wire geometries. The values for $T_j(\bar{\varrho})$ and for $a_1 = R_j E_j \partial \Delta_j^{-1}/\partial T$ are taken from Gorsky relaxation strength measurements.

The result for D is shown in Fig. 2.12. With the concentration near its critical value, T_W and T_F differ by about 100 K, but the tracer diffusion coefficient D as obtained from D_F^* agrees within experimental accuracy with the D as obtained from D_W^*. Furthermore, the effect of critical slowing down is indeed absent in D which follows Arrhenius' law $D = D_0 \exp(-U/k_B T)$. Finally, the results demonstrate that D_j^* obeys the conventional theory of critical slowing down.

In the computation of D from D_j^*, the use of data from relaxation strength measurements may be avoided if we eliminate $a(T,\bar{\varrho})$ from the first line of (2.146). One finds that

$$D = \frac{k_B T}{\bar{\varrho}} \cdot \frac{D_F^* - D_W^*}{E_W - E_F} \quad (2.149)$$

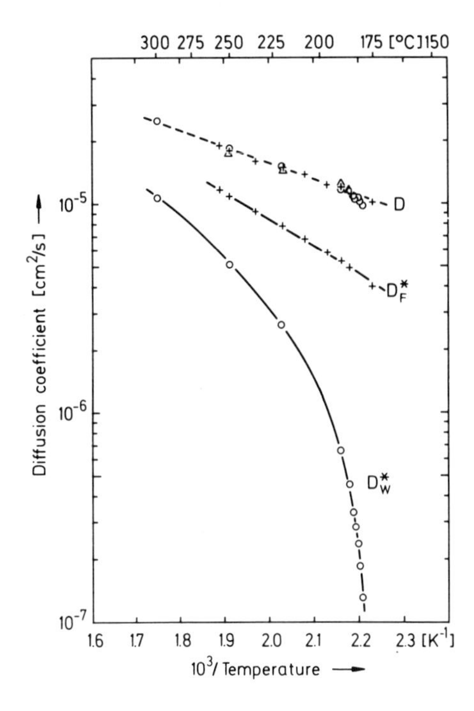

Fig. 2.12. Comparison of the tracer diffusion coefficient D of hydrogen in niobium with the effective diffusion coefficients for the wire (D_W^*) and for the foil (D_F^*). Tracer diffusion coefficient D: ○ : D from D_W^*, $c=0.34$ [H/Nb], + : D from D_F^*, $c=0.32$ [H/Nb], △ : D according to (2.149). The data for D_W^* and D_F^* are those given in [2.43]. See also Chapter 1

with $E_W - E_F = P^2[\kappa_T - 2/(C_{11} + C_{12})]$ in the plane-stress approximation. A practical disadvantage of (2.149) is that for a pointwise computation of D, knowledge of foil and wire data at the same concentration and temperature is required. The D as obtained from (2.149) at some suitable temperatures is shown in Fig. 2.12 and is seen to be consistent with the previous results.

2.6.4 Inhomogeneous Phase: Local Lattice Parameter

Consider a hydrogen-loaded crystal at critical concentration. After quenching the sample to a temperature $T_f < T_c$, we expect an inhomogeneous phase to develop, as described in Sections 2.3.3 and 2.5.3. The spinodal formation of the macroscopically varying density distribution $\varrho(\boldsymbol{x})$ induces a similar variation in the local lattice parameter $b_L(\boldsymbol{x})$ which should be observable by x-ray diffraction methods [2.49].

If a narrow beam of x-rays is diffracted from a small localized area of the coherent sample, we expect to see a Bragg reflection corresponding to an average local lattice parameter. The Bragg peak will in general be shifted from its position in the one-phase region, the shift being proportional to the local deviation $\varrho(\boldsymbol{x}) - \bar{\varrho}$. However, the occurrence of a well-defined Bragg peak is only to be expected for temperatures T_f not far below T_c where only a few macroscopic modes contribute to $\varrho(\boldsymbol{x})$. At lower temperatures, the variation of

$\varrho(\boldsymbol{x})$ within the scattering region is no longer negligible and should cause a significant increase in the width or even a splitting of the Bragg reflection. As long as the crystal remains coherent, the changes in the Bragg reflection below T_c are predicted to be reversible: after heating the sample to a few degrees above T_c, one should recover the original Bragg reflection corresponding to the uniform lattice parameter of the homogeneous phase. The formation of the inhomogeneous phase after quenching, as well as the restoration of the homogeneous phase after heating, is predicted to occur on macroscopic time scales, typically on the order of hours [2.49]. If the sample has been quenched to temperatures sufficiently far below T_c, the growing coherency stresses will be removed by incoherent phase separation. In this case we expect to observe again well-defined Bragg peaks corresponding to the lattice parameter of either the high- or low-density phase, depending on the location of the scattering region. Since there will still be residual coherency stresses, the local lattice parameter will in general not coincide with either $b_L^{(\alpha)}(T_f)$ or $b_L^{(\alpha')}(T_f)$. Furthermore, after heating the incoherent sample to temperatures above T_c we expect to observe hysteresis effects due to stresses arising from dislocations created during the transition at T_f to the incoherent state.

The above qualitative predictions will now be compared with *Zabel*'s recent x-ray diffraction measurements on a disc-shaped Nb single crystal (diameter: 13.5 mm, thickness: 0.5 mm, orientation: 110) [2.52]. The scattering area was $1.5 \times 10\,\mathrm{mm}^2$.

In the first run, after in situ loading from the gas phase at about 870 K ($c = 0.32$ H/Nb $\approx c_c$), the sample was cooled in steps down to 433 K. At each step T_n, starting with $T_1 = 455$ K, the local lattice parameter was measured after a waiting time extending up to 12 h. Below $T_c(\mathrm{exp}) = 444$ K the lattice parameter in the exposed region starts to increase gradually with decreasing temperature (see Fig. 2.13a). One also observes a slight broadening of the Bragg peak. After heating from 433 K back to 455 K and waiting for 17 h, the lattice parameter went back to its original value at 455 K within an accuracy of 0.025%.

The second run consisted of cooling in steps from 455 K to 423 K. Below 430 K the Bragg reflection splits symmetrically and the peak intensities are strongly reduced (see Fig. 2.13b). After heating back to 455 K, the splitting vanished after 11 h and the peak went again to its original position, having nearly its original intensity.

The third run was a cooling in steps from 455 K to 417 K. Splitting started at 421 K with both peaks shifted towards values corresponding to lower local concentration $c < c_c$. The width of the splitting was observed to fluctuate with time and temperature. After heating to 455 K and a waiting time of 10 h, a single peak at the original position was found; the half-width had increased by 50% however.

The fourth run involved cooling in steps from 455 K to 398 K. After broadening (417 K) and splitting (409 K), one observes at 398 K initially a single, asymmetrically broadened peak. With increasing time the asymmetry decreases and the peak is simultaneously shifted towards the α phase boundary

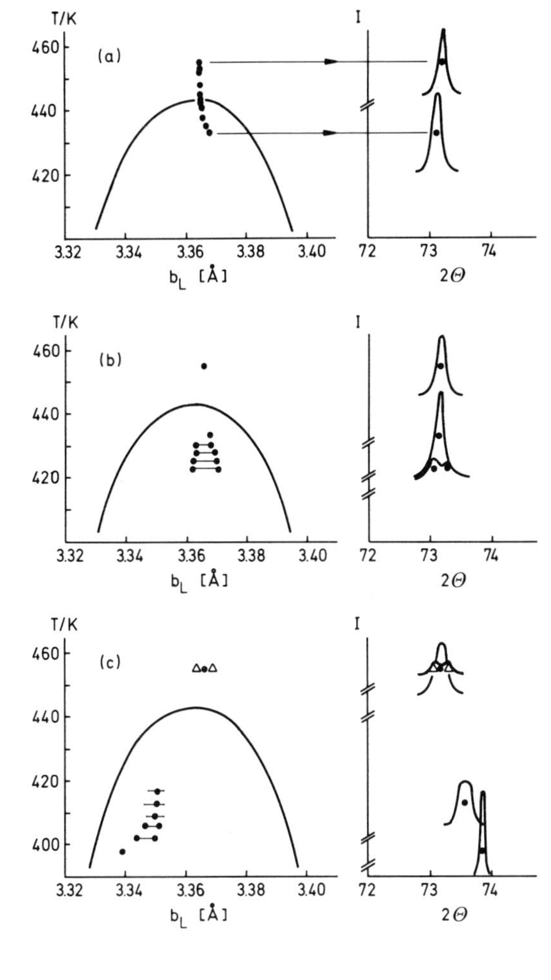

Fig. 2.13a–c. Local *Bragg* reflection in the miscibility gap of the $\alpha-\alpha'$ phase transition of hydrogen in niobium (after *Zabel* [2.52]). Left part: local lattice parameter (●) b_L observed after waiting up to 12 h; curves: coexistence boundary of the incoherent $\alpha-\alpha'$ phase separation [2.57]. Right part: Angular dependence of the *Bragg* intensity I (arbitrary units). (a): first cooling, (b): second cooling after first reheating to 455 K, (c): fourth cooling after third reheating to 455 K. △: lattice parameter after fourth reheating

(see Fig. 2.13c). After heating to 455 K and waiting for 25.2 h, two peaks were found, lying symmetrically with respect to the starting position at 455 K (see Fig. 2.13c). The intensity of the peaks was strongly reduced.

A comparison with the predictions discussed above shows that the mode picture appears to provide a framework in which many of the qualitative features of these experimental findings can be understood. Taken together with the other experimental results mentioned in this section, they strongly support the theory of the $\alpha-\alpha'$ phase transition based on the elastic interaction.

Acknowledgement. Many of the theoretical results presented in this chapter have been worked out together with *R. Bausch* and *H. Horner* to whom I am deeply indebted for a stimulating period of collaboration. I would like to thank *G. Alefeld*, his co-workers, and *H. Zabel* for allowing me to show some of their experimental results prior to publication. Thanks are also due to *J. Völkl* for help in the preparation of some of the drawings.

References

2.1 G. Alefeld, J. Völkl (eds.): *Hydrogen in Metals II, Application-Oriented Properties*. Topics in Applied Physics, Vol. 29 (Springer, Berlin, Heidelberg, New York 1978) in preparation
2.2 G. Alefeld: Phys. Stat. Sol. **32**, 67 (1969)
2.3 R. Fowler, E. A. Guggenheim: *Statistical Thermodynamics* (Cambridge University Press, Cambridge 1960)
2.4 H. Buck, G. Alefeld: Phys. Stat. Sol. **49**, 317 (1972)
2.5 Y. de Ribaupierre, F. D. Manchester: J. Phys. C: Solid State Phys. **7**, 2126 (1974); **7**, 2140 (1974); **8**, 1339 (1975)
2.6 J. D. Eshelby: In *Solid State Physics*, Vol. 3, ed. by F. Seitz, D. Turnbull (Academic Press, New York, London 1956) p. 79
2.7 M. A. Krivoglaz: Fiz. Tver. Tela **5**, 3439 (1963)
[English transl.: Soviet Phys.—Solid State **5**, 2526 (1964)]
2.8 A. G. Khachaturyan: Fiz. Tver. Tela **8**, 2709 (1966)
[English transl.: Soviet Phys.—Solid State **8**, 2163 (1967)]
2.9 J. R. Hardy, R. Boullough: Phil. Mag. **15**, 237 (1967)
2.10 R. Siems: Phys. Stat. Sol. **42**, 105 (1969)
2.11 H. E. Cook, D. de Fontaine: Acta Met. **17**, 915 (1969); **19**, 607 (1971)
2.12 R. Burch: Trans. Faraday Soc. **66**, 749 (1970) and references therein
2.13 G. Alefeld: Ber. Bunsenges. Phys. Chem. **76**, 335 (1972); **76**, 746 (1972)
2.14 H. Wagner, H. Horner: Advan. Phys. **23**, 587 (1974)
2.15 J. D. Eshelby: Acta Met. **3**, 487 (1955)
2.16 H. Horner, H. Wagner: J. Phys. C: Solid State Phys. **7**, 3305 (1974)
2.17 H. Kanzaki: J. Phys. Chem. Solids **2**, 24 (1957)
2.18 V. K. Tewary: Advan. Phys. **22**, 757 (1973)
2.19 J. Buchholz, J. Völkl, G. Alefeld: Phys. Rev. Lett. **30**, 318 (1973)
2.20 J. W. Cahn: Acta Met. **9**, 795 (1961)
2.21 R. Bausch, H. Horner, H. Wagner: J. Phys. C: Solid State Phys. **8**, 2559 (1975)
2.22 I. N. Sneddon, D. S. Berry: "The Classical Theory of Elasticity". In *Elastizität und Plastizität*. Handbuch der Physik, Band VI. Hrsg. S. Flügge (Springer, Berlin, Göttingen, Heidelberg 1958)
2.23 E. Kröner: *Kontinuumstheorie der Versetzungen und Eigenspannungen*. Ergebnisse der angewandten Mathematik, Heft 5 (Springer, Berlin, Göttingen, Heidelberg 1958)
2.24 J. W. Cahn: Acta Met. **10**, 179 (1962); **10**, 907 (1962); Trans. Met. Soc. AIME **242**, 166 (1968)
2.25 P. H. Dederichs, G. Leibfried: Phys. Rev. **188**, 1175 (1969)
2.26 K. H. G. Lie, J. S. Koehler: Advan. Phys. **17**, 421 (1968)
2.27 J. W. Flocken, J. R. Hardy: Phys. Rev. B **1**, 2447 (1970)
2.28 R. A. Masumara, G. Sines: J. Appl. Phys. **41**, 3930 (1970)
2.29 P. H. Dederichs, J. Pollmann: Z. Physik **255**, 315 (1972)
2.30 F. Bitter: Phys. Rev. **37**, 1526 (1931)
2.31 J. D. Eshelby: J. Appl. Phys. **25**, 255 (1954)
2.32 F. J. Wegner: J. Phys. C: Solid State Phys. **7**, 2109 (1974)
2.33 B. May: Dissertation, Universität des Saarlandes (1976)
2.34 S. P. Timoshenko, J. N. Goodier: *Theory of Elasticity* (McGraw-Hill, New York 1970)
2.35 H. A. Goldberg: J. Phys. C: Solid State Phys. **10**, 2059 (1977)
2.36 V. A. Somenkov, V. A. Gurskaya, M. G. Zemlyanov, M. E. Kost, N. A. Chernoplekov, A. A. Chertkov: Fiz. Tver. Tela **10**, 1355 (1968)
[English transl.: Soviet Phys.—Solid State **10**, 1076 (1968)]
2.37 N. G. Van Kampen: Phys. Rev. **135**, A 362 (1964)
2.38 T. W. Burkhardt: Z. Physik **269**, 237 (1974)
2.39 C. K. Hall, G. Stell: Phys. Rev. B **11**, 224 (1973)
2.40 L. D. Landau, E. M. Lifshitz: *Statistical Physics* (Pergamon Press, London-Paris 1959)
2.41 H. Conrad, G. Bauer, G. Alefeld, T. Springer, W. Schmatz: Z. Physik **266**, 239 (1974)
2.42 J. Völkl: Ber. Bunsenges. Phys. Chem. **76**, 797 (1972)
2.43 J. Völkl, G. Alefeld: Nuovo Cimento **33** B, 190 (1976)

2.44 J.Tretkowski: Jül-Bericht Jül-1049-FF (Kernforschungsanlage Jülich)
J.Tretkowski, J.Völkl, G.Alefeld: Z. Physik B **28**, 259 (1977)
2.45 G. Alefeld, J.Völkl, G.Schaumann: Phys. Stat. Sol. **37**, 337 (1970)
2.46 H.K.Janssen: Z. Physik B **23**, 245 (1976)
2.47 M.Hillert: D.Sc. Thesis. MIT, Cambridge, Mass., (1956)
J.W.Cahn, J.E.Hilliard: J. Chem. Phys. **28**, 258 (1958)
M.Hillert: Acta Met. **9**, 525 (1961)
2.48 J.S.Langer: Acta Met. **21**, 1649 (1973) and references therein
2.49 T.W.Burkhardt, W.Wöger: Z. Physik B **21**, 89 (1975)
2.50 W.Kappus, H.Horner: Z. Physik B **27**, 215 (1977)
2.51 M.A.Pick: Jül-Bericht Jül-951-FF (Kernforschungsanlage Jülich)
2.52 H.Zabel: Dissertation, Universität München (1977) to be published
2.53 W.M.Albrecht, W.D.Goode, M.W.Mallett: J. Electrochemical Soc. **106**, 981 (1959)
2.54 J.Pryde, C.Titcomb: Trans. Faraday Soc. **65**, 2758 (1969)
2.55 E.Veleckis, R.K.Edwards: J. Phys. Chem. **73**, 683 (1969)
2.56 R.J.Walter, W.T.Chandler: Trans. AIME **233**, 762 (1965)
2.57 H.Zabel, H.Peisl: Phys. Stat. Sol. (a) **37**, K67 (1976)
2.58 D.Schnabel: Jül-Bericht Jül-878-FF (Kernforschungsanlage Jülich)
2.59 R.I.Sharp: J. Phys. C: Solid State Phys. **2**, 421 (1969)
2.60 H.Metzger, H.Peisl, J.Wanagel: J. Phys. F: Metal Phys. **6**, 2195 (1976)
2.61 D.I.Bolef: J. Appl. Phys. **32**, 100 (1961)
2.62 A.Magerl, B.Berre, G.Alefeld: Phys. Stat. Sol. **36**, 161 (1976)
2.63 W.Münzing, N.Stump. G.Göltz: To be published
2.64 H.C.Bauer, J.Völkl, J.Tretkowski, G.Alefeld: Z. Physik B **29**, 17 (1978)
2.65 G.Leibfried, N.Breuer: *Point Defects in Metals I*, Springer Tracts in Modern Physics, Vol. 81 (Springer, Berlin, Heidelberg, New York 1978)

3. Lattice Strains due to Hydrogen in Metals

H. Peisl

With 14 Figures

3.1 Overview

Hydrogen dissolves in many metals and occupies interstitial sites in the host lattice. In all known metal-hydrogen alloys the dissolved hydrogen expands the crystal lattice of the host metal. Each hydrogen interstitial causes displacements of the metal atoms from their regular sites, and the resulting crystal lattice distortions (described as strain or stress fields) give rise to a series of physical property changes which have attracted both fundamental and applied research activities.

Many of the metals which dissolve large quantities of hydrogen are technological materials of present or future importance. Iron, steel, Nb, Ta, V, and Pd and its alloys are such materials. Nb and V are possible candidates as construction materials in fusion reactors, and using hydrogen as an energy carrier depends on storage containers and transport lines.

Typical relative volume expansions due to the solution of one hydrogen atom per metal atom are of the order of 20%. This fact could cause severe construction problems if such metals are used in a hydrogen environment owing to the drastic change in the dimensions of the construction materials.

A variety of disordered and ordered phases are observed over the wide range of composition in which metal-hydrogen alloys exist (see [Ref. 3.1, Chap. 2]). As the lattice distortions depend on the hydrogen concentrations in the various phases, the formation of a different phase is connected with coherency stresses. The phase transitions may depend on the sample geometry due to these coherency stresses (Chap. 2). If the coherency stresses exceed the critical yield stress they are released by the formation of dislocations which themselves change the properties of the material.

On an atomistic scale the hydrogen atoms interact via their distortion fields ("elastic interaction"). According to *Alefeld* [3.2], this elastic interaction is the relevant interaction for the $\alpha-\alpha'$ phase transition in the hydrogen-niobium and hydrogen-palladium system. An elastic interaction between the hydrogen atoms and impurities may also exist and influence the nucelation of a different phase. The elastic interaction of hydrogen with dislocations and internal (or external) stress fields, e.g., close to a crack in the material, plays an important role in the most serious mechanical property change, the hydrogen embrittlement of metals (see [Ref. 3.1, Chap. 9]).

In the following we describe the main experimental methods which have been used to study the lattice distortions due to hydrogen in metals and summarize the results for some typical metal hydrogen systems.

3.2 Lattice Distortions

3.2.1 Strain Field of a Point Defect

The introduction of an interstitial atom into a crystal lattice in general is accompanied by an increase of the lattice volume. The additional volume Δv needed by one interstitial can be determined from the change of the mean atomic distance in the lattice, i.e., from the lattice parameter change Δa. The so-called size factor $\lambda = a^{-1}(\Delta a/\Delta c)$ is related to $\Delta v = 3\lambda$. c is the interstitial concentration (c = number of interstitials/number of host lattice atoms).

In general the displacements $\boldsymbol{u}$ of the lattice atoms from their regular sites $\boldsymbol{r}$ may be anisotropic and a tensor is necessary to describe the strain field ε_{ij}^{d} of the interstitial [3.3]. For an interstitial described by λ_{ij}^{ν} various equivalent orientations ν may exist in the crystal lattice. ε_{ij}^{d} is then obtained by summing over all n_d possible orientations.

$$\varepsilon_{ij}^{d} = \sum_{\nu=1}^{n_d} \lambda_{ij}^{\nu} c \varrho_{\nu} . \tag{3.1}$$

ϱ_{ν} is the fraction of interstitials having the orientation ν. Although λ_{ij} is determined in most experiments there is some advantage in describing the lattice distortions of an interstitial by its stress field.

3.2.2 Stress Field of a Point Defect

Kanzaki [3.4] has introduced the following very useful concept for a theoretical description of the displacement field. The actual displacements of the lattice atoms by the defect can be simulated in a defect-free lattice by applying virtual forces f_j^m ("Kanzaki forces") to each lattice atom m (distance from the defect site x_i^m) so that these forces cause the same displacements as the defect does. The force distribution can be described by a multipole expansion in analogy to a charge distribution in the electric case. It turns out that for most cases it suffices to take only the "dipole part" of the force distribution

$$P_{ij} = \sum_{m} f_j^m x_i^m , \tag{3.2}$$

the so-called double force tensor. Furthermore only forces on a few neighbor atoms (e.g., only next nearest neighbors of the defect) are necessary to give a good description of the displacements.

A distribution of c defects with possible orientations (occupation ϱ_ν) each described by a double force tensor P^ν_{ij} gives a stress field

$$\sigma^d_{ij}=1/\Omega \sum_{\nu=1}^{n_d} P^\nu_{ij} c \varrho_\nu . \tag{3.3}$$

Stress and strain are related by Hooke's law and we get a relation between λ_{ij} and P_{ij}

$$\lambda^\nu_{ij}=\frac{1}{\Omega}\sum_{kl} S_{ijkl} P^\nu_{kl} . \tag{3.4}$$

Ω is the mean atomic volume of a lattice atom and S_{ijkl} are the elastic compliances in the four index notation.

For defects randomly distributed and oriented in a cubic crystal, the volume change measured, for example, by lattice parameter change is given by

$$3\Delta a/a \simeq \Delta V/V=\sum_i \varepsilon^d_{ii}=c\,\mathrm{Trace}\,\lambda_{ij}=(c/3\Omega)K\,\mathrm{Trace}\,P_{ij} . \tag{3.5}$$

$K=3(S_{11}+2S_{12})=3/(C_{11}+2C_{12})$ is the compressibility. A lattice defect with its distortion field has been named an *elastic dipole*. In analogy to an electric dipole, an elastic dipole may also lower its energy in an external field by reorientation. For an external strain field ε^e_{ij} the energy of a defect is changed by $u=-\varepsilon^e_{ij}P_{kl}$. This is the basis of all anelastic relaxation processes which thus give information on the P_{ij} and hence the λ_{ij}.

3.3 Experimental Methods

3.3.1 Change of Macroscopic Dimensions

Dissolving n hydrogen atoms in a metal changes the volume V of the metal by

$$\Delta V=n\Delta v , \tag{3.6}$$

where Δv is the characteristic volume change per hydrogen atom, the quantity we are looking for. [Δv is directly related to the mean partial molar volume $V_M=\Delta v\cdot L$ (L is Avogadro's number).] A metal crystal with the volume V contains N metal atoms. If the mean atomic volume of a metal atom is Ω, we have $V=N\cdot\Omega$ and the relative volume change due to an atomic fraction $c=n/N$ hydrogen atoms is

$$\Delta V/V=c(\Delta v/\Omega) . \tag{3.7}$$

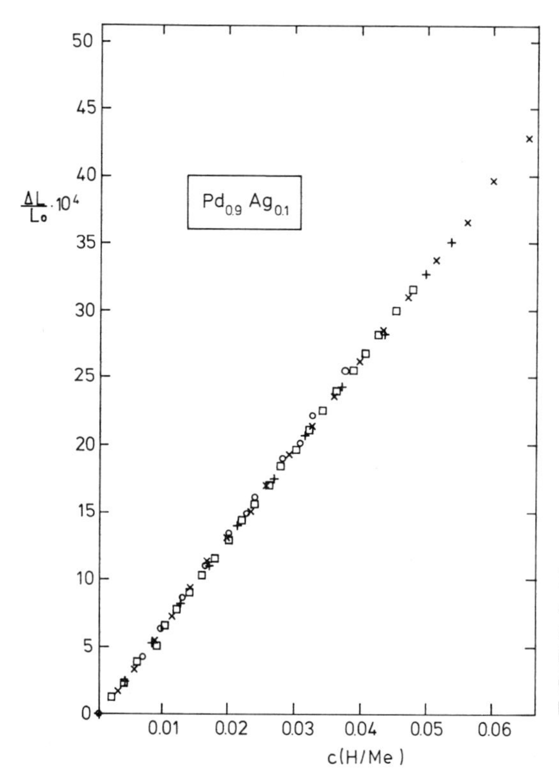

Fig. 3.1. Relative length change $\Delta L/L_0$ vs hydrogen concentration c(H/Me) for a $Pd_{0.9}Ag_{0.1}$ alloy (*Loos* [3.7]) at various temperatures (○ 150 °C, + 170 °C, × 190 °C, □ 210 °C)

Measuring the relative volume change as a function of the hydrogen concentration yields $(\Delta v/\Omega)$, which is related by (3.5) with the quantities used to describe the distortion field.

For a cubic crystal and random interstitial site occupancy it suffices to measure the change of one sample dimension, e.g., the length change $\Delta L/L$. For small changes $\Delta V/V = 3\Delta L/L + 0[3(\Delta L/L)^2] + \dots$ and

$$\Delta L/L = 1/3 \cdot c(\Delta v/\Omega)\,. \tag{3.8}$$

This method is well suited for relatively large samples if care is taken that the sample volume does not change by any other influence. Length changes can be measured with a vernier gauge with an accuracy $d(\Delta L) = 10^{-6}$ m [3.5, 6]. Using samples of $L = 10^{-2}$ m, this leads to $\Delta L/L = 10^{-4}$, a relative length change typically caused by a hydrogen concentration $c \approx 2 \cdot 10^{-3}$. Optical interferometers can give $\Delta L \approx 10^{-7}$ m. Figure 3.1 shows a typical result [3.7]. The relative length change $\Delta L/L$ of $Pd_{0.9}Ag_{0.1}$ was measured at various temperatures as a function of the hydrogen pressure, and thus for different hydrogen concentrations in the crystal.

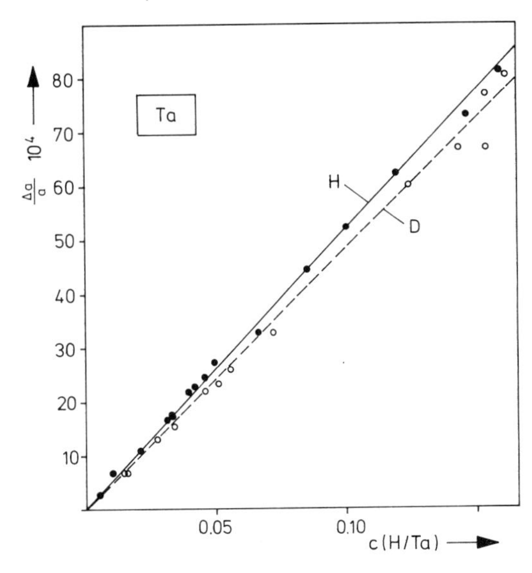

Fig. 3.2. Relative lattice parameter change $\Delta a/a$ of tantalum vs hydrogen (●) and deuterium (○) concentration (*Pfeiffer* and *Peisl* [3.9])

3.3.2 Lattice Parameter Change

The same information can be obtained from precision lattice parameter measurements by x-ray or neutron diffraction studies as from the above measurements. Here one determines the volume change of the mean elementary cell and thus does not depend on the volume of the sample not being changed during the hydrogen solution process by some process like evaporation of metal atoms, loss of part of a powder sample, etc. For a cubic crystal and a random interstitial site occupancy we have

$$\frac{a_H^3 - a_0^3}{a_0^3} = c\frac{\Delta v}{\Omega} = 3\frac{\Delta a}{a} + 0\left[3\left(\frac{\Delta a}{a}\right)^2\right] + \dots \,. \tag{3.9}$$

In the case of cubic crystals with interstitial sites of type v having different fractional occupations ϱ_v and a distortion field of lower than cubic symmetry, the lattice parameter changes $\Delta d/d$ depend on the orientation of the sample

$$(\Delta d/d)_{e_i} = \sum_{v=1}^{n_d} c\varrho_v \lambda_{ij}^v e_i e_j \,. \tag{3.10}$$

e_i is the unit vector giving the lattice direction under consideration.

Small lattice parameter changes are measured from the shift $\Delta\theta_B$ of a high angle Bragg reflection (Bragg angle θ_B)

$$\Delta a/a = -\mathrm{ctg}\,\theta_B \cdot \Delta\theta_B \,. \tag{3.11}$$

Given a suitable x-ray or neutron diffraction setup, the accuracy with which θ_B and $\Delta\theta_B$ can be determined depends mainly on the width of the Bragg peaks.

For powder or polycrystalline samples, $\Delta a/a$ can be determined with an accuracy of $d(\Delta a/a) \geqq 10^{-4}$. Using single crystals and high resolution x-ray diffractometry [3.8, 9] gives $d(\Delta a/a) \geqq 10^{-5}$. Neutron and x-ray backscattering diffractometers [3.10, 11] can achieve $d(\Delta a/a) \geqq 10^{-7}$. These techniques exceed the accuracy with which the hydrogen concentration can be determined even under favorable conditions (see Sects. 3.3.5 and 3.3.6). Figure 3.2 shows a typical result [3.9]. The relative lattice parameter change $\Delta a/a$ of tantalum was measured as a function of the hydrogen and deuterium concentration.

3.3.3 Diffuse Scattering of X-Rays and Neutrons

The ideal scattering intensity distribution of a metal crystal is altered threefold by the distortion field of dissolved hydrogen (see, e.g., [3.12, 13]. In Section 3.3.2 we have used 1) the shift of the diffraction peaks to determine an average new lattice parameter $a_0 + \Delta a$. Local deviation from this average expanded lattice gives rise to 2) a static Debye-Waller factor which causes an attenuation of the Bragg intensities and 3) a diffuse scattering intensity distribution close to the Bragg peaks (Huang diffuse scattering) and between the Bragg peaks ("Zwischenreflexstreuung").

The elastic coherent diffuse scattering intensity I_D is given by

$$I_D(\boldsymbol{K}) \sim c \sum_{\nu} \varrho_\nu |F_I^\nu + iF_M \boldsymbol{K} \cdot \tilde{\boldsymbol{u}}^\nu(\boldsymbol{K})|^2 , \tag{3.12}$$

$$\tilde{\boldsymbol{u}}^\nu(\boldsymbol{K}) = \sum_{n} \boldsymbol{u}^\nu(\boldsymbol{r}_m) \exp(i\boldsymbol{K} \cdot \boldsymbol{r}_m) . \tag{3.13}$$

$\boldsymbol{K}$ is the scattering vector, F is the scattering amplitude for x-rays or the coherent scattering length for neutrons, respectively. Index I denotes the value for the interstitial and M for the metal host atoms. Debye-Waller factors are included in F. $\tilde{\boldsymbol{u}}^\nu(\boldsymbol{K})$ is the Fourier transform of the displacement field $\boldsymbol{u}^\nu(\boldsymbol{r}_m)$ caused by an interstitial of type ν.

Huang Diffuse Scattering of X-Rays

Huang diffuse scattering of x-rays has been successfully applied to determine the double force tensor P_{ij} of hydrogen in metals by *Metzger* et al. [3.14, 15]. In the case of x-rays, the scattering from the interstitial hydrogen can be neglected compared with that from the metal atoms ($F_M \gg F_I$, $F_I \approx 0$). The Huang scattering is observed for small distances $\boldsymbol{g} = \boldsymbol{K} - \boldsymbol{G}$ from a Bragg reflection given by the reciprocal lattice vector $\boldsymbol{G}$. In this case the Fourier transformed

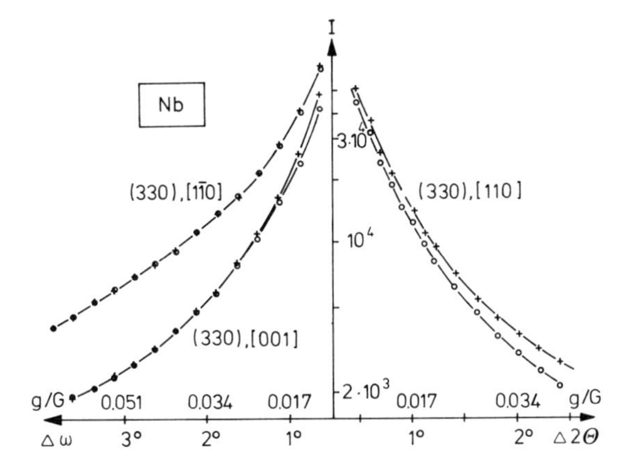

Fig. 3.3. Diffuse scattering intensity from hydrogen-doped niobium near the (330) reflection in [110], [1$\bar{1}$0], and [001] direction (*Metzger* et al. [3.14]) ○ pure Nb; + $c=0.023$

$\boldsymbol{u}^{\nu}(\boldsymbol{K})$ can be expressed by the double force tensor P_{ij} and elastic continuum theory [3.16]. For an anisotropic distortion field and a random distribution of the defect orientations, one gets for a cubic crystal

$$I_{\text{HDS}} \sim cF_{\text{M}}^2(G/g)^2 \Pi \,. \tag{3.14}$$

Π contains the components of P_{ij} and the elastic constants C_{ijkl} and depends on the orientation of $\boldsymbol{G}$ and $\boldsymbol{g}$. For a double force tensor P_{ij} with tetragonal symmetry

$$P_{ij} = \begin{pmatrix} A & 0 & 0 \\ 0 & B & 0 \\ 0 & 0 & B \end{pmatrix}, \tag{3.15}$$

all possible information is obtained from the scattering intensity distribution around a ($hh0$) Bragg reflection. Figure 3.3 shows a typical result [3.14] for hydrogen in niobium. The diffuse scattering intensity was measured close to the (330) reflection of Nb in the three perpendicular directions [110], [1$\bar{1}$0], and [001] on a pure Nb single crystal and on the same crystal after $c=0{,}023$ hydrogen was dissolved. Additional scattering intensity due to the lattice distortions caused by the hydrogen is observed only in the [110] direction. An absence of additional scattering intensity in the [001] direction is expected for a double force tensor with no off-diagonal elements. Zero intensity in the [1$\bar{1}$0] direction means that $(A-B)^2=0$, i.e., the distortion field of hydrogen in niobium (described by P_{ij}) has cubic symmetry. The remaining component of P_{ij} could be determined from the absolute scattering intensity in the [110] direction and $P_{ij}=\delta_{ij}\,(3.37\pm0.1)\,\text{eV}$. The corresponding $\Delta v/\Omega$ value is included in Table 3.1.

Diffuse Scattering of Neutrons

Diffuse scattering of neutrons has been used by *Bauer* et al. [3.17] to study the lattice distortions due to deuterium in niobium. In practice only deuterium can be studied because hydrogen has a very high incoherent scattering cross section. Now F_{I}, the scattering length of deuterium, can no longer be neglected. The Fourier transform $\boldsymbol{u}^{\mathrm{v}}(\boldsymbol{K})$ can no longer be replaced by P_{ij} and C_{ijkl} because the measurements were done between the Bragg peaks, i.e., for large $\boldsymbol{g}$. Using the concept of the Kanzaki forces (see Sect. 3.2.2), the Fourier transform $\boldsymbol{u}^{\mathrm{v}}(\boldsymbol{K})$ can be calculated by

$$\tilde{\boldsymbol{u}}^{\mathrm{v}}(\boldsymbol{K})=\phi^{-1}(\boldsymbol{K})\tilde{f}^{\mathrm{v}}(\boldsymbol{K}), \tag{3.16}$$

where ϕ^{-1} is the inverse dynamical matrix of niobium known from lattice dynamics (phonon dispersion relations) and $\tilde{f}^{\mathrm{v}}(\boldsymbol{K})$ is the Fourier transform of the Kanzaki forces f_j^m.

The experimental results for the cross section of diffuse neutron scattering from deuterium in niobium as a function of the scattering vector $\boldsymbol{K}$ are shown in Fig. 3.4 for four different directions in reciprocal space [3.17]. The experimental points are compared with the following model calculations. Assuming deuterium located on tetrahedral interstitial sites and applying radial forces of equal magnitude $\boldsymbol{f}_1$ on the four next nearest neighbors, one gets $B-A=-1/5\,\mathrm{Trace}\,P_{ij}$ and $A=2B=a\cdot 2f_1/\sqrt{5}$ and expects the curves α given as dashed lines. A value of $\mathrm{Trace}\,P_{ij}=10\,\mathrm{eV}$ and $B-A=-2\,\mathrm{eV}$ has been used.

To improve the agreement with the experimental data and to be consistent with other experiments (see Huang Diffuse Scattering of x-rays and Snoek Effect), the tetragonality was varied. This can be achieved by applying additional forces of equal magnitude $\boldsymbol{f}_2$ to the four second nearest neighbors. In this case $A=a\cdot f_1/\sqrt{5}+9af_2/\sqrt{13}$ and $B=2af_1/\sqrt{5}+2af_2/\sqrt{13}$. $B-A=0$ can be achieved if $f_2=0.23f_1$. The results for this model are shown in Fig. 3.4 as curves β (solid lines). Satisfactory agreement between measured and calculated data is found.

To demonstrate that the diffuse scattered intensity also depends on the interstitial sites, calculated data are also given for the octahedral interstitial sites. Again the two cases $\mathrm{Trace}\,P_{ij}=10\,\mathrm{eV}$, $B-A=+2\,\mathrm{eV}$ (curves γ, dotted lines) and $B-A=0$ (curves δ, dash-dotted lines) have been calculated.

3.3.4 Mechanical Relaxation Methods

Defects which distort the crystal lattice have been described as elastic dipoles (see Sect. 3.2). The reorientation of an anisotropic elastic dipole in an external

Fig. 3.4. Diffuse neutron scattering cross section from deuterium in niobium as a function of the scattering vector $|\boldsymbol{K}|$ for different directions in reciprocal space (*Bauer* et al. [3.17]). (For explanation of the various calculated curves see text) ▶

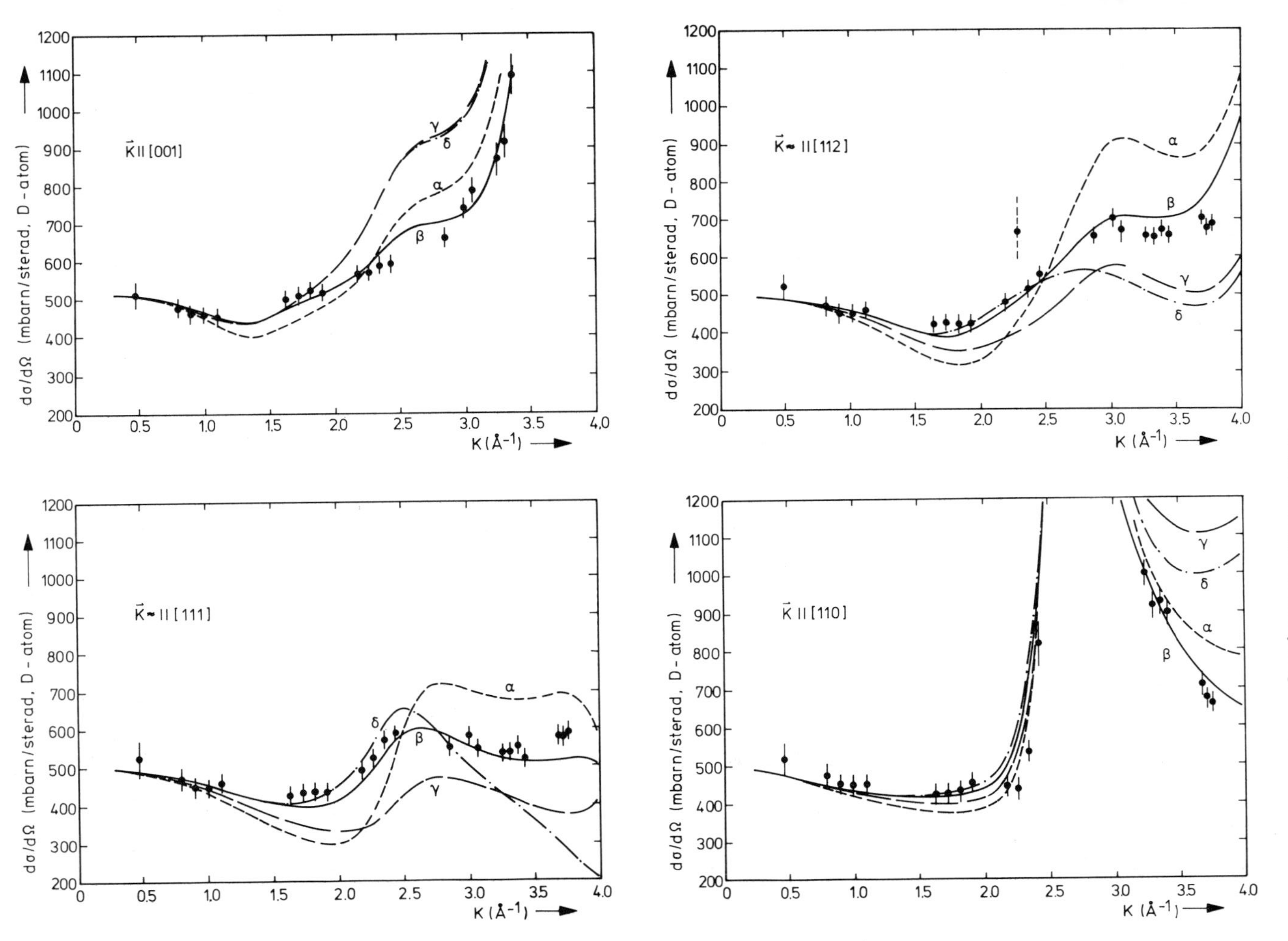
K⃗ ∥ [001]
K⃗ ≈ ∥ [112]
K⃗ ≈ ∥ [111]
K⃗ ∥ [110]
dσ/dΩ (mbarn/sterad, D - atom)
K (Å⁻¹)
α
β
γ
δ

strain field gives rise to an anelastic relaxation the *Snoek effect* [3.18]. If the defects are highly mobile like H in metals, they cause a diffusional relaxation in a strain gradient, the *Gorsky effect* [3.19, 20].

Hydrogen in bcc metals seems to fulfill all the conditions to observe these relaxation processes. Hydrogen expands the lattice; in the bcc structure no interstitial site exists with full cubic symmetry. It has been demonstrated experimentally that hydrogen occupies tetrahedral sites in Nb and Ta [3.21, 25]; therefore the elastic dipole is expected to be anisotropic and hydrogen is highly mobile. Typical diffusion constants at room temperature are $D \approx 5 \cdot 10^{-5}\,\mathrm{cm^2\,s^{-1}}$.

Gorsky Effect

The Gorsky relaxation is commonly measured as an elastic aftereffect [3.26, 27]. A dilatation gradient is applied to a sample by bending it. The lattice is expanded on one side and contracted on the opposite side of the sample. The hydrogen atoms will follow this gradient and migrate towards the expanded side, setting up a concentration gradient in the sample. (The high mobility is necessary so that the migration over sample dimensions occurs in reasonable times.) The concentration gradient causes an additional time-dependent anelastic strain.

The relaxation time of the anelastic strain is the relaxation time with which the concentration gradient is established and is closely connected with the diffusion coefficient. Precise measurements of the diffusion coefficients is the main application of the Gorsky effect [3.28]. But there is another observable quantity, the magnitude of the anelastic strain ε^{a} which is finally set up. Normalized to the applied elastic strain ε^{e}, it is called the relaxation strength $\Delta_E = \varepsilon^{\mathrm{a}}/\varepsilon^{\mathrm{e}}$ and for low hydrogen concentration it is given by

$$\Delta_E = \frac{\varepsilon^{\mathrm{a}}(t \to \infty)}{\varepsilon^{\mathrm{e}}} = \Omega_{\mathrm{G}} \frac{a^3 c (\Delta v/\Omega)^2}{18 k_{\mathrm{B}} S_{11} T} . \tag{3.17}$$

Ω_{G} is an orientation-dependent factor which contains the elastic coefficients and direction cosines, k_{B} is the Boltzmann constant, T is temperature and S_{11} are elastic compliances. The Gorsky relaxation has been measured on a number of hydrogen-metal systems, the relaxation strength Δ_E determined, its $1/T$ temperature dependence verified [for higher concentrations the hydrogen interstitials interact and the temperature dependence goes over to a Curie-Weiß type behavior $\Delta_E \sim (T - T_{\mathrm{G}})^{-1}$], and the magnitude $(\Delta v/\Omega)$ determined. The values have been recently collected by *Völkl* [3.26] and are included in the tables.

Snoek Effect

The application of an uniaxial stress, e.g., along one of the main crystal axis, causes a time-dependent ordering of anisotropic elastic dipoles. This anelastic relaxation is proportional to $(\lambda_1 - \lambda_2)$ or $(A - B)$ for a tetragonal distortion field. It can be measured as a change of the elastic constants. The shear moduli $C' = (C_{11} - C_{12})/2$ and C_{44} and the bulk modulus $B = (C_{11} + 2C_{12})/3$ have been investigated. For low defect concentrations and tetragonal distortions one expects a change of C'

$$\Delta C' = \frac{2\Omega c}{3k_B T}(\lambda_1 - \lambda_2)^2 . \tag{3.18}$$

Experiments on hydrogen-metal systems have been performed to determine $(\lambda_1 - \lambda_2)$. Careful measurements of the Snoek relaxation by one group of researchers performed by two different techniques (*Buchholz* [3.29, 30], *Magerl* et al. [3.5]) gave an unexpected small value for $\lambda_1 - \lambda_2$, which was essentially governed by their experimental error. Another group (*Fisher* et al. [3.31]) found a $(\lambda_1 - \lambda_2)$ to be as large as expected but have not yet checked so far whether the change of the elastic modulus shows the expected temperature dependence.

3.3.5 Concentration Determination

All the methods discussed to determine $\Delta v/\Omega$ or λ_{ij} or P_{ij} depend on an exact knowledge of the hydrogen concentration. The usual procedure is to start with a pure reference sample. This must be produced by a proper high-temperature ultrahigh vacuum degassing treatment [Ref. 3.1, Chap. 2]. Doping with hydrogen can be achieved in a hydrogen gas atmosphere or in liquid electrolyte containing protons. As hydrogen is highly mobile in the metal, the rate of hydrogen doping or hydrogen loss of a sample depends essentially on the surface of the sample. In most cases the surface is poisoned by impurities. This has the advantage that after doping, the samples remain sealed and the hydrogen loss is negligible at temperatures $T < 600$ K. On the other hand, doping is more complicated. It has to be done in the UHV container right after the degassing procedure or at elevated temperatures (~ 800 K) where the permeation of hydrogen through the surface layer is high. At these temperatures, however, other gases like oxygen and nitrogen may be dissolved, at least in a surface layer. This unknown amount of impurities may cause serious errors. The characteristic volume change Δv is about twice as big for oxygen and nitrogen as for hydrogen. Also, if the concentration is determined from the mass change of the sample, an impurity concentration of oxygen or nitrogen may appear as a hydrogen concentration more than ten times greater.

Measuring the *mass change* of the sample is indeed a widely used, fast and simple method to determine the concentration of dissolved hydrogen. As the influence of impurities is mainly within a surface layer (thickness a few microns [3.32]), the accuracy of this method increases with decreasing surface to volume ratio of the sample. For large single-crystal samples the surface layer may be neglected, whereas for powder samples one may easily investigate only surface layers and thus have a large rather unknown error source.

The pressure increase during a *high-temperature vacuum extraction* of hydrogen is often used to determine the hydrogen content. Other gases are detected with the same sensitivity as hydrogen.

Once a reliable calibration exists, measuring the *relative lattice parameter change* by x-ray or neutron diffraction or the relative length change gives the hydrogen concentration. The penetration depth of the x-rays or neutrons in most cases easily exceeds the thickness of the perturbed surface layer.

An unique method has been applied by *Metzger* et al. [3.14] to cross check the hydrogen concentration of their samples. Combining relative lattice parameter change and Huang diffuse scattering intensity gives the interstitial concentration independent from any other information. The results agreed very well with the results from two other methods ($\Delta a/a$ and mass change).

3.4 Experimental Results: Lattice Expansion

In the following sections the experimental results will be collected for a number of typical and interesting metal-hydrogen systems. The most reliable data for each of the rather well-investigated systems are collected in figures from which a best value can be deduced. Furthermore a more complete but less critically selected collection of experimental data is given in tables. Independent of the experimental method and the result given in the paper, we have always calculated the relative volume change $\Delta v/\Omega$ per unit concentration of hydrogen in order to have one value which can be compared. Each table also gives the experimental method. The concentration and temperature region are given in order to see to what phase the experimental value belongs. Many results could not be accepted because they were obtained in an undefined coexistence region of two phases. The form of the samples used is also noted.

As discussed in the preceding chapter, the strain field of hydrogen has cubic symmetry, at least $|A-B|$ seems to be extremely small. In the case of niobium, *Pick* et al. [3.33] deduced from the orthorhombic cell dimensions in the β-phase values $|A-B|=0.04\,\text{eV}$ and $0.13\,\text{eV}$ which are consistent with the values mentioned in Sections 3.3.3 and 3.3.4. Only one component of the double force tensor $P_{ij}=\delta_{ij}P$ remains and all information is deduced from $\text{Trace}\,P_{ij}=3P$.

3.4.1 Niobium

The niobium-hydrogen system has recently attracted most attention. Lattice strains have been studied by a variety of methods to test the concept of elastic interaction; this has been proposed as the most relevant interaction to explain

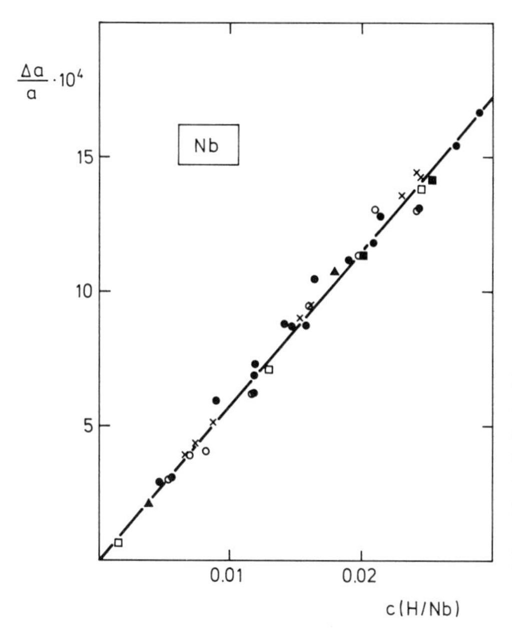

Fig. 3.5. Relative lattice parameter change $\Delta a/a$ of niobium single crystals vs hydrogen concentration in the α-phase. ● Nb/H ○ Nb/D $\Delta a/a$, *Pfeiffer* [3.9]; ▲ Nb/H $\Delta L/L$, *Buck* et al. [3.6]; ■ Nb/H □ Nb/D $\Delta L/L$, *Magerl* et al. [3.5, 48]; × Nb/H Huang scattering, *Metzger* et al. [3.14]

the $\alpha - \alpha'$ phase transition [3.2]. In Fig. 3.5 the results obtained at room temperature in the pure α-phase are collected. All experiments were performed on single crystals. At these low concentrations the accuracy would hardly be good enough to measure on powder samples. All data scatter about a straight line with a slope which practically coincides with the value given by *Pfeiffer* et al. [3.9]. As a best value we propose

$$\Delta v/\Omega = 0.174 \pm 0.005 .$$

No systematic difference between hydrogen and deuterium has ever been observed.

In Fig. 3.6 the relative volume change for higher concentrations is given. Measurements are included which were performed at high temperatures in order to have a homogeneous phase. An extrapolation of the straight line from Fig. 3.5 is also given. Almost all high concentration values seem to give a lower slope, close to the value given in [Ref. 3.1, Chap. 2], $\Delta v/\Omega = 0.14 \pm 0.08$. In this high concentration region there are, however, only two data points (×) from large single-crystal samples, which lie exactly on the straight line extrapolated from low concentrations. We therefore propose that the smaller slope is typical for powder samples. Because of surface effects (see Sect. 3.3.5) the concentration given may be too high and, due to trapping of hydrogen by impurities in the surface layer, $\Delta v/\Omega$ may actually be lower [3.34]. Table 3.1 summarizes the results for the niobium hydrogen (deuterium) system.

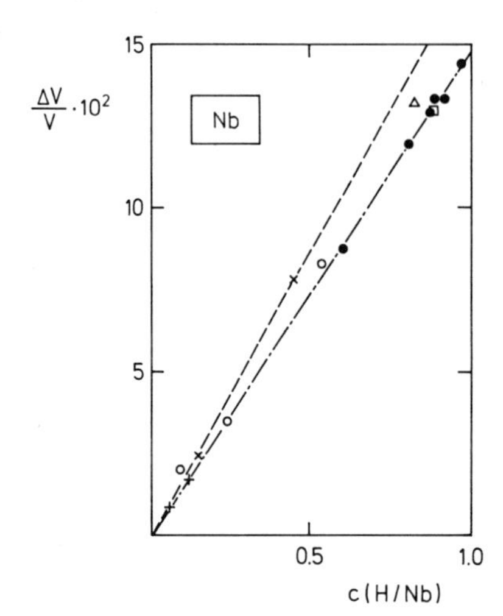

Fig. 3.6. Relative volume change $\Delta V/V$ of niobium vs hydrogen concentration. ○ *Albrecht* et al. [3.36], △ *Amato* and *Negro* [3.38], □ *Brauer* and *Herman* [3.35], —·—·—● *Pick* et al. [3.33], + *Walter* and *Chandler* [3.37] (Nb/H powder samples); × *Rowe* et al. [3.40] Nb/D, single crystal; — — — — extrapolation of straight line of Fig. 3.5

Table 3.1. Relative volume change $\Delta v/\Omega$ due to hydrogen in niobium

$\Delta v/\Omega$	Method	Sample[a]	Concentration	Temperature	Author(s)
0.15	$\Delta a/a$ x-ray	p; H, D	β-phase	RT	*Brauer* and *Herman* [3.35]
0.154[b]	$\Delta a/a$ x-ray	b; H	0…0.54	$T > T_c$	*Albrecht* et al. [3.36]
0.135	$\Delta a/a$ x-ray	b, p; H	0.057…0.864	RT…93° C	*Walter* and *Chandler* [3.37]
0.16	$\Delta a/a$ x-ray	p; H	β(0.83)	RT	*Amato* and *Negro* [3.38]
0.195	Gorsky relax	b; H, D	α-phase		*Schaumann* et al. [3.39]
0.18	$\Delta L/L$	sc; H	0.02; 0.09	RT	*Buck* et al. [3.6]
0.17	$\Delta a/a$ neutron	sc; D	α'(0.15; 0.45)	475 K	*Rowe* et al. [3.40]
0.171	$\Delta L/L$	sc; H	α-phase	RT	*Fisher* et al. [3.31]
0.142	$\Delta a/a$ x-ray	p; H	$c \geqq 0.6$	160° C	*Pick* and *Bausch* [3.33]
0.168	$\Delta L/L$	sc; H, D	α-phase	RT	*Magerl* et al. [3.5]
0.175	HDS x-ray	sc; H, D	α-phase	RT	*Metzger* et al. [3.14]
0.174	$\Delta a/a$ x-ray neutron	sc; H, D	α-phase	RT	*Pfeiffer* and *Peisl* [3.9]

[a] p powder; b bulk; sc single crystal.

[b] The value $\Delta a = 0.23 \cdot c\,(\Delta v/\Omega = 0.21)$ given in the paper [3.36] is in obvious disagreement with the data given as experimental results (see also [3.33]). A proper evaluation of the experimental data gives the value included here.

3.4.2 Tantalum

In the tantalum-hydrogen (deuterium) systems the α-phase extends to higher concentrations at room temperature. The results for the α-phase and for higher concentrations are collected in Fig. 3.7 and Table 3.2. The results of *Pfeiffer* et al. [3.9] have already been shown in Fig. 3.2. These results show an isotope dependence of the lattice expansion; hydrogen expands the lattice about 8%

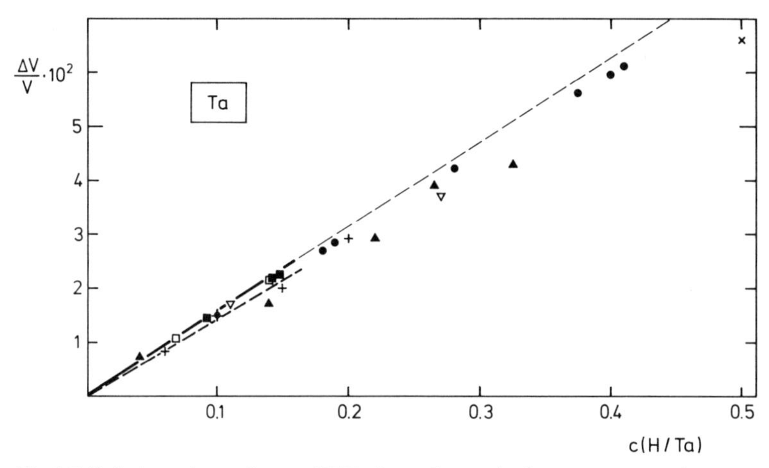

Fig. 3.7. Relative volume change $\Delta V/V$ of tantalum vs hydrogen concentration. —— H, — — D (*Pfeiffer* et al. [3.9]); ■ H, □ D *Magerl* et al. [3.5, 48]; ● H *Zierath* [3.44]; ▲ H *Stalinski* [3.41]; ▽ D *Slotfeld-Ellingsen* and *Pedersen* [3.47]; × D *Wallace* [3.43]; + H *Ducastelle* et al. [3.46]

Table 3.2. Relative volume change $\Delta v/\Omega$ due to hydrogen in tantalum

$\Delta v/\Omega$	Method	Sample[a]	Concentration	Temperature	Author(s)
0.129	$\Delta a/a$ x-ray	p; H	0.04...0.75	RT	*Stalinski* [3.41]
0.227	$\Delta a/a$ x-ray	?	$\leqq 0.4$	$-145°$ C..70° C	*Waite* et al. [3.42]
0.132	$\Delta a/a$ neutrons		Ta_2H	RT	*Wallace* [3.43]
0.15	$\Delta a/a$ x-ray	p, H	$\leqq 0.41$	RT	*Zierath* [3.44]
0.135	Gorsky rel.	b; H	α-phase		*Schaumann* et al. [3.39]
0.132	Gorsky rel.	b; D	α-phase		*Schaumann* et al. [3.39]
0.136			0.58; 0.61	RT	*Ducastelle* et al. [3.46]
0.143	$\Delta a/a$ x-ray	p; D	0.11...0.665	RT	*Slotfeld-Ellingsen* and *Pedersen* [3.47]
0.18	$\Delta L/L$ $\Delta L/L$		0.0115; 0.11	RT	*Fisher* et al. [3.31]
0.15		sc; H, D	α-phase	RT	*Magerl* et al. [3.5]
0.155	$\Delta a/a$ x-ray	sc; H	α-phase	RT	*Pfeiffer* and *Peisl* [3.9]
0.143	$\Delta a/a$ x-ray	sc; D	α-phase	RT	*Pfeiffer* and *Peisl* [3.9]

[a] p powder; b bulk; sc single crystal.

more than deuterium. In the low concentration α-phase, a number of results for hydrogen are quite consistent with the value

$$\Delta v/\Omega = 0.155 \pm 0.005 .$$

3.4.3 Vanadium

Results for the vanadium-hydrogen system in the low concentration α-phase are shown in Fig. 3.8 [3.48]. All results are collected in Fig. 3.9 and Table 3.3.

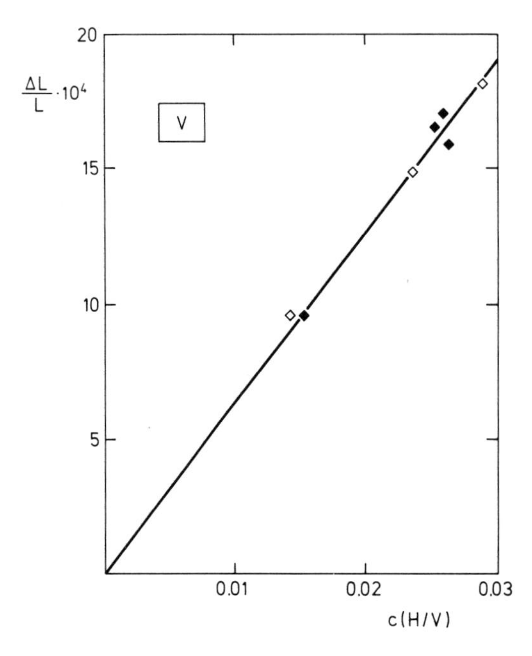

Fig. 3.8. Relative length change $\Delta L/L$ of vandium vs hydrogen concentration in the α-phase (*Magerl* [3.48]), ◆V/H, ◇V/D

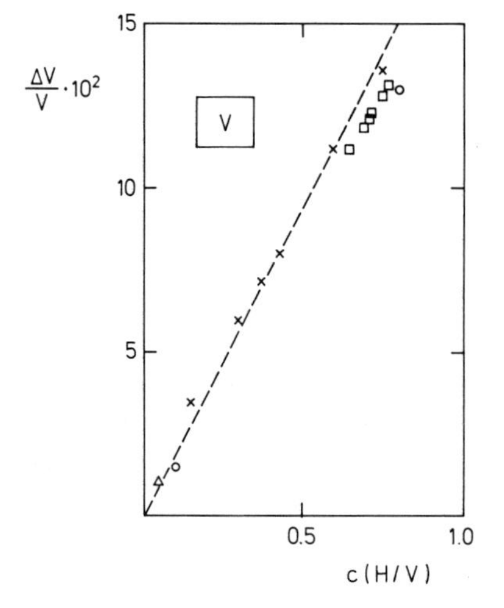

Fig. 3.9. Relative volume change $\Delta V/V$ of vanadium vs hydrogen concentration. ——— *Magerl* [3.48]; × *Mealand* [3.49]; ○ *Hardcastle* and *Gibb* [3.50]; □ *Asano* and *Hirabayashi* [3.51]

Especially in the low concentration region the results agree with a value [3.5]

$$\Delta v/\Omega = 0.19 \pm 0.01 .$$

No definitive evidence of isotope dependence has as yet been detected.

Table 3.3. Relative volume change $\Delta v/\Omega$ due to hydrogen in vanadium

$\Delta v/\Omega$	Method	Sample[a]	Concentration	Temp.	Author(s)
0.2	$\Delta a/a$ x-ray	p; H	α-phase (<0.05)	RT	*Maeland* [3.49]
0.185[b]			0…0.75	200° C	
0.177	Gorsky rel.	b; H	α-phase		*Schaumann* et al. [3.39]
0.183		b; D			
0.149	$\Delta a/a$ x-ray	p; D	α-phase (<0.1)	RT	*Hardcastle* and *Gibb* [3.50]
0.163			α′-phase (0.7..0.8)		
0.172[b]	x-ray	p; D	α,α′-phase	RT	*Asano* and *Hiradayashi* [3.51]
0.189	$\Delta a/a$ neutron $\Delta L/L$	sc; H, D	α-phase (0.03)	RT	*Magerl* et al. [3.5]

[a] p powder; b bulk; sc single crystal.
[b] Value obtained from own evaluation of the experimental data given in the paper.

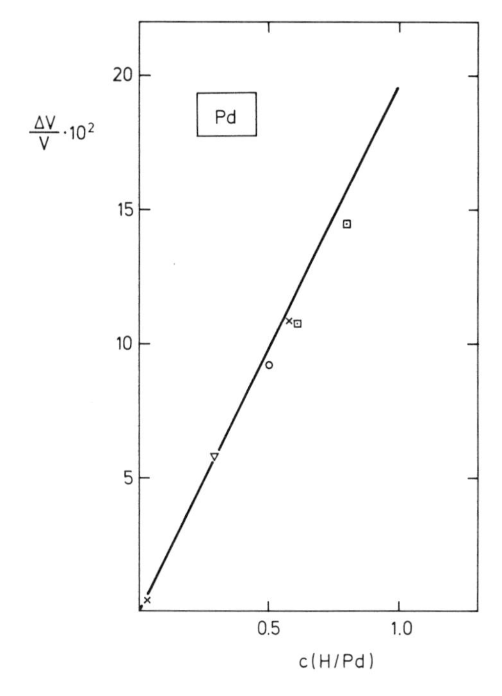

Fig. 3.10. Relative volume change $\Delta V/V$ of palladium vs hydrogen concentration. —— *De Ribaupierre* and *Manchester* [3.59]; ○ *Hanawalt* [3.53]; ▽ *Yamada* [3.52]; × *Maeland* and *Flanagan* [3.57]; ⊡ *Aben* and *Burgers* [3.55]

3.4.4 Palladium

Although palladium-hydrogen is one of the most investigated systems, reliable experimental data on lattice expansion are scarce. Most experiments were performed in the two-phase $\alpha+\beta(\alpha')$ region, so that one could only determine the lattice expansion at the upper phase boundary of the α-phase ($c_\alpha^{\max}$) and at the lower boundary of the $\beta(\alpha')$ phase ($c_\beta^{\min}$). The results for the pure palladium-hydrogen system are collected in Fig. 3.10 and Table 3.4. The experiments in

Table 3.4. Relative volume change $\Delta v/\Omega$ due to hydrogen in palladium

$\Delta v/\Omega$	Method	Sample[a]	Concentration	Temperature	Author(s)
0.2	$\Delta a/a$ x-ray	w, H	0.29	RT	*Yahmada* [3.52]
0.185	$\Delta a/a$ x-ray	b, H	0.5	RT	*Hanawalt* [3.53]
0.141	$\Delta a/a$ x-ray	b, H	0.77	RT	*Krüger* and *Gehm* [3.34]
0.177	$\Delta a/a$		$C_\beta^{\min}=0.61$	RT	*Aben* and *Burgers* [3.55]
0.182	$\Delta a/a$		0.8	RT	*Wicke* and *Nernst* [3.56]
0.155 / 0.187	$\Delta a/a$ x-ray	b; H, D	$C_\alpha^{\max}$ / $C_\beta^{\min}$	RT	*Maeland* and *Flanagan* [3.57]
0.198	Gorsky rel.	b; H	α-phase		*Völkl* et al. [3.58]
0.198	$\Delta L/L$	b, H	$f(c)$	$f(T)$	*De Ribaupierre* and *Manchester* [3.59]
0.132	$\Delta a/a$		0.8 ... 0.89	77 K	*Schirber* and *Morosin* [3.60]

[a] w wire; b bulk.

which the concentration dependence was measured over a wide range [3.58, 59] suggest a value for palladium-hydrogen of

$$\Delta v/\Omega = 0.19 \pm 0.01\,,$$

which is in quite good agreement with the rest of the data and fits to Baranowski's plot (see Sect. 3.4.5) quite well. *Maeland* and *Flanagan* [3.57] report an isotope effect. Deuterium expands the lattice about 5% more than hydrogen.

3.4.5 Other Metals and Metal Alloys

For pure metals there exist data for Ni and a few rare earth metals. *Bauer* et al. [3.61] report x-ray lattice parameter change measurements for Ni at room temperature. From their data a value $\Delta v/\Omega = 0.28$ can be deduced.

The relative volume change per unit concentration of hydrogen was determined from x-ray lattice parameter change at high temperatures (450 °C, 500 °C) by *Bonnet* [3.62] for erbium, thulium, and lutetium. The values are $\Delta v/\Omega = 0.1$ (Er); 0.12 (Tm); 0.092 (Lu). For Lu an isotope effect was detected; deuterium expands the lattice about 8% more than hydrogen. *Beaudry* and *Spedding* [3.63] obtained from measurements at room temperature for hydrogen $\Delta v/\Omega = 0.13$ (Er); 0.12 (Tm); 0.11 (Lu); 0.1 (Y).

Loos [3.7] investigated the length change of a palladium (90%)—silver (10%) alloy at various temperature and hydrogen pressures. He shows that the relative volume change per hydrogen $\Delta v/\Omega = 0.197 \pm 0.004$ does not depend on concentration ($0 \leqq c \leqq 0.06$) and temperature ($150\,°\mathrm{C} \leqq T \leqq 210\,°\mathrm{C}$) variations. The experimental results for hydrogen are shown in Fig. 3.1. Figure 3.11 shows the relative length change versus deuterium concentration in the $Pd_{0.9}Ag_{0.1}$ alloy. For comparison reasons a straight line fitted to the experimental data for

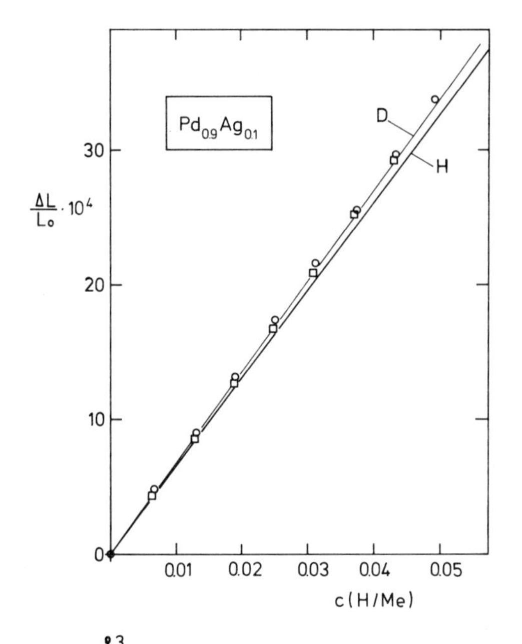

Fig. 3.11. Relative length change $\Delta L/L_0$ vs deuterium (D) concentration c(H/Me) for a $Pd_{0.9}Ag_{0.1}$ alloy at 170 °C (*Loos* [3.7]). For comparison the straight line for hydrogen (H) from Fig. 3.1 is also given

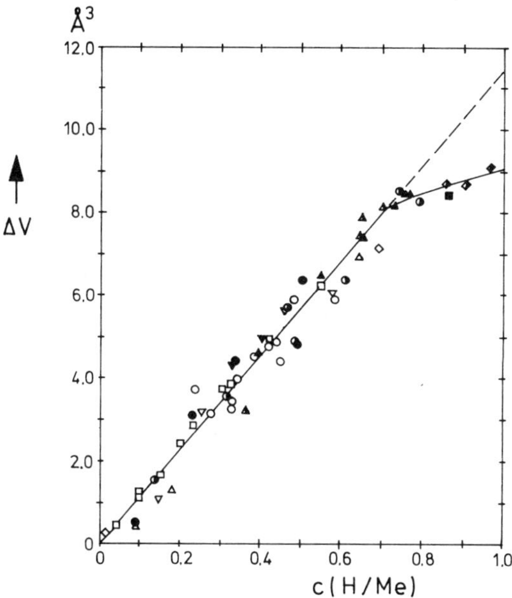

Fig. 3.12. Volume change ΔV vs hydrogen concentration c for a series of fcc metals and alloys in single-phase regions (25 °C) (*Baranowski* et al. [3.64]). ◇ ◆ palladium, ■ nickel, ▽ irridium-palladium, □ gold-palladium, ○ ◑ silver-palladium, △▲ platinum-palladium, ▼ copper-palladium, ● copper-nickel

hydrogen (Fig. 3.1) is also shown. In this case deuterium expands the lattice about 3% more than hydrogen.

Baranowski et al. [3.64], and *Krukowski* and *Baranowski* [3.65] have measured and collected the volume change due to hydrogen in a large series of fcc metals and alloys. The results are given in Figs. 3.12 and 3.13. Especially

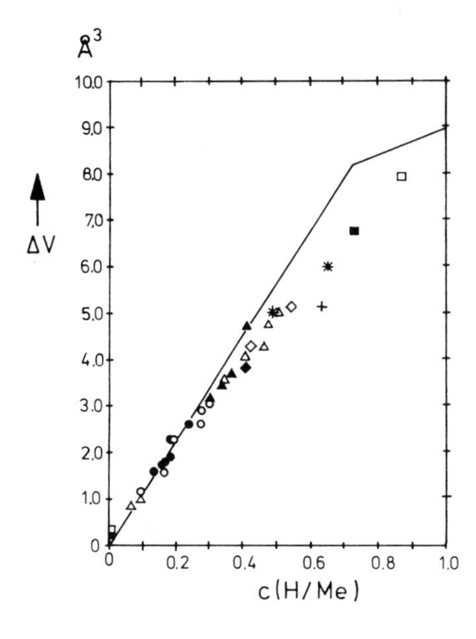

Fig. 3.13. Volume change ΔV vs hydrogen concentration c for various fcc nickel-manganese alloys in single-phase regions (25 °C) (*Krukowski* and *Baranowski* [3.65])

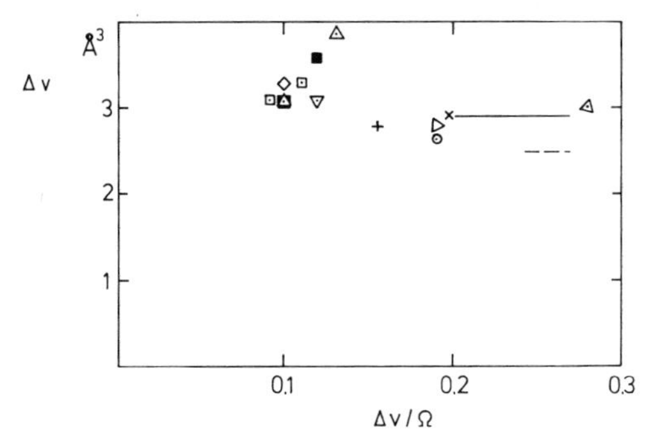

Fig. 3.14. Volume change ΔV vs relative volume change $\Delta v/\Omega$ for various hcp, bcc, and fcc metals and alloys. ——— Pd alloys; – – – NiMn alloys; □ Lu, △ Er, ■ Tm hcp; ▷ Nb, + Ta, ○ V bcc; × $Pd_{0.9}Ag_{0.1}$, ◁ Ni fcc

from Fig. 3.12 a unique slope $\Delta v/c = 11.5$ Å^3 is observed for $0 \leqq c \leqq 0.7$ for quite different metals and alloys. This surprising result means that addition of one hydrogen atom increases the volume of the crystal by $\Delta v = 2.9$ Å^3 independent of the material.

To check the above statement for other materials not included in Baranowski's plot we have plotted in Fig. 3.14 the additional volume per hydrogen atom Δv versus the relative volume change $\Delta v/\Omega$ for various hcp, bcc, and fcc metals. $\Delta v/\Omega$ varies almost by a factor of three, whereas Δv is very close to 2.9 Å^3. No explanation for this unique "size" of a hydrogen atom in many materials has been given so far.

Acknowledgement. The author wishes to thank *H. Pfeiffer* and *H. Zabel* for helpful discussions and critical reading of the manuscript. Special thanks are due to Dr. *J. Stott* for a revision of the English manuscript.

References

3.1 G. Alefeld, J. Völkl (eds.): *Hydrogen in Metals II, Application-Oriented Properties*. Topics in Applied Physics, Vol. 29 (Springer, Berlin, Heidelberg, New York 1978) in preparation
3.2 G. Alefeld: Phys. Stat. Sol. **32**, 67 (1969)
3.3 A.S. Nowick, B.S. Berry: *Anelastic Relaxation in Crystalline Solids* (Academic Press, New York-London 1972)
3.4 H. Kanzaki: J. Phys. Chem. Solids **2**, 24 (1957)
3.7 H. Loos: Thesis, Technische Universität, München (1976)
3.8 H. Pfeiffer: Dissertation, Universität München (1977)
3.9 H. Pfeiffer, H. Peisl: Phys. Lett. **60** A, 363, 1977
3.10 B. Alefeld: Z. Physik **222**, 155 (1969)
3.11 B. Sykora, H. Peisl: Z. Angew. Phys. **30**, 320 (1970)
3.12 P.H. Dederichs: J. Phys. F: Metal Phys. **3**, 471 (1973)
3.13 H. Peisl: "Structural Information and Defect Energies Studied by x-Ray Methods", in *Defects and Their Structure in Nonnectallic Solids*, ed. by B. Henderson, A.E. Hughes (Plenum Press, New York 1976)
3.14 H. Metzger, H. Peisl, J. Wanagel: J. Phys. F: Metal Phys. **6**, 2195 (1976)
3.15 H. Metzger, H. Peisl: J. Phys. F: Metal Phys. **8**, 391 (1978)
3.16 H. Trinkaus: Phys. Stat. Sol. **51**, 307 (1972)
3.17 G. Bauer, W. Schmatz, W. Just: Proc. Second International Congress on Hydrogen in Metals, (Paris 1977)
3.18 L. Snoek: Physica **8**, 711 (1941)
3.19 W.S. Gorsky: Z. Physik SU **8**, 457 (1935)
3.20 G. Schaumann, J. Völkl, G. Alefeld: Phys. Rev. Lett. **21**, 891 (1968)
3.21 H.D. Carstanjen, R. Sizmann: Ber. Bunsenges. Phys. Chem. **76**, 1223 (1972)
3.22 H.D. Carstanjen, R. Sizmann: Phys. Lett. **40** A, 93 (1972)
3.23 M. Antonini, H.D. Carstanjen: Phys. Stat. Sol. (a) **34**, K 153 (1976)
3.24 N. Stump, W. Gissler, R. Rubin: Phys. Stat. Sol. **54**, 295 (1972)
3.25 V.A. Somenkov, A.V. Gurskaya, M.G. Zemlyanov, M.E. Kost, N.A. Chernoplekov, A.A. Chertkov: Soviet Phys.-Solid State **10**, 1076 (1968)
3.26 J. Völkl: Ber. Bunsenges. Phys. Chem. **76**, 797 (1972)
3.27 J. Völkl, G. Alefeld: Nuovo Cimento **33** B, 190 (1976)
3.28 J. Völkl, G. Alefeld: "Hydrogen Diffusion im Metals", in *Diffusion in Solids: Recent Developments*, ed. by A.S. Nowick, J.J. Burton (Academic Press, New York 1975) p. 231
3.29 J. Buchholz, J. Völkl, G. Alefeld: Phys. Rev. Lett. **30**, 318 (1973)
3.30 J. Buchholz: Jül-962-FF (Kernforschungsanlage Jülich 1973)
3.5 A. Magerl, B. Berre, G. Alefeld: Phys. Stat. Sol. (a) **36**, 161 (1976)
3.31 E.S. Fisher, D.G. Westlake, S.T. Ockers: Phys. Stat. Sol. (a) **28**, 591 (1975)
3.32 H. Pfeiffer, H.D. Carstanjen: Private communication
3.33 M.A. Pick, R. Bausch: J. Phys. F: Metal Phys. **6**, 1751 (1976)
3.34 U. Freudenberg, J. Völkl, W. Münzing, G. Alefeld: Verhandl. DPG (VI) **12**, 317 (1977)
3.35 G. Brauer, R. Herman: Z. Anorg. Allg. Chemie **274**, 11 (1953)
3.36 W.M. Albrecht, W.D. Goode, W.M. Mallet: J. Electrochem. Soc. **106**, 981 (1959)
3.37 R.J. Walter, W.T. Chandler: Trans AIME **233**, 762 (1965)
3.38 I. Amato, A. Negro: J. Less-Common Metals **16**, 468 (1968)
3.39 G. Schaumann, J. Völkl, G. Alefeld: Phys. Stat. Sol. **42**, 401 (1970)
3.6 O. Buck, D.O. Thompson, C.A. Wert: J. Phys. Chem. Solids **32**, 2331 (1971)
3.40 J.M. Rowe, N. Vagelatos, J.J. Rush, H.E. Flotow: Phys. Rev. B **12**, 2559 (1975)
3.41 B. Stalinski: Bull. Acad. Polen. Sci. Cl. III, **2**, 245 (1954)
3.42 T.R. Waite, W.E. Wallace, B.S. Craig: J. Chem. Phys. **24**, 634 (1956)
3.43 W.E. Wallace: J. Chem. Phys. **35**, 2156 (1961)
3.44 J. Zierath: Dissertation Universität Münster (1969)
3.45 E. Wicke, A. Obermann: Z. Phys. Chem. N. F. **77**, 163 (1972)
3.46 F. Ducastelle, R. Courdan, P. Costa: J. Phys. Chem. Solids **31**, 1247 (1970)

3.47 D.Slotfeld-Ellingsen, B.Pedersen: Phys. Stat. Sol. (a) **25**, 115 (1974)
3.48 A.Magerl: Private communication
3.49 A.J.Maeland: J. Phys. Chem. **68**, 2197 (1964)
3.50 K.I.Hardcastle, T.R.P.Gibb, Jr.: J. Phys. Chem. **76**, 927 (1972)
3.51 H.Asano, M.Hirabayashi: Phys. Stat. Sol. (a) **15**, 267 (1973)
3.52 M.Yamada: Phil. Mag. **45**, 241 (1923)
3.53 J.D.Hanawalt: Phys. Rev. **33**, 444 (1929)
3.54 F.Krüger, G.Gehm: Ann. Physik **16**, 174 (1933)
3.55 P.C.Aben, W.G.Burgers: Trans. Faraday Soc. **58**, 1989 (1962)
3.56 E.Wicke, G.H.Nernst: Ber. Bunsenges. Phys. Chem. **68**, 224 (1964)
3.57 A.J.Maeland, T.B.Flanagan: J. Phys. Chem. **68**, 1419 (1964)
3.58 J.Völkl, G.Wollenweber, K.-H.Klatt, G.Alefeld: Z. Naturforsch. **26**a, 922 (1971)
3.59 Y.de Ribaupierre, F.D.Manchester: J. Phys. C: Solid State Phys. **7**, 2126 (1974)
3.60 J.E.Schirber, B.Morosin: Phys. Rev. B**12**, 117 (1975)
3.61 H.J.Bauer, G.Berninger, G.Zimmermann: Z. Naturforsch. **23**a, 2023 (1968)
3.62 J.E.Bonnet: J. Less-Common Metals **49**, 451 (1976)
3.63 B.J.Beaudry, F.H.Spedding: Met. Trans. **6**B, 419 (1975)
3.64 B.Baranowski, S.Majchrzak, T.B.Flanagan: J. Phys. F: Metal Phys. **1**, 258 (1971)
3.65 M.Krukowski, B.Baranowski: J. Less-Common Metals **49**, 385 (1976)

4. Investigation of Vibrations in Metal Hydrides by Neutron Spectroscopy

T. Springer

With 16 Figures

The dynamics of hydrogen or deuterium dissolved in heavy metals can be discussed in terms of three kinds of motions with widely differing time scales so that, from a physical point of view, the motions can be discussed separately. First there are diffusive motions with characteristic times which are, normally, longer than 10^{-12} s. Secondly, vibrations appear in the region of the acoustic spectrum of the host lattice with frequencies up to 10^{13} Hz, and, finally, optic phonons or localized modes with frequencies in the region of 10^{14} Hz which are well separated from the acoustic band. Here we deal with vibrations, and diffusion will be treated in other chapters of this book (see Chap. 10).

At present, neutron spectroscopy is the only available method for the investigation of such vibrations. In particular, the localized modes of hydrides can be studied very easily because of the large incoherent scattering cross section of the proton. To our knowledge, light-reflection spectroscopy on hydrides has so far been unsuccessful for such investigations. At present, there is a general interest in the dynamics of hydrogen or deuterium in metals, since these are linked with many physical properties of such systems, in particular superconductivity, electronic properties, and hydrogen diffusion, and a great abundance of experimental research exists in this field. However, the status of theoretical interpretation is quite unsatisfactory at the present.

4.1 General and Experimental Aspects

The low-energy vibrations of hydrogen are linked to the acoustic phonons of the host lattice. For *long wavelength* phonons, the hydrogen amplitudes of such vibrations agree with the host atom amplitudes because the dissolved hydrogens move in phase with the host metal atoms. On approaching the zone boundary there is, in general, a slow decrease of the *amplitude* (see [4.1, 2]). These modes are often called the *band modes* of the dissolved hydrogen. The *optic modes* are caused by the additional degrees of freedom which are introduced by the dissolved hydrogen or deuterium. For stoichiometric hydrides these vibrations can be considered as collective vibrations of the light H-atom against the heavy and only slightly moving host atoms. Owing to the small hydrogen mass, the amplitude of these components in the spectrum is relatively large.

Table 4.1. Scattering cross sections

Element	σ_s [barns][a]	σ_{inc}[barns][b]
H	81.96(6)[c]	80.20(6)
D	7.64(3)	2.04(3)
T	2.3(7)	$0.0^{+0.1}_{-0.8}$
Nb	6.311(6)	0.0024(3)
V	5.01(2)	4.99(2)
Ta	6.01(1)	0.011(3)
Ce	4.0(2)	1.0(3)
Th	13.9(2)	0.00
Pd	5.1(3)	0.093(10)

From *L. Koester*, personal communication and [4.3].

[a] Total bound scattering cross section per atom ($\sigma_{coh}+\sigma_{inc}$) with $\sigma_{coh}=4\pi b^2$ (b=bound coherent scattering length).

[b] Incoherent bound scattering cross section per atom (spin+isotope contributions).

[c] Number in parentheses indicates mean standard deviation of the last figure quoted.

1 barn $=10^{-24}$ cm^2.

For *dilute* hydrides with randomly distributed hydrogen atoms, the optic vibrations should be considered as "*localized modes*" (instead of collective excitations) where the isolated H impurity, together with the neighboring host atoms, can be considered as forming a "macromolecule" embedded in the undisturbed host lattice. Obviously, there may be a gradual transition with concentration from localized modes to optic phonons: for instance in PdD_x, the observed optic phonon peaks shift strongly if the wave vector is changed, even at rather large deviations from a stoichiometric composition, and the vibrations still have collective features.

The experimental information obtained from neutron spectra of *hydrides* and *deuterides* is basically different. This is a consequence of the scattering properties of the proton and the deuterium. For hydrides, the spectrum is essentially determined by incoherent scattering processes from protons ($\sigma_{inc} \gg \sigma_{coh}$)[1]. Under these circumstances the host lattice scattering is more or less negligible, except for low hydrogen concentrations; here it can be corrected for by an experiment with an unloaded sample. Since there is practically no interference of the waves scattered from the protons, the quasi-momentum selection rule does not come into play, and all vibrational modes appear in the spectrum, provided that their polarization vector has a component in the direction of the momentum transfer during the scattering process of the neutron.

If the hydrogen atoms were forming a simple cubic Bravais lattice, the cross section per solid angle and energy interval would read [4.4, 5]

$$(d^2\sigma/d\Omega dE_1)_{inc}=(\sigma_{inc}/4\pi)(k_1/k_0)S_{inc}(\boldsymbol{Q},\omega). \tag{4.1}$$

[1] A number of important scattering cross sections are presented in Table 4.1.

In a harmonic one-phonon approximation one obtains

$$S_{\text{inc}}(\boldsymbol{Q}, \omega) = \left[\frac{\hbar F(\hbar\omega/k_{\text{B}}T)}{2M|\omega|}\right] Q^2 g(\omega) \mathrm{e}^{-Q^2\langle u^2\rangle} . \tag{4.2}$$

$g(\omega)$ is the phonon density of states. The quantities

$$\hbar\omega = E_0 - E_1$$

and

$$\hbar\boldsymbol{Q} = \hbar(\boldsymbol{k}_0 - \boldsymbol{k}_1)$$

are the energy and momentum transfer during scattering; the quantities E_0, E_1, $\boldsymbol{k}_0$, and $\boldsymbol{k}_1$ are neutron energy and wave vector before and after scattering, respectively. $\boldsymbol{Q}$ is called the scattering vector. The thermal population factor is given by the function

$$F(x) = (\mathrm{e}^x - 1)^{-1} + \tfrac{1}{2}(1 \pm 1) , \tag{4.3}$$

where the upper and lower sign corresponds to phonon creation and phonon annihilation, respectively. The function $\exp(-Q^2\langle u^2\rangle)$ is the usual Debye-Waller factor which also determines the intensity of the "zero phonon" processes (see [4.6] and Chap. 10). $\langle u^2\rangle$ is the mean-square amplitude of the proton. In (4.1) the one-phonon processes dominate if $Q^2\langle u^2\rangle$ is sufficiently small ($\lesssim 0.1$), which requires Q values in the region of 2–4 Å^{-1}. The quantity in brackets in (4.2) is just the amplitude of a certain mode with frequency ω. In general, the hydrogen atoms do not form a Bravais lattice. Nevertheless, (4.1) can be considered as a useful *definition* of an apparent phonon density of states if M is identified with the hydrogen mass for the optic part of the spectrum, and with the metal atom mass for the acoustic part ("pseudo-phonon density")[2].

In this discussion we have neglected the diffusive motions. This is only allowed if ω is sufficiently large compared to the jump rate of hydrogen diffusion. The influence of diffusion on the Debye-Waller factor will be discussed more generally in Chapter 10. It should be mentioned that, for a dilute system, the Debye-Waller factor deviates from the simple form quoted above [4.8, 9]. However, this deviation seems to be small and has not yet been studied experimentally.

For a metal *deuteride*, the neutron spectrum contains more detailed information on the lattice dynamics. In general, sharp peaks appear caused by

---

[2] More generally, $g(\omega)$ can be related to the Fourier transform of the velocity autocorrelation function of the scattering proton [4.7].

coherent scattering on phonons which are selected by the quasi-momentum rule $\boldsymbol{Q}=\boldsymbol{q}+\boldsymbol{G}$ where $\boldsymbol{G}$ is a reciprocal lattice vector. For scattering from a single crystal, the intensity of the spectral line due to scattering from a phonon with frequency $\omega_{\boldsymbol{q}s}$, wave vector $\boldsymbol{q}$, and polarization s, is given by [4.10].

$$I_{\boldsymbol{q}s}=\left|\sum_i \frac{b_i}{M_i^{1/2}\omega_{\boldsymbol{q}s}^{1/2}}(\boldsymbol{Q}\boldsymbol{e}_{i\boldsymbol{q}s})\mathrm{e}^{\mathrm{i}\boldsymbol{Q}\boldsymbol{r}_i}\mathrm{e}^{-W_i}\right|^2 F(\hbar\omega_{\boldsymbol{q}s}/k_\mathrm{B}T). \qquad (4.4)$$

b_i is the coherent scattering amplitude of atom i (deuterium or metal), and $\boldsymbol{r}_i$ and $\boldsymbol{e}_{i\boldsymbol{q}s}$ are its position vector and unit polarization vector, respectively. $\exp\{-W_i\}$ is the Debye-Waller factor for atom i. A special situation occurs in the case of VD_x since σ_{coh} for V is very small, and the phonon line is only determined by deuterium scattering.

Disorder in nonstoichiometric deuterides may cause a line broadening, in particular for optic and short wavelength acoustic phonons (see Sect. 4.2). Since deuterium contributes also a finite amount of incoherent scattering, weak density-of-state peaks from $g(\omega)$ appear in addition to the coherent scattering lines. For vanadium deuteride, an incoherent distribution is also caused by the host lattice itself, due to its large incoherent scattering cross section.

In general, optic and acoustic dispersion branches $\omega_{\boldsymbol{q}s}$ of *deuterides* were measured in symmetry directions with conventional triple-axis spectrometers. Optic modes in *hydrides* were mainly investigated by time-of-flight spectroscopy, mostly using small incident energies (a few 10^{-3} eV), and studying phonon annihilation or "up-scattering" at fixed scattering angles. For typical optic modes, one obtains $\hbar\omega_\mathrm{H}=0.1...0.2$ eV. Consequently, $k_1 \gg k_0$, so that roughly

$$Q^2=(\boldsymbol{k}_0-\boldsymbol{k}_1)^2 \simeq k_1^2 = 2m\omega_\mathrm{H}/\hbar^2\,. \qquad (4.5)$$

The ω-dependence of Q as described by (4.5) causes a deformation of the apparent phonon spectrum. Furthermore, it follows that Q values occur between 7 and 10 Å^{-1}. This is a drawback because the probability for two-phonon processes $w_2/w_1 \simeq Q^2\langle u^2\rangle/2$ becomes rather high (typically $\langle u^2\rangle = 0.01...0.03$ Å^2). This enhances multiphonon processes, i.e., *combinations* of optic transitions with transitions caused by the low-energy modes. If Q is not too large, the multiphonon or "pure" or single phonon optic line still appears, and is superimposed on the broadened multiphonon line. At large Q, however, the single phonon optic line practically disappears completely, and the spectrum is dominated by the broad multiphonon line which is *approximately* centered at the energy of the optic transition. At very large Q, its width approaches the Doppler width $\Gamma \simeq (\hbar^2 Q^2 k_\mathrm{B}T/M)^{1/2}$. To avoid this broadening, the scattering experiments should be carried out at small Q, which requires large incident energies E_0 and small scattering angles (see, e.g., [4.11]). Such experiments can be performed successfully with a triple-axis spectrometer on a hot source, or on a pulsed neutron source where the epithermal flux may be relatively high.

In the following sections, a number of experiments will be discussed dealing with the optic vibrations and the phonons in various metal-hydride systems. Unfortunately, the theoretical interpretation of the results is in a rather underdeveloped state, and no successful lattice dynamics calculations exist at present. Numerous calculations for pure transition metals and for their nitrides or carbides were carried out [4.12–15], the main motivation being the understanding of phonon anomalies and their relation to the superconductivity. These calculations should be considered as a first step towards performing calculations for transition metal hydrides where there is the additional difficulty of accounting for the very strong interaction between the dissolved proton and the conduction electrons.

For palladium hydride, a few calculations exist [4.16, 17] based on its electronic structure, mainly in view of the high transition temperature for superconductivity (for literature on the electronic structure of palladium hydride see [4.18–20]). The electronic structure of the hydrides in general will be discussed in the Chapter 5 in more detail. At present, most interpretations of the experiments are based on phenomenological Born-von Karman models. The optic vibrations were treated in terms of nearest and next-nearest neighbor interactions [4.21, 22] leading to H-metal and H–H force constants. For dilute impurities the localized mode frequencies correspond to a simple Einstein oscillator if the host lattice is considered as rigid. The influence of a nonrigid host lattice on these vibrations can be treated [4.23] in terms of a displacement Greens' function of the lattice [4.24].

4.2 Optic Modes in bcc Metal Hydrides

In the transition metals niobium and tantalum, the hydrogen or deuterium atoms occupy tetrahedral sites for the disordered phases (usually called α). This holds also in certain ordered phases which reveal deviations from cubic symmetry. The tetrahedral sites in the bcc lattice have tetragonal point symmetry. Consequently, the optic vibrations split into two lines or two branches (actually into two bundles of $6+12=18$ branches because of the 6 interstitial sublattices). For vanadium at higher hydrogen concentrations, also octahedral sites become occupied. The different kinds of sites can be inferred from [Ref. 4.24a, Fig. 2.2].

Figure 4.1 presents typical time-of-flight spectra for a $\alpha NbH_{0.05}$ *single crystal* [4.25] showing the split optic line and, at lower energy, the band spectrum which comes from the host lattice phonon spectrum. After transforming the time-of-flight spectrum into an energy spectrum $g(\omega)$, the optic peak positions were found to be at $\hbar\omega=0.11$ and 0.18 eV (see Table 4.2). Only slightly different energies were observed for $NbH_{0.95}$ in the disordered α' phase

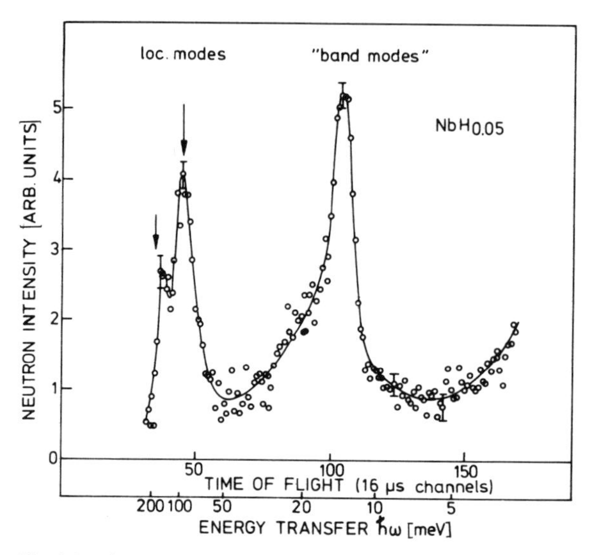

Fig. 4.1. Time-of-flight spectrum for neutrons scattered from a $\alpha NbH_{0.05}$ single crystal. The incident spectrum is centered at about 4 meV. The flight time is given in channel numbers (from *Verdan* et al. [4.25])

at elevated temperature and, at room temperature, in the ordered orthorhombic β phase [4.25a]. Figure 4.2 shows the corresponding pseudofrequency distribution $g(\omega)$ using (4.2). We assume that the same mass was used by the authors for the optic *and* for the acoustic part of the spectrum which obviously exaggerates the area under the optic contribution. The cutoff of the acoustic spectrum appears at 25 meV, in agreement with the phonon spectrum of pure niobium [4.26]. Time-of-flight spectra were also measured for $\beta NbH_{0.85}$ with a pulsed accelerator neutron source at relatively high incident energies (0.24 eV) as phonon creation processes [4.27]. The quality of the results from these experiments demonstrates the advantage of using high incident energies. Recently, the widths of the optic lines were determined at a hot-source triple-axis spectrometer between 20 and 210 °C, yielding values from 24 to 30 meV for the 0.11 eV level, and 36 to 50 meV for the 0.18 eV level (*Magerl* [4.27a]). For curiosity we mention very early measurements of the total scattering cross section vs neutron energy for $NbH_{0.85}$ where the 0.17 eV transition is clearly visible as a kink in the cross-section curve [4.28].

More recently, *coherent* scattering experiments on *niobium deuteride* were carried out with the triple-axis spectrometer operated on the hot source at the Grenoble High Flux Reactor, in order to obtain the optic branches $\omega(\boldsymbol{q})$ as a function of phonon wave vector $\boldsymbol{q}$. Such measurements are difficult because high incident energies are required (above 0.2 eV). This leads to low intensities, small Bragg angles, and relatively high background. Figure 4.3 presents typical optic dispersion branches [4.29]. Within the experimental accuracy (5...8 meV),

Table 4.2. Hydrogen and deuterium in niobium, tantalum, vanadium, and palladium. Optic mode frequencies $\hbar\omega_H$ (approximate energy of band center) and activation energy for self-diffusion (low concentration values from [4.37]) E_{act}

Sample	Lattice type; proposed H-sites	$\hbar\omega_H$ [a] [eV]		E_{act} [eV]	Lattice parameter [Å]
$\alpha NbH_{0.05}$	bcc; tetrahedral	0.11 and 0.18	[4.25]	0.10	Nb: 3.30
$\alpha NbD_{0.60}$	bcc; tetrahedral	$0.113/\sqrt{2}$ and $0.158/\sqrt{2}$	[4.29]		
$\beta NbD_{0.75}$	orthorhombic, tetrahedral	$0.120/\sqrt{2}$ and $0.170/\sqrt{2}$	[4.30]	~0.3[b]	
$\beta NbH_{0.95}$	orthorhombic, tetrahedral	0.120 and 0.170	[4.25a]		
$\alpha TaH_{0.15}$ (500 °C)	bcc; tetrahedral	0.12 and 0.17	[4.31]	0.14	Ta: 3.30
$\alpha TaD_{0.22}$	bcc; tetrahedral	$0.119/\sqrt{2}$ and $0.167/\sqrt{2}$	[4.33]		
$\beta_1 TaH_{0.7}$	monoclinic	0.13 and 0.18	[4.34]		
$\alpha VH_{0.04}$	bcc; tetrahedral	0.12 and 0.17	[4.25]	0.05	V: 3.02
$\beta VH_{0.40}$	bct; tetrahedral	0.120 and 0.175	[4.35]		
	octahedral	0.055[c]			
$\gamma VH_{1.5-1.7}$ [d]	fcc; tetrahedral	~0.16	[4.35]		
$\alpha PdH_{0.68}$	fcc; octahedral	0.056	[4.43]	0.23	Pd: 3.89
$\alpha PdH_{0.002}$	fcc; octahedral	0.066	[4.44]		
$\alpha PdD_{0.63}$	fcc; octahedral	$0.051/\sqrt{2}$ [e]	[4.40]		

Note: 1 THz = 4.15 meV; 1 cm^{-1} = 0.124 meV.

[a] The factor $\sqrt{2}$ means that $\hbar\omega_H$ was measured for the deuterided sample and transformed into the hydride value by the mass factor.

[b] For $\beta NbH_{0.9}$ [4.39].

[c] Suggested assignment from [4.35].

[d] Assignment in phase diagram not clearly established.

[e] From optic zone-center frequency [4.40]. The value of 0.056 eV in the hydride spectrum corresponds most probably to the TO modes.

1 THz = 4.15 meV; 1 cm^{-1} = 0.124 meV

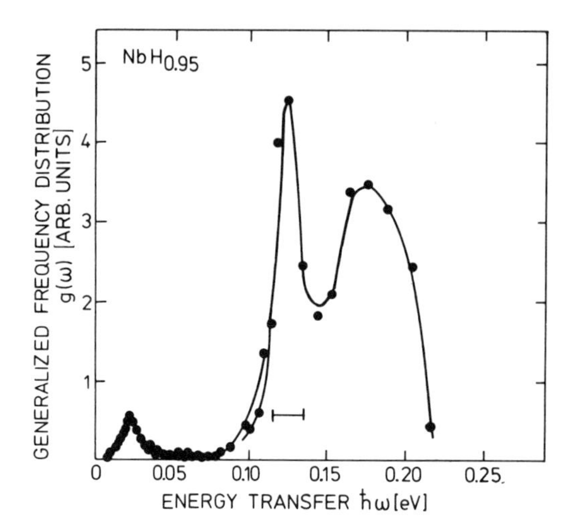

Fig. 4.2. The pseudolattice spectrum $g(\omega)$ according to (4.2) for $\beta NbH_{0.95}$, without resolution corrections. Horizontal bar: resolution width. The optic lines and the acoustic part of the spectrum are well separated (from *Chernoplekov* et al. [4.25a])

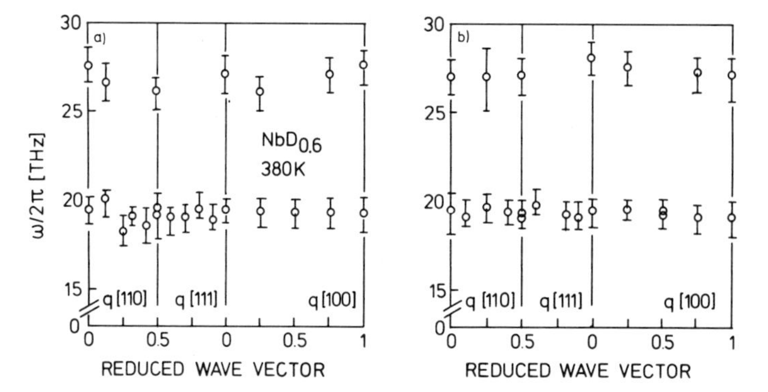

Fig. 4.3a and b. Frequency of the optic modes in $\alpha NbD_{0.6}$ vs wave vector of the phonon, for different directions of propagation and polarization. (a) transversal polarization, (b) longitudinal polarization. The bars indicate the accuracy obtained for the position of the phonon lines (from *Stump* et al. [4.29])

the frequency of the optic modes does not depend on the wave vector $\boldsymbol{q}$, and also not on the polarization of the phonons. The latter statement is based on measurements of different orientations of the scattering vector $\boldsymbol{Q}$ with respect to $\boldsymbol{q}$. Similar experiments were carried out on the partially ordered orthorhombic $\beta NbD_{0.75}$ [4.30] whose structure can be understood as a slightly distorted bcc lattice with D-atoms on rows parallel [110] forming a regular pattern. The sample was a single crystal of the niobium host metal, whereas, with regard to the orthorhombic superstructure, it was a multidomain crystal. The orthorhombic splitting of the Nb reflections causes a broadening or a splitting of the Bragg peaks which does not appreciably affect the dispersion curves. Figure 4.4 shows the results, together with the acoustic branches to be discussed later.

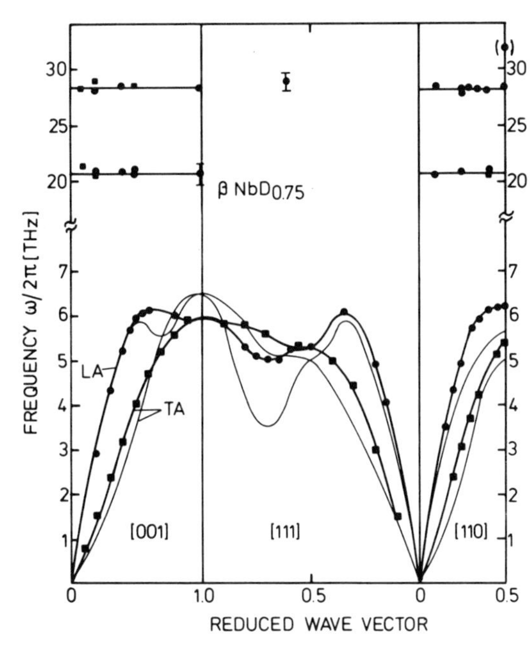

Fig. 4.4. Dispersion curves in different symmetry directions for $\beta NbD_{0.75}$ (solid line). Acoustic branches measured at 300 K, optic branches at 378 K (from *Lottner* et al. [4.30]). Thin line: pure Nb (from *Nakayawa* and *Woods* [4.26])

Again, the optic branches are completely flat and independent of polarization, with frequencies very close to those of the cubic and disordered α or α' phases. Within the range of 78...300 K, no temperature dependence of the frequencies was observed. In particular, no change in the overall dispersion curve was found when the $\alpha'\beta$ transition is approached from above within a few degrees. Additional experiments showed that no further optic peaks occur below those presented in Fig. 4.4.

For *hydrogen in tantalum*, localized mode frequencies were quoted from time-of-flight spectra on $TaH_{0.15}$ at 340 °C [4.31] and on $TaH_{0.13}$ at 500 °C [4.32] with frequencies which almost agree with those quoted before for αNbH_x. As in the case of NbD_x, *coherent* scattering experiments on $TaD_{0.22}$ [4.33] revealed two optic branches with frequencies which are independent of polarization vector and wave vector within experimental accuracy (see Fig. 4.15). Time-of-flight experiments on the ordered monoclinic β phase of tantalum hydride revealed nearly no difference for the optic frequencies with regard to the α phase, within 0.01 eV [4.34].

Broad spectral peaks in the optic phonon region were found for *hydrogen in vanadium* [4.35] for the three phases α, β, and γ. The two maxima in the spectrum of the cubic phase were again attributed to hydrogen on tetrahedral sites with tetragonal symmetry (see Table 4.2). In the tetragonal β phase, however, an additional peak appears. It can be ascribed to optic modes caused by the occupation of *octahedral* sites which have a large lattice expansion in one of the (originally cubic) axes. Figure 4.5 presents results for $\alpha VH_{0.04}$ single

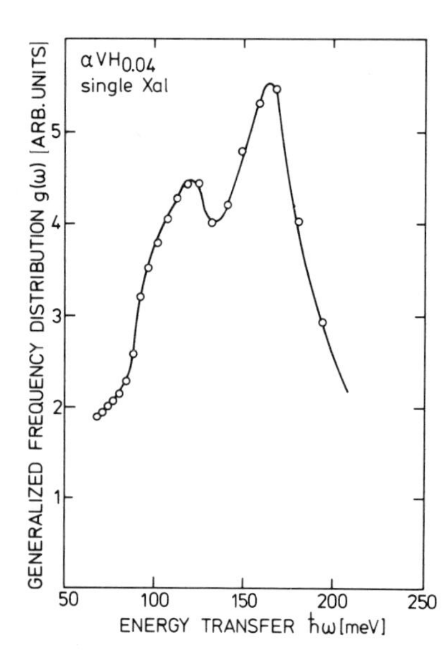

Fig. 4.5. Pseudo-frequency distribution for an $\alpha VH_{0.04}$ single crystal, showing the optic lines and the tail of the spectrum in the acoustic region. The broad wing at high energies is probably due to contamination by multiphonon processes (from *Verdan* et al. [4.25])

crystals, after converting the time-of-flight distributions into a pseudofrequency spectrum. Experiments on $VD_{0.5}$ [4.36] confirm that the center of gravity of the optic bands is lowered approximately by a factor of $\sqrt{2}$ in going from H to D.

Summarizing the results for the bcc hydrides and deuterides shown in Table 4.2, we draw the following conclusions. For niobium and tantalum, two optic branches appear in the high-symmetry directions, whose frequency does not depend on the wave vector and on the polarization within the experimental accuracy. From this we conclude that the restoring forces are determined essentially by metal-deuterium interactions, and depend only weakly on deuterium-deuterium interactions, which is in contrast to the case of PdD_x discussed later. This observation is consistent with the fact that the width and position of the optic phonon peaks depends weakly on the hydrogen concentration, and that it is also not strongly influenced by the ordering in the noncubic phases. So far, the optic branches for vanadium deuterides have not been measured. It is surprising that the frequencies are all the same, within about 0.01 eV, for hydrogen in niobium, tantalum, and vanadium, in spite of the fact that the hydrogen-metal distance changes between about 1.7 Å (for $VH_{0.04}$) to 1.9 Å (for $NbH_{0.095}$) [4.34]. The observed ratio of the frequencies for the two optic branches agrees with the value $\sqrt{2}$ as expected for hydrogen atoms in a bcc lattice bound by *central* nearest neighbor forces. The width of the optic peaks deduced from incoherent scattering spectra for the hydrides appears to be much broader than the width from the coherent scattering experiments on the deuterides. We believe that the width of the hydride spectra contains an appreciable contribution caused by multiphonon effects, as discussed in Section 4.1.

Comparing the frequencies for the hydrides and the deuterides, the ratio expected for a harmonic oscillator, $(\omega_H/\omega_D)^2 = 2$, is found in all cases, within experimental accuracy. As can be seen from Table 4.2, the activation energies for self-diffusion (see Chap. 12 and [4.37]) are obviously comparable with, or *even smaller* than $\hbar\omega_H$, which is the energy difference between the ground state and the first excited state of the observed optic vibrations. This fact suggests that the potential which provides the restoring force for the optic vibrations is not the same as the potential responsible for the motion over the saddle point. Consequently, acoustic phonons may play an important role in the diffusive step. Accurate investigations of the optic line as a function of temperature, as well as the search for higher harmonics, would be very helpful to obtain insight into this problem. This may be related to the dynamics of fast hydrogen diffusion in the transition metals (see Chap. 8).

4.3 Optic Vibrations in the Palladium Hydrides; Rare Earth and Other Hydrides

In the fcc palladium lattice, hydrogen or deuterium atoms occupy octahedral sites. For the case of *stoichiometric* PdH, this leads to a NaCl structure. In contrast to niobium, most of the optic branches of this hydride display a strong wave-vector-dependent frequency, as observed in triple-axis spectrometer experiments on $\alpha PdH_{0.63}$ [4.40, 41] in particular with respect to the longitudinal branch as shown in Fig. 4.6. The zone center frequency is more than two times lower than it is in the case of NbD_x which implies a relatively weak Pd–D interaction.

The pronounced maximum appearing in the longitudinal optic branch indicates strong second neighbor D–D interactions whose strength is comparable to the first neighbor D–D interaction. The full curves in the figure are derived from a 12-parameter Born-von Karman fit with first and second neighbor Pd–Pd and D–D interactions, and first neighbor Pd–D interactions. In these computations, the crystal was treated as if it were stoichiometric PdD, a point which will be discussed later. More recently, calculations [4.17] were carried out in which the directed *d*-electron bonds are described by first and second neighbor forces, and the Pd–H interaction in terms of screened ions with effective charges. The resulting optic branches are nearly flat, in disagreement with the experiment.

The Born-von Karman parameters form dispersion curves in $PdD_{0.63}$, shown in Fig. 4.6, were used to calculate the phonon density of states [4.40]. Figure 4.7 presents the results for the hydride, again assuming a stoichiometric structure. The strong optic peak deserves special mention since several authors have suggested [4.42] that optic phonons play a role in the superconducting properties of the system.

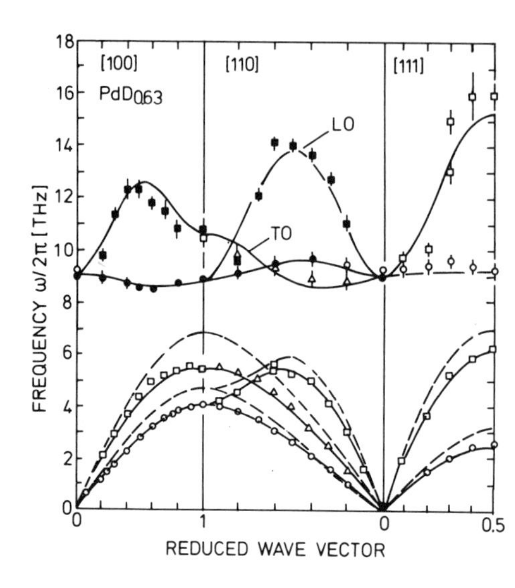

Fig. 4.6. Dispersion curve for $PdD_{0.63}$. Solid line: Born-von Karman fit (see text). Solid (open) symbols from phonon peaks at 150 K (295 K). Dashed: pure Pd [4.38] (from *Rowe* et al. [4.40])

The spectra in Fig. 4.7 can also be related to pseudo-frequency spectra obtained from *incoherent* neutron scattering on palladium hydride [4.43]. At room temperature these spectra reveal an optic line at 56 meV in reasonable agreement with the pronounced TO peak in Fig. 4.7. An additional broad line appears in the incoherent spectrum at about twice this energy which may be due to double phonon processes. We point out that it is not obviously clear whether this line is caused by a transition from the ground state to the second excited state of the oscillator (energy $\hbar\omega_{0-2}$), or if it is a two-phonon transition ($2\hbar\omega_{0-1}$). In addition, the possibility of double scattering in the sample cannot be excluded completely.

Recent experiments on palladium single crystals with low hydrogen concentrations [4.44] show an optic line whose energy changes from 63 to 66 meV if the H-concentration is lowered from 2.7 to 0.2 at.%. For the dilute hydride, the optic lines are considered as localized (or impurity) modes, in contrast to the LO optic peak in high concentration samples, discussed above, where the center of the broad line appears about 20% lower. For low temperatures and a well-annealed sample with 0.2 at.% concentration, the observed width of the localized-mode peak was below the experimental resolution (4 meV). At higher concentrations and temperatures, and for insufficient annealing of the sample, the width appeared to be significantly larger. Also in these experiments a second line was found at about 135 meV which is believed to be a second harmonic of the 63 meV transition.

Experiments with different orientations of the scattering vector $\boldsymbol{Q}$ relative to the sample orientation revealed an unexpected anisotropy of the localized mode: on changing $\boldsymbol{Q}$ from [100] to [110], the intensity drops by a factor of about 10. This could be interpreted in terms of an anomalous anisotropy of the localized modes with a large amplitude in the [100] direction. An anisotropy

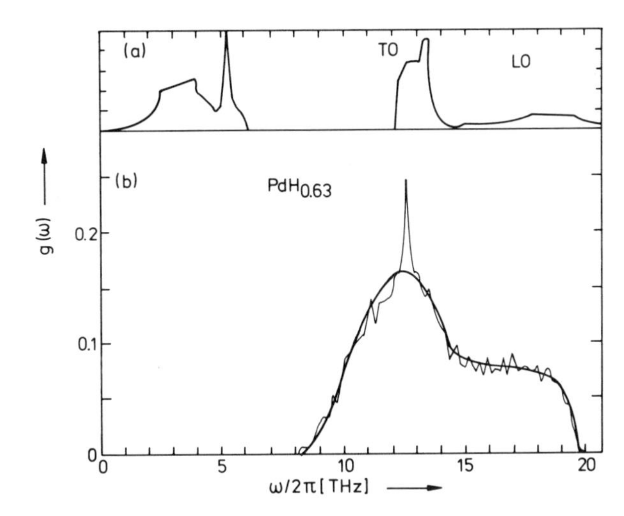

Fig. 4.7. (a) Phonon density of states for stoichiometric PdH calculated with the force constants obtained from $\omega(q)$ in Fig. 4.6 (from [4.40]). (b) Density of states calculated for a pseudo-random lattice $Pd_{108}H_{68}$ (thin line, from *Raman* et al. [4.47]. The solid line is drawn as a guideline for the eye to smooth the statistical fluctuations from the calculations)

was also observed in the Debye-Waller factor determined from measurements of the quasielastic line on palladium hydride [4.45], with mean-square amplitudes $\langle u^2 \rangle [110] = 0.05 \,\text{Å}^2$ and $\langle u^2 \rangle [100] = 0.09 \,\text{Å}^2$. Further investigations, in particular concerning the temperature dependence of the proton amplitudes, are needed to obtain a better understanding of this observation. The influence of quenching and annealing on the optic spectra of palladium hydride down to low temperatures was also studied, and, in general, no drastic changes of the spectra were observed in such experiments [4.46].

The observed localized mode line for the very dilute sample, as well as the optic peak for the high concentration sample (essentially due to TO modes, see Fig. 4.7), are believed to be strongly determined by Pd–H interactions. Consequently, the overall decrease of their energy with increasing hydrogen concentration may be understood (at least partly) in terms of an expansion of the palladium lattice. On the other hand, the H–H interaction is also important, and increasing the concentration tends to increase the frequency of the longitudinal modes.

The optic phonon peaks for coherent scattering from $PdD_{0.63}$ whose dispersion branches are shown in Fig. 4.6, were found to be broad, the longitudinal being broader than the transversal peaks, and the phonon energies were taken to be the position of the line centers. It is natural to assume that this width is caused by the disorder of the nonstoichiometric lattice structure. In view of the strong D–D interaction, the disorder is responsible for large fluctuations of certain force constants (in contrast to NbD_x where the D–D interaction is quite small). The dynamics of such a disordered lattice was studied theoretically in terms of a *pseudo-random approximation* [4.41, 47] in the following way: The Born-von Karman equations were formulated for a lattice which is subdivided into "supercells" with a cubic lattice parameter na_0, where a_0 is the cubic lattice parameter of palladium. Each supercell includes

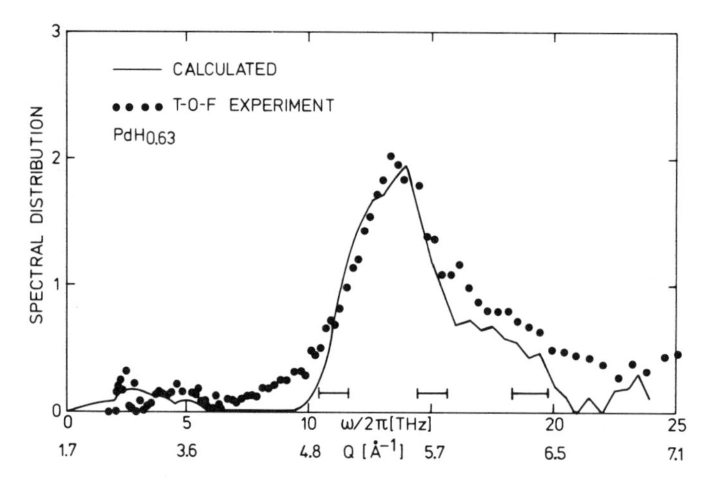

Fig. 4.8. Points: time-of-flight spectrum from incoherent scattering on hydrogen in $PdH_{0.63}$ as a function of frequency, carried out at a *fixed* scattering angle. Consequently, the scattering vector Q varies with frequency. Solid line: calculations on the basis of the pseudo-random lattice model (from *Raman* et al. [4.47])

$4n^3$ palladium and $4xn^3$ deuterium atoms, where $x=1$ corresponds to the stoichiometric case. The disorder was simulated by randomly distributing the D-atoms over the octahedral sites, and by solving the dynamic equations for an ensemble of such supercells having different distributions of the D-atoms (i.e., the effects of disorder were taken into account with respect to short-range forces). In most of the calculations a cell size of $n=3$ was used with 108 Pd and 68 H- or D-atoms. With the force constant as obtained from the $PdD_{0.63}$ dispersion curves [4.40], the density of states for $PdH_{0.63}$ was calculated, averaging over a sufficiently large number of supercells. The resulting spectrum is also shown in Fig. 4.7. Obviously, the disorder causes extensive broadening of the optic spectrum, and an enhancement of the high-energy wing of the distribution which, as we assume, is caused mainly by the longitudinal vibrations. The density of states in Fig. 4.7 was also applied to calculations of the incoherent scattering spectrum $S_{inc}(\boldsymbol{Q}, \omega)$ for $PdH_{0.63}$, including the multiphonon processes. The results were compared to time-of-flight experiments on a pulsed-source spectrometer. In order to achieve good agreement between the calculations of S_{inc} and the experimental data, the Pd–H force constant was taken to be 20% larger than the corresponding Pd–D force constant (Fig. 4.8). The authors suggest that this effect could be related to the inverse isotope effect of the superconductivity in these hydrides [4.16, 48]. The position of the broad spectral peak in Fig. 4.8 agrees roughly with the energies of the peaks in the incoherent spectra quoted before for $PdH_{0.68}$ [4.43].

Finally the pseudo-random lattice model was also applied to calculate the shape of the *coherent* phonon lines for $PdD_{0.63}$, by averaging (4.4) over many of the supercells in the pseudo-random lattice. The resulting phonon lines agree reasonably well with those for the transversal optic phonon lines obtained originally for the $PdD_{0.63}$ sample. To obtain agreement for the longitudinal optic peaks, the D–D force constant had to be readjusted[3].

As discussed in [Ref. 4.24a, Chap. 6], the transition temperature for superconductivity in palladium hydride reaches about 9 K, and a further increase was observed (up to 16 K) if Ag was added [4.49]. This observation has stimulated several measurements of the vibrational spectra on the system Pd–Ag–H. Only a slight shift of the optic peak to higher energies occurred (56 to about 58 meV) [4.50] with some disorder broadening. The spectral band in the acoustic region is similar to that of pure palladium [4.51], except at small energies where a new spectral peak appears (at about 6 meV). The authors suggest that this may be related to a Kohn anomaly, i.e., enhanced electron-phonon interactions at low energies. Such an anomaly was also reported for spectra on niobium hydride [4.52]. As a matter of fact, such an anomaly should appear in the dispersion branches. However, the energy quoted above is quite small and q is near the center of the Brillouin zone. Consequently, this effect might have been lost experimentally. In contrast to silver, the addition of cerium causes a considerable shift and broadening of the optic lines to higher energies [4.46]. This may be related to the relatively large size of the cerium atom which produces a strong lattice distortion.

In view of the superconducting properties, special attention was also paid to the vibrational spectra of the *thorium hydrides* [4.53]. For face-centered tetragonal ThH_2, which is not superconducting above 1 K, a single narrow optic line was found, which is well separated from the acoustic spectrum. On the other hand, for Th_4H_{15}, with a transition temperature of 8 K, a broad optic spectrum appears with two maxima. These may be attributed to the two different kinds of H sites in the complex bcc structure of Th_4H_{15}, having different force constants. As can be seen from Fig. 4.9, these peaks are rather broad, which is attributed to the strong q-dependence of the optic branches, i.e., due to strong H–H interactions as discussed earlier for palladium hydride. These statements have a quantitative basis in lattice dynamics calculations performed for thorium hydrides [4.54]. An interesting feature in Fig. 4.9 is the appearance of a shoulder in the Th_4H_{15} spectrum at rather low energies. The authors relate it to a soft optic branch which may play a role with respect to the high transition temperature for superconductivity.

In *rare earth dihydrides* the optic lines and the acoustic bands from hydrogen scattering were measured in the case of lantanum, praesodynium, holmium [4.46], and cerium [4.22, 55]. For most of the other rare earth metals, neutron absorption is very high, and the measurement of the spectra is difficult. Apart from the poor resolution of the time-of-flight spectra at high energies, the optic peaks were found to be wide and structured, which implies a dispersion of

[3] This inconsistency results from the fact that the Born-von Karman constants were evaluated under the assumption of a stoichiometric lattice, instead of the $PdD_{0.63}$ which contains a great number of D-vacancies, thus overestimating the D–D interactions.

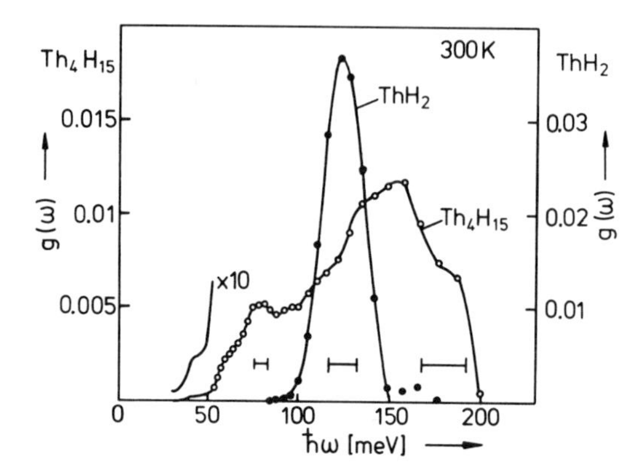

Fig. 4.9. Optic part of the pseudo-lattice spectrum for scattering on hydrogen in ThH_2 and Th_4H_{15}, as obtained from time-of-flight experiments. The bars indicate the resolution (it is poor at large $\hbar\omega$, i.e., for small flight times) (from *Dietrich* et al. [4.53])

the optic branches. The measured spectra were fitted to a four-parameter central force model, developed for CaF_2-type lattices [4.21], with nearest and next-nearest H–RE, RE–RE, and H–H interactions [4.22]. From these calculations, a systematic increase of the rare earth-hydrogen forces was observed with decreasing metal-hydrogen distance, following the sequence La, Ce, Pr, and Ho.

Special attention was paid to nonstoichiometric rare earth hydrides. The compound CeH_x exists with relatively large deviations from stoichiometry. For $x=2$, the hydrogen atoms are believed to occupy the tetrahedral sites in the fcc lattice. For x larger than 2, the hydrogen atoms begin to occupy also the octahedral sites [4.56, 57]. Time-of-flight spectra measured on $CeH_{1.98}$ show an optic line at an energy of 106 meV which is nearly as high as for the bcc hydrides. In addition, a very weak peak appears at 75 meV that is attributed to hydrogens which are already sitting on octahedral sites [4.55]. The binding is weaker at these sites, largely due to the greater H–Ce distance. For $CeH_{2.72}$ this low-energy octahedral-site peak is rather intense, and shifted to smaller energies. The peak ascribed to the tetrahedral sites appears at a higher energy than for the low concentration case, which is probably due to stronger H–H interactions. The measured spectra were fitted to the four-parameter model mentioned before. The density of states was evaluated, and the calculated specific heat was in good agreement with the experiments in the case of the dihydride. For PrH_x at different concentrations there is a similar situation as for CeH_x. It seems that a large range of concentrations between $x=2$ and 3 can be covered without two-phase precipitation. At lower concentrations, the two types of sites can be easily identified by the appearance of two optic lines in the spectrum [4.46]. Also for this system, the phonons may reveal the typical features of a crystal with disorder caused by nonstoichiometry.

4.4 Acoustic Phonons

The influence of dissolved hydrogen or deuterium on the acoustic phonon dispersion in metals can be understood in terms of two effects. *First*, the

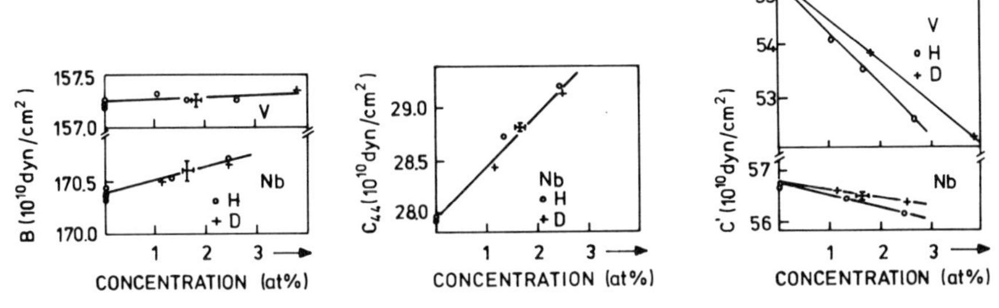

Fig. 4.10. Change of the elastic constants due to the dissolved hydrogen or deuterium. B = bulk modulus, c_{44} = shear modulus, $c' = (c_{11} - c_{12})/2$ = shear constant (from *Magerl* et al. [4.58])

hydrogen atoms are believed to add part of their electrons to the conduction band of the host metal. For palladium this expands the Fermi surface, but there may be more complicated implications, for instance a change of the shielding of the host atoms, or the appearance of new electronic states, as in palladium hydride. In this connection it is noteworthy that, for the transition metals Nb, Ta, and V, the influence of the electron concentration on the Fermi surface is just compensated by the lattice expansion due to the dissolved hydrogen. *Secondly*, there is a simple effect, namely the tendency to soften the force constants due to the lattice expansion caused by the dissolved hydrogen atoms. These two contributions, responsible for the change of the acoustic phonons, can be separated [4.58]: The expansion effect can be determined in terms of a negative hydrostatic pressure produced by the dissolved H-atoms which changes the elastic constants in a predictable manner. Concerning the change of the electronic properties, the corresponding effects were often discussed in terms of adding *molybdenum* which also introduces one additional electron. This point will be treated later.

The subsequent discussion will be divided into two parts, one dealing with acoustic modes at relatively large q, and the other treating modes close to the zone center $q=0$. From this point of view, we begin with ultrasonic experiments (see [4.58, 59]) before dealing with the large q features of the dispersion curves determined by neutron spectroscopy. The influence of the lattice disorder in nonstoichiometric hydrides on the acoustic phonon lines is weak (see Sect. 4.3).

For hydrogen and deuterium in the bcc metals Nb, Ta, and V, the behavior of the elastic constants as measured by ultrasonics [4.58] is qualitatively similar: For concentrations below 10 or 15 at.%, the bulk modulus $B=(c_{11}+2c_{12})/3$ and the shear modulus c_{44} reveal a nearly linear dependence on H- or D-concentration, with practically no isotope effect. On the other hand, the shear constant $c'=(c_{11}-c_{12})/2$ *decreases* with concentration and reveals an isotope effect, with c'(H) smaller than c'(D) for all three bcc metals. Figure 4.10 presents typical results from ultrasonic experiments. The concentration dependence is largest for c' in V, and for c_{44} in Nb (about 2% per 1 at.%) and it is several times smaller in all other cases. The increase of B and c_{44} is an indication that strong "electronic effects" exist: They oppose the *decrease* of the

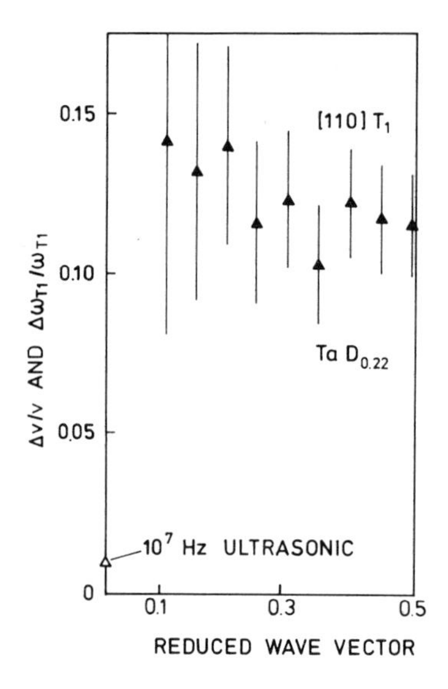

Fig. 4.11. Solid triangles: Relative change of the phonon frequency of the T_1 branch for $TaD_{0.22}$ from coherent neutron scattering (frequencies in the region of 10^{12} Hz). Open triangle at $q=0$: Relative change of sound velocity, corresponding to the c' mode, as obtained from ultrasonic experiments (10^7 Hz) (from *Magerl* et al. [4.60])

elastic constants which would be expected if there were only the influence of the volume expansion.

It is tempting to explain the anomalous decrease of the shear constant c' and its isotope effect by a relaxation process. At first sight, this point of view is supported by a comparison between ultrasonic measurements at 10^7 Hz and coherent neutron scattering experiments in the region above 10^{12} Hz carried out on $TaD_{0.22}$ with a triple-axis spectrometer at relatively small q [4.60]. As can be recognized from Fig. 4.11, for a given D- or H-concentration the relative change of the sound velocity at low frequency is about ten times smaller than the change obtained at high frequencies from the transversal dispersion branch ω_{T_1} [4].

The strain field created by an interstitial on a tetrahedral site has tetragonal symmetry. Consequently, the strain field couples to the shear waves characterized by the modulus c'. Under the influence of such a shear wave, a relaxation process could occur somewhere between ultrasonic and neutron scattering experiments (i.e., between 10^7 and 10^{12} Hz) which may be caused by deuterium atoms jumping between tetrahedral sites (Snoek effect), with a characteristic time τ. This would lead to a "relaxed" shear constant for ω *below* the characteristic rate $1/\tau$, and a higher (nonrelaxed) modulus *above* $1/\tau$. As a matter of fact, such jumps of the D-atoms to exist; however, the *strength* of the Snoek relaxation [4.61] is by far too small to allow such an interpretation. Moreover, this explanation contradicts the temperature dependence observed in the ultrasonic experiments. At present, no relaxation (or

[4] A similar observation was also made for NbD_x [4.60]. No such discrepancy between low- and high-frequency measurements was observed for c_{44}.

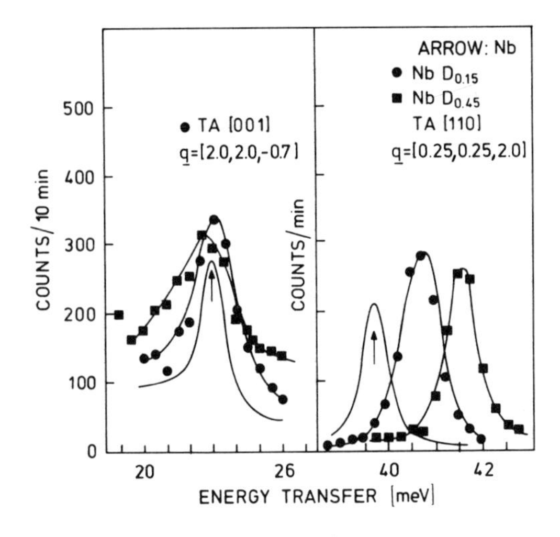

Fig. 4.12. Typical acoustic phonon lines: For pure Nb (arrow), $NbD_{0.15}$, and $NbD_{0.45}$ (from *Rowe* et al. [4.64])

any other) process can be identified which might account for these observations.

Also for $\alpha VD_{0.7}$ the slope of the acoustic branches in the three high symmetry directions was investigated by means of triple-axis spectrometry [4.62]. The slopes were found to be higher than those for pure V [4.63]. To our knowledge, no ultrasonic experiments exist on V at high D-concentrations to facilitate a comparison. In summary, apart from the anomalous behaviour of c', hydrogen loading in Nb, Ta, and V tends to increase the elastic constants, and, as well, the slope ω/q in the long wavelength limit. This means that the "electronic effects" overcompensate the softening of the restoring forces to be expected from expanding the lattice. This conclusion is in contrast to the case of palladium where deuterium loading causes a lowering of the slope of ω/q, as can be seen from Fig. 4.6. Quantitatively the following statement can be made: Using the pressure coefficient of the elastic constants and calculating the average negative pressure by subtracting the eigenvolume of the defect from the macroscopic lattice expansion (see, for example, [4.58]), only a softening of 6% in $PdD_{0.63}$ can be explained, whereas the experimental value is four times larger.

Regarding now the acoustic branches $\omega(\boldsymbol{q})$ at larger wave vectors $\boldsymbol{q}$, one finds the following qualitative features. As already stated, in the case of palladium deuteride there is a general lowering of the acoustic frequencies, but the *shape* of the branches does not change (Fig. 4.6). This is surprising since, at least, a change of the Fermi surface should occur due to the electrons from the deuterium atoms. The situation is quite different for the bcc metals Nb and Ta. Typical acoustic phonon lines for pure Nb, and for Nb with $x=0.15$ and 0.45 deuterium concentration are presented in Fig. 4.12, measured on a single crystal with a mosaic spread of 0.4° (pure Nb) and 0.6° (after loading) [4.64]. No increases of the phonon width can be observed with the available resolution.

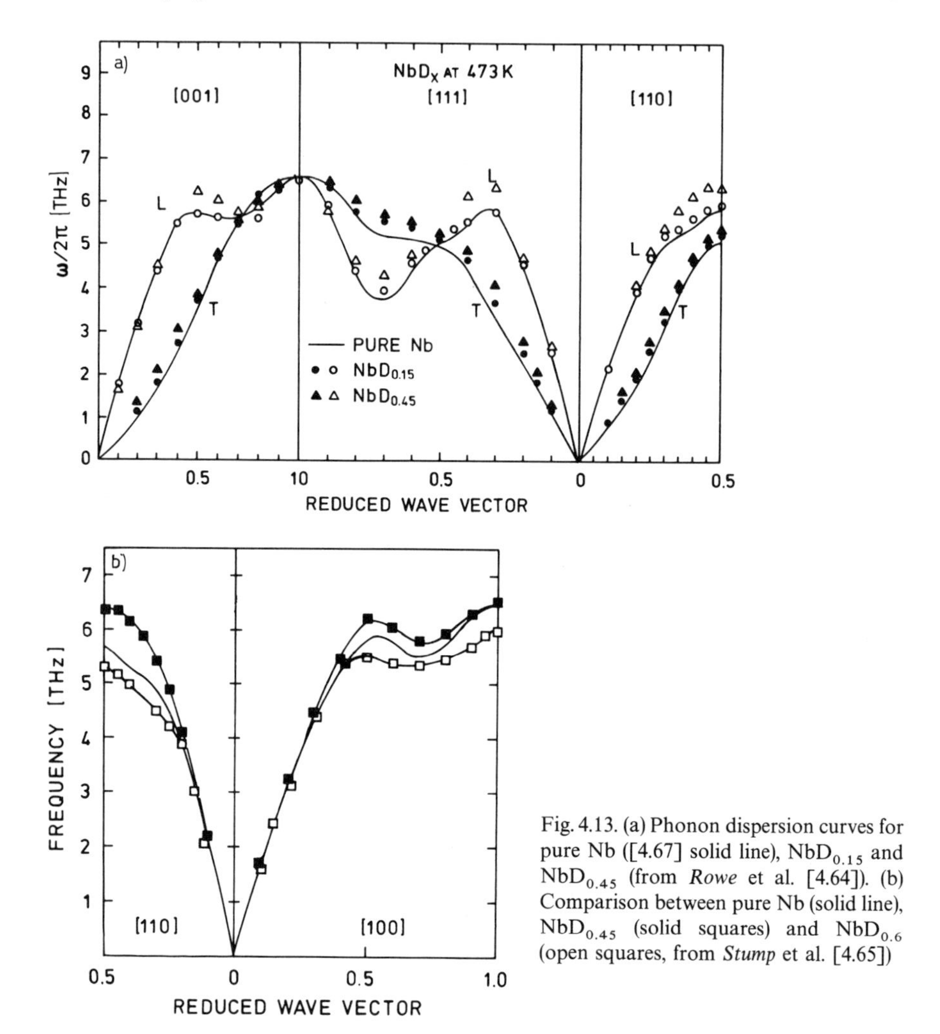

Fig. 4.13. (a) Phonon dispersion curves for pure Nb ([4.67] solid line), $NbD_{0.15}$ and $NbD_{0.45}$ (from *Rowe* et al. [4.64]). (b) Comparison between pure Nb (solid line), $NbD_{0.45}$ (solid squares) and $NbD_{0.6}$ (open squares, from *Stump* et al. [4.65])

The increased background which appears at higher concentration is mainly caused by incoherent scattering from the D-nuclei. The resulting dispersion curves are shown in Fig. 4.13a and, for experiments at higher hydrogen concentration [4.65], in Fig. 4.13b. As mentioned already, there appears an increase of the slope at smaller wave vectors, and, at higher concentrations the negative curvature of the transverse branches disappears. Secondly, there is a remarkable change of the "anomaly" in the region of about 0.7 [100] L. Obviously, there is an enhancement of the anomaly for concentrations $x=0.15$ and 0.45. At $x=0.60$, however, the anomaly flattens again. This was also found from observations at $x=0.75$ in the β phase [4.30] (see Fig. 4.4). On the other hand, the anomaly near 0.3 [110] L gets weaker at $x=0.40$.

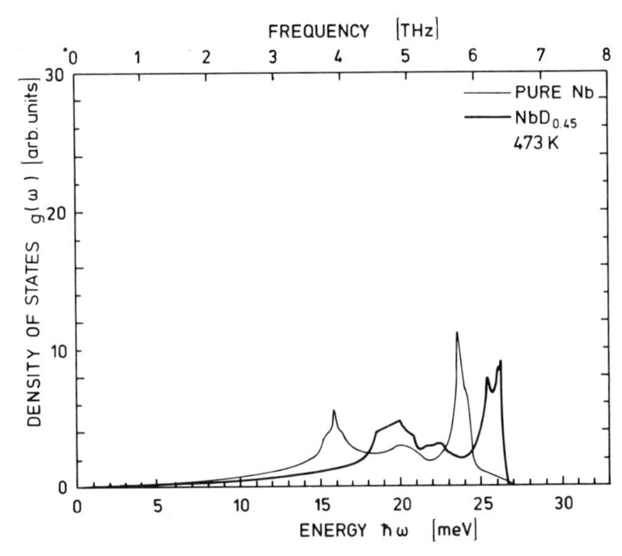

Fig. 4.14. Phonon density of states for pure Nb (thin line) and $NbD_{0.45}$ (solid line). (The optic part of the spectrum is omitted) [4.64]

For $x \simeq 0.50$ this behavior (change of the "anomalies"; disappearance of the negative curvature for small q; increase of the elastic constants except c') can be tentatively related to the niobium-molybdenum alloys. An increase of the molybdenum concentration leads to a decrease of the electron density of states N_F at the Fermi energy [4.66, 67]. This yields an increase of longitudinal sound velocities ($\propto N_F^{-1/2}$), which might also be related to an overall stiffening of the phonon branches. The decrease of the frequencies in the region of the anomalies, which seems to appear above $x=0.50$, again bears a similarity to the behavior of certain modes in $Nb_{1-x}Mo_x$ [4.67]. One observes an increase of the acoustic frequencies up to $x=0.50$, whereas above that concentration the frequency remains constant, or decreases again.

The dispersion curves measured for $NbD_{0.45}$ and discussed in the preceding paragraph were fitted to a Born-von Karman model with 6 nearest-neighbor tensor forces [4.64]. The resulting parameters were applied to represent the phonon energies in the regions of reciprocal space which were not covered experimentally, and the density of states was calculated on this basis. It is difficult to judge the reliability of this extrapolation procedure, but the general features of the spectrum may be reproduced reasonably well. Figure 4.14 shows the result. (The optic modes are not presented. They would appear as two "Einstein peaks" at high energies). The reduction of the density of states in the acoustic region obviously follows from the stiffening of the dipersion branches discussed before.

The change of the phonon spectra due to the solution of hydrogen yields an important contribution to the entropy of solution. This effect was investigated

Table 4.3. Excess entropies for H in Nb and Pd at low concentration (from *Magerl* et al. [4.68])

	Site	S_0	S_a	S_e	$\ln\beta$	S^{exc}_{calc}	S^{exc}_{exp}
Pd	tetrahedral	1.47	1.20	−0.49	0.69	2.87	2.06
	octahedral	1.47	1.20	−0.49	0.00	2.18	
Nb	tetrahedral	0.24	−0.88	−0.22	1.79	0.93	1.20
	octahedral	0.36	−0.88	−0.22	1.10	0.36	

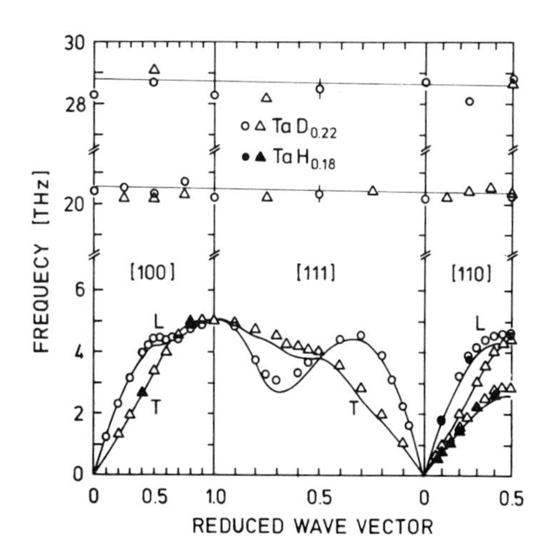

Fig. 4.15. Dispersion curves for $TaD_{0.22}$ and $TaH_{0.18}$ showing the acoustic and the optic branches. Circles: longitudinal modes. Triangles: transversal modes (from *Magerl* et al. [4.33])

systematically [4.68] for a series of metal-hydrogen systems. For small hydrogen concentrations x, the partial entropy can be written as

$$S = S^{exc} - k_B \ln x,$$

with

$$S^{exc} = S_0 + S_e + S_a + k_B \ln\beta, \tag{4.6}$$

where S^{exc} does not depend on x. The term

$$S_0 = k_B \sum_1^3 \left\{ u_i[\exp(u_i) - 1]^{-1} - k_B \ln[1 - \exp(-u_i)] \right\}$$

is caused by the local modes with frequency ω_{H^i} where $u_i = \hbar\omega_{H^i}/k_B T$. The term $S_a = -k_B d\langle \ln\omega \rangle / dx$ follows from the change of the acoustic modes by the dissolved H- or D-atoms ($\langle \dots \rangle$ means an average over all these modes), and $S_e = (d\gamma/dc) k_B T$ is the electronic contribution, where $k_B T\gamma$ is the electronic specific heat per atom, also depending on x. The last term gives the con-

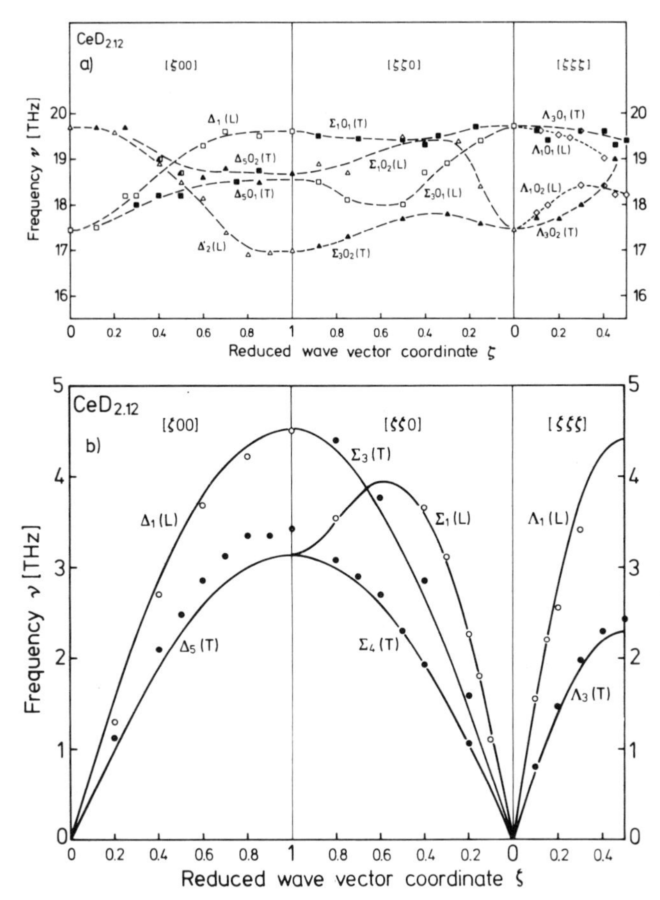

Fig. 4.16. (a) Acoustic and (b) optic dispersion curves for $CeD_{2.12}$. The dots are experimental values. Solid line: central-force model with force constants derived from incoherent scattering on $CeH_{1.98}$. Dashed line: guideline for the eye (from *Vorderwisch* et al. [4.73])

figurational entropy where β is the number of interstitial sites per metal atom. For a bcc lattice one obtains $\beta=3$ (octahedral sites) or $\beta=6$ (tetrahedral sites), and for a fcc lattice $\beta=1$ (octahedral) and $\beta=2$ (tetrahedral). The terms S_0 and S_a can be calculated from the vibrational spectra discussed in the preceding sections. In the case of Nb, the quantity $d\gamma/dc$ was approximately extrapolated from values existing for Ta. Typical results are shown in Table 4.3. From these data a clear decision can be obtained concerning the type of occupied interstitial sites, namely octahedral for Pd, tetrahedral for Nb, and also for Ta, in the α phase, in agreement with diffraction [4.69, 70] and channeling experiments [4.71]. The decisive contribution S_a caused by the acoustic modes should be noticed: In particular, for H in niobium, a neglection of this term would obviously suggest the occupancy of octahedral rather than tetrahedral sites.

Finally, we may speculate about the relation between features of the phonon dispersion and superconductivity. In contrast to Pd, the solution of hydrogen in niobium tends to hinder the superconducting state. Adding hydrogen, T_c remains practically constant at low concentration, and then drops rapidly below 4 K as soon as an H-concentration of 70 at.% is reached [4.72]. In view of the NbH_x phase diagram, the solution of H at low temperatures immediately leads to the precipitation of the ordered β phase (or of higher ordered phases). It can be supposed that the ordered phase is nonsuperconducting. Consequently, at about 70% the nonsuperconducting phase occupies most of the sample, and the "percolation" through the remaining superconducting regions of the low-concentration αNbH is interrupted. Regarding Fig. 4.4, one finds that the anomaly at 0.6 [001] L, which may be related to strong electron-phonon coupling, actually disappears in the ordered phase.

For completeness, we show also dispersion curves for tantalum [4.33] with deuterium and hydrogen (Fig. 4.15), including the flat optic branches already discussed. Qualitatively, this system reveals features very similar to those discussed for the niobium system, namely an increase of the slope, a reduction of the negative curvature of the [100] T and [110] T_2 branches at smaller q, and a particularly strong change of the frequencies in the region where the host lattice curves show the anomalies.

Recently, dispersion curves on $CeD_{2.12}$ were determined [4.73]. Figure 4.16 presents unpublished results for the branches in the symmetry directions. As compared to the palladium hydride, the Q-dependence is relatively weak, but stronger than in NbD_x. We therefore conclude that the D–D interaction is small as compared to the metal-D interaction, in contrast to the case of PdD_x.

Acknowledgement. We would like to thank *V. Lottner* and *A. Magerl* for helpful remarks and for leaving us unpublished results, and *S. Lovesey* for critically reading the manuscript.

References

4.1 W. Kley: Z. Naturforsch. **21**a, (Suppl.) 1770 (1966)
4.2 P.G. Dawber, R.J. Elliott: Proc. Roy. Soc. London A **273**, 222 (1963)
4.3 L. Koester: "Neutron Scattering Lengths and Fundamental Neutron Interactions", in *Neutron Physics*, Springer Tracts in Modern Physics, Vol. 80 (Springer, Berlin, Heidelberg, New York 1977)
4.4 A. Sjölander: Arkiv Fysik **14**, 315 (1958)
G. Placzek, L. Van Hove: Phys. Rev. **93**, 1207 (1954)
4.5 P. Egelstaff (ed.): *Thermal Neutron Scattering* (Academic Press, London 1965)
4.6 T. Springer: *Quasi-Elastic Scattering of Neutrons for the Investigation of Diffusive Motions in Solids and Liquids*. Springer Tracts in Modern Physics, Vol. 64 (Springer, Berlin, Heidelberg 1972)
4.7 A. Rahman, K.S. Singwi, A. Sjölander: Phys. Rev. **126**, 986 (1962)
P.A. Egelstaff, P. Schofield: Nucl. Sci. Eng. **12**, 260 (1962)
4.8 B. Kaufmann, H.J. Lipkin: Ann. Physik **18**, 294 (1962)
4.9 M.A. Krivoglaz: Soviet Phys. JETP **19**, 432 (1964)
4.10 G. Placzek, L. van Hove: Phys. Rev. **93**, 1207 (1954)

4.11 J.G.Couch, O.K.Harling, L.C.Clune: Phys. Rev. B **4**, 2675 (1971)
4.12 B.A.Oli, A.O.E.Animalu: Phys. Rev. B **13**, 2398 (1976)
4.13 S.T.Chui: Phys. Rev. B **9**, 3300 (1974)
4.14 W.Weber, H.Bilz, U.Schroeder: Phys. Rev. Lett. **28**, 600 (1972)
W.Hanke: Phys. Rev. B **8**, 4585, 4591 (1973)
4.15 W.Hanke, J.Hafner, H.Bilz: Phys. Rev. Lett. **37**, 1560 (1976)
4.16 B.N.Ganguly: Z. Physik **265**, 433 (1973); Phys. Rev. B **14**, 3848 (1976)
4.17 P.Hertel: Z. Physik **268**, 111 (1974)
4.18 A.C.Switendick: Ber. Bunsenges. Phys. Chemie **76**, 535 (1972); J.Less-Common Metals **49**, 283 (1976)
D.E.Eastman, J.K.Cashion, A.C.Switendick: Phys. Rev. Lett. **27**, 35 (1971)
4.19 H.Buchold, G.Sicking, E.Wicke: J. Less-Common Metals **49**, 85 (1976)
J.Zbasnik, M.Mahnig: Z. Physik B **23**, 15 (1976)
4.20 M.Mahnig, E.Wicke: Z. Naturforsch. A **24**, 1258 (1969)
4.21 E.L.Slaggie: J. Phys. Chem. Solids **29**, 923 (1968)
4.22 P.Vorderwisch, S.Hautecler, H.Deckers: Phys. Stat. Sol. (b) **65**, 171 (1974), **66**, 595 (1974)
4.23 G.Blaesser, J.Peretti, G.Toth: Phys. Rev. **171**, 665 (1968)
4.24 E.W.Montroll, R.B.Potts: Phys. Rev. **100**, 525 (1955)
4.24a G.Alefeld, J.Völkl (eds.): *Hydrogen in Metals II, Application-Oriented Properties*. Topics in Applied Physics, Vol. 29 (Springer, Berlin, Heidelberg, New York 1978) in preparation
4.25 G.Verdan, R.Rubin, W.Kley: *Neutron Inelastic Scattering*, Vol. 1, Proc. IAEA, Vienna (1968), p. 223; see also:
W.Gissler, G.Alefeld, T.Springer: J. Chem. Phys. Solids **31**, 2361 (1970)
4.25a N.A.Chernoplekov, M.G.Zemlyanov, V.A.Somenkov, A.A.Chertkov: Soviet Phys.—Solid State **11**, 2343 (1970)
4.26 Y.Nakagawa, A.D.B.Woods: Phys. Rev. Lett. **11**, 271 (1963)
4.27 S.S.Pan, M.L.Yeater, W.E.Moore: NBS Spec. Publ. **301**, 315 (1969)
4.27a A.Magerl: Unpublished
4.28 M.Sakamoto: J. Phys. Soc. Japan **19**, 1862 (1964)
4.29 N.Stump, G.Alefeld, D.Tochetti: Solid State Commun. **19**, 805 (1976), see also:
H.Conrad, G.Bauer, G.Alefeld, T.Springer, W.Schmatz: Z. Physik **266**, 239 (1974)
4.30 V.Lottner, A.Kollmar, T.Springer, W.Kress, H.Bilz, W.D.Teuchert: *Lattice Dynamics*, ed. by M.Balkanski (Flammarion Sciences, Paris 1978) p. 247
4.31 J.J.Rush, R.C.Livingston, L.A.de Graaf, H.E.Flotow, J.M.Rowe: J. Chem. Phys. **59**, 6570 (1973)
4.32 V.Lottner, A.Heim: Unpublished (1976)
4.33 A.Magerl, N.Stump, W.D.Teuchert, V.Wagner, G.Alefeld: J. Phys. C. Solid State Phys. **10**, 2783 (1977)
4.34 R.Yamada, N.Watanabe, K.Sato, H.Asano, M.Hirabayashi: J. Phys. Soc. (Japan) **41**, 85 (1976)
4.35 J.J.Rush, H.E.Flotow: J. Chem. Phys. **48**, 3795 (1968)
4.36 J.M.Rowe: Solid State Commun. **11**, 1299 (1972)
4.37 J.Völkl, G.Alefeld: *Diffusion in Solids: Recent Development*, ed. by A.S.Nowick, J.J.Burton Academic Press, New York 1975) p. 231
4.38 A.P.Miller, B.N.Brockhouse: Can. J. Phys. **49**, 704 (1971)
4.39 B.Alefeld: Unpublished; IFF Bulletin, **4**, Heft 2, KFA Jülich (1973)
4.40 J.M.Rowe, J.J.Rush, H.G.Smith, M.Mostoller, H.E.Flotow: Phys. Rev. Lett. **33**, 1297 (1974)
4.41 C.J.Glinka, J.M.Rowe, J.J.Rush, A.Rahman, S.K.Sinha, H.E.Flotow: *Proc. Int. Conf. Neutron Scattering*, Gatlinburg, Tenn. (1976)
4.42 A.Eichler, H.Wühl, B.Stritzker: Solid State Commun. **17**, 273 (1975)
P.J.Silverman, C.V.Briscoe: Phys. Lett. A **53**, 221 (1975)
4.43 M.R.Chowdhury, D.K.Ross: Solid State Commun. **13**, 229 (1973); see also
J.Bergsma, J.A.Goedkoop: In "*Inelastic Scattering of Neutrons in Solids and Liquids*", Proc. IAEA Vienna (1961) p. 501

4.44 W. Drexel, A. Murani, D. Tocchetti, W. Kley, I. Sosnowska, D. K. Ross: J. Phys. Chem. Solids **37**, 1135 (1976); see also
Proc. Gatlinburg Conf. on Neutron Scattering, Gatlinburg, Tenn. (1976)
4.45 J. M. Rowe, J. J. Rush, L. A. de Graaf, G. A. Ferguson: Phys. Rev. Lett. **29**, 1250 (1972)
4.46 D. G. Hunt, D. K. Ross: J. Less-Common Metals **49**, 169 (1976)
4.47 A. Rahman, K. Sköld, G. Pelizarri, S. K. Sinha, H. Flotow: Phys. Rev. B **14**, 3630 (1976); see also B. Splettstösser: Z. Physik B **26**, 151 (1977)
4.48 R. J. Miller, C. B. Satterwaite: Phys. Rev. Lett. **34**, 144 (1975); see also
A. Eichler, H. Wühl, B. Stritzker: Solid State Commun. **17**, 213 (1975)
4.49 W. Buckel, B. Stritzker: Phys. Lett. A **43**, 403 (1973)
4.50 M. R. Chowdhury: J. Phys. F: Metal Phys. **4**, 1657 (1974)
4.51 A. P. Miller, B. N. Brockhouse: Can. J. Phys. **49**, 704 (1971)
4.52 D. K. Ross: Personal communication (1977)
4.53 M. Dietrich, W. Reichardt, H. Rietschel: Solid State Commun. **21**, 603 (1977)
4.54 H. Winter, G. Ries: Z. Physik B **24**, 279 (1976)
4.55 P. Vorderwisch, S. Hautecler: Phys. Stat. Sol. (b) **64**, 495 (1974)
4.56 C. G. Titcomb, A. K. Cheetham, B. E. F. Fender: J. Phys. C: Solid State Phys. **7**, 2409 (1974)
4.57 G. G. Libowitz: Ber. Bunsenges. Phys. Chemie **76**, 837 (1972)
4.58 A. Magerl, B. Berre, G. Alefeld: Phys. Stat. Sol. (a) **36**, 161 (1976)
4.59 H. A. Wriedt, R. A. Oriani: Scripta Metall. **8**, 203 (1974)
E. S. Fisher, D. G. Westlake, S. T. Ockers: Phys. Stat. Sol. (a) **28**, 591 (1975)
4.60 A. Magerl, W. D. Teuchert, R. Scherm: J. Phys. C.: Solid State Phys. **11**, 2175 (1978)
4.61 J. Buchholz, J. Völkl, G. Alefeld: Phys. Rev. Lett. **30**, 318 (1973)
4.62 J. J. Rush, J. M. Rowe, C. J. Glinka, N. Vagelatos: *Proc. Gatlinburg Conf. on Neutron Scattering*, Gatlinburg, Tenn. (1976)
4.63 G. A. Alers: Phys. Rev. **119**, 1532 (1960)
4.64 J. M. Rowe, N. Vagelatos, J. J. Rush, H. E. Flotow: Phys. Rev. B **12**, 2959 (1975)
4.65 N. Stump, A. Magerl, G. Alefeld, E. Schedler: J. Phys. F: Metal Phys. **7**, L1 (1977)
4.66 L. F. Mattheiss: Phys. Rev. **139** A, 1893 (1965)
4.67 B. M. Powell, P. Martel, A. D. B. Woods: Phys. Rev. **171**, 727 (1968); see also
Neutron Inelastic Scattering, Proc. IAEA, Vienna (1969) p. 113
4.68 A. Magerl, N. Stump, H. Wipf, G. Alefeld: J. Phys. Chem. Solids **38**, 683 (1977)
4.69 G. Nelin: Phys. Stat. Sol. (b) **45**, 527 (1971)
4.70 V. A. Somenkov, A. V. Gurskaya, M. G. Zemlyanov, M. E. Kost, N. A. Chernoplekov, A. A. Chertkov: Soviet Phys.—Solid State **10**, 2133 (1969)
4.71 M. Antonini, H. D. Carstanjen: Phys. Stat. Sol. (a) **34**, K153 (1976)
H. D. Carstanjen, R. Sizmann: Phys. Lett. **40**a, 93 (1972)
4.72 J. H. Welter, H. Wenzl: IFF Bulletin, KFA Jülich **9**, Heft 1, p. 27 (1976)
4.73 P. Vorderwisch, S. Hautecler, G. G. Libowitz, W. D. Teuchert: Personal communication (1977); see also G. J. Glinha, J. M. Rowe, J. J. Rush, G. G. Libowitz, A. Maeland: Solid State Comm. **22**, 541 (1977)

5. The Change in Electronic Properties on Hydrogen Alloying and Hydride Formation*

A. C. Switendick

With 14 Figures

5.1 Properties of Hydrogen-Metal Systems

This chapter will deal with differences between the electronic properties of metal systems and the system resulting from various amounts of hydrogen in association with the metal. As the title indicates, there are two (or more) regimes to this association and there is little reason to expect similar behavior in both regimes or for that matter from one metal to the other. We hope to show that the most consistent behavior might be expected for the low concentration regime. On the other hand, as has been indicated elsewhere [5.1], usually very little hydrogen can be accommodated within the host metal lattice before a phase change to a different structure (with varying hydrogen composition as a function of temperature and pressure) occurs. Therefore, any change upon accommodation of the hydrogen must take into account the possible occurrence of a new structure and the change engendered by change of the metal lattice itself, aside from any effects of the hydrogen per se. Indeed perhaps the strongest evidence of significant hydrogen perturbation on the properties of the metal is this structure change from a system which can accommodate only modest amounts of hydrogen, on the order of one percent or so, to one which prefers sizeable numbers, $x \approx 1-3$, of hydrogens in association with each metal atom and indicates both metal-hydrogen and hydrogen-hydrogen interactions.

The first system we shall treat is the old and venerable one [5.2]—the palladium-hydrogen system and alloy systems involving palladium and its neighbors in the periodic table. The second system we shall discuss in some detail is the Ti–V–H ternary system, in which the metal lattice change involved is hexagonal (hex) to face centered cubic (fcc) in the case of titanium and body centered cubic (bcc) to fcc in the case of vanadium, and where hydrogen ratios approaching $x=2$ are achieved. The relationship of these systems to other dihydride (and trihydride) systems, including those of the rare earth series, will be made. The third system again involves vanadium, but the metal lattice remains bcc and somewhat lower concentrations of hydrogen are attained. The final system we shall discuss is the Cr–H system where again a metal lattice change occurs (bcc to hex).

* This work was supported by the United States Energy Research and Development Administration, ERDA, under contract AT(29-1)789.

Applying the techniques and principles of contemporary solid-state physics [5.3], we shall endeavor to present a consistent (but *not* the same for every metal) picture of the metal (hydrogen)-hydrogen (hydrogen) interactions and the manifestations of these on the properties of these systems. In particular, we shall examine a few systems in this context and examine the changes due to the change of metal lattice and the changes due to the influence of the hydrogen, and to the changes as the metal itself is varied. The question thus becomes which changes are due to the addition of hydrogen and which changes are due to the lattice change and what are their manifestations and what extrapolations are logical and correct.

We shall dwell primarily on experimental techniques which measure some aspect of the density of electronic states, $N(E)$, and compare and correlate these experimental results with theoretical results derived from electronic energy band calculations. A brief discussion of the connection between the experiment and the theory is given in Sections 5.1.1–5.1.4. We shall largely neglect magnetic properties as they are discussed elsewhere in this volume (Chap. 7) and shall only touch on the superconducting behavior as it directly relates to the systems above. For the same reasons, we shall not consider the electronic aspects of hydrogen motion as they too are treated elsewhere in this volume (Chaps. 8 and 9).

Our discussion, by necessity, will primarily cover the high hydrogen concentration regime since it is there that most experimental results exist, effects manifest themselves and, in fact, are also most amenable to theoretical treatment. However, where possible, we shall try to indicate the applicability to the low concentration regime. In discussing each system we shall first give the experimental results which we feel are most germane to elucidating the origin of the changes, and apologize for the omission of much work which could not be included due to restrictions of the length of this chapter and refer the reader to recent reviews and conference proceedings [5.1, 4, 5].

5.1.1 Electronic Specific Heat

At low temperatures the heat capacity at constant volume is given by

$$c_v = \alpha T^3 + \gamma T,$$

where α is governed by the lattice vibrations and determines the Debye temperature. The term linear in temperature predominates at sufficiently low temperatures in metals and is due to the heat capacity of the conduction electrons within $\sim 2kT$ of the Fermi energy E_F, which are the only ones available to be excited. The electronic specific heat coefficient γ is given by

$$\gamma = 2\pi^2 k^2 N(E_F)(1+\lambda_\gamma)/3, \tag{5.1}$$

where $N(E_F)$ is the density of spin states per formula unit, k is Boltzmann's constant, and λ_γ is a factor which represents enhancements of $N(E_F)$ from the simple noninteracting electron system, due to interactions with phonons, paramagnons, and a variety of other excitations. Its value varies from 0.05 for alkali metals to 2–3 for anomalously behaving intermetallic compounds such as Nb_3Sn. Values of λ_γ of the order of 0.50 ± 0.20 are typical of the transition metal hosts such as we shall be considering. Thus measurements of the electronic specific heat coefficient give a measure of the electronic density of states at the Fermi energy $N(E_F)$, and more accurately give (if λ_γ does not change much) its variation with composition and/or hydrogen addition.

5.1.2 Magnetic Susceptibility

The magnetic susceptibility of a solid is given by the sum of a variety of contributions

$$\chi = \chi_{\mathrm{ORB}} + \chi_{\mathrm{DIA}} + \chi_{\mathrm{L}} + \chi_{\mathrm{P}},$$

where χ_{ORB}, often called the Van Vleck susceptibility, is given by a matrix element of orbital angular momentum l_z between states below the Fermi energy and states above the Fermi energy divided by the energy difference. It generally is large for transition metals [5.6] and transition metal compounds [5.7]. Its value is difficult to calculate realistically. χ_{DIA} reflects the Lenz's law type response of the closed-shell core currents. Its value is negative and generally small due to the smallness of the core orbits. χ_{L} is the Landau diamagnetism of the conduction electrons and is generally considered proportional to the Pauli paramagnetic susceptibility

$$\chi_{\mathrm{L}} \cong -\chi_{\mathrm{P}}/3 .$$

The final term, the Pauli paramagnetism, depends directly on the electronic density of states at the Fermi energy and is due to the differing numbers of spin-up electrons and spin-down electrons caused by their different energies in a magnetic field. Its value is positive and typically represents 25–50% of the total χ. It is given by

$$\chi_{\mathrm{P}} = 2\mu_{\mathrm{B}}^2 N(E_F)(1+\lambda_\chi), \tag{5.2}$$

where $N(E_F)$ is as before, μ_{B} is the Bohr magneton, and λ_χ reflect enhancements in the susceptibility of the density of states above its noninteracting value $N(E_F)$. Similar effects which contribute to γ may contribute to χ although their magnitudes may not be equal. In any case, since the susceptibility consists of a sum of terms, one must not attribute all changes due to changes in $N(E_F)$. In particular, one must not assume

$$\chi = \chi_P$$

and infer variations in $N(E_F)$ or values of λ therefrom as is sometimes done.

5.1.3 Spectral Measures of Density of States

The transition probability for the absorption or emission of a photon of angular frequency ω and hence energy $\hbar\omega$ is proportional to

$$N(\hbar\omega)\,|\boldsymbol{r}_{\mathrm{if}}|^2 \tag{5.3}$$

with

$$E_{\mathrm{f}}(\boldsymbol{k}) - E_{\mathrm{i}}(\boldsymbol{k}) = \hbar\omega$$

when only direct (k-conserving) transitions are considered, where $\boldsymbol{r}_{\mathrm{if}}$ is the matrix element of the coordinate between the initial and final states. From the product involved, one can see that the transition probability will be large when there are a large number of initial states $N(E_{\mathrm{i}})$, or a large number of final states $N(E_{\mathrm{f}})$, all other things being equal. Matrix element variation with energy and initial and final state may be weak or strong. For ultraviolet photoemission, the variation of spectra with photon energy is generally taken to represent final state effects. For x-ray emission (absorption) involving a core level of angular momentum l, the matrix element becomes important in that initial (final) states having angular momentum character $l \pm 1$ are important. Thus different l-characters of the density of states may be sampled by different processes to provide a more complete picture and test of a theoretical model.

5.1.4 Superconductivity

The occurrence of superconductivity is governed by the strength of the electron-phonon interaction constant λ_{ep}, which also contributes to the enhancement of the electronic specific heat and the Pauli susceptibility. According to *McMillan*'s [5.8] solution of the Eliashberg equations, the superconducting transition temperature of a solid is given by

$$T_{\mathrm{c}} = \frac{\Theta}{1.45} \exp\left[-\frac{1.04(1+\lambda_{\mathrm{ep}})}{\lambda_{\mathrm{ep}} - \mu^*(1+0.62\lambda_{\mathrm{ep}})}\right],$$

where Θ is the Debye temperature, and $\mu^* \approx 0.10$ reflects electron-electron interactions. Recently *Gaspari* and *Gyorffy* have given a simplified, but apparently fairly accurate prescription for calculating λ_{ep} and hence estimating the superconducting transition temperature T_c. Following *McMillan*,

$$\lambda_{ep} = N(E_F)\langle I^2\rangle / M\langle\omega^2\rangle,$$

where the numerator depends on the density of states at the Fermi energy and the square of the electron-phonon matrix element averaged over the Fermi surface, and the denominator depends on the mass of vibrating ion and the square of an appropriately averaged phonon frequency. *Gaspari* and *Gyorffy* [5.9] showed that the numerator could be calculated quite readily from the information at hand from band structure calculations as (in atomic units)

$$\eta \equiv N(E_F)\langle I^2\rangle = \frac{2E_F}{\pi^2 N(E_F)} \sum_l \frac{(l+1)\sin^2(\delta_{l+1}-\delta_l)\eta_l\eta_{l+1}}{\eta_l^{(1)}\eta_{l+1}^{(1)}}, \tag{5.4}$$

where δ_l is the phase shift and $\eta_l/\eta_l^{(1)}$ is the lth component of the solid density of states divided by the single scatterer value. This gives a fairly detailed test of an energy band model and goes a considerable way in establishing the nature of the mechanism leading to large values of the electron phonon coupling constant λ_{ep}.

5.1.5 Theoretical Approach

To understand the causes of the various changes observed upon absorption of hydrogen and hydride formation, one needs some sort of theoretical model with which to compare, whose predictive capability and understanding can guide further experiments. Our approach will be the one-electron band theory of solids [5.3] whose basis and applicability have been well established in the areas of semiconductor physics and metal physics. The periodic nature of crystalline solids gives rise to electronic energy levels which are quasicontinuous *bands* (energy separations $\sim 10^{-21}$ eV) separated by regions of forbidden energies, *gaps*. It is the calculation of the bands and gaps and their change with hydrogen concentration which will form the basis of our theoretical understanding. Previous models of no change, in which the hydrogen electrons fill up states of the host metal (the *proton model*), or in which the addition of low-lying hydrogen states empties states of the host metal (the *anion model*), have been shown to be too naive and simple and to only represent the extremes of the actual case.

To determine the electronic states of the solid one must solve Schrödinger's equation

$$H\psi_{\boldsymbol{k}} = E_{\boldsymbol{k}}\psi_{\boldsymbol{k}}$$

in which H is the Hamiltonian containing kinetic and potential energy, $\psi_{\boldsymbol{k}}$ is the wave function, and $E_{\boldsymbol{k}}$ is the energy whose label $\boldsymbol{k}$ describes the symmetry of the solution. To go further one needs a prescription for the potential. There is considerable experience available in selecting an initial guess based on atomic [5.10] and solid-state calculations [5.11]. The crystal structure and lattice parameters are assumed as given data (although one may vary these). The limitations of the procedure depend on the choice of initial potential, although one can implement self-consistency checks and procedures. A further limitation is based on how many (of the $\sim 10^{23}$) $\boldsymbol{k}$ values in the Brillouin zone one can afford to calculate. To obtain detailed results one needs a great many k-vectors which requires very expensive and/or time-consuming calculations or the application of interpolation procedures.

Since the effects and phenomena we wish to explore depend more on symmetry than on anything else, the choice of initial potential is not critical for a qualitative understanding of the changes and we shall assess the value of the quantitative understanding. Our technique [5.12] and many of our results on hydride systems are published elsewhere [5.13–18], and we shall here present a different aspect of the results, namely the comparison of the density of states information, $N(E)\Delta E$, the number of states in the region between $E-\Delta E$ and E with energy E. We shall indicate the effects of the limitations where appropriate.

5.2 The Palladium-Hydrogen and Related Systems

The studies of the influence and interaction of palladium with hydrogen go back over a century and probably constistute the most heavily studied [5.1, 2] metal/gas system. This is probably because the equilibrium pressure for hydride formation at room temperature is less than one atmosphere. Its practical use as diffusion membranes for purifying hydrogen gas and the recent discovery of superconductivity in this system make it of wide ranging interest.

5.2.1 Lattice Structure

Upon the absorption of hydrogen at room temperature up to a ratio of 0.025–0.030, the lattice parameter of palladium metal increases from 3.891 Å to 3.894 Å. Further absorption of hydrogen occurs with the formation of the beta (β) phase of palladium hydride in equilibrium with the low concentration (α) phase. This former phase retains the face centered cubic metal lattice of palladium atoms; however, there occurs a discontinuous change to a new lattice parameter of 4.026 Å characteristic of a first-order phase transition. The hydrogen concentration at this phase change is $x \approx 0.60$ with the hydrogens apparently randomly occupying the octahedral interstices in the lattice. Thus macroscopic concentration ratios $0.03 < x < 0.60$ only reflect a mixture of α and β phase palladium hydrides and the conversion of α phase to β phase according to the lever rule. With higher pressures or by other techniques, a stoichiometric concentration of $x=1$ can be approached in which all the octahedral sites are filled and the ideal sodium chloride structure is attained. The fact that the metal lattice structure does not change and that the increase in the lattice parameter is less than 5% leads one to hope that the changes observed are directly attributable to the addition of hydrogen. Some of the more significant ones will be discussed in terms of the underlying electronic structure. We shall first present a few salient results, discuss these results in terms of modern theoretical calculations, and then indicate the extension of these results to allied systems.

5.2.2 Electronic Specific Heat Coefficient

The low-temperature extrapolation of the plot of the molar heat capacity of PdH_x divided by T vs T^2 yields at the $T^2=0$ intercept the specific heat coefficient which is related to the density of states at the Fermi energy (5.1). Recent work [5.19] in the regime of $0.78 \leqq 0.96$ shows within the experimental accuracy a continuous decrease of the electronic specific heat coefficient from a value of $1.49\,\mathrm{mJ\,mol^{-1}\,K^{-2}}$ for $x=0.78$ to $0.78\,\mathrm{mJ\,mol^{-1}\,K^{-2}}$ for $x=0.96$. This is contrasted with values for palladium and silver metal of $9.48\,\mathrm{mJ\,mol^{-1}\,K^{-2}}$ and $0.65\,\mathrm{mJ\,mol^{-1}\,K^{-2}}$, respectively, with values in between for the alloy system. Again, the addition of hydrogen decreases the electronic specific heat coefficient by over a factor of 12, but the detailed quantitative analogy with addition of silver fails.

5.2.3 Susceptibility

Measurements of the magnetic susceptibility, going back as far as 1916 [5.20] and spanning more than five decades [5.21] show the susceptibility decreasing from the large value of $\sim 600 \times 10^{-6}$ emu/g-at. to essentially zero at the β phase hydrogen composition of 0.60. The analogy of this behavior with Pd–Ag alloys, where the addition of the extra electrons (silver being one greater in atomic number than palladium), was explained by filling up the holes in the *d*-band [5.22], the system becoming diamagnetic when the postulated 0.6 holes are filled. There are flaws in this analogy in that while silver forms a *continuous substitutional* alloy $Pd_{1-y}Ag_y$ with palladium, *interstitial* hydrogen forms a second phase and the susceptibility decrease reflects the low susceptibility of $PdH_{0.60}$ and the manifestation of the lever rule as stated above. We are, however, left with the low susceptibility of this phase and its dependence on hydrogen concentration for $x>0.60$. To the extent that the susceptibility reflects the density of electronic states at the Fermi energy, we can say that the addition of hydrogen (like that of silver) decreases this density of states. Furthermore, it is known both experimentally [5.23] and theoretically [5.24] that palladium has only 0.36 holes in the *d*-band and this fact must be accounted for.

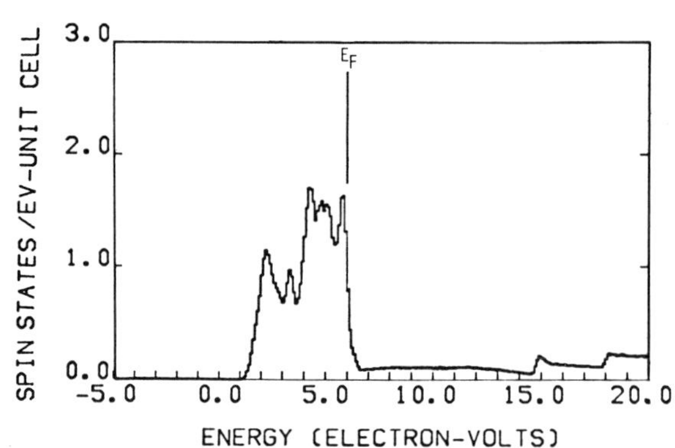

Fig. 5.1. Density of states of one spin for palladium obtained from the LCAO model of [5.16]

5.2.4 Photoemission from the PdH_x System

Eastman et al. [5.15], upon exposing a palladium foil to an environment of hydrogen under conditions which should lead to the formation of β phase palladium hydride, observed the growth of a low-energy peak below the peak associated with *d*-bands of palladium. This peak was associated with the presence of hydrogen-induced states in the hydride. Similar evidence of such states has been seen by *Antonangeli* et al. [5.25], but *Gilberg* [5.26] has failed to find any evidence in his x-ray emission results, a fact which lacks a satisfactory explanation.

5.2.5 Superconductivity

The specific heat data and resistivity data [5.27] show that for hydrogen concentrations above $x \approx 0.84$ the system becomes superconducting, with the superconducting transition temperature T_c increasing from 3 K and achieving a value in excess of 8 K near the upper end of the hydrogen concentration regime. Two more relevant results are the decrease in T_c with decreasing lattice constant (pressure) [5.28] and the higher (approximately 2 K) T_c of the deuteride than the hydride of comparable concentration [5.29]. These results are discussed more fully in Ref. [5.30], Chapter 6 and provide severe tests for various models as discussed below.

5.2.6 Theoretical Calculations of the Electronic Structure of PdH_x

Probably no system has been dealt with theoretically by so many researchers in the past five years as PdH_x [5.15, 16, 31–33]. The original work by this author was a calculation of the band structure of the stoichiometric compound PdH [5.15, 16]. Densities of states for palladium and PdH are shown in Figs. 5.1 and 5.2, respectively. Subsequent calculations have included various approxi-

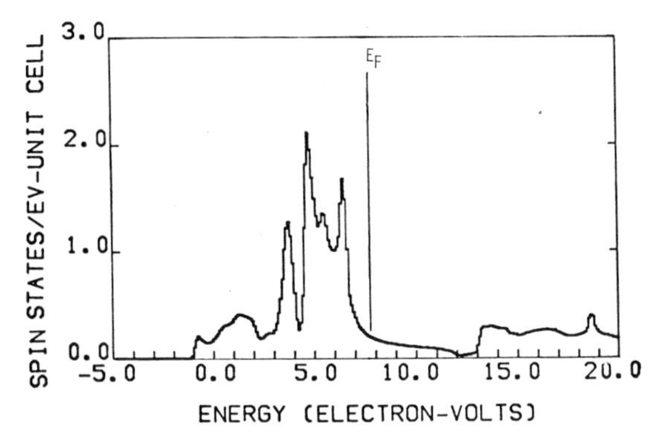

Fig. 5.2. Density of states of one spin for stoichiometric PdH obtained from the LCAO model of [5.16]

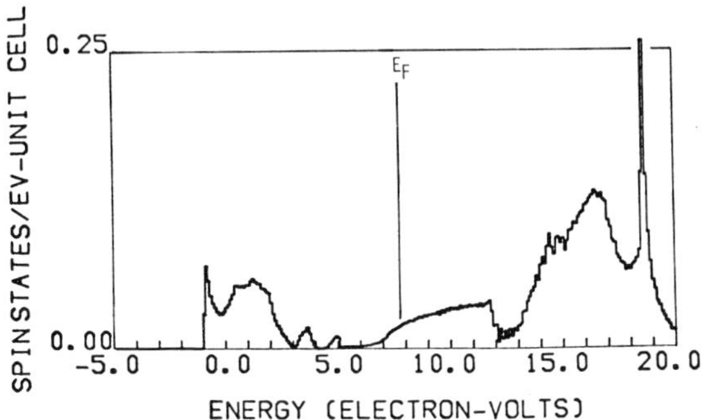

Fig. 5.3. Density of states of one spin of *s*-like character about the hydrogen site for PdH obtained from the LCAO model of [5.16]

mations and refinements to extend these results and to address specific phenomena such as the influence of nonstoichiometry and the superconductivity as discussed above. All these calculations have the following features in common which were obtained from a variety of potentials and using a variety of techniques:

1) The lowering of filled palladium states by the influence of the hydrogen in the unit cell.

2) The lowering of empty states in palladium below the Fermi energy.

3) The filling of the latter states and the holes in the *d*-band.

4) The filling of states above the top of the *d*-band.

Each of these features and how they manifest themselves experimentally will be discussed in turn.

The lowered filled palladium states lie below the *d*-bands of Pd and are a manifestation of bonding between the hydrogen and the palladium and between the hydrogens themselves. The states are clearly shown in Fig. 5.2 in the energy range -1.0 to $+2.0$ eV. The bonding is both with the 5*s*- and 4*d*-electrons of the palladium. The hydrogenlike *s* character is shown in Fig. 5.3. In stoichiometric PdH, these levels are shifted as much as 2–5 eV depending on the level and on the calculation. There is clearly a large electronic effect of the hydrogen. These levels are the ones observed in the photoemission studies [5.15].

The lowered states (which were empty in Pd) likewise are those that reflect s-like character about the hydrogen[1] site; however, their character about the palladium site is $5p$-like, being the top of the $s-p$ conduction band in the $\langle 111 \rangle$ Brillouin zone directions. The p-like states in the $\langle 100 \rangle$ directions are not lowered since they are also p-like about the hydrogen. The states at K in the $\langle 110 \rangle$ directions are lowered over 5 eV, but since they started over 8 eV above the Fermi energy, they do not get filled. (Of course, some states in this k-direction near the Fermi energy do get filled.) Thus the majority of these types of states come from the states associated with the $\langle 111 \rangle$ directions and their end points L. This uniqueness is associated with their being the *lowest* states with s-like character about the hydrogen sites while being p-like about the palladium. We have estimated that these states can hold ~ 0.5 electrons/hydrogen which, with the 0.36 electrons accommodated in the d-bands, can account for the β phase boundary at $x = 0.6$. These states are lowered over 5 eV, again a manifestation of strong electronic effects due to the hydrogen. A similar lowering of these states would be expected in copper where they are already filled and give rise to the electron neck orbits on the Fermi surface in planes normal to the $\langle 111 \rangle$ direction. A very sensitive measure of this effect is the change of the Fermi surface scattering properties for very dilute hydrogen in copper as observed by *Lengeler* et al. [5.34]. Attempts to observe this effect in palladium have proven negative [5.35] although both the electronic specific heat coefficient [5.36] and the susceptibility [5.21] measurements for *dilute* hydrogen indicate a rapidly decreasing density of states and hence should yield a quantitative modification of the Fermi surface. The fact that these states are already filled in platinum due to a relativistic lowering is most likely associated with the lack of formation of a hydride phase of platinum.

Analysis of this energy lowering shows that it is due to a large second like neighbor hydrogen-hydrogen interaction. There is also experimental evidence for this interaction in the phonon spectra analysis [5.37]. Calculations for $Pd_{32}H$ [5.38] (the hydrogens are now 8th like nearest neighbors) and the non-stoichiometric models [5.31, 33] show that the added electron goes into d-states at the Fermi level for low concentration. These results are consistent with the dilute solution α phase results where the susceptibility [5.21] and specific heat coefficient change [5.36] faster with hydrogen concentration that the change from α phase to β phase with concentration and are consistent with the band structure results for palladium [5.24].

Palladium hydride therefore differs from palladium by the strong perturbing influence of hydrogen in the octahedral interstice in the lattice. This hydrogen is strongly attractive for states (which may or may not be filled) having s-like character about *this* site; the character (or number) of states about the palladium is of minor importance except as far as they were originally filled or empty. There exists strong hydrogen-hydrogen interactions which can lower states sufficiently to be filled. The filling of these states as well as states

[1] The hydrogen atom consists of one $1s$-electron with binding energy 13.6 eV.

Table 5.1. Density of states and derived quantities compared with experiment for Pd and PdH

	$N(E)$[a] [electrons/ eV-cell]	γ_{calc} [mJ K^{-2} mol^{-1}]	γ_{expt} [mJ K^{-2} mol^{-1}]	$1+\lambda_\gamma$	$\chi_{P\,calc}$ [10^{-6} emu mol^{-1}]	χ_{expt} [10^{-6} emu mol^{-1}]
Pd	2.27	5.35	9.48[b]	1.77	73.5	735.0[c]
PdH	0.45	1.06	1.50[d]	1.42	14.6	0 ± 20[c]
			0.78[e]	0.74		

[a] Ref. [5.16]. [b] Ref. [5.36]. [c] Ref. [5.21]. [d] Extrapolation of Ref. [5.27]. [e] Ref. [5.19].

characteristic of the "host" palladium raises the Fermi level to a region of low density of states (not precisely related to those of palladium), and accounts for the low susceptibility and electronic specific heat coefficient observed. A comparison of the values of the specific heat and susceptibility derived from the calculated values of the densities of states from Figs. 5.1 and 5.2 with λ set equal to one is given in Table 5.1. Clearly, the decrease in the specific heat and susceptibility reflect the 5-fold decrease in the density of states. The 77% enhancement of the palladium density of states for the specific heat is reasonable and in good accord with other calculations [5.24]. The susceptibility results clearly show the presence of the other contributions for the susceptibility. The calculated value of the density of states for PdH clearly shows the presence of either other contributions or enhancements of the susceptibility. The calculated values of the density of states for PdH, if anything, is smaller than that obtained by others [5.32, 33] and gives reasonable results when compared with the extrapolated values [5.27] and the observed electron phonon enhancement. Worse agreement is obtained with [5.19] which extends to higher ($\chi=0.96$) and agrees with [5.27] in the range of common x, indicating the extrapolation may be in error. It is unusual for a calculation to overestimate the density of states. Two of the calculations err even more in this regard. One would infer from the calculated and measured values of the susceptibility for PdH that the contributions of χ_{ORB}, χ_{DIA}, and χ_L are small and $\approx \pm\chi_P$.

While the total density of states is falling in Fig. 5.2 at the Fermi energy, the hydrogen s-like character is increasing as Fig. 5.3 shows. It is this s-like ($l=0$) character [see (5.4)] coupled with the optical vibrations of the hydrogens which give rise to the superconducting behavior of PdH_x [5.29, 32]. This variation at the Fermi energy is relevant to the compositional alloy and pressure dependence of the superconducting properties of palladium hydride [5.39].

5.2.7 Other Palladium Based Systems

There has been extensive discussion of palladium alloy systems elsewhere [5.2] and in Ref. [5.30], Chapter 3, but some discussion in the light of modern theory of electronic properties seems called for. Work by *Wise* et al. [5.40] on the change in lattice constant at the β phase boundary for various palladium-metal alloy hydride systems shows that the change in lattice spacing over the α phase value can be linearly related to the solute-valency concentration product, with the change extrapolating to the zero value when the product equals 0.37. This strongly indicates a rigid bandlike behavior with the solute valence electrons filling up the holes in the palladium d-band. This work combined with the volume change versus x, H/M, ratio correlation of *Baranowski* et al. [5.41] shows that the hydrogen-to-metal ratio at the β phase boundary decreases from a value of ~ 0.55 for no solute in palladium to a value of 0.16–0.20 for the critical valency-solute concentration of 0.37, indicating that some of the electrons are being accomodated in states other than d-band holes. We maintain that these are of the second type above. An apparent contradiction to the above correlation is the Pd–Fe system where a valency of 2.0 must be assumed for iron. Similar behavior for this system has been observed by a variety of techniques [5.42, 43] where other valencies are required.

Particularly telling results are those for the Pd–Fe systems where it is found that the hydrogen absorption for ordered Pd_3Fe is 4.5–9.0 times larger than that of the disordered alloy of the same system [5.44]. This again shows that local effects are important since in the ordered alloy there are octahedral sites with all (6) palladium neighbors while in the disordered system 1.5 neighbors would be iron on the average. This behavior takes on further significance when the electronic specific heat coefficient measurements are considered. It is found that the coefficient for ordered $FePd_3$ is 8.17 mJ mol^{-1} K^{-1} while that for the disordered is even larger, 11.28 mJ mol^{-1} K^{-1} [5.45, 46]. These results are both in contradiction to any rigid band behavior *and* with any *simple* screening model of hydrogen solubility in metal. Clearly there are profound local effects, and the magnitude of the *total* density of states which reflects things associated with the d-metal may have little to do with the s-like hydrogen atom and its site preference.

5.3 The Titanium-Hydrogen and Related Systems

Titanium possesses one of the highest absorptive capabilities for hydrogen of metal hydride systems, with densities of hydrogen atoms per cubic centimeter in excess of 9×10^{22} being attainable. The influence of hydrogen is quite pronounced and again manifests itself in the phase diagram as changes in the lattice structure.

5.3.1 Lattice Structure

At high temperatures one can achieve concentrations as large as 10% in the α phase parent hexagonal close-packed titanium lattice [5.47]. Increasing the hydrogen concentration results in a two-phase region of α phase and hydrogen

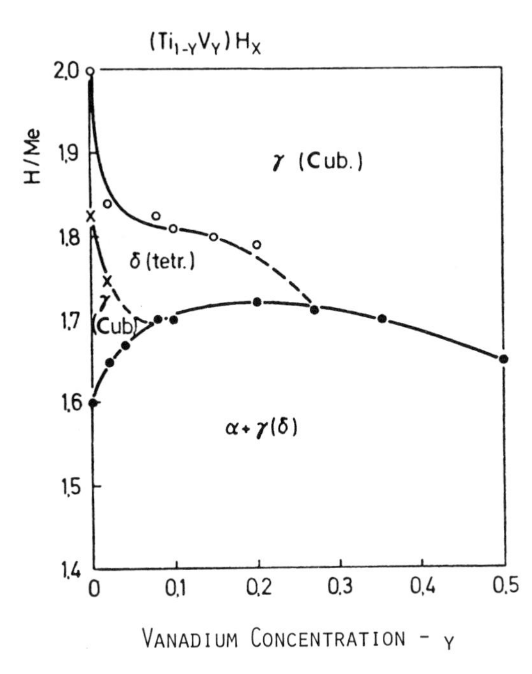

Fig. 5.4. A portion of the room temperatures titanium-vanadium-hydrogen phase diagram according to [5.48]

in the high-temperature bcc β phase of titanium. It should be noted that increasing the concentration of hydrogen (up to 40%) permits stabilization of this phase at temperatures 600 K lower than normal. Further additions of hydrogen result in another two-phase region of β and γ, a face centered cubic metal lattice with the hydrogens in the tetrahedral interstices in the lattice. This fluorite (CaF_2) structure has the stoichiometric composition of titanium dihydride, TiH_2. At room temperatures the lattice changes from α to γ with a two-phase $\alpha+\gamma$ region for $0.01<x<1.6$. This dihydride structure is important because it is the structure found for the rare earth and actinide metals as well as the nearest neighbors in the periodic table of zirconium, Y, Nb, Hf, and Ti, as well as Sc and V (see Ref. [5.1], Fig. 1.1). However, in the case of the Group IV transition metals Ti, Zr, and Hf, as the ideal composition of two is approached, the lattice continuously distorts, shrinking along the z-direction and expanding along the x- and y-directions. This tetragonal, δ phase has a volume per metal atom which differs by less than 1% from that of the undistorted structure with c/a varying from 0.98 for titanium to 0.90 for hafnium. More remarkable is the influence of substituting vanadium for titanium on the phase diagram. This system has been investigated in considerable detail by *Nagel* and *Perkins* [5.48]. A portion of their 300 K diagram is given in Fig. 5.4. For no vanadium ($y=0$), the change to the two-phase $\alpha+\gamma$, cubic γ and tetragonal δ phases as x goes from 1.4–2.0 is apparent. With the addition of vanadium, $\gamma\to\delta\to\alpha+\gamma\to\gamma$ ($x\approx1.7$), $\gamma\to\delta\to\gamma$ ($x=1.75$), and $\delta\to\gamma$ ($x=1.85$) transitions are all observed. The effect of these transformations on various physical properties again includes effects due to the implicit lattice change and effects due to the explicit variation of hydrogen and vanadium additions. The fact that changes due to substitution of vanadium can be reversed by removal of hydrogen suggests a complementary role which will be discussed below.

5.3.2 Specific Heat Results

Specific heat results as well as some transport data have been obtained by *Ducastelle* et al. [5.49]. For the two samples $TiH_{1.75}$ and $TiH_{1.97}$, they obtained specific heat coefficients of 3.8 mJ K^{-2} metal atom and 4.5 mJ K^{-2} metal atom, respectively. Their results for the zirconium hydrides which should all be δ phase indicate a continuously decreasing susceptibility with hydrogen content while the specific heat goes through a maximum, again indicating differing contributions to the susceptibility than just density of states variation. They conclude in agreement with *Nagel* and *Goretzki* [5.50] that in the cubic phase everything indicates that the density of states at the Fermi level $N(E_F)$ increases as H/Ti increases. The δ phase transition is accompanied by a lowering of $N(E_F)$. This seems to indicate the presence of a peak near the Fermi level which is lowered by a redistribution of states to lower energies by the distortion. *Gesi* et al. [5.51] proposed an interpretation of the same type although they put the Fermi level near a minimum in the density of states, a situation not supported by theoretical [5.18] or experimental results.

5.3.3 Susceptibility

Measurements of the magnetic susceptibility were first made by *Trzebiatowski* and *Stalinski* [5.52] and more recently by *Nagel* and *Goretzki* [5.50] whose results are generally in good agreement with the earlier ones. *Nagel* and *Goretzki* also considered the effects of vanadium substitution for titanium as well as variation with hydrogen content and temperature. A variety of fairly complex behavior was observed which is clouded by the $\gamma-\delta$ phase change as well as variation with all the parameters, x, y, and T. By extrapolating the high-temperature results to low-temperature, susceptibility vs vanadium content and susceptibility vs hydrogen content plots of Fig. 5.5a and 5.5b, respectively, were obtained. Although the susceptibility consists of several terms, *Nagel* and *Goretzki* conclude that χ_{vv}, the Van Vleck paramagnetism, should not vary strongly with x or y and that the *dependence* of the susceptibility is principally determined by the variation of the density of states. Since the lattice constants in the δ phase depend upon both the composition and temperature no statements can be made about the form of the $N(E)$ curve in the δ phase. They concluded that $N(E)$ is lowered by the $\gamma \rightarrow \delta$ transition.

From Fig. 5.5a it can be concluded that the density of states initially rises with increasing vanadium content and that a two-peaked structure exists. This rising density of states is generally reflected in Fig. 5.5b with increasing x except

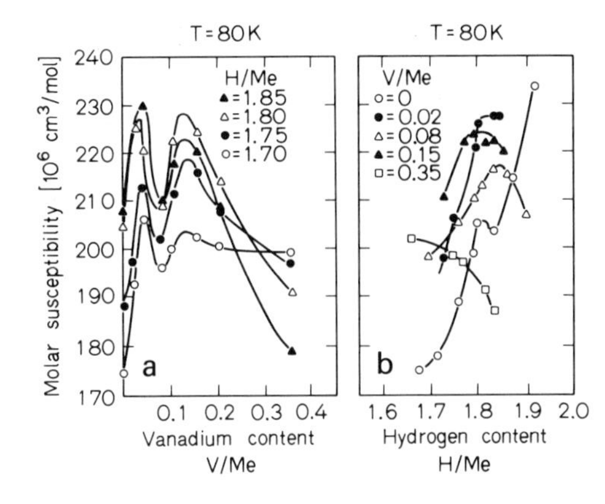

Fig. 5.5a and b. Magnetic susceptibility of titanium-vanadium-hydrogen system (a) as a function of vanadium content-y, (b) as a function of hydrogen content-x according to [5.50]

for the largest vanadium concentrations. There seems to be much less sensitivity to hydrogen variation than to vanadium variation. The peaks in Fig. 5.5a are not varying in position with x while Fig. 5.5b shows variations of smaller magnitude. This greater sensitivity to vanadium variation than to hydrogen variation is generally born out by the slopes of the phase boundary lines in Fig. 5.4. We shall discuss these results further below in the light of our theoretical calculation.

5.3.4 Spectral Studies of Titanium-Hydrogen Systems

There have been several studies of the titanium dihydride system which give complementary information about the density of states of this material. *Nemnonov* and *Kolobova* [5.53] have studied the K emission and absorption of a series of titanium compounds. The K emission of titanium metal consists of a structureless peak, K_{β_5}, about 1.5 eV below the Fermi energy and about 5 eV wide. This peak is skewed toward the Fermi energy. The absorption consists of a sharp peak 1 eV above the Fermi energy followed by a smooth continuously rising absorption. For the dihydride the peak below the Fermi energy is diminished in intensity and a much larger peak occurs 6.0 eV below the Fermi energy of overall width greater than 5 eV and again skewed towards the Fermi energy. In absorption the sharp peak above the Fermi energy disappears and a shoulder on the continuously rising background occurs 3 eV above the Fermi energy. Since this radiation is associated with the K-shell of *titantium*, it reflects density of states having p-like character about the *titanium* site.

In contrast, *Fischer* [5.54] in studying $L_{\text{II,III}}$ emission and absorption found a large peak 2.0 eV below the Fermi energy and a peak about half this height 6.3 eV below the Fermi energy for the hydride. On absorption he found a peak 2 eV above the Fermi energy with further peaks at 4.5 and 7.5 eV. There is also a shoulder on the low-energy side of the first peak at about 1 eV above E_F. Since this radiation is associated with the $L_{\text{II,III}}$ subshell it reflects density of states

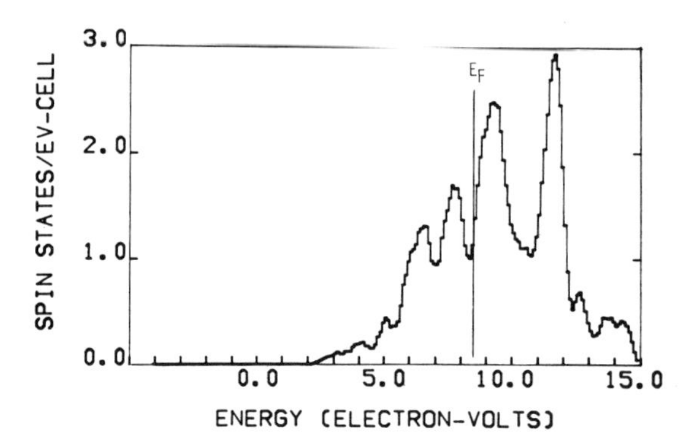

Fig. 5.6. Density of states of one spin for hexagonal titanium

having *s*- and *d*-character about the titantium, with large *d*-character remaining at the Fermi level.

Eastman [5.55], on exposing a titanium sample to hydrogen, saw the growth of a photoemission peak 5.0 eV below the Fermi energy as well as a 2–3 eV wide peak associated with the *d*-bands. Although the characterization and composition of his sample is unknown, the results are consistent with the x-ray results in the addition of intensity ~5 eV below the Fermi energy.

5.3.5 Superconductivity in Dihydride Phases

There has been no observation of superconductivity in any of the dihydride phases although the experimental verification is far from complete. Based on theoretical considerations discussed below, its occurrence is highly unlikely.

5.3.6 Theoretical Calculations of the Electronic Structure of $Ti_{1-y}V_yH_x$

In accord with our emphasis on changes, we first compare the change engendered in going from hexagonal close-packed titanium to a close-packed face centered cubic array of titaniums appropriate to the dihydride structure. The density of states for hexagonal titanium is shown in Fig. 5.6. Although there are more detailed calculations based on more points in the Brillouin zone [5.56], the gross structure is the same—four peaks of increasing height with energy—two below and two above the Fermi energy, with the Fermi energy just on the leading edge of the third peak. The x-ray emission results [5.53] for Ti metal are unable to resolve the ~1 eV splitting of the lower peaks, although a shoulder is observed on the low-energy side of the main peak. The overall width of the filled states of about 5–6 eV is in good agreement. The absorption spectra for Ti metal are also unable to resolve the two peaks above the Fermi energy although there is a peak of about 5 eV wide riding on a continuously rising background. This then establishes the level of agreement expected between the band structure results and the spectral measures.

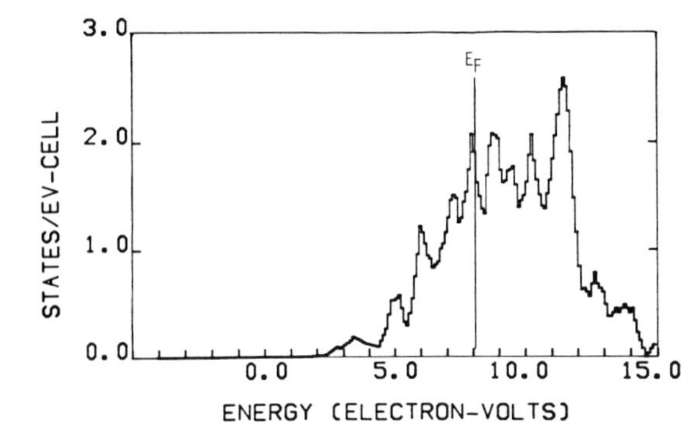

Fig. 5.7. Density of states for fcc titanium

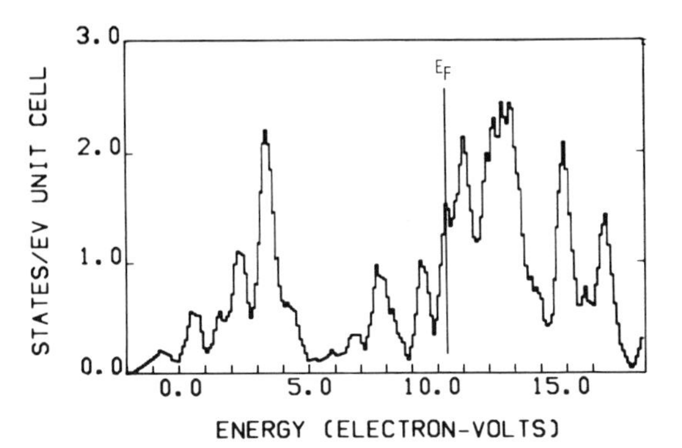

Fig. 5.8. Density of states for cubic titanium dihydride from calculation of [5.17] (256 points in Brillouin zone)

Figure 5.7 now shows the density of states for face centered cubic titanium metal. The density of states shows much less distinctive structure—the peak and valley structure is scarcely believable within the statistical sampling. The units of Fig. 5.7 have been changed to facilitate comparison with Fig. 5.6. Both are essentially in terms of electron states/titanium atom since there are two electron states/spin and two titaniums in the hexagonal unit cell. Thus the density of states for fcc titanium is larger than that of hexagonal titanium. The overall bandwidth of ~10 eV with filled and empty regions of about 5 eV is similar to the hexagonal one of Fig. 5.6.

The results for titanium dihydride are shown in Fig. 5.8. These results are derived from the calculation of [5.18]. The major change from Fig. 5.7 is the addition of a peak below 5 eV. This peak is associated with the bonding and antibonding hydrogenlike states characteristic of having two hydrogens in the unit cell. As in palladium these states consist of various types. The hydrogen bonding states were already filled in fcc titanium but are lowered 4–6 eV by the hydrogen potential. The antibonding states consist of states that were near or above the Fermi energy in fcc titanium and are lowered by the presence of the hydrogens and can be filled. It is estimated [5.18] that these states can hold ~1.15 of the two additional electrons introduced by the hydrogens in the unit cell. The remaining 0.85 electrons are accommodated in *d*-states of titanium at

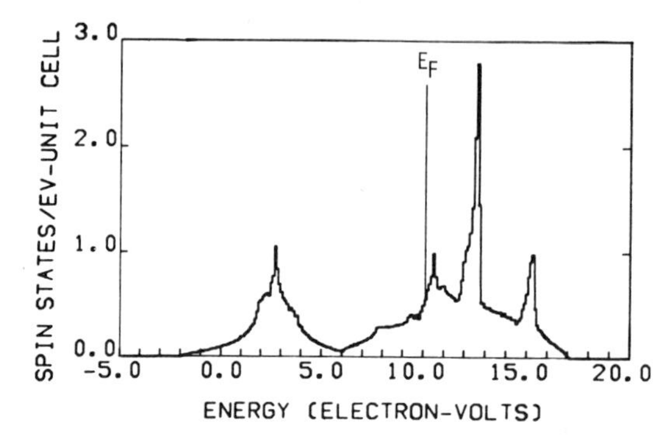

Fig. 5.9. Density of states for cubic titanium dihydride using LCAO interpolation scheme (0.5 M points in Brillouin zone)

the Fermi energy. These bonding and antibonding states are discussed in terms of the titanium and fcc dihydride structures as these form the extremes of the TiH_x $(0 \leqq x \leqq 2)$ system respectively. These states mix with the *s, p, d* states of the transition metal as their already filled nature and proximity to the Fermi energy show.

The *K*-emission thus reflects these states strongly while the *L*-emission still has its major intensity resulting from the higher lying *d*-bands of the transition metal. The presence of the hydrogen *s* character principally below the Fermi energy indicates that the addition of hydrogen and the vibrational modes associated therewith will not contribute to superconducting behavior of these systems. The density of hydrogen *s*-like states as the Fermi energy is one-fortieth the value for palladium hydride. The Fermi energy falls on the low-energy side of a peak. The densities of states for ScH_2, VH_2, and YH_2 are qualitatively very similar, with the major variation being the location of the low-energy peak relative to the Fermi energy. The balance between the accommodation of hydrogen electrons in new low-lying states below the Fermi energy and in states at the Fermi energy has been shown to have relevance to the stability of the di- and trihydride structures [5.14]. In tungsten dihydride, many more electrons must be accommodated at the Fermi energy due to insufficient lowering of the hydrogen antibonding states. The result correlates with the absence of formation of WH_2. Similar conclusions hold with regards to the absences of a trihydride structure of scandium (see [Ref. 5.1, Fig. 1.1]) and [5.14].

A density of states derived from many more points, based on an interpolation scheme, is shown in Fig. 5.9 and gives an indication of the statistical noise present in Fig. 5.8. The major features of Fig. 5.8 are reproduced; however: low-energy hydrogenlike peak, Fermi energy at increasing density of states, peak above Fermi energy. The peak just above the Fermi energy seen in Fig. 5.8 is not seen in Fig. 5.9 and its reality would appear questionable on that basis. The second peak is clearly real and corresponds in magnitude (change in units) and approximately in position to that in Fig. 5.8. In [5.18] we show that the peak at the Fermi energy in Fig. 5.8 is associated with an orbitally degenerate state (in the absence of spin-orbit coupling) which is split by the

Table 5.2. Densities of states and derived quantities compared with experiment for titanium-vanadium hydrides

	$N(E)$ [electrons/ eV-cell]	γ_{calc} [mJ K^{-2} mol^{-1}]	γ_{expt} [mJ K^{-2} mol^{-1}]	$1+\lambda_\gamma$	$\chi_{P\,calc}$ [10^{-6} emu mol^{-1}]	χ_{expt} [10^{-6} emu mol^{-1}]
Ti (hcp)	2.28[a]	2.69	3.32[c]	1.23	36.8	152[e]
	1.82[b]	2.15		1.54	29.4	
Ti (fcc)	1.91[a]	4.48			61.4	
TiH_2	1.52[a]	3.58	4.5[d]	1.26	49.1	215[f]
VH_2	2.03[a]				65.8	160[h]
	2.08[g]					

[a] This work.
[b] Ref. [5.56].
[c] Ref. [5.57].
[d] Ref. [5.49].
[e] Ref. [5.72].
[f] Ref. [5.50].
[g] Based on rigid band and Fig. 5.9.
[h] Ref. [5.70].

tetragonal distortion lowering the energy of the filled states. Adding vanadium fills (in a rigid band sense) both states and results in no net energy lowering and hence the tetragonal structure is not favored as Fig. 5.4 shows. Removal of hydrogen also appears consistent with the removal of some electrons at the Fermi energy in terms of the phase diagram but does not appear consistent with the lack of peak position shift of Fig. 5.5a, although the susceptibility clearly rises with hydrogen addition as Fig. 5.5b shows.

Finally, we discuss the relationship between the density of states results and the specific heat and susceptibility derived therefrom and the experimental results in Table 5.2. Again it is clear that the susceptibility is only partly given by the Pauli contribution both for titanium and its dihydride, and its variation only remotely reflects density of states information. In fact, the calculations predict an increase in the paramagnetic susceptibility in going from TiH_2 to VH_2 (the peak at 11 V corresponds to one more electron), whereas the measured experimental susceptibility decreases. The specific heat coefficient is in good accord with the calculations, and the enhancement due to the *d*-electrons agrees very well with that obtained for titanium metal. The model appears to give overall qualitative results and quantitative results where its application is most straightforward.

5.4 Vanadium Hydride and Related Systems

In addition to the dihydride discussed above, vanadium as well as the other Group V elements, niobium and tantalum, form a variety of other hydride phases with lesser hydrogen content. There has been considerable structural work [5.58–65] although there is far from universal agreement as indicated in the next section.

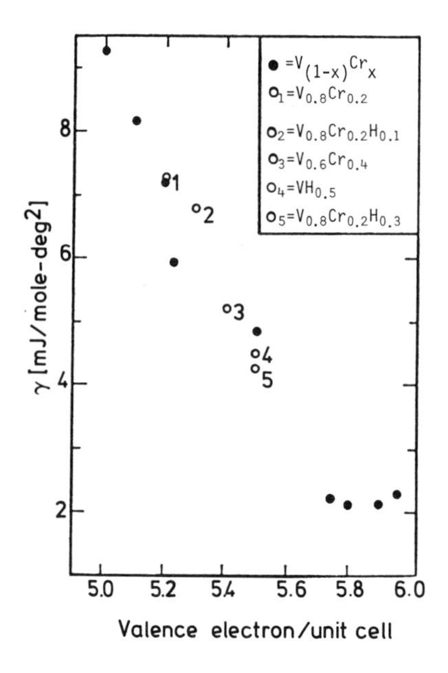

Fig. 5.10 Electronic specific heat coefficient for $V_{1-y}Cr_yH_x$ systems from [5.66]

5.4.1 Structures

Vanadium is body centered cubic with cubic lattice constant of $a = 3.028$ Å. Upon absorption of hydrogen a hydride phase of varying composition occurs. The vanadium atoms remain on a body centered lattice or a distortion thereof. Since the bcc lattice has 3 octahedral and 6 tetrahedral sites per vanadium and the hydride phases have <1 hydrogen per vanadium, true cubic symmetry is destroyed and random or ordered occupancy of these sites can occur. Ordered occupancies lead to compositions such as V_2H, V_4H_3, and VH. Both sites appear to be occupied with transitions occurring between octahedral and tetrahedral sites occurring as temperature changes. Curiously enough the hydride and deuteride systems are not completely analogous [5.62, 65].

5.4.2 Electronic Specific Heat

The electronic specific heat coefficient was measured by *Rohy* and *Cotts* [5.66] for a series of $V_{1-y}Cr_yH_x$ systems. These results are shown in Fig. 5.10. The decrease from 9.25 mJ mol^{-1} K^{-2} for V to 4.55 mJ mol^{-1} K^{-2} for $VH_{0.5}$ is consistent with the decrease in the susceptibility discussed next and appears to justify a rigid band approach. Similar conclusions appear justified based on the NMR results [5.67]. Qualitative and quantitative comparison with our theoretical models will be made below.

5.4.3 Magnetic Susceptibility

The susceptibility of all Group V subhydrides, $x<1$, decreases from that of the host bcc element. Values for V of 300.6×10^{-6} emu/g-at. to 208.8×10^{-6} for $VH_{0.7}$ [5.63]; for Nb of 201.6×10^{-6} emu/g-at. to 65×10^{-6} emu/g-at. for $NbH_{0.86}$ [5.68]; and for Ta of 152.0×10^{-6} emu/g-at. to 76×10^{-6} emu/g-at. for TaH [5.69]. These results are in reasonable agreement with those of *Aronson* et al. [5.70]. These latter authors attempt to infer the number of electrons donated to the *d*-bands by comparing susceptibility of bcc subhydrides with fcc dihydrides *superimposed* (in ordinate) on χ vs electron per atom ratio for bcc alloys systems, a dubious procedure due to the structure change as indicated above and the varying terms contributing to the susceptibility. The 30–60% change within the bcc based structures is a point to be addressed by any model, however.

5.4.4 Superconductivity

Although the resistivity V_2H is lower than that of vanadium metal [5.71], no evidence of superconductivity has been found down to 1.7 K [5.66] although vanadium is superconducting with $T_c\approx5$ K. T_c is lowered for all Group V metals upon hydrogen uptake although tantalum appears to remain superconducting at a composition of Ta_2H. Detailed application of the microscopic theory as outlined in Section 5.1.4 is needed to ascertain if this behavior can be understood.

5.4.5 Spectral Measures

Brytov et al. [5.73] give curves of the K_{β_5} emission ($4p\rightarrow1s$) and $L_\alpha(3d, 4s\rightarrow2p)$ for vanadium metal, VH_{cub}, VH_{tetr} as well as some mixed carbohydride phases. They conclude that there is very little change in shape and position from V metal to the hydrides for the *L*-emission, indicating that the majority-3*d*-electrons are *not* largely involved in "screening" the hydrogen. In the *K*-emission there appeared new low-energy bands which reflects *p*-like character about the vanadiums, which appears larger in the tetragonal phase. This peak is about 8 eV below the Fermi energy. The *sd* manifold appears about 5 eV wide.

Similarly *Fukai* et al. [5.74] have measured the L_3-emission from VH_x and VD_x and V. They conclude that there is a hump on the low-energy tail of the *d*-band peak accompanied by a depression of the low-energy side of the *d*-band and a bulging out on the high-energy side without any change in the energy of the peak. This hump is 7 eV below E_F irrespective of composition and phase. They also suggest that the Fermi energy rise accompanying the high-energy bulge ~0.3 eV is sufficient to contain 0.6 electrons. These authors support the proton model at the Fermi energy but by no means rigid band behavior below the peak of the *d*-band. Similar results and conclusions were obtained by *Gilberg* [5.75] in the case of NbH_x.

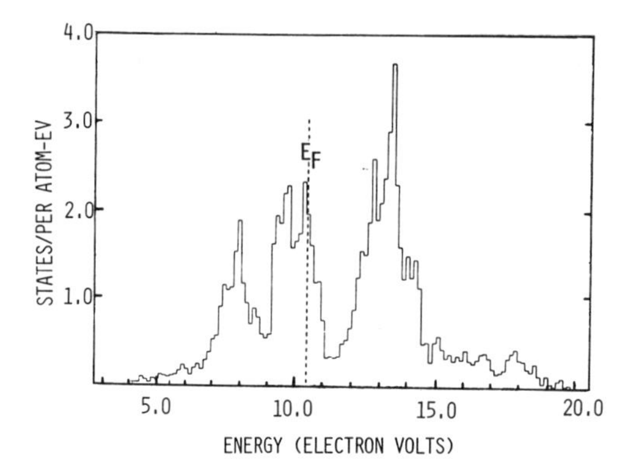

Fig. 5.11. Density of states for body centered cubic vanadium according to [5.76]

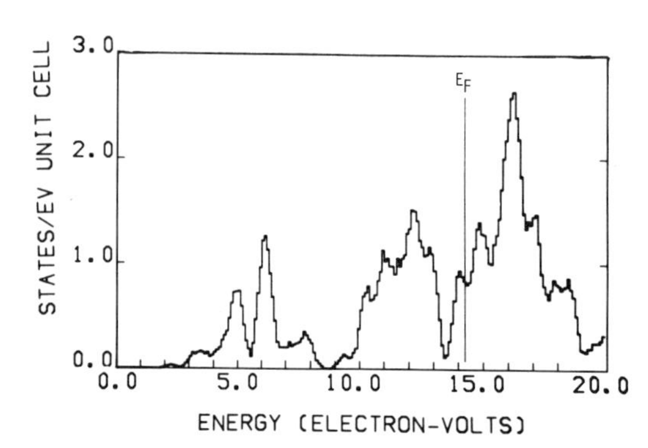

Fig. 5.12. Density of states for VH derived from the calculations of [5.17]

5.4.6 Theoretical Calculations of Electronic Structure of VH_x Systems

There is less than universal agreement on the structures of the vanadium hydrides and deuterides [5.59–65], and some of the proposed structures are of quite low symmetry and hence fairly difficult to calculate. We have chosen a fairly simple structure which retains the body centered cubic vanadium lattice and places a hydrogen above every vanadium at $(0, 0, a/2)$ [5.17]. This structure is the one actually proposed for VH by *Asano* et al. [5.65]. The density of states calculations for body centered cubic vanadium from results of energy band calculations [5.76] are shown in Fig. 5.11. These results show the Fermi energy falling on a peak in the density of states with a deep minimum $\sim$0.5 eV above it. The occupied bandwidth is about 4.0 eV and agrees reasonably well with the spectral measures cited above. Our results for VH are shown in Fig. 5.12. Again weight has been shifted out of the lower peak and a peak occurs 9 eV below the Fermi energy. This peak is made up of states which have *s*-like character around the hydrogen site and *p*- and *d*-like character around the vanadium site. This is in good qualitative and quantitative agreement with the experimental results [5.73, 74]. The calculations [5.17] indicate that with the exception of

Table 5.3. Densities of states and derived quantities compared with experiment for V–H systems

	$N(E)$ [electrons/ eV-cell]	γ_{calc} [mJ K^{-2} mol^{-1}]	γ_{expt} [mJ K^{-2} mol^{-1}]	$1+\lambda_{\gamma}$	$\chi_{P\,calc}$ [10^{-6} emu mol^{-1}]	χ_{expt} [c] [10^{-6} emu mol^{-1}]
V	1.96 [a]	4.61	9.25 [b]	2.00	63.5	310
$VH_{0.5}$	—	—	4.55 [b]		(62.8) [d]	260
VH	0.84 [e]	1.98	—	—	27.2	190

[a] Ref. [5.76].
[b] Ref. [5.66].
[c] Ref. [5.70].
[d] Derived from specific heat.
[e] This work.

states associated with the point N in the ⟨110⟩ directions the lowest states of s-like symmetry about the hydrogen are filled in bcc vanadium. However, these states are strongly modified by the attractive hydrogen potential with shifts of up to 5 eV resulting. The calculations indicate that somewhat less than one-half an electron can be accommodated in the states associated with N which are empty in vanadium and are lowered in VH and can be filled and that the remainder must be accommodated at the Fermi energy. The states at N have p-like character about the vanadium and appear 6–7 eV below the Fermi energy in good agreement with the K_{β_5} spectra [5.73].

A comparison of the specific heat and susceptibility derived from the calculations is given in Table 5.3. The factor of two decrease in the specific heat coefficient can be accounted for by the decrease in the density of states. Again the susceptibility is largely due to orbital contributions and its decrease and change cannot be accounted for in terms of the change of the spin paramagnetism. The variation is specific heat coefficient with hydrogen and chromium content cannot be given by any rigid band filling of the states of Fig. 5.12. The comparison of Figs. 5.11 and 5.12 shows that strict rigid band considerations are inappropriate. Furthermore, the number of N-like states which are lowered depends on the number of hydrogens as does the number of added electrons. The agreement between rigid band filling of Fig. 5.11 and the specific heat results should not be taken as substantiation of that picture since this viewpoint is incapable of explaining the spectral results [5.73–75] while the approach leading to the results of Fig. 5.12 can. More experimental results and calculations are needed to assess the appropriateness of the model. The spectral measures and the specific heat changes seem in reasonable accord.

5.5 Chromium-Hydrogen and Related Systems

Chromium exhibits extremely low solubility for hydrogen. *Snavely* [5.77] has reported a hexagonal monohydride CrH and a cubic dihydride CrH_2.

5.5.1 Lattice Structure

The hexagonal hydride has the nickel arsenide structure with the chromium forming a close-packed hexagonal lattice and the hydrogens occupying the octahedral intersites. The cubic dihydride is reported to have the fluorite structure discussed above in Section 5.3.1. Attempts to repeat this preparation technique have not succeeded [5.78] and on the basis discussed below we doubt its existence. *Baranowski* [5.79] has prepared the hexagonal form using high-pressure synthesis and has recently reported the synthesis of MnH [5.80].

5.5.2 Electronic Specific Heat and Superconductivity

Viswnathan et al. [5.81] have measured the electronic specific heat of the same sample—$CrH_{0.97}$—whose magnetic susceptibility was measured. They found a value for γ of 4.5 mJ g-mol^{-1} K^{-2} as compared with a value of 1.4 mJ g-mol^{-1} K^{-2} for bcc chromium. A value of 10.2 mJ g-mol^{-1} K^{-2} for $CrH_{0.84}$ has been reported by *Albrecht* and *Wolf* [5.82], but this value may be in error due to magnetic impurities and extrapolation below 12 K. In [5.81] two measurements are cited which claim no superconductivity down to 0.007 K.

5.5.3 Magnetic Susceptibility

Hanson et al. [5.83] have measured the magnetic susceptibility of $CrH_{0.97}$. They found a value of 220.9×10^{-6} emu g-mol^{-1} for $CrH_{0.97}$ as compared with 181.8×10^{-6} emu g-mol^{-1} for Cr. The susceptibility of $CrH_{0.97}$ exhibits considerable temperature and field dependence and rises to a value in excess of 265×10^{-6} emu g-at.$^{-1}$ below 10 K. From this and a comparison with the electronic specific heat coefficient they concluded that CrH has a strongly enhanced Pauli susceptibility with the Fermi energy falling near a peak in the density states and drew considerably on an analogy with palladium and speculated about the possible occurrence of giant moments.

5.5.4 Spectroscopic Measures

Brummer et al. [5.84] have measured the K-absorption spectra of cubic Cr and hexagonal CrH and they show considerable differences in the spectra. They found that the K-edge of hexagonal CrH is shifted 3.8 eV higher in energy than the K-edge of cubic Cr. As compared with shifts of 7–20 eV in chemical compounds of chromium, they conclude that the binding of hydrogen in "hexagonal chromium" is loose. This shift can be due both to the increase of the chromium Fermi level corresponding to the filling up of the chromium d-states and to the increased binding of the chromium K-shell due to transfer of charge to the hydrogen.

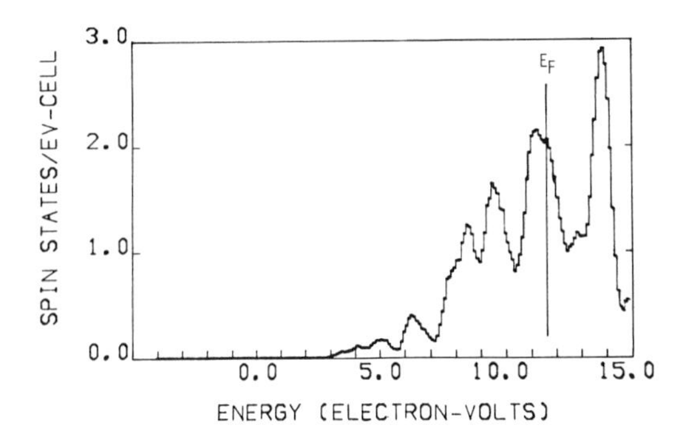

Fig. 5.13. Density of states for hexagonal chromium using the same chromium potential as in CrH

Table 5.4. Density of states and derived quantities compared with experiment for Cr–Mn–H systems

	$N(E)$ [electrons/eV-cell]	γ_{calc} [mJ K^{-2} mol^{-1}]	γ_{expt} [mJ K^{-2} mol^{-1}]	$1+\lambda_\gamma$	$\chi_{\text{P calc}}$ [10^{-6} emu mol^{-1}]	χ_{expt} [10^{-6} emu mol^{-1}]
Cr (bcc)	0.96[a]	2.27	2.90[b]	1.28	31.1	182[c]
Cr (hcp)	4.10[d]	4.83	—	—	66.4	—
CrH	2.86[d]	3.37	4.5[e]	1.34	46.3	265[c]
MnH	1.82[d]	2.14	—	—	—	—

[a] Ref. [5.85]. [b] Ref. [5.86]. [e] Ref. [5.81]. [c] Ref. [5.83]. [d] This work.

5.5.5 Theoretical Calculation for CrH and MnH Systems

The low electronic specific heat of chromium metal is accounted for by the low density of states in the body centered cubic structures for six electrons per atom and is clearly shown in Fig. 5.11. Although this figure was calculated for vanadium, direct calculation for chromium [5.85] shows that rigid band behavior is justified. Again concentrating on change, we first display the results of calculations for hexagonal chromium using the same potential for chromium as in CrH but with no hydrogens in the unit cell in Fig. 5.13. The similarity to Fig. 5.6 for hexagonal titanium is apparent; however, the Fermi energy is higher because of the two extra electrons per atom. The density of states is quite high as Table 5.4 indicates. In fact the density of states for hexagonal Cr is more than twice that of body centered cubic Cr on a per chromium atom basis.

Detailed analysis of the band structure calculations which led to Fig. 5.13 shows that the states which have *s*-like symmetry about the hydrogens are already filled in hcp Cr. The added electrons will be added at the Fermi energy, leading to a decrease in the density of states at the Fermi energy. Although one might expect rigid bandlike behavior at the Fermi energy, this is not true everywhere as Fig. 5.14 shows for CrH. The last three peaks are largely unaffected but the lowest energy peak becomes separated from the main structure. These states again are largely associated with the hydrogens although they contain sizeable *d*, *p*, and *s* contributions from the chromiums. The shift in

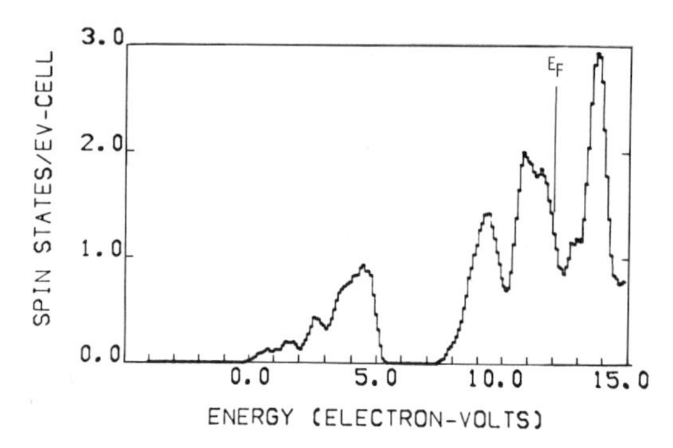

Fig. 5.14. Density of states for hexagonal chromium hydride in the nickel arsenide structure

Fermi level on hydrogen addition is 0.5 eV. The involvement of chromium electrons in bonding with the hydrogen should be counteracted by addition of *d*-like electrons at the Fermi energy. Without detailed self-consistent calculations the 3 eV shift of the *K*-absorption does not appear accessible to our model.

The specific heat results can be understood on the basis of Figs. 5.13 and 5.14 and Table 5.4. Agreement with the specific heat results can be obtained with an enhancement $(1+\lambda)$ of 1.34. This is comparable to the value for bcc chromium and can be credited as good agreement. Direct calculation and rigid band predictions place the Fermi energy at the minimum and should yield a correspondingly lower specific heat coefficient and susceptibility for MnH.

The susceptibility increase cannot be accounted for by the spin paramagnetic contribution. Calculations of the orbital susceptibility for body centered cubic chromium by *Mori* [5.87] show over 75% of the susceptibility is due to orbital contributions. With the Fermi energy falling between two-*d*-band peaks in CrH separated by less than 3 eV which is comparable to the situation for bcc metals shown in Fig. 5.11, a similar orbital contribution might be expected to be present for CrH.

5.6 Summary

In this chapter we have presented the results of the application of modern energy band theory and its calculational implementation to the study of the incorporation of hydrogen into a variety of metal lattices. This incorporation may or may not result in a change in the host metal structure. The change in the host metal structure may induce changes in the electronic structure which are as great as or greater than the effect of hydrogen incorporation, and one should study the influence of the hydrogen in this context. In all systems discussed for concentrated hydrogen $0.5 \leqq x \leqq 3.0$ there is evidence for strong hydrogen influence on the metal electronic structure in that states having *s*-like character about the hydrogen are lowered 3–5 eV. This lowering is due to the attractive potential at the hydrogen site and interactions between the hydrogens and the metal atoms and each other. The position and number of lowered states have been correlated with the stability of hydride structures and heats of formation [5.14, 31]. These lowered states have been seen experimentally by a variety of spectroscopic techniques. To the extent that these states were already filled by electrons from the host metal, the electrons from the hydrogen must be accommodated in the empty states at the Fermi energy which are characteristic of the host *d*-metal. This results in 0.5 (PdH) $\sim$ 0.4 (TiH_2), $>$0.5 (VH), and 1.0 (CrH) electrons/hydrogen being added at the Fermi energy. Thus aspects of both simplistic pictures—1) *Proton Model:* Electrons added at the Fermi level and 2) *Anion Model:* Low-lying states associated with electronic charge in the vicinity of the hydrogen—are common to all hydride systems. For dilute hydrogen in metals where hydrogen-hydrogen interactions vanish, one could still expect charge to pile up around the proton but with no significant lowering of any states and consequently the incorporation of the hydrogen electron in states characteristic of the host metal at the Fermi energy.

As far as giving an overall picture of the electronic properties, electronic specific heat coefficient, susceptibility, and superconductivity, this theory is on a par—no better, no worse—than its application to the simple transition metal host, giving reasonable results for the electronic specific heat coefficient; not being strictly applicable to an estimation of the total magnetic susceptibility; and capable of elucidating the terms important to the occurrence of superconductivity.

Further extensions such as self-consistency and different treatments of exchange and correlation effects are of course desirable to enable one to refine the picture of these systems, but the above results are not expected to change except in detail. Now after over twenty years of the successful application of band theory to the elemental semiconductors, one has been able to gain a full and fairly complete picture of these materials and to reduce the physical picture to a simple interaction between the atomic orbitals in the unit cell. Such a picture will eventually evolve for the hydrogen systems of this chapter and we have attempted as in Figs. 5.3 and 5.9, as well as in the text, to indicate this viewpoint. Much more needs to be treated in detail, but the basic physical picture has advanced significantly in the ten years since the last comprehensive review of this subject [5.1] as indicated by conference proceedings on this subject [5.4, 5] and this volume itself.

Acknowledgement. It is a pleasure to acknowledge the contributions made by the following people which enabled this chapter to be prepared: *Lesley* and *Suzanne Switendick*, *Shelia Guynes*, *Dana Spoeneman*, and *Maria Bargsten.*

References

5.1 W.M. Mueller, J.P. Blackledge, G.G. Libowitz: *Metal Hydrides* (Academic Press, New York-London 1968)
5.2 F.A. Lewis: *The Palladium Hydrogen System* (Academic Press, New York-London 1967)
5.3 J. Callaway: *Energy Band Theory* (Academic Press, New York-London 1964)

5.4 J. Less-Common Metals **49**, 1/2 (1976); *Conference Proceedings Hydrogen in Metals*, Jan. 5–6, 1976 Birmingham, UK
5.5 Ber. Bunsenges. Physik. Chem. **76**, 705–863 (1972)
5.6 J. Butterworth: Proc. Phys. Soc. **83**, 71 (1964)
5.7 R. Eibler, A. Neckel: Monatsschr. Chemie **106**, 577 (1975)
5.8 W. L. McMillan: Phys. Rev. **167**, 331 (1968)
5.9 G. D. Gaspari, B. L. Gyorffy: Phys. Rev. Lett. **28**, 801 (1972)
5.10 F. Herman, S. Skillman: *Atomic Structure Calculations* (Prentice Hall, Englewood Cliffs, New Jersey 1963)
5.11 L. F. Mattheiss: Phys. Rev. **133**, A1399 (1964)
5.12 L. F. Mattheiss, J. H. Wood, A. C. Switendick: "The Augmented Plane Wave Method", in *Methods in Computational Physics*, Vol. 8, ed. by B. Alder, S. Fernbach, M. Rotenberg (Academic Press, New York 1968) pp. 63–147
5.13 A. C. Switendick: Solid State Commun. **8**, 1463 (1970); **8**, XXXIV (erratum) (1970)
5.14 A. C. Switendick: Int. J. Quantum Chem. **5**, 459 (1971)
5.15 D. E. Eastman, J. K. Cashion, A. C. Switendick: Phys. Rev. Lett. **27**, 35 (1971)
5.16 A. C. Switendick: Ber. Bunsenges. Physik. Chemie **76**, 535 (1972)
5.17 A. C. Switendick: "Hydrogen in Metals—A New Theoretical Model", in *Hydrogen Energy, Part B*, ed. by T. N. Veziroğlu (Plenum Press, New York 1975) pp. 1029–1042
5.18 A. C. Switendick: J. Less-Common Metals **49**, 283 (1976)
5.19 M. Zimmerman, G. Wolf, K. Bohmhammel: Phys. Stat. Sol. (a) **31**, 511 (1975)
5.20 H. F. Biggs: Phil. Mag. **32**, 131 (1916)
5.21 H. C. Jamieson, F. D. Manchester: J. Phys. F: Metal Phys. **2**, 323 (1972)
5.22 N. F. Mott, H. Jones: *The Theory of the Properties of Metals and Alloys* (Dover Publications, New York 1958) pp. 198–200
5.23 J. J. Vuillemin, M. G. Priestly: Phys. Rev. Lett. **14**, 307 (1965)
5.24 F. M. Mueller, A. J. Freeman, J. O. Dimmock, A. M. Furdyna: Phys. Rev. B **1**, 4617 (1970)
5.25 F. Antonangeli, A. Balzarotti, A. Bianconi, E. Burattini, P. Perfetti, N. Nistico: Phys. Lett. **55** A, 309 (1975)
5.26 E. Gilberg: Extended abstracts, International Conference on the Physics of X-Ray Spectra, Aug. 30–Sept. 2. National Bureau of Standards, Gaithersburg, Maryland 1976
5.27 C. A. Mackliet, D. J. Gillespie, A. I. Schindler: J. Phys. Chem. Solids **37**, 379 (1976)
5.28 J. E. Schirber: Phys. Lett. **46** A, 285 (1973)
5.29 T. Skośkiewicz, A. W. Szafrański, W. Bujnowski, B. Baranowski: J. Phys. C: Solid State Phys. **7**, 2670 (1974)
5.30 G. Alefeld, J. Völkl: *Hydrogen in Metals II, Application-Oriented Properties*. Topics in Applied Physics, Vol. 29 (Springer, Berlin, Heidelberg, New York 1978) to be published
5.31 C. D. Gelatt, Jr., J. A. Weiss, H. Ehrenreich: Solid State Commun. **17**, 663 (1975)
5.32 D. A. Papaconstantopoulos, B. M. Klein: Phys. Rev. Lett. **35**, 110 (1975)
5.33 J. Zbasnik, M. Mahnig: Z. Physik B **23**, 15 (1976)
J. S. Faulkner: Phys. Rev. B **13**, 2391 (1976)
5.34 B. Lengeler, W. R. Wampler: In Proceedings of 13th International Low Temp. Phys. Conf., Aug. 20–26 1972, Boulder, Colorado
D. Dye, D. H. Lowndes, B. Lengeler: Phys. Stat. Sol. (b) **60**, 399 (1973)
5.35 J. Jacobs, R. Griessen, F. D. Manchester, Y. de Ribaupierre: Bull. Am. Phys. Soc. **21**, 409 (1976)
5.36 U. Mizutani, T. B. Massalski, J. Bevk: J. Phys. F: Metal Phys. **6**, 1 (1976)
5.37 J. M. Rowe, J. J. Rush, H. G. Smith, M. Mostoller, H. E. Flotow: Phys. Rev. Lett. **33**, 1297 (1974)
5.38 A. C. Switendick: Unpublished
5.39 A. C. Switendick: Bull. Am. Phys. Soc. **20**, 420 (1972)
5.40 M. L. H. Wise, J. P. G. Farr, I. R. Harris: J. Less-Common Metals **41**, 115 (1975)
5.41 B. Baranowski, S. Majchrzak, T. B. Flanagan: J. Phys. F: Metal Phys. **1**, 258 (1971)
5.42 M. Mahnig, E. Wicke: Z. Naturforsch. **24**a, 1258 (1969)
5.43 A. Obermann, W. Wanzl, M. Mahnig, E. Wicke: J. Less-Common Metals **49**, 75 (1976)
5.44 T. B. Flanagan, S. Majchrzak, B. Baranowski: Phil. Mag. **25**, 257 (1972)
5.45 C. A. Beckman, W. E. Wallace, R. S. Craig: Phil. Mag. **27**, 1249 (1973)

5.46 P. Merker, G. Wolf, B. Baranowski: Phys. Stat. Sol. (a) **26**, 167 (1974)
5.47 A. D. McQuillan: Proc. Roy. Soc. (London) A **204**, 309 (1950)
5.48 H. Nagel, R. S. Perkins: Zeitschr. Metallkunde **66**, 362 (1975)
5.49 F. Ducastelle, R. Caudron, P. Costa: J. de Phys. **31**, 57 (1970)
5.50 H. Nagel, H. Goretzki: J. Phys. Chem. Solids **36**, 431 (1975)
5.51 K. Gesi, Y. Takagi, T. Takeuchi, S. Noguchi: "Nuclear Metallurgy", in *International Symposium on Compounds of Interact in Nuclear Reactor Technology*, Aug. 3–5, 1964, Boulder Colorado. Vol. X, p. 45, AIME, New York
5.52 W. Trzebiatowski, B. Stalinski: Bull. Acad. Pol. Sci. Class III **1**, 131 (1953)
5.53 S. A. Nemnonov, K. M. Kolobova: Fiz. Metal. Metalloved. **22**, 680 (1966)
5.54 D. Fischer: Private communication
5.55 D. E. Eastman: Solid State Commun. **10**, 933 (1972)
5.56 O. Jepsen: Phys. Rev. B **12**, 2988 (1975)
5.57 J. G. Daunt: *Progress in Low Temperature Physics* (North-Holland, Amsterdam 1955)
5.58 B. W. Roberts: Phys. Rev. **100**, 1257 (1955)
5.59 A. Yu. Chervyakov, I. R. Éntin, V. A. Somenkov, S. Sh. Shil'stein, A. A. Cherktov: Soviet Phys.—Solid State **13**, 2172 (1972)
5.60 V. A. Somenkov, I. R. Éntin, A. Yu. Chervyakov, S. Sh. Shil'stein, A. A. Cherktov: Soviet Phys.—Solid State **13**, 2178 (1972)
5.61 V. A. Somenkov: Ber. Bunsenges. Phys. Chem. **76**, 733 (1972)
5.62 D. G. Westlake, M. H. Mueller, H. W. Knott: J. Appl. Cryst. **6**, 206 (1973)
5.63 R. L. Zanowick, W. E. Wallace: J. Chem. Phys. **36**, 2059 (1962)
5.64 H. Asano, M. Hirabayashi: Phys. Stat. Sol. (a) **15**, 267 (1973)
5.65 H. Asano, Y. Abe, M. Hirabayashi: J. Phys. Soc. (Japan) **41**, 974 (1976)
5.66 D. Rohy, R. M. Cotts: Phys. Rev. B **1**, 2484 (1970)
5.67 D. Rohy, R. M. Cotts: Phys. Rev. B **1**, 2070 (1970)
5.68 W. Trzebiatowski, B. Stalinski: Bull. Acad. Polan. Sci. Class III **1**, 317 (1953)
5.69 B. Stalinski: Bull. Acad. Polan. Sci. Class III **2**, 245 (1954)
5.70 S. Aronson, J. J. Reilly, R. H. Wiswall, Jr.: J. Less-Common Metals **21**, 439 (1970)
5.71 D. G. Westlake, S. T. Ockers: Phys. Rev. Lett. **25**, 1618 (1970)
5.72 E. W. Collings, J. C. Ho: Phys. Rev. B **2**, 235 (1970)
5.73 I. A. Brytov, E. Z. Kurmaev, K. I. Konashenok, M. M. Antonova: Izvestiya Akademii Nauk SSSR—Neorganicheskie Materialy (Trans.) **9**, 137 (1973)
5.74 Y. Fukai, S. Kazama, K. Tanaka, M. Matsumoto: Solid State Commun. **19**, 507 (1976)
5.75 E. Gilberg: Phys. Stat. Sol. (b) **69**, 477 (1975)
5.76 D. A. Papaconstantopoulos, J. R. Anderson, J. W. McCaffrey: Phys. Rev. B **5**, 1214 (1974)
5.77 C. A. Shavely, D. A. Vaughan: J. Am. Chem. Soc. **71**, 313 (1949)
5.78 J. T. Waber: Private communication
5.79 B. Baranowski: Ber. Bunsenges. Physik. Chem. **76**, 714 (1972)
5.80 M. Krukowski, B. Baranowski: J. Less-Common Metals **49**, 385 (1976)
5.81 R. Viswanathan, H. R. Khan, A. Knoedler, Ch. J. Raub: J. Appl. Phys. **46**, 4088 (1975)
5.82 G. Albrecht, G. Wolf: Phys. Stat. Sol. **18**, K 119 (1966)
5.83 M. Hanson, H. R. Khan, A. Knödler, Ch. J. Raub: J. Less-Common Metals **43**, 93 (1975)
5.84 von O. Brummer, G. Dräger, H. Baum: Z. Naturforsch. **18**a, 1102 (1963)
5.85 M. Yasui, E. Hayashi, M. Shimizu: J. Phys. Soc. (Japan) **29**, 1446 (1970)
5.86 F. Heiniger, E. Bucher, J. Muller: Phys. Kondens. Mater. **5**, 243 (1966)
5.87 N. Mori: J. Phys. Soc. (Japan) **20**, 1383 (1965)

6. Mössbauer Studies of Metal-Hydrogen Systems

F. E. Wagner and G. Wortmann

With 13 Figures

6.1 Background

The possibility of applying the Mössbauer method in studies of metal-hydrogen systems was recognized more than a decade ago [6.1, 2]. By now, a variety of applications has proven the usefulness of Mössbauer spectroscopy in this field. The probes used by Mössbauer spectroscopy to obtain information on solid-state properties are nuclei with low-energy transitions to the ground state. For γ ray energies below about 150 keV, there is a reasonable probability for emission and absorption without recoil energy loss if the nuclei are bound in a solid. Since such emission and absorption lines are unaffected by thermal broadening, it becomes possible to observe the natural linewidth and to scan the resonance absorption spectra by means of a small Doppler shift produced by moving either the source or the absorber with velocities of the order of a few $\mathrm{cm\,s^{-1}}$.

The most valuable information gained from Mössbauer spectra is primarily on the hyperfine interactions between the probe nuclei and their electronic environment. Three different types of interactions can be observed. Two of these, namely, the interaction between the nuclear magnetic dipole moment and the magnetic hyperfine field produced by the electrons in magnetic materials, and the interaction between the nuclear quadrupole moment and the electric field gradient arising from a noncubic environment, result in a splitting of the nuclear energy levels, and hence of the γ ray emission and absorption patterns. The information thus obtained corresponds to that from NMR, NQR, perturbed $\gamma-\gamma$ angular correlation, or nuclear orientation experiments. The third type of hyperfine interaction, the isomer shift, is specific to Mössbauer spectroscopy. It manifests itself as a shift of the Mössbauer pattern and arises from the electrostatic monopole interaction between the finite volume of the nuclear charge distribution and the electron density inside the nucleus. Causing only a shift but no splitting of the nuclear levels, this interaction can be studied by Mössbauer spectroscopy, but remains unseen in the other methods in use for hyperfine interaction studies in solids. The width of the Mössbauer line is another parameter that often bears valuable information on unresolved hyperfine interactions, diffusion, or relaxation processes. Finally, the Mössbauer-Lamb f-factor, i.e., the fraction of γ rays that are emitted or absorbed without loss or gain of energy through phonon interactions, can be obtained from Mössbauer spectra and give insight into the dynamic properties of the lattice.

In virtually all its practicable applications, Mössbauer spectroscopy is a strictly microscopic method that senses only the local environment of the Mössbauer nuclei but is insensitive to long-range correlations. If different Mössbauer nuclei have different local environments, as would usually be the case in nonstoichiometric hydrides, the Mössbauer spectrum will be a mere superposition of the spectra for the individual sites. Fluctuating hyperfine interactions result in more complicated Mössbauer patterns, whose interpretation may become rather difficult. In metal-hydrogen systems the diffusion of the interstitial hydrogen is an important source of such fluctuations. Less specific fluctuation phenomena may arise, for instance, from paramagnetic relaxation or superparamagnetism.

The parameters obtained from Mössbauer spectra of metal-hydrogen systems have yielded information on phase diagrams, on magnetic properties, on the electronic structure as seen by the isomer shift, on the diffusion rates of interstitial hydrogen, on the local hydrogen environments of substitutional solute atoms in nonstoichiometric metal hydrides, and on changes in the vibrational spectrum ensuing from hydrogenation.

A general introduction to Mössbauer spectroscopy would exceed the scope of the present chapter and appears superfluous in view of the numerous texts on this subject at large and on major aspects of it [6.3–9]. Moreover, a complete bibliography of the Mössbauer literature is being carried on by *Stevens* and *Stevens* [6.10]. In the following we shall give a concise introduction only to those aspects of Mössbauer spectroscopy that seem particularly relevant in the present context. Similarly, the properties of metal hydrides will be discussed only to the extent that seems indispensable for the understanding of the Mössbauer work. In the second half of the chapter we shall then review the Mössbauer studies performed so far on metal-hydrogen systems. Finally, an outlook on future perspectives will be attempted.

6.2 Some Relevant Properties of Metal-Hydrogen Systems

6.2.1 Hydride Phases of the *d*-Transition Metals

Exothermal hydride formation is typical for the Group IIIa, IVa, and Va transition elements (Fig. 6.1) and for Pd. Ni and some alloys, like stainless steels, can be hydrogenated with some difficulty. Certain intermetallic compounds also readily take up large amounts of hydrogen, well-known examples being those of the RNi_5 and RCo_5 type, where R stands for a rare earth element (see Sect. 6.5).

The dissolved hydrogen usually occupies interstitial sites in the metal lattice. Interstices of different symmetry are available in most matrices, but

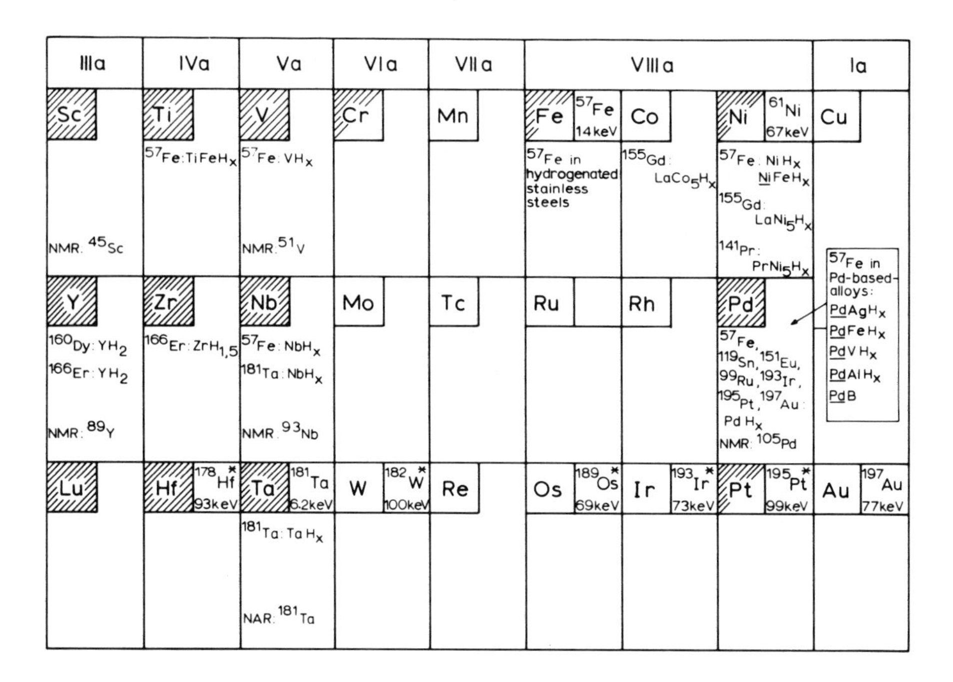

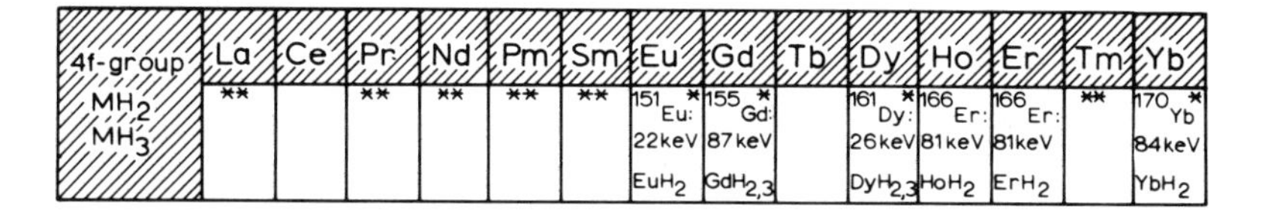

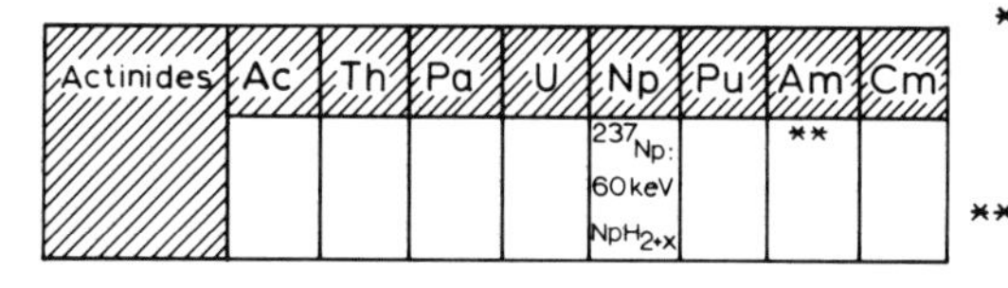

* Other Mössbauer isotopes or transitions suitable for hyperfine interactions exist for this element (see, e.g., ref. [10]).

** Suitable Mössbauer resonances exist (see, e.g., ref. [10]).

Fig. 6.1. Periodic table covering the *d*-transition elements, rare earths and actinides. Shaded squares designate the exothermal formation of hydrides. Half-shaded squares indicate that some hydrogen can be dissolved either in the pure element or in alloys thereof. For the *d* elements, the isotopes and γ ray energies of Mössbauer resonances that are suitable for hyperfine interaction studies are given besides the element symbols. For the rare earths and actinides, only those resonances that have actually been used for studying hydrides are included. In the space below the symbols for the individual elements the hydride systems studied so far and the Mössbauer isotopes used for this are summarized. Finally, some isotopes suitable for NMR and NAR experiments in hydride systems are listed

often only specific sites can be occupied. Of the metals that have attracted specific interest from Mössbauer spectroscopists, the fcc lattices of Pd and Ni accommodate the hydrogen in octahedral sites, whereas in bcc Nb and Ta the hydrogen occupies tetrahedral interstices. In bcc V, the solute hydrogen may occupy both tetrahedral and octahedral sites [6.11]. The question which type of conceivable hydrogen location is actually populated in individual hosts can be answered by neutron diffraction or channeling techniques. In principle, the hyperfine spectra as observed in Mössbauer spectroscopy can also be expected to be sensitive to the site occupancy. Difficulties in the quantitative interpretation of the hyperfine parameters, e.g., the isomer shift, have so far precluded such applications for the *d*-transition metals. Mössbauer spectra of rare earth hydrides have, however, been interpreted in terms of the hydrogen location (see Sects. 6.2.2 and 6.5.1).

In nonstoichiometric hydride phases, the hydrogen is highly mobile, owing the diffusion via empty interstitial sites with jump rates as high as $10^{12}\,s^{-1}$ at room temperature [6.12]. Being strongly temperature dependent, the time between jumps becomes very long—of the order of hours or more—at cryogenic temperatures, where quantum diffusion rather than thermally stimulated processes may become important. The way in which such diffusion processes affect the Mössbauer spectra, and how Mössbauer spectroscopy can be used to obtain information on them, will be discussed in Section 6.3.7.

Hydrogen in transition metals often gives rise to phase diagrams that resemble those of single-component gas-liquid-solid systems. Figure 6.2 shows, in a somewhat simplified manner, the phase diagrams for three cases that have been studied by Mössbauer spectroscopy, as will be discussed in detail in Section 6.4. Hydrogen in Nb is a good example for the gas-liquid-solid analogy. In the α phase of such systems the dissolved hydrogen can be considered as a lattice gas. The gas-liquid phase transition to the α' phase goes along with an expansion of the host lattice and the onset of short-range order among the dissolved hydrogen. Metal hydride phases corresponding to the solid state are generally characterized by a change of the symmetry of the host lattice and by long-range order of the hydrogen atoms. The existence of more than one phase at certain hydrogen concentrations and temperatures will result in a superposition of different hyperfine patterns in the Mössbauer spectrum. It is important to keep this in mind in the interpretation of the data even when the contributions from the individual phases are not clearly resolved, as is often the case. While Mössbauer spectroscopy is insensitive to the long-range nature of the order occurring in the β phases of transition metal hydrides, the short-range effects of the ordering phenomena in both α' and β phases may affect the hyperfine parameters, since they influence the hydrogen distribution around the Mössbauer atoms. So far, however, practically no results in this direction have been obtained.

The electronic properties of metal-hydrogen systems like Pd–H have, for a long time and with considerable success, been described in the framework of the rigid band model that assumes that each dissolved hydrogen atom donates one

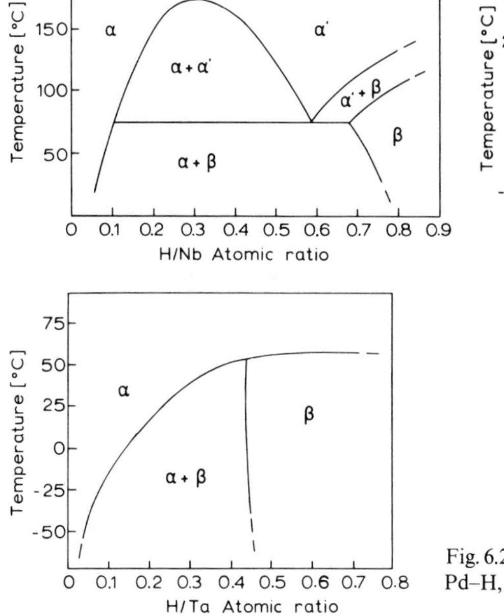

Fig. 6.2. Schematic phase diagrams for the Nb–H, Pd–H, and Ta–H systems (adapted from [6.12])

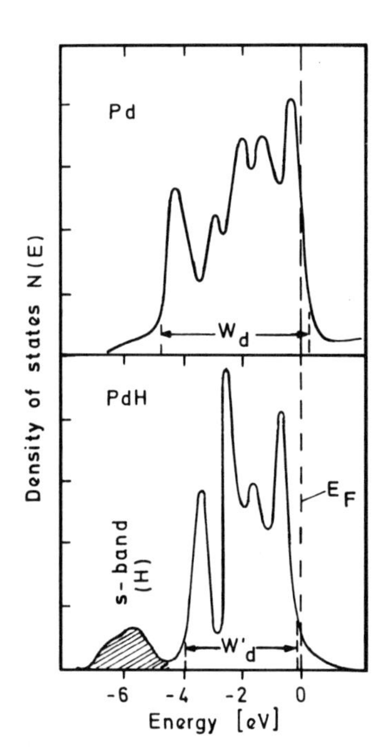

Fig. 6.3. Density of states curves for Pd and PdH (after [6.15])

electron into the otherwise unaltered conduction band of the host. This model explains, at least qualitatively, for instance, the transition from strong paramagnetism to diamagnetism in the PdH_n system at $n \approx 0.6$. De Haas-van Alphen experiments [6.13] and band structure calculations [6.14] showing that there are only 0.36 holes in the *d*-band of pure Pd have, however, revealed its shortcomings. By now, UPS and XPS measurements [6.15, 16] as well as band structure calculations [6.15, 17, 18] for the Pd–H system have established the more consistent picture that the hydrides are characterized by a high density of $s-p$ band states below the Pd *d*-band (Fig. 6.3). It is obvious that the band structure of PdH_n, and of metal hydrides in general, directly affects the Mössbauer parameters, notably the isomer shift (see Sect. 6.3.3) and, for instance in **Pd**FeH_n, the magnetic hyperfine interactions. With more accurate approaches still in their infancy, the rigid band model is often quite useful at least for a qualitative understanding of experimental results. Moreover, it gave rise to the idea that alloying with electron-donating partners should fill the band in the same way as hydrogenation. Comparisons between hydrides (e.g., PdH_n) and alloys (e.g., $Pd_{1-x}Ag_x$ [6.19–21]) have therefore aroused some interest as tests of the rigid band model, and as ways to reveal its shortcomings. Some such Mössbauer studies will be described in Section 6.4.2.

6.2.2 Hydrides of the Rare Earths

The rare earths are the second major group of elements whose hydrides have been studied by Mössbauer spectroscopy. Rare earth hydrides form readily when the metals are exposed to hydrogen at elevated temperatures (600–800 °C). Schematic phase diagrams for the hydrides of both the lighter and the heavier rare earths are shown in Fig. 6.4. The α phase solid solutions of hydrogen in the hcp lattice of the rare earth metals are unstable at ambient temperature. The RH_2 phase has the CaF_2-type fcc structure. Nonstoichiometric RH_{2+x} hydrides having this lattice exist over a wide range of compositions. For $x<0$, part of the regular tetrahedral hydrogen sites (see Fig. 6.4) remain empty. For $x>0$, the excess hydrogen goes to the octahedral interstices. For the lighter rare earths (La, Ce, Pr, Nd), the RH_{2+x} fcc phase persists up to the composition RH_3, in which all octahedral interstices are filled. The heavier rare earths form a separate hexagonal RH_{3-x} phase. The divalent rare earth metals are exceptional in forming orthorhombic dihydrides [6.22].

For the interpretation of Mössbauer spectra of rare earth isotopes in such hydride lattices it is important whether the protonic or the anionic (hydridic) model applies, i.e., whether the hydrogen has to be considered as positively or negatively charged. This is so because the hyperfine interactions in rare earths are determined by the splitting of the electronic ground state of the free ion in the crystalline electric field which, in its turn, depends on the charges on the surrounding atoms. Experimental evidence, including the Mössbauer experi-

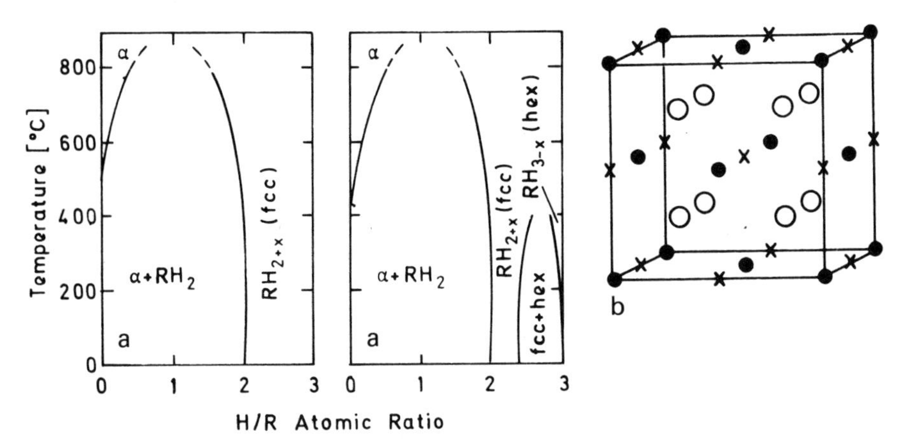

Fig. 6.4. (a) Schematic phase diagrams of the hydrides of the lighter (left) and heavier (right) rare earths. (b) Unit cell of the fcc fluorite-type rare earth hydrides. Full circles represent the rare earth atoms, open circles the regular (tetrahedral) hydrogen sites, and crosses the octahedral hydrogen sites that are populated in the RH_{2+x} phases with $x>2$ (after [6.22])

ments discussed in Section 6.5, favors the hydridic model. The crystalline field ground states expected in both the hydridic and the protonic model for the RE^{3+} ions in fcc REH_2 hydrides can readily be calculated [6.23] and have been compiled by *Shenoy* et al. [6.24].

In nonstoichiometric hydrides, the individual rare earth atoms will have different hydrogen surroundings. This complicates the interpretation of the Mössbauer spectra, particularly because the crystalline electric field can no longer be considered as cubic. On the other hand, it means that the Mössbauer spectra contain information, for instance on short-range order effects. At low temperatures the environment can be considered as static, but at higher temperatures jump diffusion of the hydrogen becomes sufficiently fast [6.25] to cause rapidly fluctuating crystalline electric fields and electric quadrupole interactions, and hence relaxation phenomena in the Mössbauer spectra (see Sect. 6.3.7).

6.3 Mössbauer Spectroscopy on Metal-Hydrogen Systems

6.3.1 Mössbauer Resonances for the Investigation of Metal Hydrides

The usefulness of a Mössbauer resonance depends on the resolution for hyperfine interactions and the ease of observation. In this sense, only one of the *d*-transition elements that readily form hydrides, namely Ta, has a suitable Mössbauer transition (Fig. 6.1). Hf has several Mössbauer resonances [6.10], but these are all virtually useless for isomer shift measurements because the observable shifts are only a minute fraction of the natural linewidth [6.26]. For the same reason ^{61}Ni is of limited value. With this isotope, problems arise also from the necessity to correct for a relatively large second-order Doppler shift, and from the short half-life of the source isotopes [6.27]. ^{195}Pt would, from the Mössbauer point of view, be a reasonably good candidate, but the capability of Pt metal to take up hydrogen is very limited. Thus the 6.2 keV resonance in ^{181}Ta is the only resonance suitable for Mössbauer studies of "pure" transition metal hydride systems. This resonance has very good resolution for all types of hyperfine interactions. The major difficulty in its use is actually the inhomogeneous line broadening arising from this high sensitivity [6.28, 29]. Results obtained with it will be discussed in Sections 6.4.5 and 6.4.6.

Transition metal hydrides other than the Ta–H system can be studied by Mössbauer spectroscopy only when an appropriate Mössbauer isotope is introduced into their lattice as an impurity. Then, however, the hydrogen distribution around the Mössbauer atom will usually be different from the environment of a host atom because the impurity will attract or repel hydrogen in a way that is not a priori known in most cases. This is a serious complication for the interpretation of the Mössbauer data, but on the other hand it opens a way to study the local environment of substitutional solute atoms in metal hydrides.

For such studies, any Mössbauer isotope with sufficient resolution for hyperfine interactions can be used unless there are insurmountable metallurgic problems. So far the majority of experiments have, of course, been performed with the 14.4 keV resonance of ^{57}Fe. Other transition elements with suitable resonances for such work are ^{99}Ru (90 keV), ^{191,193}Ir (82 keV; 73 keV), ^{197}Au (77 keV), and with lower resolution, ^{189}Os (69 keV; 36 keV) and ^{195}Pt (99 keV, 130 keV). The impurity Mössbauer isotopes need not themselves be transition elements. For instance, ^{119}Sn (23 keV) and ^{151}Eu (21 keV) have been used for studies of the Pd–H system. The work actually performed so far will be reviewed in Sections 6.4.1–6.4.4. Data, experimental details, and references on other Mössbauer isotopes can be found in [6.4, 10, 30].

As far as the availability of Mössbauer isotopes is concerned, the hydrides of the rare earths can readily be studied owing to the large number of resonances available in this region of the periodic table. In Fig. 6.1, only the Mössbauer isotopes actually used so far have been included for the sake of clarity. Detailed information on these and all others can be found in the literature (e.g., [6.4, 10, 30, 31]).

6.3.2 Experimental Techniques

General reviews of the experimental techniques of Mössbauer spectroscopy can, for instance, be found in [6.3–9, 32]. Here we shall only mention a few specific problems connected with the use of metal hydrides as Mössbauer sources or absorbers.

Usually one will first prepare an alloy of the Mössbauer source or absorber isotope with the host metal and then convert this to the hydride. In absorbers, a few percent of the Mössbauer element are usually required. The use of enriched isotopes may help to keep the impurity concentrations low. Thus about 0.1 atomic percent of highly enriched ^{57}Fe in Pd will give an efficient Mössbauer absorber. When the hydrides are studied as sources, the impurity concentration may often be kept much lower if the radioactive parent is available with high specific activity or, preferably, carrier-free. Sources of ^{57}Co in Pd for ^{57}Fe Mössbauer spectroscopy, for instance, have been made with ^{57}Co concentrations down to 0.6 ppm [6.33]. Whatever the impurity concentration, clustering of the Mössbauer atoms may be a problem in certain systems and should be paid some attention. Particularly in sources, autoradiographic methods are of great help in checking the distribution of the Mössbauer atoms in the matrix [6.34].

The Mössbauer sources or absorbers are usually loaded with hydrogen electrolytically or from the gas phase. The amount of dissolved hydrogen can be determined gravimetrically, by measuring the uptake from the gas phase, or by observing the evolution of hydrogen from the sample after the experiment. The time necessary to acquire a Mössbauer spectrum usually ranges from several hours to several days. During this time the hydrogen content of the samples must be kept constant despite the tendency of many hydrides to lose hydrogen in the course of time. The easiest way to prevent the release of hydrogen is to cool the sample to liquid nitrogen temperature or below. For Mössbauer transitions with energies above about 30 keV, one must cool the samples anyhow in order to obtain reasonably high *f*-factors. In other cases, e.g., in work with ^{181}Ta, ^{57}Fe, ^{119}Sn, ^{151}Eu or ^{161}Dy, measurements up to room temperature or even above are feasible, and will often be of physical interest. It may then be necessary to take special measures against hydrogen loss, e.g., sealing by deposition of iodine or copper onto the sample surface, maintenance of an appropriate hydrogen pressure, or sustained electrolytic loading during the Mössbauer measurement.

Specific problems may result from the requirement to use very thin samples for measurements with low-energy γ rays. Thus, for 14.4 keV γ rays in Pd, the mean penetration depth for attenuation by photoeffect is only 20 μ. This enforces the use of thin samples and thus aggravates the problems of hydrogen loss. On the other hand, Mössbauer transitions of higher energy like those in ^{99}Ru (90 keV), ^{193}Ir (73 keV), ^{197}Au (77 keV), ^{166}Er (81 keV), or ^{170}Yb (84 keV) often call for absorber thicknesses of the order of 0.5–1 mm. Then the homogeneous hydrogenation may become problematic, for instance in the case of Ni hydride, which forms electrolytically only as a thin surface layer of about 30 μ thickness [6.35].

6.3.3 The Isomer Shift

The isomer shift of the Mössbauer pattern in velocity units is given by [6.3–9]

$$S=(c/E_\gamma)(2\pi/3)\cdot Z\cdot e^2\cdot \Delta\langle r^2\rangle\cdot \Delta\varrho(0)\,, \tag{6.1}$$

where c, Z, e, and E_γ are the velocity of light, the atomic number, the unit charge and the γ ray energy, respectively. The isomer shift thus measures the product of a nuclear parameter, namely, the change $\Delta\langle r^2\rangle$ of the mean square nuclear charge radius accompanying the Mössbauer transition, and an electronic quantity, namely, the difference $\Delta\varrho(0)$ of the electron density at the nucleus between two different materials. The isomer shift must therefore always be given relative to a standard source or absorber material. Since hardly any $\Delta\langle r^2\rangle$ values are presently known to better than 30%, and many much worse [6.9], the $\Delta\varrho(0)$ values extracted from isomer shift measurements are also not very accurate.

The electron density $\varrho(0)$ at the nuclear site arises from s and, to a lesser extent, from relativistic $p_{1/2}$ electrons. Its changes $\Delta\varrho(0)$ due to different chemical environments are of the order of $10^{-4}\varrho(0)$ or even less. They arise either directly from changes of the density of valence shell s electrons or in an indirect way via the shielding effect of valence shell nd electrons or of $4f$ electrons on the s electrons of the core, a higher nd or $4f$ population inducing a decrease of $\varrho(0)$.

As a rule, one finds that $\varrho(0)$ is smaller in metal hydrides than in the hydrogen-free metal or alloy. The question arises, how much of this reduction is attributable to the volume expansion going along with hydrogenation? To answer this question, one may, at least for small changes of the hydrogen/metal ratio n, use the phenomenological equation

$$\frac{dS}{dn}=\left(\frac{\partial S}{\partial \ln V}\right)_n\cdot\frac{d\ln V}{dn}+\left(\frac{\partial S}{\partial n}\right)_V, \tag{6.2}$$

where V is the macroscopic volume. Since the volume coefficient of the isomer shift, $(\partial S/\partial \ln V)_n$, can be obtained from Mössbauer experiments under high pressure [6.36] and $d\ln V/dn$ from x-ray diffraction data, one can solve (6.2) for $(\partial S/\partial n)_V$. This quantity represents the explicit effect of the interstitial hydrogen on the electronic structure of the Mössbauer atom. It is expected to reflect the filling and distortion of the host conduction band, as well as the local distribution of hydrogen around the Mössbauer atom. This distribution will usually be unknown for Mössbauer atoms that are impurities in a metal-hydrogen matrix, but in certain cases it can be studied by the comparison of source and absorber experiments (see Sect. 6.3.6). The determination of the volume term in (6.2) from pressure isomer shift data and the macroscopic lattice dilatation $d\ln V/dn$ as obtained from x-ray diffraction does not take into account any local lattice distortions in the vicinity of hydrogen interstitials [6.37]. The effect of such distortions on S is thus also included in the explicit term $(\partial S/\partial n)_V$. This manner of separation is mainly prompted by the fact that then the first term in (6.2) can be determined separately with relative ease.

Qualitative and phenomenological interpretations of this type are not utterly satisfactory, but still most interpretations of isomer shifts, particularly in metallic systems, are based on such considerations. More rigorous interpretations based on band structure calculations are still in their infancy. Such approaches have recently been discussed by *Freeman* and *Ellis* [6.38].

6.3.4 Electric Quadrupole Interactions

The electric quadrupole splitting of the Mössbauer line arises from the interaction of the nuclear electric quadrupole moment Q and the electric field gradient produced at the site of the Mössbauer nucleus by the surrounding electronic charges. The electric field gradient tensor is usually described by two scalar quantities, q and η, called the field gradient and asymmetry parameter, respectively. The quadrupole patterns expected for the individual Mössbauer resonances have been discussed, for instance, in [6.3–9]. Notably, the electric field gradient vanishes for cubic point symmetry, and η is zero whenever the site of the Mössbauer atom has an axis of trifold or higher symmetry.

Traditionally, the electric field gradient in metallic systems is rationalized as a sum of a "lattice" contribution and a "local" contribution [6.39–41]

$$q = q_{\text{latt}}(1-\gamma_\infty) + q_{\text{loc}}(1-R), \tag{6.3}$$

where R and γ_∞ are the Sternheimer shielding and anti-shielding factors (see, e.g., [6.42]). The first term in (6.3) actually comprises all contributions from the charge distribution outside the core of the Mössbauer atom, i.e., from the ionic cores in the lattice and the cloud of conduction electrons in between. The second term describes the electric field gradient arising from the nonspherical charge distribution inside the core of the Mössbauer atom. For rare earths and actinides, q_{loc} would thus contain the $4f$ or $5f$ contributions in addition to those from valence shell p and d electrons.

As far as their electric quadrupole interactions are concerned, it is reasonable to distinguish between cubic metal hydrides (e.g., α-TaH_n, α'-PdH_n, RH_{2+x}) and hydrides with a noncubic arrangement of the metal atoms (e.g., EuH_2, YbH_2, hcp–RH_{3-x}, $LaNi_5H_x$). In the latter type of materials, electric quadrupole interactions can often be observed (see Sect. 6.5). In the cubic hydride phases, however, electric field gradients arise only from the more or less randomly distributed interstitial hydrogen, at least if one disregards local distortions of the metal lattice. As a consequence, one expects a distribution of q and η values resulting in complicated Mössbauer patterns.

In practice, it has turned out that the resolution of Mössbauer spectroscopy is insufficient, at least in the cases studied so far, for more than some line broadening to be observed. This shows that the electric field gradients in these hydrides are much smaller than the values calculated on the basis of a simple point charge model, assuming one positive unit charge on the hydrogen in protonic hydrides. This may be due to a strong shielding of the protonic charge or to other cancellation effects. In nonstoichiometric hydrides, the Mössbauer line may, however, also be broadened by a distribution of isomer shifts. Observed line broadenings therefore cannot unambiguously be interpreted in terms of unresolved quadrupole interactions except for Mössbauer resonances that are insensitive to isomer shifts because of their small $\Delta\langle r^2\rangle$ value.

6.3.5 Magnetic Hyperfine Interactions

Magnetic hyperfine interactions can be observed by Mössbauer spectroscopy both in magnetically ordered systems and, under certain conditions, in paramagnetic atoms. In the former case, the magnetic hyperfine interaction can be described by an effective hyperfine field B_{hf} acting on the nuclei and causing a simple Zeeman splitting of the nuclear levels and, consequently, of the Mössbauer line [6.3–8]. Mössbauer spectroscopy can yield information on the magnitude and temperature dependence of B_{hf} and thus on the electronic structure of the system and the magnetic ordering temperature.

Magnetic ordering is frequent in rare earth hydrides at low temperatures [6.24, 43–45], but the pure hydrides of the *d* transition metals are nonmagnetic. Often, however, the addition of small amounts of magnetic impurities, e.g., of Fe to PdH_n, causes magnetic ordering at low temperatures. This can readily be studied by Mössbauer spectroscopy (see Sect. 6.4.3).

Magnetic hyperfine interactions in paramagnetic ions can be observed in Mössbauer spectra either when the relaxation times of the ionic spins are long compared to—or at least comparable with—the typical nuclear precession time, or when the paramagnetic moments are polarized by a sufficiently strong external field at low temperatures. In the latter case, the Mössbauer hyperfine pattern can again be described by an effective hyperfine field B_{hf}. From measurements of B_{hf} as a function of the applied field, one can derive both the magnitude of the magnetic moment of the paramagnetic ion and the saturation value of B_{hf}. Such measurements have, for instance, been performed on ^{57}Fe in Pd [6.33], and similar work on hydrogenated materials would seem very promising.

The Mössbauer spectra observed in the slow-relaxation regime with no or only a very weak external field applied cannot be described in the effective field approximation, and further complications arise if the requirement of slow relaxation of the electronic spin is not rigorously fulfilled [6.46–49]. The requirement of slow relaxation rates calls for a high dilution of the paramagnetic spins to prevent spin-spin relaxation and for low temperatures in order to slow down spin-lattice relaxation and relaxation by exchange with the spins of the conduction electrons. For the 4*f* spins of rare earths, the coupling to the conduction electrons is relatively weak, and liquid helium temperature is often sufficient for the slow-relaxation limit to be reached [6.24, 50–52]. Results on rare earth hydrides will be discussed in Section 6.5.1. For paramagnetic 3*d* impurities in metallic systems, much lower temperatures are usually needed to slow down relaxation. ^{57}Fe in Pd, for instance, shows clearcut zero-field

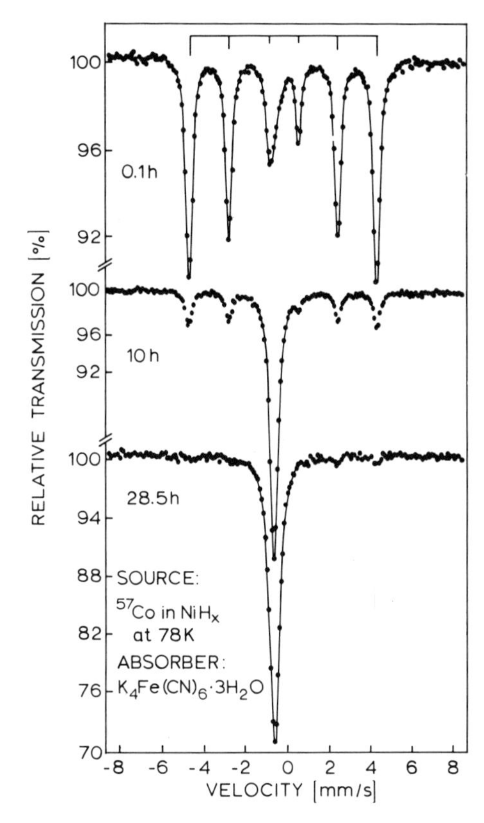

Fig. 6.5. ^{57}Fe Mössbauer emission spectra of a source of ^{57}Co in Ni as a function of the time of electrolytic loading with hydrogen (after [6.2]). The single line from the nonmagnetic $NiH_{0.7}$ phase is clearly resolved from the Zeeman pattern of ferromagnetic Ni. According to the generally accepted convention, we use a positive sign for the Doppler velocity v whenever the source and absorber are moving towards one another. This implies that the sign of an observed isomer shift is different for a source and an absorber experiment. In the discussion of isomer shift data, we always use the sign proper for absorber experiments, however the data have been obtained. It should be noted that in [6.2] that the data are presented in a different sign convention

relaxation Mössbauer spectra only below about 50 mK [6.33]. With the Korringa relation predicting a strong dependence of the relaxation times on the density of states at the Fermi level, such experiments in hydrogenated systems will be an interesting field of future Mössbauer work.

6.3.6 Phase Analysis, Hydrogen Distribution, and Short-Range Order

Coexisting phases in metal-hydrogen systems can be identified by Mössbauer spectroscopy whenever their hyperfine parameters are sufficiently different. A fine example for this is nickel hydride (Fig. 6.5), which is nonmagnetic and therefore exhibits an unsplit ^{57}Fe Mössbauer line, whereas pure Ni is ferromagnetic and gives rise to a well-split Zeeman pattern. Other cases where the isomer shift [6.53, 54] or quadrupole interaction [6.55] are used to distinguish different phases will be discussed in Sections 6.4.4 and 6.5.1.

Quantitative phase analyses will be based on the assumption that the area under each of the superimposed Mössbauer patterns is proportional to the number of Mössbauer atoms in the respective phase. However, several other considerations are important in this context: 1) The *f*-factors in the individual phases will generally be different. This may often be a small effect for ^{57}Fe because of the low γ ray energy, but with other Mössbauer resonances it can become important. For the 77 keV resonance of ^{197}Au, for instance, the *f*-factor at 4.2 K is 20% lower in the β phase of Pd–H than in the α phase [6.56]. 2) In absorber experiments the finite absorber thickness must be taken into account because it may distort the intensity ratios. A straightforward way to evaluate such spectra is to fit transmission integral line shapes rather than a sum of *Lorentzians* [6.57]. 3) Clustering of the Mössbauer impurity atoms may influence the domain distribution in the mixed-phase hydride. As a consequence, the concentration of Mössbauer atoms need not be the same in all phases. 4) Mössbauer spectroscopy is insensitive to the distribution of the phases within the sample. An imperfectly hydrogenated sample may give the same spectrum as an equilibrium mixture of two phases.

It has already been pointed out that the hyperfine interactions, in principle, bear information on the hydrogen distribution around the Mössbauer atom, although an actual interpretation, e.g., of the magnitude of the isomer shift, in such terms may often be difficult. Sometimes the absence of a distribution of hyperfine parameters itself is a valuable piece of information indicating the existence of a well-defined hydrogen environment. In rare earth hydrides, Mössbauer spectra often yield information on the crystalline electric field, and hence on the hydrogen distribution (see Sect. 6.5.1 for examples).

The Mössbauer atom will often be an impurity, whose environment may differ substantially from that of a matrix atom. Moreover, different impurity elements will have different hydrogen environments. As a consequence, emission and absorption spectra for the same Mössbauer resonance and matrix may display different hyperfine parameters because the source isotope—e.g., ^{57}Co for Mössbauer spectroscopy with ^{57}Fe—may favor a hydrogen environment other than the Mössbauer element itself. Immediately after the nuclear decay, the hydrogen environment will therefore be in a nonequilibrium state, and the question arises whether the Mössbauer emission spectrum will be typical for this "old" hydrogen configuration, or whether it will already see the "new" equilibrium typical for the Mössbauer element itself. Here, the decisive parameters are the mean time τ_R needed for the hydrogen distribution to adjust to the new situation, and the mean time elapsing between the nuclear transformation and the emission of the Mössbauer gamma quanta. The latter usually is simply the lifetime τ_n of the excited Mössbauer level. For the limiting cases $\tau_R \gg \tau_n$ and $\tau_R \ll \tau_n$, Mössbauer spectroscopy sees the old and new hydrogen configuration, respectively. For the intermediate case, $\tau_R \approx \tau_n$, one obtains complicated emission spectra that cannot be represented by a superposition of Lorentzian lines. The theory of emission spectra under such nonequilibrium relaxation conditions has, at least for simple cases, been worked out by various authors [6.47, 58–61]. The lifetimes τ_n of Mössbauer levels of practical importance range from a few ns to a few μs. The relaxation time τ_R will be closely related to the characteristic time for single jumps of hydrogen atoms towards or away from the nearest neighborhood of the Mössbauer atom during the transition to equilibrium. Little is known on these, the use of a single relaxation time τ_R may be an oversimplification, and additional complications arise from the fact that the equilibrium hydrogen environment both before and after the nuclear decay need not be unique and static. Therefore, the observation of such nonequilibrium transitions appears as an interesting field of future research.

Differences in isomer shift between source and absorber experiments observed at 4.2 K, e.g., with the ^{197}Au resonance [6.56, 62] (see Sect. 6.4.2 and Fig. 6.8), seem to represent the limit of slow relaxation, $\tau_R \gg \tau_n$. Around room temperature, on the other hand, one expects to observe the fast relaxation limit, $\tau_R \ll \tau_n$, because of the fast jump diffusion then occurring in metal-hydrogen systems like Pd–H [6.12].

6.3.7 Hydrogen Diffusion

The nonequilibrium diffusion described in the previous section is by no means the only way in which the mobility of solute hydrogen affects the Mössbauer spectra. Normal equilibrium diffusion of interstitial hydrogen results in a fluctuating environment of the Mössbauer atom and hence in fluctuating hyperfine interactions. The effect of these on the Mössbauer spectrum depends on the correlation time τ_c of the fluctuating environment and the characteristic hyperfine interaction time τ_{hf}. The correlation time τ_c is expected to differ from the mean residence time τ_r between hydrogen jumps only by a statistical factor. The interaction time is $\tau_{hf} = \hbar/E_{hf}$, where E_{hf} is the energy involved in the fluctuating hyperfine interactions. For an isomer shift that fluctuates between two values S_1 and S_2 [(6.1)], for instance, one has $E_{hf} = (E_\gamma/c)/(S_1 - S_2)$. The nuclear lifetime τ_n of the Mössbauer level is of minor importance in stationary relaxation phenomena, except as the parameter determining the natural width $W = 2\hbar/\tau_n$ of the Mössbauer line and hence the resolution that can be obtained.

In the quasistatic limit $\tau_c \gg \tau_{hf}$, one gets a mere weighted superposition of the hyperfine interaction patterns for the different hydrogen environments that occur as a consequence of the hydrogen diffusion. In the limit of rapid fluctuations $\tau_c \ll \tau_{hf}$, one observes a single hyperfine pattern with time-averaged hyperfine parameters. The intermediate case $\tau_c \approx \tau_{hf}$ can only be described by a complicated analysis [6.47] similar to that applicable to paramagnetic relaxation phenomena (Sect. 6.3.5). This intermediate case has so far not been identified in the Mössbauer spectra of metal hydrides, but isomer shifts fluctuating in this regime due to electron hopping have been reported [6.31, 63].

The case of rapid fluctuations has been studied [6.64] with the ^{181}Ta resonance in the α phase of Ta–H (see Sect. 6.4.6). In this case, one observes a broadening of the Mössbauer line that is proportional to τ_c [6.64–66], and thus disappears for infinitely fast diffusion ($\tau_c \to 0$). It should be pointed out that this motional narrowing of the Mössbauer line due to the diffusion of the lattice gas is utterly different from cases where the *Mössbauer atoms themselves* diffuse in the lattice. In the latter case, the diffusion leads to a broadening of the Mössbauer line that is proportional to the diffusion jump rate τ_r^{-1} *of the Mössbauer atom* [6.67, 68]. This means that the linewidth increases when the diffusion jumps become faster, until finally the Mössbauer effect becomes altogether unobservable.

6.3.8 Lattice Dynamics

Mössbauer absorption spectroscopy can be used to study the vibrational behavior of the emitting or absorbing atom by means of the Mössbauer-Lamb f-factor [6.3–8]

$$f = \exp(-k^2 \langle x^2 \rangle), \tag{6.4}$$

where k is the wave vector of the γ ray and $\langle x^2 \rangle$ the mean-square motional amplitude of the Mössbauer atom, or by means of the second-order Doppler shift [6.3–8, 69, 70]

$$S_{\rm SOD} = -\langle v^2 \rangle / 2c, \tag{6.5}$$

where $\langle v^2 \rangle$ is the mean-square velocity of the emitting or absorbing nucleus. Since the second-order Doppler shift is always observed together with the isomer shift [(6.1)] of the Mössbauer line, it is difficult to obtain, particularly since it is often small compared to the isomer shift [6.69]. Favorable cases for the observation of the second-order Doppler shift are, for instance, ^{57}Fe, ^{119}Sn, and ^{61}Ni.

So far, hardly any Mössbauer work on the lattice dynamics of metal hydrides has been performed. The few existing data seem to indicate that the solution of hydrogen in d-transition metals lowers the f-factor, as is indeed expected as a consequence of the softening of the vibrational modes that results from the lattice expansion. The f-factor of ^{197}Au at 4.2 K, for instance, is 20% lower in $PdH_{0.7}$ than in pure Pd [6.56]. Superconducting hydrides are a promising field for f-factor measurements. First results for ^{119}Sn in Pd–H, which reveal a softening of the lattice modes on hydrogenation, have recently been reported [6.71].

An interesting effect is expected to arise from the diffusion of solute hydrogen: The time-dependent distortion fields induced in the host lattice by the diffusing lattice gas will increase $\langle r^2 \rangle$ and $\langle v^2 \rangle$, and therefore are expected [6.65, 72] to affect the f-factor, the second-order Doppler shift, and in certain cases the linewidth. Recent observations [6.56] indicate that such effects can indeed be observed.

6.4 Review of the Mössbauer Work on Hydrides of *d*-Transition Metals

In this section and in Section 6.5 we shall briefly review the results of Mössbauer studies of individual metal-hydrogen systems. A table of the hydride systems studied by Mössbauer spectroscopy can be found as an Appendix. A survey of this work has recently also been given by *Wortmann* [6.73].

6.4.1 ^{57}Fe in Hydrides of Ni and Ni–Fe Alloys

In one of the first Mössbauer studies of metal hydrides, *Wertheim* and *Buchanan* [6.2] investigated the electrolytic hydrogenation of a ^{57}Co**Ni** source. As in the Pd–H system (Fig. 6.2), in Ni–H there exists a phase containing virtually no hydrogen and a hydride phase having a composition close to $NiH_{0.7}$. The former is ferromagnetic and thus gives rise to the typical six-line Zeeman pattern of the ^{57}Fe resonance. The hydride is nonmagnetic and therefore exhibits an unsplit Mössbauer line. The Mössbauer spectra (Fig. 6.5) of a ^{57}Co**Ni** source show how this unsplit line grows with the time of hydrogenation at the expense of the Zeeman pattern. However, Mössbauer experiments with a strong external magnetic field applied show [6.2] that the iron atoms in the hydride phase do not altogether lose their magnetic moment, a situation reminiscent of the behavior of dilute iron in Cu and Ag [6.74]. Moreover, the magnetic moment of the Fe atoms in Ni hydride has been found [6.2] to depend on the number of nearest hydrogen neighbors. At 77 K, the Mössbauer line of the hydride has an isomer shift of $S = +0.44\,\mathrm{mm\,s^{-1}}$ [6.2] with respect to the Zeeman pattern of pure Ni (see Fig. 6.5). This corresponds to a lower electron density $\varrho(0)$ at the Fe nuclei in the hydride. Only part of this decrease can be attributed to the 15% volume expansion going along with hydrogenation [6.22]: From the volume coefficient of the isomer shift in pure Ni, $(\partial S/\partial \ln V)_T = 1.1\,\mathrm{mm\,s^{-1}}$ [6.75], one expects the volume contribution to the shift between Ni and $NiH_{0.7}$ to be only $+0.16\,\mathrm{mm\,s^{-1}}$. The large difference from the experimental value may be considered to reflect the explicit effect of the hydrogenation on the electronic structure of the iron atoms. *Wertheim* and *Buchanan* discussed this in terms of the filling of the host *d*-band [6.2]. They also found indications of a distribution of isomer shifts and, possibly, quadrupole interactions due to different hydrogen configurations in the neighborhood of the Mössbauer atoms.

Janot and *Kies* [6.35] have used Mössbauer spectroscopy to study the time dependence of the hydrogen evolution from electrolytically produced $(Ni_{1-x}Fe_x)H_n$ for $x = 0.01$ and 0.004. More recently, *Mizutani* et al. [6.76] have studied such hydrides with $0.005 \leqq x \leqq 0.1$ down to temperatures of 1.5 K. They observed magnetic ordering of the hydride phases, with critical temperatures that decrease when the iron content is lowered, and have fallen below 1.5 K for $x=0.005$. Moreover, experiments with an external field applied to $(Ni_{0.995}Fe_{0.005})H_n$ yield larger values for both the iron magnetic moment and the hyperfine field than those deduced by *Wertheim* and *Buchanan* [6.2]. *Mizutani* et al. [6.76] conclude that substitutional Co has a greater number of hydrogen neighbors in Ni hydride than substitutional Fe, which perhaps has no H neighbors at all, and that this difference in the local environment causes the different observations in source [6.2] and absorber [6.76] experiments.

This conclusion is supported by the isomer shift data: The shift between the hydrogenated and pure $Ni_{1-x}Fe_x$ phases is only about 0.2 mm s^{-1} [6.35, 76], as compared to 0.44 mm s^{-1} in the ^{57}Co**Ni** sources [6.2]. The interpretation of this difference in terms of the local hydrogen distribution implies, that, ceteris paribus, $\varrho(0)$ decreases when the number of hydrogen neighbors to the Mössbauer atom increases.

Further corroboration for these views comes from a Mössbauer and x-ray investigation of the hydrogenation products of concentrated $Ni_{1-x}Fe_x$ alloys with $x=0.25$ and 0.55 [6.77]. The authors find that in these systems there exist two hydride phases, designated as γ' and β. In the γ' phase, they observe nearly no reduction of the hyperfine field and a small isomer shift with respect to the hydrogen-free γ phase, while in the β phase the hyperfine field is reduced by 15% and a larger isomer shift of about 0.4 mm s^{-1} is seen. These results, as well as x-ray data, have been explained by the assumption [6.77] that in the γ' phase the iron atoms repel the hydrogen, while in the β phase long-range order forces hydrogen into the vicinity of iron.

6.4.2 Isomer Shifts of ^{57}Fe, ^{99}Ru, ^{193}Ir, ^{195}Pt, ^{197}Au, ^{151}Eu and ^{119}Sn in Hydrides of Pd and Pd-Based Alloys

Although Pd has no Mössbauer resonance of its own, the PdH_n system has attracted much interest from Mössbauer spectroscopists. Indeed, the metallurgic properties of Pd facilitate the introduction of many Mössbauer isotopes as substitutional solutes.

Loading a source of ^{57}Co in Pd electrolytically with hydrogen, *Bemski* et al. [6.1] were the first to report that the electron density at the ^{57}Fe nuclei is lowered upon hydrogenation. This result was subsequently confirmed by more careful experiments [6.78–83], which were, however, all performed with absorbers of ^{57}Fe in $(Pd_{1-x}M_x)$ H_n hydrides where M stands for iron and, in some cases [6.82, 83], other additional alloying components like Ag, Al, or V. It turns out that in $(Pd_{1-x}Fe_x)H_n$ hydrides with $x=0.003$, the isomer shift for the β phase at the phase limit ($n \approx 0.55$) is only about +0.02 mm s^{-1} with respect to the hydrogen-free alloy [6.80]. This is less that the shift of about +0.07 mm s^{-1} expected to arise merely from the 11% volume increase and the isomer shift data for ^{57}Fe**Pd** under pressure [6.75]. In the region of the β phase, however, one finds a steep increase of the isomer shift S with the hydrogen concentration

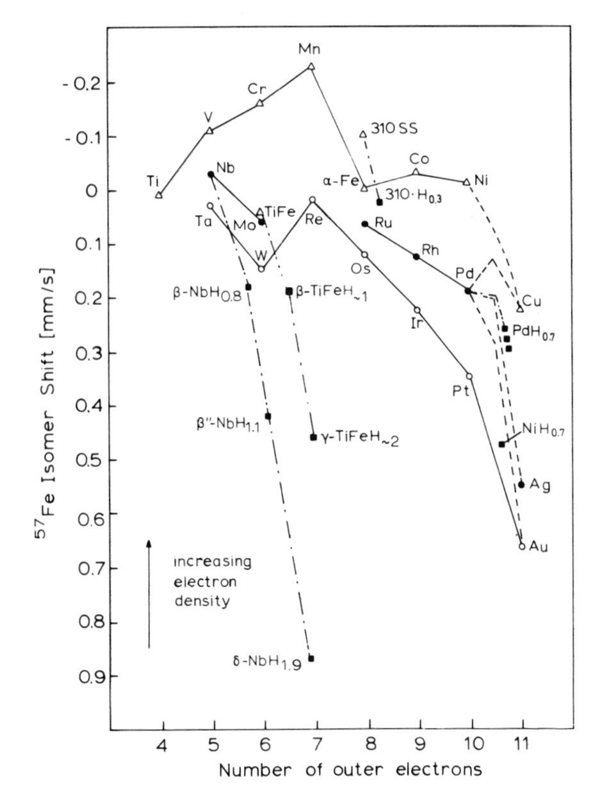

Fig. 6.6. ^{57}Fe isomer shifts in *d*-transition metals as well as some alloy and hydride systems: The solid lines connect the isomer shifts for ^{57}Fe as a dilute impurity in pure 3*d*, 4*d*, and 5*d* metals, respectively. The dashed lines represent, in a slightly idealized way, the variation of the ^{57}Fe isomer shift in $Pd_{1-x}M_x$ (M=Cu, Ag, Au) and $Ni_{1-x}Cu_x$ alloys with composition. The dot-dashed lines connect different hydride phases of the same metal, n_e [see (6.6)] being the increment on the abscissa. The isomer shift data were adapted from [6.2, 53, 54, 80, 85–88]. The concept of this plot is discussed in more detail in [6.26, 28, 29]

n, corresponding to a slope of $dS/dn = 1.1\,\mathrm{mm\,s^{-1}}$ [6.80]. The resulting kink in the S vs n curve is indicated in Fig. 6.6. *Mahnig* and *Wicke* [6.80] were the first to observe that this behavior, as well as the dependence of S on the hydrogen concentration in alloys with a higher iron concentration, can be rationalized in a unified way if one considers S as a function of the additional electron concentration

$$n_e = x \cdot Z_M + n \cdot Z_H \tag{6.6}$$

stemming from the substitutional solute M *and* the interstitial hydrogen in $(Pd_{1-x}M_x)H_n$ systems. Z_M and Z_H stand for the number of electrons donated by the constituent M (e.g., $Z_{Fe}=3$) and the hydrogen ($Z_H=1$). The more or less abrupt change of the slope of the S vs n_e curve is then generally found to occur near $n_e=0.55$. *Wanzl* [6.82] has extended this concept to ternary alloys containing substitutional Ag ($Z_{Ag}=1$), Al ($Z_{Al}=3$), and V ($Z_V=5$) in addition to iron. Moreover, he has replaced hydrogen by boron ($Z_B=3$) as the interstitial solute. Except for alloys containing high concentrations of V ($x \geqq 0.05$), all systems studied by *Wanzl* exhibit very similar S vs n_e curves with a change of slope at $n_e \approx 0.55$ [6.82, 83]. In the framework of the rigid band model

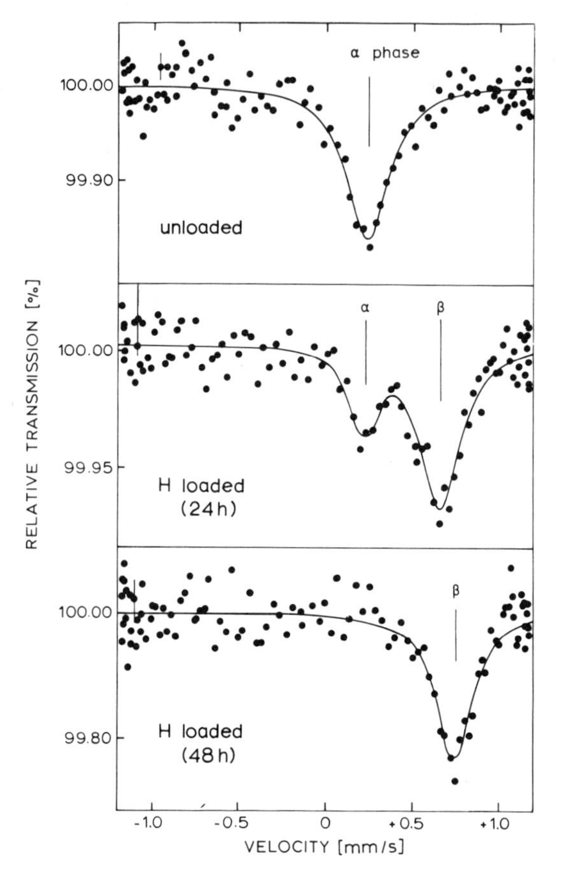

Fig. 6.7. ^{99}Ru Mössbauer spectra taken at 4.2 K with a source of ^{99}Rh in $(Pd_{0.995}Ru_{0.005})H_n$ and a Ru metal absorber after different times of electrolytic hydrogenation of the source [6.56]. The spectrum in the middle shows the coexistence of the α and the β phase. The shift between the β phase lines in the middle and bottom spectra reveals a dependence of the isomer shift on n in the β phase region

this behavior can be understood as a consequence of the filling of the Pd d-band near $n_e = 0.55$. The rapid decrease of the electron density at the ^{57}Fe nuclei has been attributed [6.81] to the filling of localized $3d$ states of iron once the Pd d-band is filled. This model is also quite useful in explaining the magnetic properties of Fe–Pd–H systems (see Sect. 6.4.3). Its success is quite unexpected if one considers that the actual band structures of PdH_n and Pd-based alloys deviate considerably [6.15–20] from the rigid band model. The reason for this success probably is that in many Pd-based systems the d-band is filled at $n_e \approx 0.55$, irrespective of the detailed composition and band structure.

As has been pointed out in Section 6.4.1, there are strong indications that in Ni–Fe–H the local distribution of hydrogen around the iron atoms affects the Mössbauer parameters. The same is expected in Pd-based systems. Indeed, the sudden increase of the isomer shift near $n_e \approx 0.55$ in Pd–H might also reflect changes in this distribution. Such a redistribution might, in its turn, be driven by the electronic structure. A comparison of Mössbauer source and absorber experiments could probably help to clarify this point.

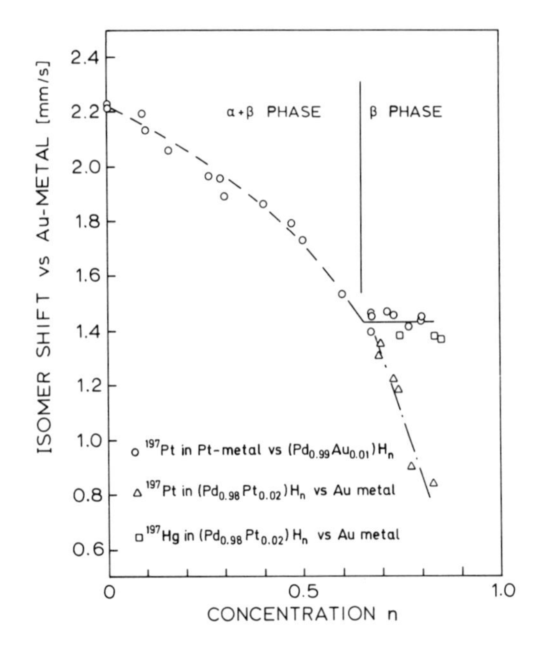

Fig. 6.8. ^{197}Au isomer shift data for different sources and absorbers at 4.2 K. All shifts are given in the absorber convention and relative to metallic Au. In the mixed $(\alpha+\beta)$ phase region $(n \lesssim 0.65)$, only the average center shifts could be determined due to the limited resolution. The dashed line drawn to these data represents the concentration dependence of the center shift expected if one takes into account that the f-factor is 20% lower in the β phase than in the α phase [6.56, 62]

The isomer shifts of the Mössbauer resonances in ^{99}Ru, ^{193}Ir, ^{195}Pt, and ^{197}Au in Pd hydrides have been studied by *Iannarella* et al. [6.84]. In all these cases the electron density $\varrho(0)$ at the Mössbauer nucleus was found to be lower in $PdH_{\approx 0.8}$ than in pure Pd. The resolution for ^{99}Ru, ^{193}Ir, and ^{197}Au is sufficiently good to facilitate accurate isomer shift measurements. As an example, some spectra for ^{99}Ru are shown in Fig. 6.7. So far, however, only the ^{197}Au isomer shifts have been studied in some detail as a function of the hydrogen content [6.62]. The 77 keV Mössbauer transition in ^{197}Au can be studied in absorber experiments as well as in source experiments in which the Mössbauer level is populated either after the β decay of ^{197}Pt or after the electron capture decay of ^{197}Hg. Results for all three types of experiment are shown in Fig. 6.8. At the β phase limit $(n \approx 0.65)$, both the source and the absorber experiments yield a decrease of the isomer shift by about $-0.8\,\mathrm{mm\,s^{-1}}$ with respect to pure Pd, corresponding to a lowered electron density $\varrho(0)$ in the hydrides. This shift is close to the volume contribution of $-0.9\,\mathrm{mm\,s^{-1}}$ expected from (6.2) and the volume coefficient of the isomer shift of ^{197}Au in pure Pd, $(\partial S/\partial \ln V)_T = -(8.0 \pm 1.5)\,\mathrm{mm\,s^{-1}}$ [6.62]. Within the β phase region the volume effect is expected to cause a slower but continuous decrease of the isomer shift. This is not observed in the absorber experiments and in the experiments with the ^{197}Hg sources (Fig. 6.8), which yield a virtually constant shift for $n \geqq 0.65$. With the sources of ^{197}Pt in $(Pd_{0.98}Pt_{0.02})H_n$, however, one finds the isomer shift to decrease more rapidly with n than is expected for the volume effect alone. These differences in the behavior of the different samples must be attributed to different hydrogen environments of Pt,

Au, and Hg atoms in the hydride matrix. At 4.2 K, where all these experiments have been performed, these environments are expected to persist in the sources until the time of the emission of the Mössbauer γ ray. Conceivably, the interstitial sites are empty next to Au and Hg impurities, while they become progressively occupied next to Pt when the hydrogen concentration rises above $n \approx 0.65$. The ensuing conclusion that hydrogen nearest neighbors reduce $\varrho(0)$ agrees with the interpretation of the data for ^{57}Fe in NiH_n (Sect. 6.4.1). The constant shift in the absorbers for $n \geqq 0.65$ indicates that there is a positive contribution to compensate the volume effect. This could reflect the filling of *s*-band states, an interpretation that seems to be supported by isomer shift data for ^{197}Au in $Pd_{1-x}Ag_x$ alloys, which also exhibit an increase of $\varrho(0)$ above the band-filling point at $x \approx 0.55$ [6.62].

The Pd–H system seems to undergo a phase transition near 50 K (see, e.g. [6.89]), in the course of which at least part of the hydrogen is supposed to migrate from octahedral to tetrahedral sites. While *Phillips* and *Kimball* [6.79] found no effect of this transformation in their ^{57}Fe Mössbauer spectra, recent ^{197}Au measurements [6.56] with sources of ^{197}Pt in PdH_n show a small isomer shift difference between sources cooled rapidly and slowly through the critical region near 50 K.

Isomer shifts of ^{151}Eu in PdH_n ($n \leqq 0.8$) have been measured by *Meyer* et al. [6.90], who observed a decrease of $\varrho(0)$ in the hydrogenated samples. This decrease is very close to the estimate of the pure volume effect.

The only case where hydrogen loading of the Pd matrix has been found to increase $\varrho(0)$ is that of ^{119}Sn [6.91]. The authors explained this, as well as a similar increase for ^{119}Sn in $Pd_{1-x}Ag_x$ alloys, by the filling of *s*-band states. However, subsequent pressure experiments for ^{119}Sn in Pd showed [6.36, 92], that in this matrix the electron density at the tin nuclei *decreases* under pressure. As a consequence, the shifts in the hydrides can, within the limits of error, be explained by the volume contribution alone. This seems to support the view that volume effects are, in many cases, the dominant contribution to the isomer shifts in metal-hydrogen systems. One should not forget, however, that the separation of the isomer shift into a volume and an intrinsic part [(6.2)] is a very crude picture to describe the change between the α and the β phase of PdH_n.

6.4.3 Magnetism and ^{57}Fe Hyperfine Fields in the Pd–Fe–H System

At low concentrations, the iron atoms in **Pd**Fe alloys possess localized "giant" moments that interact ferromagnetically owing to a long-range exchange coupling mechanism [6.74, 93–95]. This behavior is closely connected with the high polarizability of the Pd matrix and hence with the band structure of Pd. Hydrogen loading or alloying, for instance with Ag, is therefore expected to have a profound influence on the magnetic properties of such alloys.

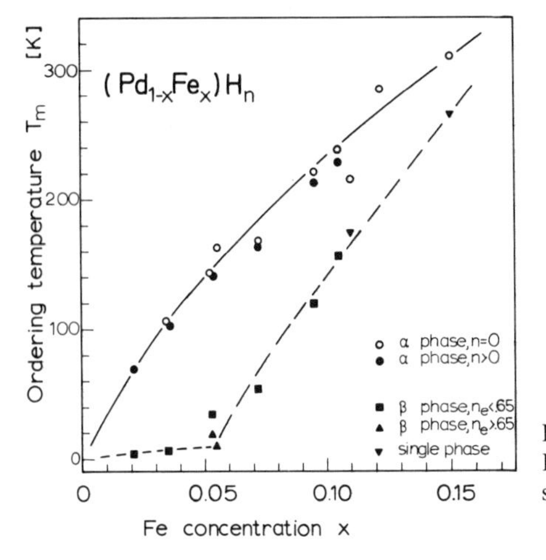

Fig. 6.9. Magnetic ordering temperatures in the Fe–Pd–H system obtained from Mössbauer studies [6.78, 79, 81, 96]

The first Mössbauer investigations of the magnetic properties of the $(Pd_{1-x}Fe_x)H_n$ system were performed by *Jech* and *Abeledo* [6.78] at iron concentrations of $x=0.11$ and 0.15, and by *Phillips* and *Kimball* [6.79] at $x=0.02$ and 0.05. This work revealed the decrease of the ordering temperature on hydrogenation (see Fig. 6.9), which amounts to more than an order of magnitude for $x \lesssim 0.05$. In an extensive Mössbauer study [6.81, 96], *Carlow* and *Meads* found the phase diagram of the Pd–Fe–H system to be rather similar to that of Pd–Ag–H: With increasing iron content the gap between the α and the β phase becomes narrower until the phase separation disappears altogether for iron concentrations above $x=0.115$. Where the α and β phase coexist, the β phase forms with $n_e \approx 0.55$ and exhibits the same low-temperature saturation value for the hyperfine field as the α phase despite its drastically lowered ordering temperature. At higher hydrogen contents ($n_e \gtrsim 0.6$), however, the saturation field drops to lower values (Fig. 6.10). This goes along with a further decrease of the ordering temperature and the rapid increase of the isomer shift described in Section 6.4.2. The n_e dependence of the hyperfine field is strikingly similar to the behavior of the ^{57}Fe hyperfine field in $Pd_{0.98-y}Au_yFe_{0.02}$ alloys [6.85] (see Fig. 6.10). A nearly constant hyperfine field has also been observed in $Pd_{1-x-y}Ag_yFe_x$ ($x=0.03$ and 0.07) alloys up to the band-filling point [6.86], but in this case metallurgic difficulties forbid higher silver concentrations that could bring about the expected drop of the hyperfine field values.

Carlow and *Meads* [6.81] have convincingly discussed the magnetic properties of the Pd–Fe–H system in terms of a model that implies the filling of two different types of *d*-holes at the Pd atoms: The t_{2g} orbitals are assumed to have band character and to be responsible for the long-range polarization of the Pd matrix, while the e_g orbitals are localized and carry an induced magnetic moment when an iron atom is in the nearest neighborhood. In the β phase with $n_e \approx 0.6$, only the t_{2g} holes are filled. This destroys the long-range ordering

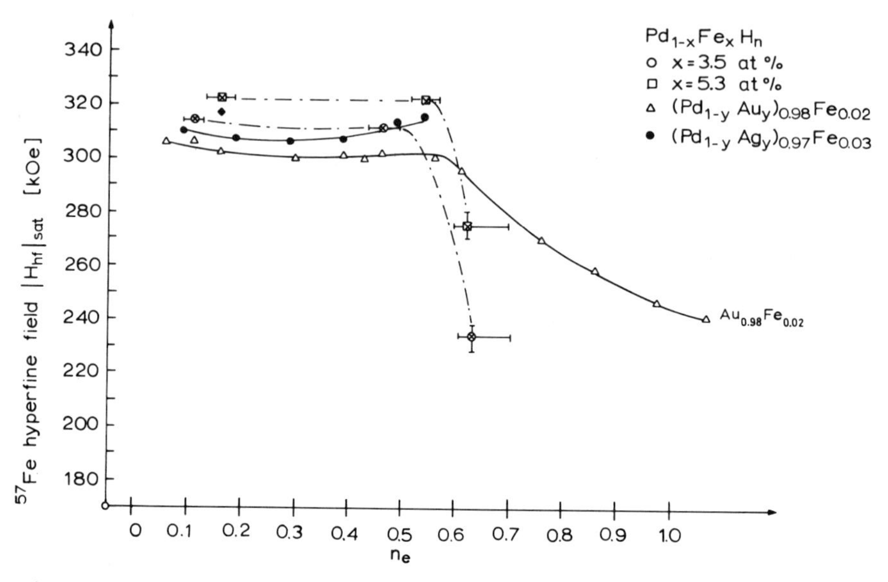

Fig. 6.10. Low-temperature saturation values of the ^{57}Fe hyperfine field in Pd–Au(Ag)–Fe alloys (solid curves) and in Pd–Fe–H hydrides (dot-dashed curves) [6.81, 85, 86]. Note that the data points for the hydrides at $n_e \approx 0.15$ refer to the α phase, while those for larger values of n_e are for the β phase

potential, but leaves the hyperfine field unchanged. The latter starts to decrease only when the e_g holes begin to be filled at $n_e \lesssim 0.6$.

The magnetic order in the β phase is then established via s electrons in a random RKKY mechanism. Indeed resistivity measurements [6.97] on $(Pd_{1-x}Fe_x)H_n$ reveal a spin-glass behavior for $x \gtrsim 0.01$ and a Kondo behavior for lower iron concentrations. These findings suggest a reinterpretation of some earlier Mössbauer data and further studies with low iron concentrations and at low temperatures.

6.4.4 ^{57}Fe in Hydrides of Nb, V, Ti, and Some Transition Metal Alloys

The NbH_n system has been studied by Mössbauer spectroscopy with absorbers containing 1.6 at. % of enriched ^{57}Fe [6.53]. Four different phases (α, β, β'', and δ) have been identified by their isomer shifts, which are large enough in this case to produce at least partially resolved lines in mixtures of different phases. Between the α phase ($n \approx 0.05$) and the δ phase ($n \approx 1.9$) an isomer shift of 0.90 mm s^{-1} is observed, the β ($n \approx 0.8$) and β'' ($n \approx 1.1$) phases lying in between (see Fig. 6.6). These isomer shifts correspond to a decrease of the electron density at the iron nuclei that by far exceeds the values expected from the lattice expansion [6.36, 75]. They have been explained by the occupation of Fe 3d holes as a consequence of the filling of the host 4d-band. The relative intensities of the Mössbauer lines for different coexisting phases have been used [6.53] to obtain information on the Nb–H phase diagram at low temperatures. Measurements on the β phase at ambient temperature in an external magnetic field have been interpreted [6.53] as evidence for a localized magnetic moment on the iron atoms in $NbH_{\approx 0.8}$. This is a tempting idea deserving a conclusive test by measurements at low temperatures, since Fe in $Nb_{1-x}Mo_x$ alloys does develop a magnetic moment for $x \geqq 0.4$ [6.74, 98].

Mössbauer measurements on ^{57}Fe in hydrogenated vanadium have been performed by *Simopoulos* and *Pelah* [6.99], who found no change in the isomer shift but a surprisingly large increase in the f-factor on hydrogenation. The V–H system seems to be a promising candidate for further, more detailed Mössbauer studies.

^{57}Fe in the $TiFeH_n$ system has recently been studied by several groups [6.54, 100, 101]. The Mössbauer spectra of hydrides of equiatomic TiFe alloys consist of partially resolved contributions from different phases (α, β, and γ) with small quadrupole splittings in the β and γ phase. The isomer shift (Fig. 6.6) increases by 0.4 mm s^{-1} between the α phase ($n \lesssim 0.1$) and the γ phase ($n \approx 2$), corresponding to a decrease of the electron density that clearly exceeds the volume effect [6.54, 100] and thus indicates the filling of iron d states on hydrogenation.

The hydrogenation of stainless steels would seem to be a promising field for Mössbauer studies. Indeed *Wertheim* and *Buchanan* [6.2] in their early work studied electrolytically charged type 310 stainless steel and found an increase of the isomer shift by 0.25 mm s^{-1} at a hydrogen to host atomic ratio of $n \approx 0.3$. Similar results have recently been obtained for some austenitic steels by *Fujita* and *Sohmura* [6.77].

The intermetallic cubic Laves phase compounds of the RFe_2 type, where R is a rare earth, form hydrides of the composition RFe_2H_4 with interesting magnetic properties [6.102, 103]. Thus in YFe_2 and $GdFe_2$ the Curie temperature decreases on hydrogenation, while the magnetic moment per iron atom increases. A ^{57}Fe Mössbauer study of these hydrides has shown that, at least for YFe_2, this increase of the iron moment goes along with an increase of the hyperfine field at the iron nuclei, while the electron density decreases [6.103].

6.4.5 ^{181}Ta Isomer Shifts in the α Phase of Ta–H

The Mössbauer studies of the Ta–H system with the 6.2 keV resonance of ^{181}Ta [6.64, 104] are remarkable for two reasons: First of all, they were performed on a pure metal-hydrogen system, and, secondly, the high resolution of the 6.2 keV resonance facilitated not only very precise isomer shift measurements at low hydrogen concentrations, but also detailed studies of the hydrogen-induced line broadening as a function of hydrogen concentration and temperature.

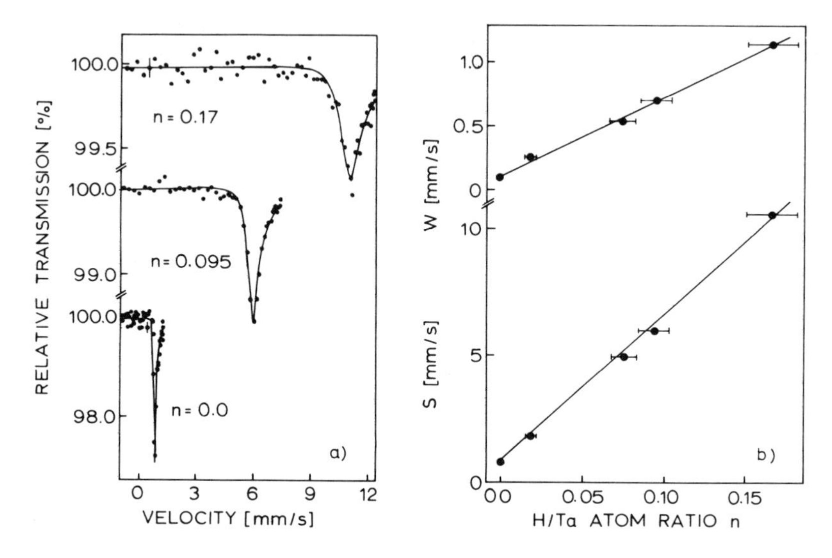

Fig. 6.11. (a) ^{181}Ta Mössbauer spectra of α-TaH_n absorbers obtained at room temperature for various hydrogen concentrations n. The asymmetry of the lines is due to the dispersion term in the absorption cross section that is typical for the 6.2 keV resonance [6.28, 29]. (b) Room temperature values of the full linewidth at half maximum W and the isomer shift S relative to ^{181}Ta in metallic tungsten as a function of the hydrogen concentration n (from [6.64])

Results obtained at room temperature [6.64, 29] are shown in Fig. 6.11. The isomer shift S increases linearly with the hydrogen concentration n with a slope of $dS/dn = +57 \pm 6\,\mathrm{mm\,s^{-1}}$. Since $\Delta\langle r^2\rangle$ is negative for the 6.2 keV transition, this corresponds [(6.1)] to a decrease of the electron density $\varrho(0)$ at the Ta nuclei. Going along with the increasing isomer shift, there is a considerable hydrogen-induced line broadening which also increases linearly with n by $dW/dn = 6.3 \pm 0.7\,\mathrm{mm\,s^{-1}}$ at room temperature. This line broadening depends strongly and reversibly on temperature (Fig. 6.12) and will be discussed in terms of the diffusion behavior of the interstitial hydrogen in Section 6.4.6.

In order to explain the concentration dependence of the isomer shift, one must again consider two main sources, namely, the lattice expansion and the explicit change of the electronic structure arising from the dissolved hydrogen. Since the hydrogen concentrations are rather small and there is no phase change, α-TaH_n should be an ideal case for the application of (6.2), except for the fact that the first term in this equation can describe only the effect of the macroscopic lattice expansion, but not the local lattice distortions expected in α-TaH_n [6.37]. From pressure isomer shift data for ^{181}Ta [6.28, 29], one finds [6.29] that the macroscopic volume expansion can, at most, account for one-third of the experimental dS/dn value. This means that the dissolved hydrogen induces a substantial explicit reduction of the electron density $\varrho(0)$ at the Ta nuclei.

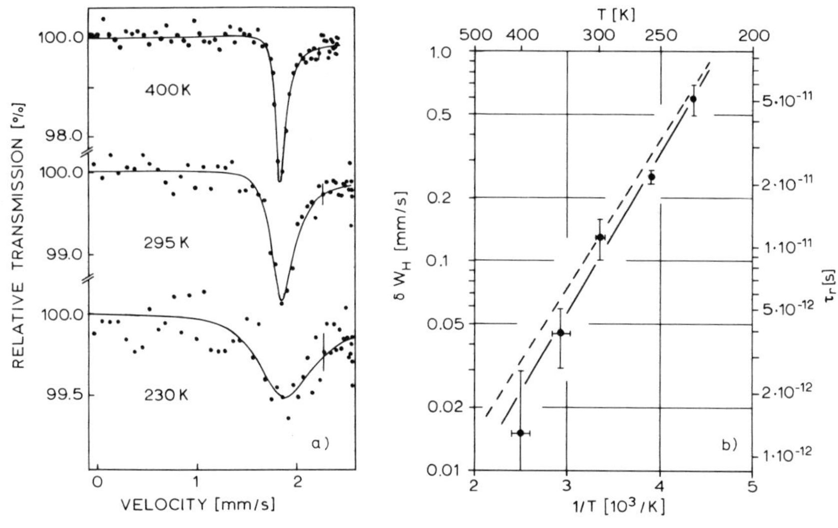

Fig. 6.12. (a) ^{181}Ta Mössbauer spectra of a $TaH_{0.018}$ absorber at different temperatures. (b) Arrhenius plot of the temperature dependence of the hydrogen-induced line broadening for a $TaH_{0.018}$ absorber. The τ_r scale follows from (6.10) and (6.11) and is then used for giving the τ_r values as calculated from macroscopic diffusion measurements [6.12] (dashed line). The Mössbauer data are from [6.64]

If one interprets this decrease of $\varrho(0)$ in terms of a rigid-bandlike filling of the Ta conduction band, one concludes [6.29] that between 50 and 100% of the electrons from the dissolved hydrogen must fill up empty Ta $5d$ states. The more realistic picture that has recently emerged for PdH_n [6.15–18] as well as photoemission studies of VH_n and VD_n [6.106] shed some doubt on this interpretation and suggest that $s-p$-like states and the formation of localized metal-hydrogen bonds might also play some part in the decrease of $\varrho(0)$.

6.4.6 Motional Narrowing of the ^{181}Ta Mössbauer Lines by Hydrogen Diffusion

The temperature and concentration dependence of the width of the ^{181}Ta Mössbauer line in α-TaH_n can be understood in terms of the motional narrowing described in Section 6.3.7. This is, indeed, the only case in which this phenomenon has so far yielded [6.64, 104] quantitative information on the diffusion process, i.e., on the activation energy and jump frequency of the diffusing hydrogen atoms.

In order to arrive at a quantitative description of the hydrogen-induced line broadening, *Heidemann* et al. [6.64, 104] assume 1) that only the isomer shift contributes to the line broadening while the effect of the electric quadrupole interaction is negligible, and 2) that only H-atoms in the nearest neighbor interstitial sites around the Mössbauer atom contribute to the fluctuating isomer shift.

The solute hydrogen in bcc α-TaH_n occupies the tetrahedral interstitial sites [6.11]. There are 6 such sites per Ta site and 24 in nearest neighbor positions around each Ta atom. Therefore, the mean number of hydrogen nearest neighbors to a Ta atom is $4n$, and the mean isomer shift induced by these is

$$\bar{S}_{\mathrm{H}} = \omega_{\mathrm{H}} \cdot 4n \tag{6.7}$$

if ω_{H} is the shift induced by every single H nearest neighbor. From the observed variation of S with n (Fig. 6.11b), one finds $\omega_{\mathrm{H}} = 14.2 \pm 1.5\,\mathrm{mm\,s^{-1}}$. The influence of the isomer shift fluctuations on the resonance line shape can be described by two parameters [6.65, 66, 107], namely the correlation time τ_c and the modulation amplitude Δ, which is closely related to the inverse of the hyperfine interaction time τ_{hf} of Section 6.3.7. By definition, Δ^2 is the mean square deviation of the isomer shift from its time average $\bar{S}_{\mathrm{H}}$. For low hydrogen concentrations, this becomes

$$\Delta^2 = \omega_{\mathrm{H}}^2 \cdot 4n\,. \tag{6.8}$$

The correlation time τ_c is proportional to the mean residence time τ_r of the H-atoms between two diffusion jumps,

$$\tau_c = \nu \cdot \tau_r\,, \tag{6.9}$$

where ν is the mean number of jumps a nearest neighbor H-atom must make before it leaves the nearest neighbor shell. Equation (6.9) thus takes into account that the isomer shift remains unchanged when the H-atom merely changes its position within the nearest neighbor shell. Actually three out of the four possible jump directions from a tetrahedral interstitial site in a bcc lattice are such irrelevant jumps. Numerical calculations for this case [6.64, 108] give a value of $\nu = 12.6$. According to stochastic theory [6.66, 107], the shape of the Mössbauer line in the limit of fast modulation ($\Delta \cdot \tau_c \ll 1$) will be a Lorentzian with a hydrogen-induced line broadening

$$\delta W_{\mathrm{H}} = 2\Delta^2 \cdot \tau_c = 8\omega_{\mathrm{H}}^2 \cdot \tau_r \cdot \nu \cdot n\,. \tag{6.10}$$

The linear dependence of δW_{H} on n is in agreement with the experimental results (Fig. 6.11b), which can be used to determine a value of $(9.8 \pm 1.5) \cdot 10^{-12}\,\mathrm{s}$ for the jump time τ_r at room temperature, since ω_{H} is known from (6.7) and the experimental dS/dn value.

The temperature dependence of the linewidth (Fig. 6.12) is contained in the temperature dependence of τ_r. For a thermally stimulated diffusion process with an activation energy U, one therefore expects

$$\delta W_{\mathrm{H}} \sim \tau_r \sim \exp(U/kT)\,. \tag{6.11}$$

Such an Arrhenius behavior is actually observed (Fig. 6.12b) and yields $U = 0.14 \pm 0.01$ eV [6.64]. This result for U is in good agreement with the value obtained from macroscopic diffusion measurements [6.12], and so are the values obtained for τ_r. The latter, in particular, is an a posteriori justification for neglecting effects of other than nearest neighbor hydrogen sites and the electric quadrupole interaction. Moreover, independent information from NAR measurements [6.109, 110] shows that the quadrupole interaction does not contribute substantially to the broadening of the Mössbauer line.

The Mössbauer studies of hydrogen diffusion in TaH_n are presently extended to lower hydrogen concentrations and to lower temperatures, where changes in the diffusion behavior are expected [6.12]. Preliminary results for TaD_n absorbers [6.111] show a stronger line broadening than in TaH_n, reflecting the slower diffusion rate of the heavier isotope.

6.5 Mössbauer Experiments on Hydrides of the Rare Earths and Actinides

6.5.1 Rare Earth Hydrides and Related Systems

The RH_2 dihydrides of the rare earths (RE) and some isostructural transition metal hydrides have attracted the interest of Mössbauer spectroscopists mainly for two reasons which are both closely related to their cubic crystal structure (see Fig. 6.4 and Sect. 6.2.2). First, there is their suitability as single-line Mössbauer sources and absorbers that will be discussed further below. Secondly, they are promising systems for studying magnetic hyperfine interactions in RE ions because, owing to the cubic point symmetry of the RE sites in the stoichiometric compounds, the theoretical treatment of the crystalline electric field (CEF) states is rather straightforward and requires only two CEF parameters, A_4 and A_6 [6.23]. The calculation of the hyperfine spectra can be simplified further by the use of the Mössbauer transitions between 2^+ and 0^+ nuclear spin states which are available in the deformed even-even nuclei of the heavier RE, notably in ^{160}Dy (87 keV), ^{166}Er (81 keV), and ^{170}Yb (84 keV). The computational benefits of the use of such resonances arise mainly from the fact that the 0^+ nuclear ground state undergoes no hyperfine splitting. They become particularly important in the calculation of paramagnetic relaxation patterns [6.47–49].

The first work of this type is the study of dilute ^{160}Dy and ^{166}Er impurities in a diamagnetic YH_2 matrix by *Stöhr* and *Cashion* [6.51], who found a Γ_7 Kramers doublet to be the ionic ground state in both cases. From this result they conclude that the hydridic model accounts properly for the CEF experienced by the RE ions. Subsequently, *Shenoy* et al. [6.24] studied ^{166}Er in ErH_2 and YH_2. They confirmed the Γ_7 ground state for Er^{3+} in YH_2, but concluded from measurements in an external field that in ErH_2 the ground state is Γ_6. Such a change can, indeed, arise from a small change of the ratio of A_4 and A_6 [6.24]. The Mössbauer data also showed that ErH_2 orders magnetically near 2.4 K. The somewhat lower (2.13 K) ordering temperature found from specific heat data [6.112] may be due to deviations from stoichiometry.

Shenoy et al. [6.50] have also studied Er^{166} in δ-zirconium hydride, $ZrH_{1.5}$, which has the RH_2 fluorite structure, with only 3/4 of the H sites occupied. They found an almost undistorted cubic CEF, and Γ_7 as the Er^{3+} ground state. The existence of a nearly cubic CEF at 4.2 K, where no motional averaging of the hydrogen distribution can take place, shows that all H sites in the nearest neighborhood of the Er impurities are occupied.

The $2^+ - 0^+$ transitions in even-even nuclei are of little use for isomer shift measurements because of their small $\Delta\langle r^2\rangle$ values [6.9]. Isomer shift data for various hydrides of Gd and Dy have, however, been obtained with the Mössbauer resonances of ^{155}Gd (87 keV) and ^{161}Dy (26 keV) [6.113–116]. These resonances have also been used to study the magnetic hyperfine interactions in DyH_2 and GdH_2 at low temperatures [6.114, 117]. The SmH_x ($2 \leqq x \leqq 3$) system has been investigated by ^{153}Eu (103 keV) Mössbauer spectroscopy [6.116, 118]. The large isomer shifts of this resonance permitted the determination of the relative amounts of Eu^{2+} and Eu^{3+} formed in this system after the nuclear β decay of ^{153}Sm as a function of x.

The usefulness of fcc RH_2 hydrides as single-line Mössbauer sources and absorbers is due to several favorable circumstances. The hydrides are quite stable, can easily be prepared from RE metals, and have reasonably high f-factors. Owing to the cubic point symmetry, they should exhibit no electric quadrupole splitting at least when they are stoichiometric. Moreover, their magnetic ordering temperatures are rather low [6.43–45, 112], such that some, like ErH_2 or TmH_2, can be used even at liquid He temperature. In others, e.g., DyH_2 or HoH_2, the ordering temperatures can be sufficiently lowered by magnetic dilution with Y or Sc. Actual applications of hydrides as single-line sources or absorbers have been described in [6.24, 50, 51, 119, 120].

Of the hexagonal RH_3 hydrides, DyH_3, GdH_3, and ErH_3 have been studied [6.113–115, 119] by Mössbauer spectroscopy. In GdH_3, a resolved quadrupole splitting was observed and interpreted in terms of the hydridic model [6.113, 114]. In ErH_3, and unsplit line was found down to 1.7 K, but dilute Er^{3+} in YH_3 exhibits a paramagnetic hyperfine structure [6.119]. Assuming the point symmetry of the Er^{3+} sites to be approximately hexagonal, *Suits* et al. [6.119] could interpret this spectrum satisfactorily and derive the components of the Er^{3+} g-tensor.

In orthorhombic EuH_2 and YbH_2, the RE are divalent. In both of them, substantial quadrupole interactions were revealed by Mössbauer measurements and could be explained in terms of the hydridic model [6.121]. EuH_2 is ferromagnetic below 16 K with a saturation hyperfine field of 305 kOe [6.121].

Recently, the hydrides of EuPd and $EuRh_2$ have been studied by ^{151}Eu Mössbauer spectroscopy [6.105]. It turns out that, while Eu is trivalent in

$EuRh_2$, it assumes a metallic divalent configuration in $EuRh_2H_x$. In EuPd, on the other hand, it is already divalent, and it remains so on hydrogenation. The electron density at the Eu nuclei decreases markedly, however. At the same time the magnetic hyperfine field at 4.2 K increases from 16.0 T in EuPd to 26.4 T in $EuPdH_x$. $EuRh_2$ also undergoes a drastic change of its magnetic properties. While is does not order magnetically in its pure state, the $EuRh_2H_x$ sample was found to become magnetic at 15.5 K, with a hyperfine field of 26.9 T at 4.2 K. These results indicate that the electronic state of Eu in both hydrides is rather similar to that of EuH_2. Very recently, the hydrides of $EuNi_5$ and $EuMg_2$ have also been studied.

Finally, a recent ^{155}Gd Mössbauer investigation of $(Gd_{0.1}La_{0.9})Ni_5H_x$ and $(Gd_{0.1}La_{0.9})Co_5H_x$ should be mentioned. In the Mössbauer spectra obtained for values of x up to 6.7 and 4.2, respectively, *Bauminger* et al. [6.55] could identify the hydride phases known from the work of *Kuipers* [6.122] by their different isomer shifts and quadrupole interactions. This is in contrast to the results of a ^{141}Pr Mössbauer study of $PrNi_5$ and $PrNi_5H_{4.3}$, for which no change of the isomer shift on hydrogenation and no quadrupole splitting was observed [6.123].

6.5.2 Mössbauer Study of NpH_{2+x}

Np is one of the heavier actinides, whose hydrides can fairly well be described by the phase diagram for the heavier rare earths (Fig. 6.4). The NpH_{2+x} system with $x=0.1$, 0.3, and 0.5 has been studied by Mössbauer spectroscopy with the 60 keV resonance in ^{237}Np. The spectra obtained by *Mintz* et al. [6.124] vary drastically with x (Fig. 6.13) and exhibit a fairly well-resolved structure. The authors attributed this to the distribution of hydrogen atoms on the partially filled octahedral interstices, and fitted the spectra assuming that the isomer shift for a Np atom with m octahedral hydrogen neighbors is $m \cdot \omega_H(x)$, where the shift $\omega_H(x)$ induced by a single octahedral neighbor is itself a function of x. Similarly, the quadrupole interaction induced by several neighbors was considered to be a linear superposition of single-neighbor contributions. Since the spectra could not be reproduced satisfactorily in this way when a statistical occupation of the octahedral sites was assumed, the site occupancy was modified by the assumption of a repulsive or attractive interaction between the hydrogen atoms on the octahedral sites. With an attractive H–H interaction for $x=0.1$ and a repulsive one for the higher concentrations, reasonable fits could then be obtained (Fig. 6.13). These yielded a negligibly small quadrupole interaction and $\omega_H(x)$ values of $+1.2$, $+2.3$, and $+3.6\,\mathrm{mm\,s^{-1}}$ for $x=0.1$, 0.3, and 0.5, respectively. This corresponds to a decrease of the electron density at the Np nuclei when the number of H neighbors increases, since $\Delta\langle r^2\rangle$ is negative [6.9]. Similarly, the shift of about $20\,\mathrm{mm\,s^{-1}}$ between Np metal and $NpH_{\approx 2}$ corresponds to a reduced electron density in the hydride. The interpretation of *Mintz* et al. [6.124] is, however, not the only one conceivable.

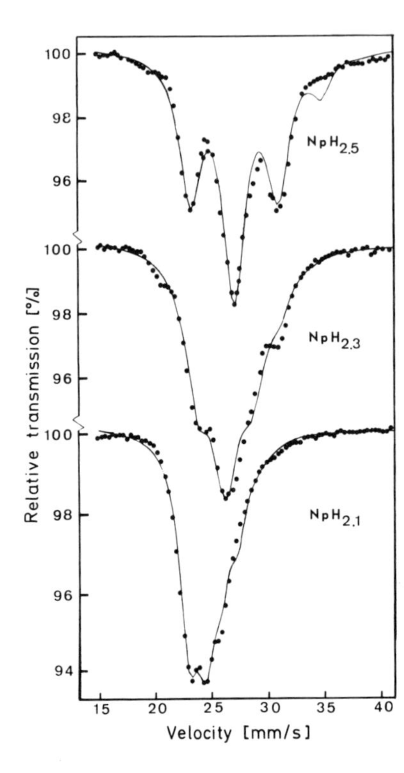

Fig. 6.13. ^{237}Np Mössbauer spectra of NpH_{2+x} at 4.2 K taken with a source of Am metal. The solid curves represent the fit based on the assumption of a distribution of isomer shifts and quadrupole splittings arising from different nearest neighbor hydrogen configurations as discussed in the text (from [6.124])

As has been pointed out [6.73], the spectrum of the $NpH_{2.5}$ sample could also be explained by a unique site with a large quadrupole interaction arising from the formation of a superstructure. Moreover, the dispersion term in the absorption cross section of the ^{237}Np resonance is rather large [6.125] and should be taken into account. In any case, the spectra presented by *Mintz* et al. clearly demonstrate the high resolution power of the ^{237}Np resonance for studying hydride systems. With recent improvements in counting speed [6.126], it should even be possibly to study the temperature region where the hydrogen becomes mobile.

6.6 Conclusions

In the preceding sections we have outlined how the Mössbauer effect can be applied to studies of metal hydrides, and we have briefly reviewed the Mössbauer results as yet obtained on such systems. Clearly, the work performed so far must often be considered as a demonstration of the capabilities of the method rather than as a novel contribution to our knowledge of metal-hydrogen systems. Some specific topics of special interest have, however, emerged already.

Thus, the isomer shift has turned out to be an interesting, perhaps even the most important, parameter in this context. Isomer shift data, and the virtually ubiquitious reduction of the electron density at the nucleus on hydrogenation in particular, call for a quantitative theoretical interpretation. This is actually an appeal to theorists rather than to experimentalists, although some progress is also expected from semiempirical work trying to unravel different contributions, like the volume effect and the influence of nearest neighbors.

Similarly, the electric quadrupole interaction and the crystalline electric field induced by interstitial hydrogen should be studied more carefully, both theoretically and experimentally. Here, even semiquantitative approaches could yield valuable information on the screening of the charge of the interstitial hydrogen, for instance as a function of the overall hydrogen content. Experimental data on quadrupole interactions may, however, often be better obtained from methods that are insensitive to the isomer shift, like the perturbed angular correlation technique.

A major field of interest will be the magnetic properties of metal hydrides containing magnetic atoms like iron or cobalt, particularly since in these systems one finds nearly the whole palette of magnetic properties, like ferromagnetism, mictomagnetism, spin-glass behavior, or Kondo effect. Studies of dilute systems of this type at temperatures in the mK region, as can be obtained with a ^{3}He–^{4}He dilution refrigerator, would seem particularly appealing. To yield conclusive results, Mössbauer studies of magnetic properties should, however, be supplemented by other techniques, e.g., by magnetization measurements, wherever this is possible.

The sensitivity of Mössbauer spectroscopy to the local hydrogen environment opens up a way to study the hydrogen distribution even around dilute substitutional solute atoms. Such experiments are still in their infancy, but they promise to purvey information that could hardly be obtained otherwise, for instance on the dynamic behavior of the hydrogen distribution after a nuclear transformation.

Diffusion effects at large are expected to play a major role in future Mössbauer studies of metal hydrides. To name just one conceivable application, it seems possible to study the dynamics of slow phase transitions like the ill-understood phase change in Pd–H near 50 K, during which diffusion jumps between octahedral and tetrahedral interstices might be of importance.

Other perhaps less ambitious but equally useful applications of Mössbauer spectroscopy will be in analyses of equilibrium phase distributions or studies of technologically promising materials like TiFe or $LaNi_5$, or of superconducting hydrides. We have named some fields of major interest only at the risk of omitting others that may be or become important, but we hope to have shown that Mössbauer spectroscopy can indeed contribute to the understanding of metal-hydrogen systems, both in basic and applied research.

Appendix

Compilation of Mössbauer Studies of Metal-Hydrogen Systems

Matrix	Mössbauer isotope	Reference
Pd	^{57}Fe	*Bemski* et al. (1965) [6.1]
		Jech and *Abeledo* (1967) [6.78]
		Phillips and *Kimball* (1968) [6.79]
		Mahnig and *Wicke* (1969) [6.80]
		Carlow and *Meads* (1969, 1972) [6.96, 81]
		Wanzl (1973) [6.82]
		Obermann et al. (1976) [6.83]
		Oliver (1976) [6.127]
	^{119}Sn	*Chekin* and *Naumov* (1967) [6.91]
		Mahnig and *Wicke* (1969) [6.80]
		Kimball et al. (1976) [6.71]
	^{151}Eu	*Meyer* et al. (1975) [6.90]
	^{99}Ru, ^{193}Ir	*Iannarella* et al. (1974) [6.84]
	^{195}Pt, ^{197}Au	
	^{197}Au	*Karger* et al. (1978) [6.62]
Ni	^{57}Fe	*Wertheim* and *Buchanan* (1966, 1967) [6.2]
		Janot and *Kies* (1972) [6.35]
Ni–Fe	^{57}Fe	*Fujita* and *Sohmura* (1976) [6.77]
		Mizutani et al. (1976) [6.76]
Stainless steels	^{57}Fe	*Wertheim* and *Buchanan* (1967) [6.2]
		Fujita and *Sohmura* (1976) [6.77]
V	^{57}Fe	*Simopoulos* and *Pelah* (1969) [6.99]
Nb	^{57}Fe	*Ableiter* and *Gonser* (1975) [6.53]
	^{181}Ta	*Heidemann* (1978) [6.129]
Ta	^{181}Ta	*Heidemann* et al. (1975, 1976) [6.104, 64, 111]
TiFe	^{57}Fe	*Swartzendruber* et al. (1976) [6.54]
		Ron et al. (1976) [6.100]
		Bläsius (1976) [6.101]
YFe_2, $GdFe_2$	^{57}Fe	*Buschow* and *van Diepen* (1976) [6.103]
SmH_2	^{153}Eu	*Mustachi* (1973) [6.118], see also [6.116]
EuH_2	^{151}Eu	*Lounasmaa* and *Kalvius* (1967) [6.128]
EuPd, $EuRh_2$	^{151}Eu	*Buschow* et al. (1977) [6.105]
$EuNi_5$, $EuMg_2$	^{151}Eu	*Oliver* et al. (1978) [6.130]
EuH_2, YbH_2	^{151}Eu, ^{171}Yb	*Mustachi* (1974) [6.121]
GdH_2, GdH_3	^{155}Gd	*Cashion* et al. (1973) [6.113]
		Lyle et al. (1975) [6.114]
ScH_2	^{155}Gd	*Prowse* et al. (1973) [6.120]
DyH_2, DyH_3	^{161}Dy	*Abeles* et al. (1969) [6.115]
DyH_2		*Hess* et al. (1971) [6.117]
DyH_2, ErH_2	^{166}Er	*Stöhr* and *Cashion* (1975) [6.51]
HoH_2	^{160}Dy	
YH_2	^{166}Er	*Stöhr* and *Cashion* (1975) [6.51]
	^{160}Dy	
ErH_2, YH_2	^{166}Er	*Shenoy* et al. (1976) [6.24]
ErH_3, YH_3	^{166}Er	*Suits* et al. (1977) [6.119]
$ZrH_{1.5}$	^{166}Er	*Shenoy* et al. (1976) [6.50]
NpH_{2+x}	^{237}Np	*Mintz* et al. (1976) [6.124]
$Gd_{0.1}La_{0.9}Ni_5$	^{155}Gd	*Bauminger* et al. (1977) [6.55]
$PrNi_5$	^{141}Pr	*Thoma* (1973) [6.123]

References

6.1 G.Bemski, J.Danon, A.M.DeGraaf, X.A.DaSilva: Phys. Lett. **18**, 213 (1965)
6.2 G.K.Wertheim, D.N.E.Buchanan: J. Phys. Chem. Solids **28**, 225 (1967); Phys. Lett. **21**, 255 (1966)
6.3 T.C.Gibb: *Principles of Mössbauer Spectroscopy* (Chapman and Hall, London 1976)
6.4 N.N.Greenwood, T.C.Gibb: *Mössbauer Spectroscopy* (Chapman and Hall, London 1971)
6.5 L.May (ed.): *An Introduction to Mössbauer Spectroscopy* (Plenum Press, New York 1971)
6.6 H.Wegener: *Der Mössbauer-Effekt und seine Anwendungen in Physik und Chemie* (Bibliographisches Institut, Mannheim 1966)
6.7 V.I.Goldanskii, R.H.Herber (eds.): *Chemical Applications of Mössbauer Spectroscopy* (Academic Press, New York 1968)
6.8 U.Gonser (ed.): *Mössbauer Spectroscopy*, Topics in Applied Physics, Vol. 5 (Springer, Berlin, Heidelberg, New York 1975)
6.9 G.K.Shenoy, F.E.Wagner (eds.): *Mössbauer Isomer Shifts* (North-Holland, Amsterdam 1978)
6.10 J.G.Stevens, V.E.Stevens: *Mössbauer Effect Data Index* (IFI/Plenum Data Corporation, New York 1970ff.); *Mössbauer Effect Data and Reference Journal*, Vol. 1 (1978) ff.
6.11 J.Hauck, H.J.Schenk: J. Less-Common Metals **51**, 251 (1977)
6.12 J.Völkl, G.Alefeld: In *Diffusion in Solids, Recent Developments*, ed. by A.S.Nowik, J.J.Burton (Academic Press, New York 1975)
6.13 J.J.Vuillemin: Phys. Rev. **144**, 396 (1966)
6.14 F.M.Mueller, A.J.Freeman, J.O.Dimmock, A.M.Furdyna: Phys. Rev. B **12**, 4617 (1970)
6.15 D.E.Eastman, J.K.Cashion, A.C.Switendick: Phys. Rev. Lett. **27**, 35 (1971)
6.16 F.Antonangeli, A.Balzarotti, A.Bianconi, E.Burattini, P.Perfetti, N.Nisticó: Phys. Lett. **55** A, 309 (1975)
6.17 J.Zbasnik, M.Mahnig: Z. Physik B **23**, 15 (1976)
6.18 J.S.Faulkner: Phys. Rev. B **13**, 2391 (1976)
6.19 C.Norris, H.P.Myers: J. Phys. F.: Metal Phys. **1**, 62 (1971)
6.20 G.M.Stocks, R.W.Williams, J.S.Faulkner: J. Phys. F: Metal Phys. **3**, 1688 (1973)
6.21 V.V.Nemoshkalenko, V.G.Aleshin, V.M.Pessa, M.G.Chudinov: Physica Scripta **11**, 387 (1975)
6.22 W.M.Müller, J.P.Blackledge, G.G.Libowitz: *Metal Hydrides* (Academic Press, New York 1968)
6.23 K.R.Lea, M.J.M.Leask, W.P.Wolf: J. Phys. Chem. Solids **23**, 1381 (1962)
6.24 G.K.Shenoy, B.D.Dunlap, D.G.Westlake, A.E.Dwight: Phys. Rev. B **14**, 41 (1976)
6.25 D.S.Schreiber, R.M.Cotts: Phys. Rev. **131**, 1118 (1963)
6.26 F.E.Wagner, U.Wagner: In Ref. [6.9]
6.27 F.E.Obenshain: In *Mössbauer Effect Data Index Covering the* 1972 *Literature*, ed. by J.G.Stevens, V.S.Stevens (IFI/Plenum Data Corporation, New York 1973)
6.28 G.Kaindl, D.Salomon, G.Wortmann: Phys. Rev. B **8**, 1912 (1972)
6.29 G.Kaindl, D.Salomon, G.Wortmann: In Ref. [6.9] and in: *Mössbauer Effect Methodology*, Vol. 9, ed. by I.J.Gurverman, C.W.Seidel (Plenum Press, New York 1974); see also International Conference on Hyperfine Interactions Studied in Nuclear Reactions and Decay, Uppsala, Sweden, June 1974. Contributed Papers, p. 238
6.30 G.M.Kalvius, F.E.Wagner, W.Potzel: J. Phys. (Paris) **37**, C6–657 (1976)
6.31 E.R.Bauminger, G.M.Kalvius, I.Nowik: In Ref. [6.9]
6.32 G.M.Kalvius, E.Kankeleit: In *Mössbauer Spectroscopy and Its Applications*, STI/PUB/304 (International Atomic Energy Agency, Vienna 1972)
6.33 W.Gierisch, W.Koch, F.J.Litterst, G.M.Kalvius, P.Steiner: J. Mag. Magn. Materials **5**, 129 (1977)
6.34 W.Koch, G.M.Kalvius: J. Phys. (Paris) **37**, C6–728 (1976)
6.35 Ch.Janot, A.Kies: In *Proceedings of the International Conference on Hydrogen in Metals*, Paris 1972 (Editions Science et Industrie, Paris 1972)
6.36 D.L.Williamson: In Ref. [6.9]

6.37 J.D.Eshelby: In *Solid State Physics*, Vol. 3, ed. by F.Seitz, D.Turnbull (Academic Press, New York, 1956)
6.38 A.J.Freeman, D.E.Ellis: In Ref. [6.9]
6.39 R.S.Raghavan, E.N.Kaufmann, P.Raghavan: Phys. Rev. Lett. **34**, 1280 (1975)
6.40 K.W.Lodge: J. Phys. F: Metal Phys. **6**, 1989 (1976)
6.41 T.Butz: Physica Scripta **17**, 87 (1978)
6.42 F.D. Feiock, W.R.Johnson: Phys. Rev. **187**, 39 (1969)
6.43 Z.Biegański, J.Opyrchal, M.Drulis, B.Staliński: Phys. Stat. Sol. (a) **31**, 289 (1975); Fiz. Nizkikh Temp. **1**, 685 (1975)
6.44 Z.Biegański, J.Opyrchal, M.Drulis: Phys. Stat. Sol. (a) **28**, 217 (1975); Solid State Commun. **17**, 353 (1975)
6.45 Z.Biegański, B.Staliński: J. Less-Common Metals **49**, 421 (1976)
6.46 H.H.Wickman, G.K.Wertheim. In Ref. [6.7]
6.47 F.Hartmann-Boutron: Ann. Phys. (Paris) **9**, 285 (1975)
6.48 G.K.Shenoy, B.D.Dunlap, S.Dattagupta, L.Asch: J. Phys. (Paris) **37**, C6–85 (1976)
6.49 C.Meyer, F.Hartmann-Boutron, D.Spanjaard: J. Phys. (Paris) **37**, C6–79 (1976)
6.50 G.K.Shenoy, B.D.Dunlap, D.G.Westlake, A.Dwight: J. Phys. (Paris) **37**, C6–129 (1976)
6.51 J.Stöhr, J.D.Cashion: Phys. Rev. B**12**, 4805 (1975)
6.52 J.Stöhr: Phys. Rev. B**11**, 3559 (1975)
6.53 M.Ableiter, U.Gonser: Z. Metallkunde **66**, 86 (1975)
6.54 L.J.Swartzendruber, L.H.Bennett, R.E.Watson: J. Phys. F: Metal Phys. **6**, L331 (1976)
6.55 E.R.Bauminger, D.Davidov, I.Felner, I.Nowik, S.Ofer, D. Shaltiel: Physica **86**–**88**B, 201 (1977)
6.56 M.Karger, F.E.Wagner, F.Pröbst: Unpublished results
6.57 G.K.Shenoy, J.M.Friedt, H.Maletta, S.L.Ruby: In *Mössbauer Effect Methodology*, Vol. 9, ed. by I.J.Gruverman, C.W.Seidel, D.K.Dieterly (Plenum Press, New York 1974) p. 277
6.58 A.M.Afanasev, V.D.Gorobchenko: Phys. Stat. Sol. (b) **73**, 73 (1976): **76**, 465 (1976)
6.59 M.Blume: J. Phys. (Paris) **37**, C6–61 (1976)
6.60 E.Kankeleit: Z. Physik A**275**, 119 (1975)
6.61 E.Kankeleit, A.Körding: J. Phys. (Paris) **37**, C6–65 (1976)
6.62 M.Karger, F.E.Wagner, J.Moser, G.Wortmann, L.Iannarella: Hyperfine Interactions **4**, 849 (1978)
6.63 O.Berkooz, M.Malamud, S.Shtrikman: Solid State Commun. **6**, 185 (1968)
6.64 A.Heidemann, G.Kaindl, D.Salomon, H.Wipf, G.Wortmann: Phys. Rev. Lett. **36**, 213 (1976)
6.65 M.A.Krivoglaz, S.P.Repetskii: Soviet Phys.—Solid State **8**, 2325 (1967)
6.66 H.Wegener: In *Proceedings of the International Conference on Mössbauer Spectroscopy*, Vol. 2, ed. by A.Z.Hrynkiewicz, J.A.Sawicki (Cracow, Poland 1975) p. 257
6.67 K.Singwi, A.Sjölander: Phys. Rev. **120**, 1093 (1960)
6.68 R.C.Knauer: Phys. Rev. B**3**, 567 (1971)
6.69 G.K.Shenoy, F.E.Wagner, G.M.Kalvius: In Ref. [6.9]
6.70 R.V.Pound, G.A.Rebka: Phys. Rev. Lett. **4**, 337 (1960)
6.71 C.W.Kimball, G.van Landuyt, J.Spillman, E.E.Chain, F.Y.Fradin: J. Phys. (Paris) **37**, C6–29 (1976)
6.72 J.P.Motte, A.El Maslout, N.N.Greenwood: J. Phys. (Paris) **35**, C6–507 (1974)
6.73 G.Wortmann: J. Phys. (Paris) **37**, C6–333 (1976)
6.74 T.A.Kitchens, W.A.Steyert, R.D.Taylor: Phys. Rev. **138**, A467 (1965)
6.75 R.Ingalls, H.G.Drickamer, G.DePasquali: Phys. Rev. **155**, 165 (1967)
6.76 T.Mizutani, T.Shinjo, T.Takada: J. Phys. Soc. (Japan) **41**, 794 (1976)
6.77 F.E.Fujita, T.Sohmura: J. Phys. (Paris) **37**, C6–379 (1976)
6.78 A.E.Jech, C.R.Abeledo: J. Phys. Chem. Solids **28**, 1371 (1967)
6.79 W.C.Phillips, C.W.Kimball: Phys. Rev. **165**, 401 (1968)
6.80 M.Mahnig, E.Wicke: Z. Naturforsch. **24**a, 1258 (1969)
6.81 J.S.Carlow, R.E.Meads: J. Phys. F: Metal Phys. **2**, 982 (1972)
6.82 W.Wanzl: Thesis, Westfälische Wilhelms-Universität, Münster, Germany (1963) (unpublished)
6.83 A.Obermann, W.Wanzl, H.Mahnig, E.Wicke: J. Less-Common Metals **49**, 75 (1976)

6.84 L. Iannarella, F.E. Wagner, U. Wagner, J. Danon: J. Phys. (Paris) **35**, C6–517 (1974)
6.85 G. Longworth: J. Phys. C: Metal Phys. Suppl. **1**, 81 (1970)
6.86 P. Brill, G. Wortmann: Phys. Lett. **56** A, 477 (1976)
6.87 G. Kaindl, D. Salomon, G. Wortmann: Phys. Rev. B**8**, 1912 (1973); see also: *Mössbauer Effect Methodology*, Vol. 8, ed. by I.J. Gruverman, C.W. Seidel (Plenum Press, New York 1973) p. 211
6.88 G.K. Wertheim, J.H. Wernick: Phys. Rev. **123**, 755 (1961)
6.89 J.K. Jacobs, F.D. Manchester: J. Phys. F: Metal Phys. **7**, 23 (1977)
6.90 M. Meyer, J.M. Friedt, L. Iannarella, J. Danon: Solid State Commun. **17**, 585 (1975)
6.91 V.V. Chekin, V.G. Naumov: Soviet Phys. JETP **24**, 699 (1967)
6.92 H.S. Möller: Z. Physik **212**, 107 (1968)
6.93 J. Crangle, W.R. Scott: J. Appl. Phys. **36**, 921 (1965)
6.94 G.G. Low, T.M. Holden: Proc. Phys. Soc. **89**, 119 (1966)
6.95 W.L. Trousdale, G. Longworth, T.A. Kitchens: J. Appl Phys. **38**, 922 (1967)
6.96 J.S. Carlow, R.E. Meads: J. Phys. C: Solid State Phys. **2**, 2120 (1969)
6.97 J.A. Mydosh: Phys. Rev. Lett. **33**, 1562 (1974)
6.98 A.M. Clogston, B.T. Matthias, M. Peter, H.J. Williams, E. Corenzwit, R.C. Sherwood: Phys. Rev. **125**, 541 (1962)
6.99 A. Simopoulos, I. Pelah: Phys. Rev. **51**, 5691 (1969)
6.100 M. Ron, R.S. Oswald, M. Ohring, G.M. Rothberg, M.R. Polcari: Bull. Am. Phys. Soc. **21**, 273 (1976)
6.101 A. Bläsius: Thesis, Universität des Saarlandes, Saarbrücken, Germany (1976) (unpublished)
6.102 K.H.J. Buschow: Solid State Commun. **19**, 421 (1976)
6.103 K.H.J. Buschow, A.M. van Diepen: Solid State Commun. **19**, 79 (1976)
6.104 A. Heidemann, G. Kaindl, D. Salomon, H. Wipf, G. Wortmann: In *Proceedings of the International Conference on Mössbauer Spectroscopy*, Vol. 1, ed. by A.Z. Hrynkiewicz, J.A. Sawicki (Cracow, Poland 1975) p. 411
6.105 K.H.J. Buschow, R.L. Cohen, K.W. West: J. Appl. Phys. **48**, 2589 (1977)
6.106 Y. Fukai, S. Kazama, K. Tanaka, M. Matsumoto: Solid State Commun. **19**, 507 (1976)
6.107 R. Kubo: In *Fluctuation, Relaxation and Resonance in Magnetic Systems*, ed. by D. Ter Haar (Oliver and Boyd, Edinburgh 1962)
6.108 R.H. Swendsen, K.W. Kehr: Solid State Commun. **18**, 541 (1976)
6.109 B. Ströbel, K. Läuger, H.E. Bömmel: Appl. Phys. **9**, 39 (1976)
6.110 R.H. Swendsen, K.W. Kehr: Phys. Rev. B**13**, 5096 (1976)
6.111 A. Heidemann, H. Wipf, G. Wortmann: Hyperfine Interactions **4**, 844 (1978)
6.112 J. Opyrchal, Z. Biegański: Solid State Commun. **20**, 261 (1976)
6.113 J.D. Cashion, D.B. Prowse, A. Vas: J. Phys. C: Solid State Phys. **6**, 2611 (1973)
6.114 S.J. Lyle, P.T. Walsh, A.D. Witts, J.W. Ross: J. Chem. Soc. Dalton, 1406 (1975)
6.115 T.P. Abeles, W.G. Bos, P.J. Ouseph: J. Phys. Chem. Solids **30**, 2159 (1969)
6.116 E.R. Bauminger, G.M. Kalvius, I. Nowik: In Ref. [6.9]
6.117 J. Hess, E.R. Bauminger, A. Mustachi, I. Nowik, S. Ofer: Phys. Lett. **37** A, 185 (1971)
6.118 A. Mustachi: Thesis, The Hebrew University, Jerusalem (1973) (unpublished)
6.119 B. Suits, G.K. Shenoy, B.D. Dunlap, D.G. Westlake: J. Mag. Magn. Materials **5**, 344 (1977)
6.120 D.B. Prowse, A. Vas, J.D. Cashion: J. Phys. D.: Appl. Phys. **6**, 646 (1973)
6.121 A. Mustachi: J. Phys. Chem. Solids **35**, 1447 (1974)
6.122 F.A. Kuipers: J. Less-Common Metals **27**, 27 (1972)
6.123 K. Thoma: Thesis, Technical University of Munich, Munich, Germany (1973) (unpublished)
6.124 M.H. Mintz, J. Gal, Z. Hadari, M. Bixon: J. Phys. Chem. Solids **37**, 1 (1976)
6.125 W. Potzel, F.E. Wagner, G.M. Kalvius, L. Asch, J.C. Spirlet, W. Müller: AIP Conference Proceedings **38**, 77 (1977)
6.126 A. Forster, N. Halder, G.M. Kalvius, W. Potzel, L. Asch: J. Phys. (Paris) **37**, C6–725 (1976)
6.127 F.W. Oliver: J. Phys. Chem. Solids **37**, 1175 (1976)
6.128 O.V. Lounasmaa, G.M. Kalvius: Phys. Lett. **26** A, 21 (1967)
6.129 A. Heidemann: Unpublished results
6.130 F.W. Oliver, K.W. West, R.L. Cohen, K.H.J. Buschow: J. Phys. F: Metal Phys. **8**, 701 (1978)

7. Magnetic Properties of Metal Hydrides and Hydrogenated Intermetallic Compounds

W. E. Wallace

With 14 Figures

7.1 Introduction and Scope

The magnetic effects associated with the well-known solubility of hydrogen in palladium [7.1–6] attracted attention many years ago because of the inferences which can be drawn from them regarding the electronic makeup of the host metal. Results obtained for this system sparked interest in studies of other *d*-transition metal-hydrogen systems. Later these studies were broadened to include rare earth and actinide hydrides and, more recently, rare earth intermetallic compounds containing dissolved hydrogen. The salient features of each of these classes of hydrides are covered in this review: the influence of hydrogenation on the *d*-transition elements—Section 7.2; results obtained in studies involving rare earth and actinide hydrides—Section 7.3; the very recent work on hydrogenated rare earth intermetallics—Section 7.4. The results obtained for the rare earth hydrides are of interest in themselves but they are additionally of interest in that they represent the important experimental confirmation of the magnetic interaction mechanism operating in metallic rare earth systems proposed some years ago by theoretical solid-state physicsts [7.7–9].

The first observations of the influence of absorbed hydrogen on the magnetic properties of a rare earth intermetallic were contained in the work of *Zijlstra* and *Westendorp* [7.10]. They discovered by accident that certain etching techniques being used on the powerful intermetallic $SmCo_5$ degraded its magnetic properties—specifically its coercivity and remanence. They surmised that the effect was a consequence of hydrogen which had been absorbed in the lattice. Results for several other hydrogenated systems have become available in recent years and at present this is the most active area in the study of the magnetic properties of metal-hydrogen systems, partly because of its relationship to problems of storing energy in the form of hydrogen (see Ref. [7.11], Chap. 5). Results to date appear to represent the early stages in the development of an important new area of inquiry in metal-hydrogen systems.

7.2 Hydrides of the Transition Metals

A number of studies have been carried out to establish the influence of hydrogenation on the magnetic properties of *d*-transition elements. Pertinent

Table 7.1. Magnetic susceptibilities[a] (at 300 K) of transition elements and their hydrides

Sc 8.1 [7.12][b]	Ti 153 [7.13]	V 303 [7.14]	Cr 166 [7.15]
ScH_2 —[c]	TiH_2 229 [7.13]	V_2H 255 [7.14]	$CrH_{0.97}$ 217 [7.15]
Y 186 [7.16–18]	Zr 125 [7.19–21]	Nb 202 [7.22]	
$YH_{2.8}$ 55 [7.18]	ZrH_2 35 [7.19]	$NbH_{0.86}$ 66 [7.22]	
La 99 [7.23]	Hf —	Ta 152 [7.24]	
LaH_3 −21 [7.23]	HfH_2 —	Ta_2H 118 [7.24]	
		Ta_4H_3 76 [7.24]	

[a] The susceptibilities are in em u/g-at. of metal.
[b] Numbers in brackets indicate the reference.
[c] *Gschneidner* [7.12] indicates that the susceptibility of Sc is diminished by hydrogenation.

information for the systems that have been studied is summarized in Table 7.1. It is to be noted that all except LaH_3 are paramagnetic. The temperature dependence of the parent metals and the hydrides, described in the references cited, makes it clear that the paramagnetism does not originate with a system of local moments. These materials are all Pauli paramagnets, i.e., the magnetism derives from the spin imbalance (and the orbital motion) of the delocalized s and d electrons. In all cases except Ti and Cr, hydrogenation of the metal produces a decrease in the susceptibility (χ). This is presumably because the introduction of hydrogen leads to a decline in the density of states. In the free electron model χ is given by the expression

$$\chi = 4/3\,\mu_B^2 N(E)\,, \tag{7.1}$$

where $N(E)$ is the density of states per mole at the Fermi limit. This expression is obviously not applicable to transition elements; however, a decline in susceptibility is anticipated with a decline in density of states. A correlation of this nature is well exemplified by the studies of *Zanowick* and *Wallace* [7.14] and *Childs* et al. [7.25]. The former investigators found that χ decreases by about 15% in the hydride V_2H compared to the parent metal (see Table 7.1). *Childs* et al. studied the susceptibilities of V–Cr solid solutions and found a progressive decrease in χ as Cr replaces V in the lattice. From electronic specific heat work it is observed that V is near a peak in the density of states versus electron concentration curve and $N(E)$ drops sharply between V and Cr (see, for example, the review by *Mott* [7.26]). The drop in χ observed by *Childs* et al. reflects the declining density of states. Likewise, the decline in χ between V and

V_2H can be ascribed to modification of the band structure to reduce $N(E)$. The simplest kind of modification to envision, which is not necessarily correct (*vide infra*), is that hydrogen merely donates its electron to the conduction band, increasing the Fermi energy and decreasing the density of states. If this is assumed for hydrogen and also if it is assumed that Cr contributes an extra electron when it replaces V, then we expect V_2H and $V_{0.7}Cr_{0.3}$ to have about the same value of χ. This is indeed observed.

The situation for Nb, and probably Ta, resembles that for V. An increase in electron concentration beyond that in Nb results in a sharp decline in the density of states [7.26]. This correlates with the effect of introduction of hydrogen, suggesting that hydrogen in Nb and Ta lowers the value of $N(E)$, probably by raising the Fermi energy. The results of *Stalinski* for Ta show that the effect is progressive in that at the concentration Ta_2H, χ has been lowered by about 20% (compared to elemental Ta) and for Ta_4H_3 there has been an additional lowering of χ by about 30% (see Table 7.1).

Similar effects are observed in Y and La and their hydrides. Elemental Y and La are moderately strong paramagnets, whereas LaH_3 is diamagnetic. *Parks* and *Bos* [7.18] have observed a threefold reduction in χ when Y is hydrogenated to the composition $YH_{2.8}$ and they surmise that diamagnetic behavior would be realized for YH_3, which they were unable to prepare. As will be indicated in some detail in Section 7.3, experiment indicates that in La and Tm rare earths (and presumably with chemically similar Y) the sea of delocalized electrons is depleted during hydrogenation to the point that in RH_3, where R = a rare earth or Y, no delocalized electrons remain. Thus in LaH_3 one has a saline hydride $La^{3+}(H^-)_3$. An assemblage consisting exclusively of ions with closed shells is, of course, diamagnetic. The saline nature of LaH_3 is consistent with its observed diamagnetism.

As noted above, increments of χ upon hydrogenation are noted in two cases—Ti and Cr. This is not altogether unexpected since density of states information [7.26] (from electronic specific heat work) shows that these two metals are in a region of electron concentration (e.c.) such that a rise in e.c. leads to an increase of $N(E)$. Thus if electrons are donated to the band by the entering hydrogen, a rise in χ is expected.

The susceptibility of TiH_2 as measured by *Trzebiatowski* and *Stalinski* [7.13] shows a maximum at about 295 K (Fig. 7.1). Initially it was thought that this was due to the existence of antiferromagnetism below the temperature of the maximum. Later *Yakel* [7.27] showed that upon cooling, TiH_2 transforms at 295 K from a cubic to a tetragonal structure. The maximum in χ appears to be a consequence of this structural distortion rather than by the formation of a magnetically ordered material.

Palladium is the transition element which has been most extensively studied as a host for hydrogen [7.1–6]. It was established at an early date that its susceptibility declined with hydrogenation until at $PdH_{0.6}$, $\chi = 0$ and at still higher hydrogen concentrations χ becomes negative. Initially it was thought that there was 0.6 holes in the Pd *d*-band (actually the *d*-*s* hybrid band) and the

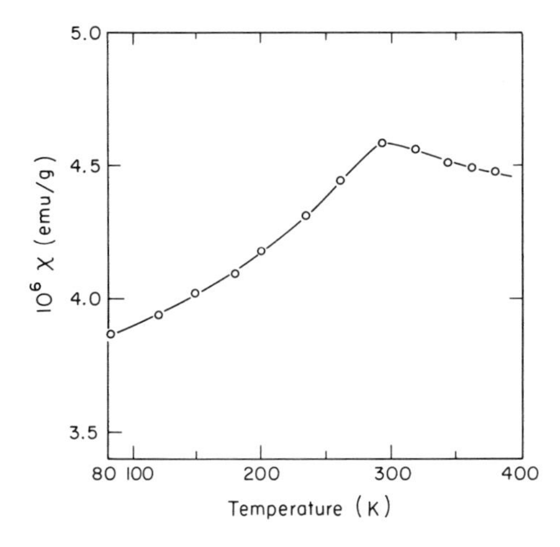

Fig. 7.1. Susceptibility (χ) versus temperature for TiH_2 [7.13]

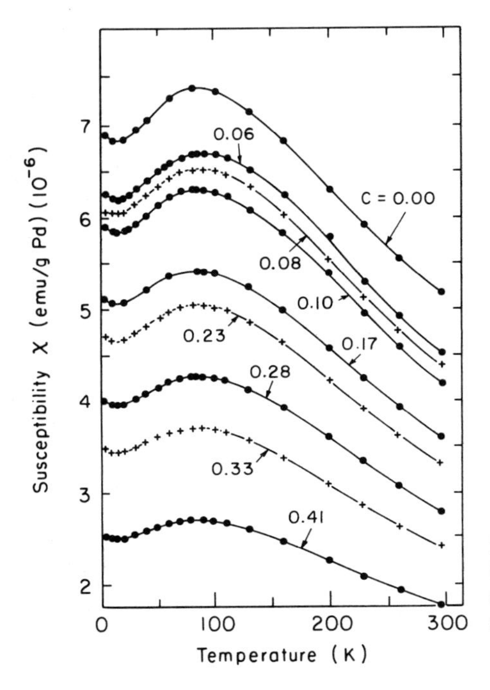

Fig. 7.2. Temperature dependence of susceptibility (χ) for Pd and Pd–H (●) and Pd–D (+) solid solutions. C is the number of H (or D) atoms per atom of Pd [7.5]

onset of diamagnetism at $PdH_{0.6}$ was ascribed to the filling of the band by the 1*s* electron supplied by hydrogen. Although the number of holes is now thought [7.28, 29] to be only 0.36, the reduction in susceptibility is still regarded as a band-filling effect. A thorough study of magnetic effects in the Pd–H and Pd–D systems was made by *Jamieson* and *Manchester* [7.5]. Their results are shown in Figs. 7.2 and 7.3.

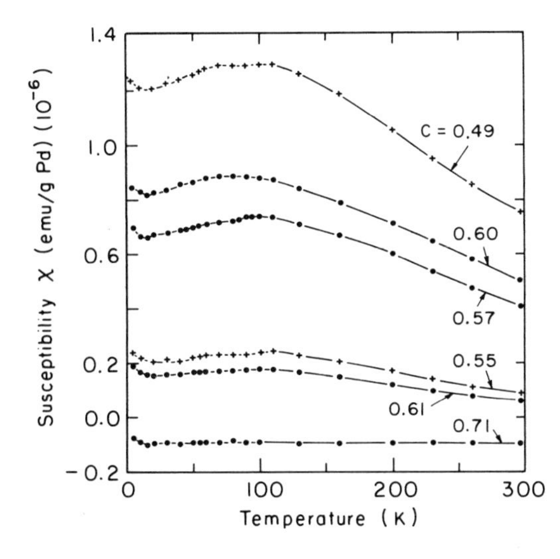

Fig. 7.3. Temperature dependence of susceptibility (χ) for Pd–H and Pd–D solid solutions showing the decline of χ with increasing C and the onset of diamagnetism for the most H-rich materials [7.5]. For definition of symbols used see caption for Fig. 7.2

The early idea of the progressive filling of the d-band as Pd is hydrogenated has given way to the sounder concept that the susceptibility behavior is primarily a consequence of the two-phase nature of the Pd–H (or Pd–D) system. The terminal solution based on Pd (usually termed the α phase) gives way to an intermediate (β) phase as the hydrogen concentration (C) is increased. Thus as C is increased the α phase material is progressively transformed into the β phase. Since χ for the α phase exceeds that for the β phase, χ diminishes with increase in C merely because of the α–β phase transformation. It is clear that the early concept of the filling of a rigid d-band as the metal is progressively hydrogenated is inapplicable in the Pd–H system. More generally, it can be stated that the rigid band model is of limited applicability for interpreting the magnetic properties of hydrogenated d-transition metals, particularly for systems having high hydrogen concentration.

Despite the limitations of the rigid band model, the physical properties of the transition metal hydrides have traditionally been interpreted on this model regarding the hydrogen as either the protonic or anionic [7.30]. Both of these have been involved above in the discussion of susceptibilities—the anionic model for the hydrides of Y and La and the protonic model in the other cases. It has been apparent for many years that these models, particularly the protonic models, represent an oversimplification of the condition which actually prevails. The incoming hydrogen according to the protonic model contributes electrons but not states to the band. Clearly H also contributes states, originating with its 1s eigenfunctions. If these states are above the Fermi energy, they are largely without effect, and the simple protonic model has validity and hence utility. If, however, the states lie below the Fermi energy, they may be very significant. The presence of hydrogen could in this case lower the Fermi energy. *Switendick* [7.31] has used the APW method to explore the

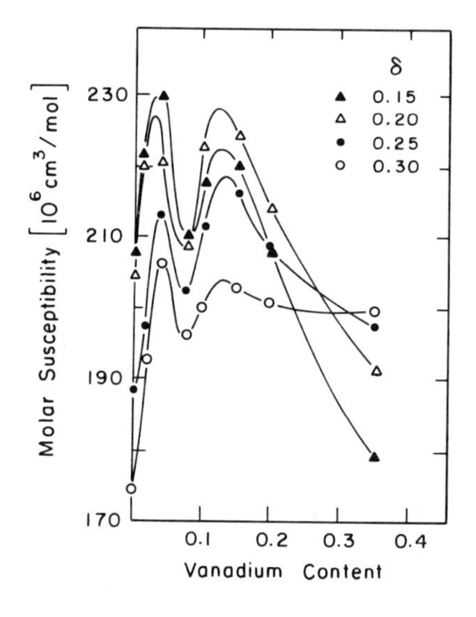

Fig. 7.4. Molar susceptibility versus x in $Ti_{1-x}V_xH_{2-\delta}$ obtained by extrapolating linearly the cubic susceptibilities to 80 K [7.32]

actual situation in a number of cases. Hybridization of the hydrogen 1s orbitals occurs to form states with energies lying below the Fermi limit. The energies of these states is concentration dependent, and in some cases their energies fall within the range of d-band so that the hybridization involves the d and s states of the host metal. *Switendick*'s work (see Chap. 5) and extensions of it to the structures observed will in time be very useful in interpreting results such as those shown in Table 7.1. At present the practice is to draw qualitative inferences from (7.1) and trends of density of states with electron concentration, the type of analysis made above for V_2H, Ta_2H, etc.

The failure of the rigid band model is well exemplified by the results of *Nagel* and *Goretzki* [7.32] on $Ti_{1-x}V_xH_{2-\delta}$. χ vs x results are shown in Fig. 7.4. The compositional dependence of χ, the double maxima in the curve, reflects the dependence of $N(E)$ at the Fermi limit on the vanadium content of the system, the Fermi energy being shifted to higher energies as x increases because the number of delocalized electrons increases by one for each vanadium atom.

It is to be noted that the positions of the maxima in Fig. 7.4 are independent of δ. This indicates that degree of occupation of the band is not affected by the hydrogen concentration as would be the case if the band were rigid with respect to changing hydrogen content.

7.3 Rare Earth and Actinide Hydrides

Extensive investigations of hydrides of the strongly magnetic rare earths have been carried out by *Wallace* and his associates [7.33–38]. As noted above, *Stalinski* [7.23] has studied the lanthanum hydrides.

The phases which form exist over a considerable composition range. They are based upon stoichiometries RH_2, the so-called dihydrides, and RH_3, the so-called trihydrides [7.39, 40]. At pressures of one or two atmospheres, all the lanthanides form dihydrides. For Eu and Yb these are the most hydrogen-rich phases formed under these conditions. The other rare earths under comparable conditions absorb hydrogen to compositions either reaching or approaching RH_3.

Almost all the experimental evidence suggests that hydrogen in the rare earth hydrides is anionic, i.e., H^-, and hence these materials essentially are saltlike or saline in nature. The trihydrides are similar to the saline hydrides of Groups Ia and IIa in that the stoichiometry corresponds to that expected from the maximum positive valence of the metal. Their saline nature is also indicated by their structures [7.40, 41] and electrical conductivity behavior [7.42]. Investigations of the magnetic characteristics of rare earth hydrides are consistent with this concept of these materials.

7.3.1 Rare Earths Which Form Trihydrides

Pr, Nd, Sm, Gd, Tb, Dy, Ho, Er, and Tm all form hydrides RH_x with x approaching or equalling 3. Electrical conductivity work on these hydrides and on LaH_3 and CeH_3 revealed that metallic conductivity disappears when x approaches 3. Early studies of the magnetic properties of cerium hydrides by *Stalinski* [7.43] and of GdH_2 by *Trombe* [7.44] showed that the paramagnetic rare earth magnetic moment is unmodified by hydrogenation, establishing that the hydrogen acquires its electron to form H^- not from the core, but instead from the conduction band. This has led to the following concept of the hydrogenation process: the hydride anion is formed by electron capture from the conduction band and this process continues as hydrogen is introduced until the conduction band is fully depopulated.

This concept has very strong implications in regard to the magnetic behavior of the rare earth hydrides. It is generally accepted that the magnetic interactions in the elemental rare earths are transmitted via the conduction electrons[1]. A rare earth ion polarizes the conduction electrons in its vicinity and the polarization propagates outward from the ion in a damped oscillatory fashion. Neighboring ions interact with the polarized conduction electrons and align either with or against the polarization wave, depending upon the sign of the effective exchange integral. The conduction electrons are thus the coupling mechanism involved in the magnetic interactions between the paramagnetic ions, the entire interaction being termed the RKKY interaction after *Ruderman-Kittel-Kasuya-Yosida* [7.7–9].

The RKKY interaction is the dominant interaction in the elemental rare earths and clearly, for it to be operative, there must be conduction electrons

[1] The f electrons centered on adjacent rare earth atoms are too localized to overlap appreciably so that direct exchange is negligible [7.45].

Table 7.2. Magnetic characteristics of polycrystalline rare earths and rare earth hydrides[a]

	μ_{eff} (μ_B per formula unit)	$g\sqrt{J(J+1)}$	θ [K]	T_c [K]	T_N [K]	Ref.
Pr	3.41	3.58	7	*n*	*n*	[7.34]
$PrH_{2.57}$	3.72	3.58	−75	*n*	*n*	[7.36]
Nd	3.33	3.62	4		19	[7.34]
$NdH_{2.75}$	3.41	3.62	−39	*n*	*n*	[7.34]
Sm	*nl*		*nl*		106	[7.34]
$SmH_{2.91}$	*nl*		*nl*	*n*	*n*	[7.34]
Gd		7.94		291		[7.37]
$GdH_{3.00}$	7.3	7.94	−3	*n*	*n*	[7.37]
Tb	9.85	9.72	230		235	[7.37]
$TbH_{2.97}$	9.7	9.72	−7	*n*	*n*	[7.37]
Dy	9.9	10.6	160		184	[7.35]
$DyH_{2.97}$	9.5	10.6	−2	*n*	*n*	[7.35]
Ho	10.4	10.6	86		135	[7.33]
$HoH_{3.06}$	10.1	10.6	−8	*n*	*n*	[7.33]
Er	9.85	9.6	56		88	[7.35]
$ErH_{3.02}$	9.54	9.6	−19	*n*	*n*	[7.35]
Tm	7.5	7.6	32		56	[7.35]
$TmH_{2.95}$	7.47	7.6	−34	*n*	*n*	[7.35]

[a] θ is the Weiss constant. μ_{eff} is the paramagnetic moment in Bohr magnetons per formula unit, obtained from the slope of the plot of inverse susceptibility versus temperature. $g\sqrt{J(J+1)}$ is the free ion paramagnetic moment. T_c and T_N represent the Curie and Néel temperatures, respectively. *nl* indicates that inverse susceptibility is nonlinear with temperature. *n* signifies that the hydrides give no indication of magnetic ordering at temperatures to 4 K. The data in this table are for the fully hydrogenated rare earth metals.

present. It therefore follows that if the conduction electrons are removed when the rare earth is hydrogenated, one expects the interactions to be greatly suppressed. In consequence, one anticipates for the fully hydrogenated metal either complete suppression of magnetic ordering at experimentally accessible temperatures or a substantial lowering of the ordering temperature in the rare earth hydride as compared to the parent metal.

Results, summarized in Table 7.2, show that these expectations in respect to magnetic ordering are borne out. The fully hydrogenated materials do *not* form a cooperative magnetic phase at temperatures extending down to 4 K. Thus the results are consistent with the notion of anionic hydrogen and a completely depopulated conduction band. Magnetic ordering does occur in the RH_2 systems (Table 7.3). That this occurs in no way invalidates the conclusions reached for the trihydrides. In the dihydrides, the conduction band has not been completely emptied (as evidenced by the occurrence of metallic conduction [7.42]) and the basis for the RKKY interaction remains.

It is to be noted that in both the trihydrides and dihydrides the observed paramagnetic moments are in reasonable agreement with $g\sqrt{J(J+1)}$, the value for a free ion. The materials are behaving in the paramagnetic region as an assemblage of free tripositive ions.

Table 7.3. Magnetic characteristics of polycrystalline rare earth dihydrides[a]

	μ_{eff} (μ_B per formula unit)	$g\sqrt{J(J+1)}$	θ [K]	T_c [K]	T_N [K]	Ref.
PrH_2	3.69	3.58	−33	*n*	*n*	[7.36]
NdH_2	3.38	3.62	−30	9.5		[7.34]
SmH_2	*nl*		*nl*	*n*	*n*	[7.34]
EuH_2	7.0	7.94	25	24	*n*	[7.38]
GdH_2	7.7	7.94	3		21	[7.37]
TbH_2	9.8	9.7	− 6		40	[7.37]
DyH_2	10.8	10.6	−16		8	[7.35]
HoH_2	9.9	10.6	− 5		8	[7.33]
ErH_2	9.75	9.6	−18	*n*	*n*	[7.35]
TmH_2	7.60	7.6	−40	*n*	*n*	[7.35]

[a] For nomenclature in this table see footnote in Table 7.2.

While the preponderance of experimental evidence supports the concept of anionic hydrogen in rare earth hydrides, there are some aspects of these systems which have been interpreted to indicate protonic hydrogen. *Heckman* has pointed out that by making some rather special assumptions about the band shape in these systems the results are susceptible to interpretation within the framework of the protonic model [7.42]. Attention was focused on the protonic model through Hall coefficient measurements.

Heckman measured [7.42] the Hall coefficient of CeH_x, with x in the range 2.18 to 2.8. He observed that the Hall coefficient was positive and became increasingly positive as x increased toward 2.8, and the conductivity diminished toward values characteristic of a poor insulator or a semiconductor. The sign of the Hall coefficient indicates a nearly filled band. From this *Heckman* concluded that CeH_x was consistent with the protonic model and that the hydrogenation of the rare earths involved band filling rather than band depopulation. However, to rationalize the conductivity, magnetism and Hall coefficients of the nearly fully hydrogenated systems, *Heckman* was forced to postulate a special kind of splitting of the conduction band. This postulate has proved to be not very convincing. Recent band structure calculations of *Switendick* [7.31] provide a framework for interpreting all the magnetic and electrical properties of the rare earth trihydrides.

In his calculations, *Switendick* finds for RH_3 that there are two bands, one characteristic of the host metal and a second, a new band, formed from the hydrogen 1*s* orbitals. This new band is lower lying than the *d*-band of the host metal and has the capacity to hold precisely the six valence electrons supplied by one R and three H. In the RH_3 compounds it is filled, and these materials are semiconductors. In slightly substoichiometric RH_3, the new band contains a few vacancies, giving rise to the positive Hall coefficient. When filled, the band cannot transmit the RKKY interaction. This accounts for the suppression of the Curie and Néel temperatures of the RH_3 systems compared to R.

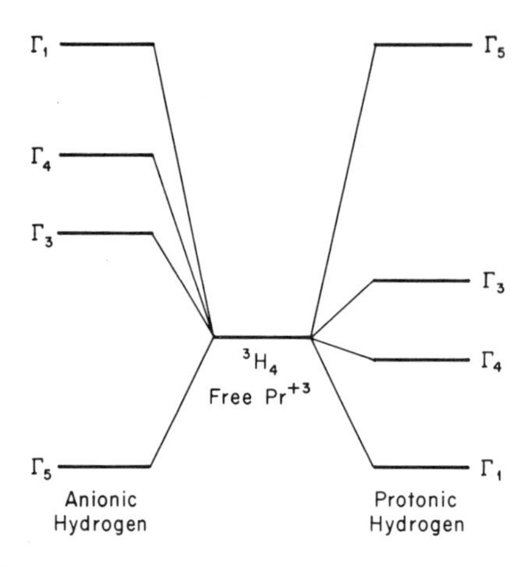

Fig. 7.5. Crystal field splitting of the Pr^{3+} 9-fold degenerate ground state multiplet in PrH_2. The patterns to the left and right are those expected if hydrogen is anionic (H^-) or protonic (H^+), respectively

At the time, when there was aggressive advocacy of the protonic model, *Wallace* and *Mader* [7.36] made a careful examination of PrH_2, whose magnetic properties sensitively depend upon the sign of the charge carried by hydrogen. Pr^{3+} in PrH_2 is surrounded by eight hydrogen ions bearing either a positive or negative charge, depending upon whether the protonic or anionic model is invoked. The free Pr^{3+} ion is in an $^3H_4$ state which is $(2J+1)$-fold degenerate where $2J+1=9$. The nine states in the free ion are pure M states with M ranging from -4 to $+4$ in steps of unity. In a crystal the ion is perturbed by the electrostatic field of the neighboring ions, and the pure M states give way to states that are linear combinations of the pure M states. The latter are called the crystal field states. These states are depicted in Fig. 7.5 in which it is shown that with anionic hydrogen a splitting pattern occurs with a Γ_5 (triplet) state as the lowest lying state. If the hydrogen is protonic, the splitting pattern is inverted and a Γ_1 state is the ground state. The magnetic behavior of an assemblage of Pr^{3+} ions is very different when distributed over sets of energy levels represented by the two splitting patterns shown. When Γ_1 is the ground state, the susceptibility becomes temperature independent at low temperatures (i.e., at temperatures such that $kT<E_{\Gamma_4}-E_{\Gamma_1}$). This is the so-called Van Vleck paramagnetism. When Γ_5 is the ground state, the Van Vleck paramagnetism does not occur [7.46]. χ for an assemblage of ions distributed over the crystal field states according to the Boltzmann factor is given by the fundamental Van Vleck expression

$$\chi=(N_0/HQ)\sum \mu_i \exp(-E_i/kT)\,, \tag{7.2}$$

where μ_i and E_i are the moment and energy, respectively, of the ith crystal field state, Q is the partition function, N_0 is the number of ions per mole, and H is the magnetic field strength. Neglecting the influence of the sixth order crystal field interaction, one can express Q, E_i, and μ_i in terms of a single parameter, the overall splitting (E_c) [7.47].

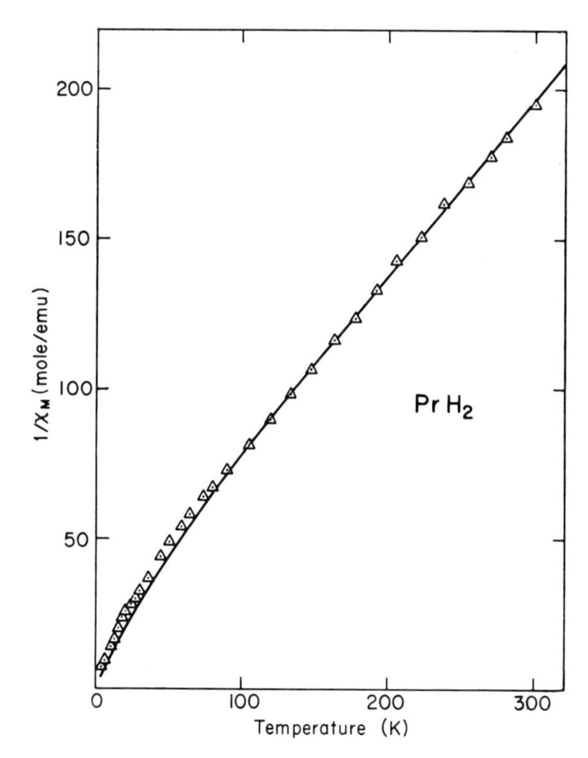

Fig. 7.6. The points give the reciprocal susceptibility of PrH_2 versus temperature for PrH_2 [7.36]. The full line is the calculated χ^{-1} versus T behavior if hydrogen is anionic and the crystal field splitting is that shown to the left in Fig. 1.5

Susceptibility data for $PrH_{2.01}$ plotted as inverse susceptibility versus temperature are shown in Fig. 7.6 taken from the work of *Wallace* and *Mader*. The χ^{-1} values computed from (7.2) using the spectrum in Fig. 7.5 for anionic hydrogen and $E_c = 657$ K are also shown. It is clear that an excellent accounting for the susceptibility is possible on the basis of the anionic model. This is not possible using the protonic model, for which, as noted above, the Van Vleck temperature-independent paramagnetism is expected at the lowest temperatures. No value of E_c can lead to a fit between observed and calculated χ^{-1} values. Thus, the paramagnetic behavior of PrH_2 is only consistent with the anionic model.

The validity of the anionic model for PrH_2 (and by extension to all the rare earth hydrides) is supported by two other lines of evidence: 1) heat capacity measurements at low temperatures [7.48–50] and 2) inelastic neutron scattering studies [7.51]. The crystal field interaction influences the heat capacity of the system as well as its susceptibility. At the lowest temperatures, the ions are in the ground crystal field state or states. With increasing temperature they occupy the excited states, giving rise to a contribution to the heat capacity [7.48, 49]. This contribution can be calculated in terms of the energy level schemes (Fig. 7.5) for both the protonic and anionic model. *Bieganski* and *Stalinski* have made [7.50] such computations for RH_2 with R = Ce, Pr, and Nd, and compared the calculated heat capacities with experiment for these dihydrides. They find that the heat capacity behavior can be accounted for reasonably well on the basis of the anionic model, whereas results computed using the protonic model cannot be brought into accord with experiment.

The crystal field splitting, such as is shown in Fig. 7.5, can be elucidated by inelastic neutron scattering experiments. Thermal neutrons impinging upon the sample can transfer energy to the ion to excite it from one crystal field state to another. This absorption can be detected, and the analysis of such an absorption spectrum gives information about the nature of the crystal field splitting. PrH_2 was studied by *Knorr* and *Fender* [7.51] using such an approach. Analysis of their results fully supported the concept of anionic hydrogen in PrH_2.

From the above discussion it is clear that there is considerable evidence indicating that hydrogen bears a negative charge. This is consistent with the known band structure of the hydride. The new band is formed from antibonding hydrogen states. In such states, charge is localized around the nucleus (rather than at points between the nuclei, which is the case with bonding orbitals). Thus when the new band is filled, all six valence electrons are in states which localize charge on the hydrogen nucleus, giving essentially the hydride ion. This accounts for the success of the anionic model in treating the properties of the system which depend upon the details of the crystal field splitting—the temperature dependence of paramagnetic susceptibility, low-temperature heat capacity behavior, and the inelastic scattering of neutrons.

From the preceding discussion it is seen that magnetic studies of the rare earth hydrides have proved to be exceedingly valuable. The information obtained has been useful in establishing the charge on hydrogen in this material, and equally importantly it has provided experimental confirmation of the RKKY interaction postulated by the solid-state theorists to account for the magnetic interactions in metallic rare earth systems.

7.3.2 Rare Earths Which Form Only Dihydrides

Europium and ytterbium under normal conditions form only the dihydride. This is a consequence of the stability of the divalent state in these elements, which is, in turn, a consequence of the special stability of the $4f^7$ and $4f^{14}$ ion core configurations. Eu^{+3} and Yb^{+3} would capture an electron from the conduction band to achieve a half-filled and filled f-shell, respectively.

YbH_2 is expected to exhibit diamagnetism in view of the closed-shell electronic configuration of Yb^{+2} and H^-. This expectation has been confirmed experimentally [7.37]. The situation is quite different for EuH_2. Eu^{+2} has seven unpaired spins and hence is strongly magnetic. Magnetization versus temperature data for EuH_2 (actually $EuH_{1.86}$) indicate [7.38] that EuH_2 orders ferromagnetically with a Curie temperature of 24 K. Magnetization versus field measurements (extending only to an applied field of 7.5 kOe) showed a saturation moment of about $6\mu_B$ per EuH_2, which approaches the expected

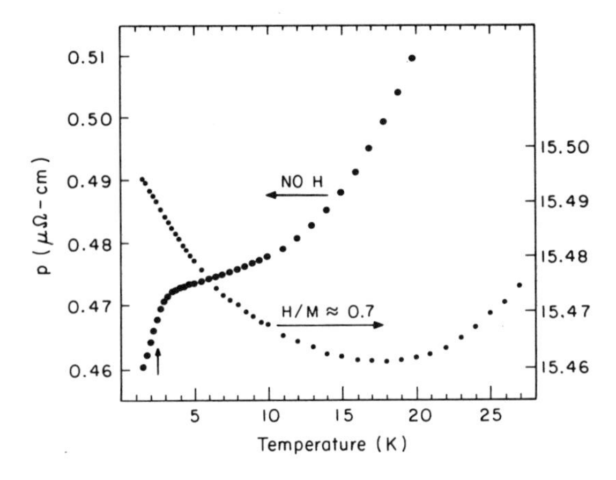

Fig. 7.7. Temperature dependence of electrical resistivity for Pd + 0.002 atomic % Fe with and without hydrogen. The arrow represents the Curie temperature of the hydrogen-free system. H/M represents the hydrogen-to-metal ratio [7.52]

value for Eu^{+2}, i.e., $7\mu_B$. Curie-Weiss dependence was observed for the variation of susceptibility with temperature. An effective moment of $7.0\mu_B$ was indicated. This is somewhat low compared to the $7.94\mu_B$ expected. It is, however, much too large to be ascribed to Eu^{+3}, which exhibits only Van Vleck susceptibility, since $J=0$ for tripositive europium.

Mydosh [7.52] in resistivity studies of hydrogenated Pd which had been doped with Fe observed behavior somewhat similar to that just described for the fully hydrogenated rare earth elements. Results for Pd containing 0.2 atomic percent Fe are shown in Fig. 7.7. The hydrogen-free alloy shows a sharp decline in resistivity near 2.5 K at which temperature the sample orders ferromagnetically. When hydrogenated to saturation, the alloy shows rather different ϱ vs T behavior. There is a resistivity minimum, which *Mydosh* ascribes to the Kondo effect, and the resistivity is still rising at the lowest temperature measured, ~1.5 K. The widely separated Fe atoms in the Pd matrix are coupled by the RKKY interaction giving rise as noted to magnetic ordering at 2.5 K in the hydrogen-free material. This interaction is obviously altered when the alloy is hydrogenated to saturation, and in this respect these systems resemble the rare earth hydrides. *Mydosh*'s results for systems more concentrated in Fe make it clear that hydrogenation weakens the Fe–Fe interaction just as hydrogenation weakens the R–R interaction in RH_3. However, studies of $PrCo_2$ and $NdCo_2$ (*vide infra*) indicate that hydrogenation strengthens Co–Co exchange, at least when hydrogen is present in small quantities.

7.3.3 Actinide Hydrides

Hydrides of the actinide elements have not been studied to any significant extent, except for UH_3. This material exists in two allotropic modifications, both cubic. βUH_3 exists in the W_3O structure. It (and αUH_3) are exceptional in that they order ferromagnetically at low temperatures with a Curie temperature

Table 7.4. Magnetic properties of UH_3

	T_c [K]	θ [K]	μ_{sat} (μ_B/formula unit)	μ_{eff} (μ_B/formula unit)
αUH_3	178 [7.56][a]	174 [7.54]	0.9 [7.54]	2.8 [7.54]
βUH_3	173 [7.53]	137 [7.55]	1.2 [7.57]	2.79 [7.55]

[a] The numbers in brackets indicate the reference.

of about 175 K, the highest Curie temperature observed for metal hydrides. The observation that βUH_3 becomes ferromagnetic at low temperatures was initially made by *Trzebiatowski* et al. [7.53]. Later *Sliwa* and *Trzebiatowski* [7.54] studied αUH_3 and found that the magnetic properties of the two allotropic modifications are very similar. Both forms exhibit Curie-Weiss behavior, above 350 K for βUH_3 [7.55] and above 200 K for αUH_3 [7.54]. The properties of the two forms of UH_3 are summarized in Table 7.4. The characteristics of these two hydrides are very similar, indicative of a similar magnetic structure of uranium in these materials.

The $5f$ electrons are not well localized in the light elemental actinides, and hence elemental U does not exhibit Curie-Weiss behavior [7.58]. In the hydrides the U–U distances have been increased (by about 10% in β and 20% in α, compared to the element) and the $5f$ electrons appear to have become more nearly localized. In this respect the α form is better behaved than αUH_3 in the sense that it conforms to Curie-Weiss behavior at all temperatures down to its Curie temperature, whereas for βUH_3 proper Curie-Weiss behavior does not occur until the temperature exceeds T_c by about 175 degrees.

The paramagnetic moments (Table 7.4) suggest two unpaired spins for which $\mu_{eff} = 2.82\mu_B$. The ordered moment is less than that expected ($2.0\mu_B$), perhaps because of the quenching effect of the crystal field interaction.

The PuH_{2+x} system has been studied by proton NMR by *Cinader* et al. [7.59]. The results obtained indicate ferromagnetic Pu–Pu interactions.

7.4 Hydrogenated Rare Earth Intermetallics

7.4.1 Early Indications of the Hydrogen Affinity of Rare Earth Intermetallics

In 1969 *Zijlstra* and *Westendorp* [7.10] observed, while studying the capabilities of $SmCo_5$ as a material for forming very powerful permanent magnets, that this intermetallic absorbed very large quantities of hydrogen rapidly and reversibly at room temperature. Almost simultaneously *Neumann* [7.60] observed the solvent power of $LaNi_5$ for hydrogen, and shortly thereafter *Van Vucht* et al. [7.61] presented information concerning the hydrogen solubility in several RCo_5 and RNi_5 compounds. *Kuijpers* and *Loopstra* provided [7.62] structural

Table 7.5. Effect of degassing on the lattice parameters of RCo_5 systems

	a_0 [Å]		c_0 [Å]		Δc_0
	Virgin	Degassed	Virgin	Degassed	
YCo_5	4.946	4.945	3.975	3.958	0.017
$LaCo_5$	5.105	5.108	3.967	3.950	0.017
$CeCo_5$	4.922	4.920	4.008	3.785	0.023
$Ce_{0.6}Y_{0.4}Co_5$	4.924	4.924	4.005	3.986	0.019
$Pr_{0.4}Y_{0.6}Co_5$	4.970	4.966	3.973	3.960	0.013
$Ce_{0.2}La_{0.8}Co_5$	5.054	5.037	3.987	3.971	0.016
$La_{0.4}Pr_{0.6}Co_5$	5.042	5.042	3.974	3.952	0.022

data for $PrCo_5D_4$, based on neutron diffraction measurements; they showed that molecular hydrogen (or deuterium) is broken down into atoms (or possibly ions) when taken into the intermetallic. The solute atom (or ion) fills 2/3 of the available octahedral sites and 1/3 of the tetrahedral sites.

In *Zijlstra* and *Westendorp*'s study the hydrogen was initially introduced by inadvertence, as a consequence of etching $SmCo_5$. Later, hydrogen was deliberately introduced to ascertain its effect on magnetic properties, primarily coercivity and remanence. The former was reduced tenfold or more and the latter by $\sim 20\%$. $SmCo_5$ was being studied as useful permanent magnet material, and both of these reductions are disadvantageous in this regard.

During measurements of electronic specific heats of RNi_2 [7.63] and RNi_5 [7.64] systems, *Neumann* communicated his observations [7.60] of hydrogen solubility in $LaNi_5$ to *Wallace*, whereupon *Wallace* and *Titcomb* initiated work in 1970 to ascertain whether rare earth intermetallics as normally prepared contained hydrogen as a contaminant. This appeared likely in view of the hydrogen affinity of the rare earth intermetallics then being revealed. If so, it was of interest to know the effect of dissolved hydrogen on the properties of the intermetallic. The compounds were prepared in the usual way, that is, by fusing together the component metals. They were then placed in an ultrahigh vacuum unit and held at about 950 °C until degassing ceased. The "degassed" material was then examined and its behavior compared with that of "virgin" material. The observations made included determinations of crystal structure and measurements of electrical resistivity and magnetization as functions of temperature. Lattice parameter determinations (Tables 7.5 and 7.6) showed in many cases a substantial difference between the virgin and degassed material, the former larger because of the hydrogen contaminant.

7.4.2 Effect of H_2 on $PrCo_2$ and $NdCo_2$

The largest differences noted were for $PrCo_2$ and $NdCo_2$, indicating these materials to be the most heavily contaminated with hydrogen. These materials

Table 7.6. Effect of degassing on the lattice parameters of RCo_2 and RFe_2 systems

	a_0 [Å]						
	$PrCo_2$	$NdCo_2$	$GdCo_2$	$DyCo_2$	$HoCo_2$	$GdFe_2$	$HoFe_2$
Virgin	7.301	7.352	7.264	7.186	7.155	7.367	7.275
Degassed	7.140	7.166	7.245	7.131	7.008	7.360	7.277
Difference	0.161	0.186	0.019	0.055	0.147	0.007	0.002

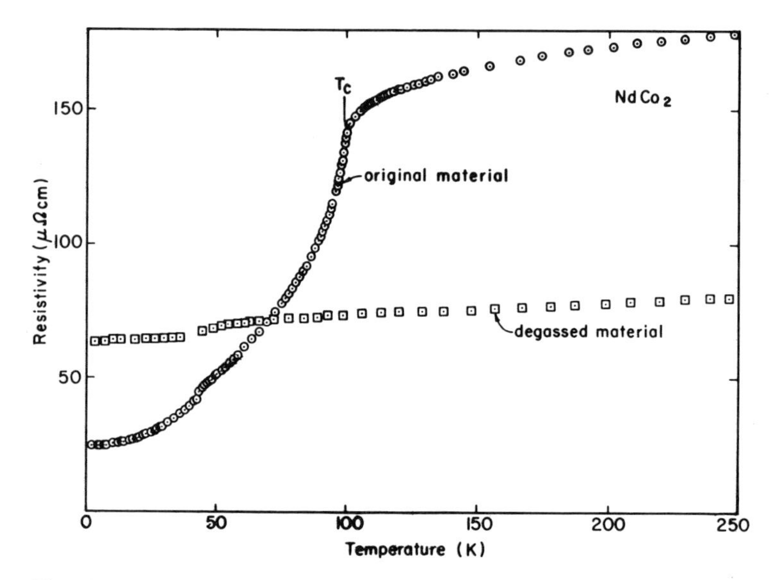

Fig. 7.8. Resistivity-temperature behavior of original, or virgin, and degassed $NdCo_2$ [7.65]. The Curie temperature (T_c) of the virgin (or hydrogen-contaminated) material is indicated at about 100 K. Degassed material gives no evidence of magnetic ordering at low temperatures

were selected for detailed study. Striking results were obtained [7.65]. Earlier work on virgin $PrCo_2$ and $NdCo_2$ indicated that these materials become ferromagnetic [7.66, 67] at 45 and 116 K, respectively. Resistivity measurements (Figs. 7.8 and 7.9) for these materials show, as expected, a sharp decline in resistance near the Curie temperature. This is ascribed to the loss of the paramagnetic scattering which occurs as the material orders magnetically. Surprisingly, degassed $PrCo_2$ and $NdCo_2$ failed [7.65] to show this decline. This strongly suggests that the degassed materials do not order magnetically, at least at temperatures in excess of 4 K, and this was confirmed by magnetic measurements (Fig. 7.10).

It therefore appears that the presence of hydrogen in $PrCo_2$ and $NdCo_2$ is necessary for the formation of the cooperative phase; the magnetic interactions are too weak for ordering to occur in the degassed materials. Although the

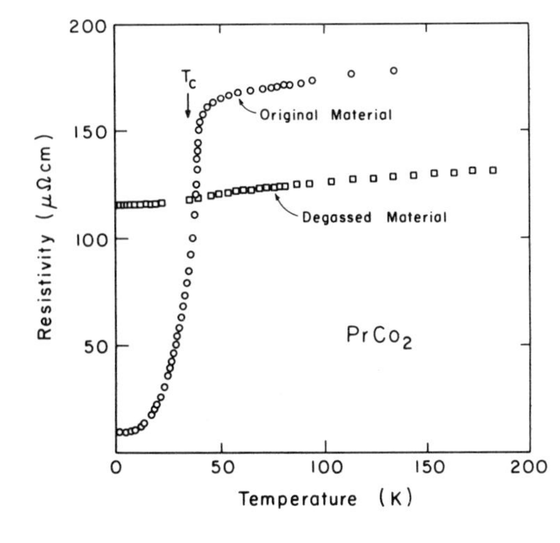

Fig. 7.9. Resistivity-temperature behavior of original, or virgin, and degassed $PrCo_2$. The normal Curie temperature is designated by the arrow and T_c [7.65]

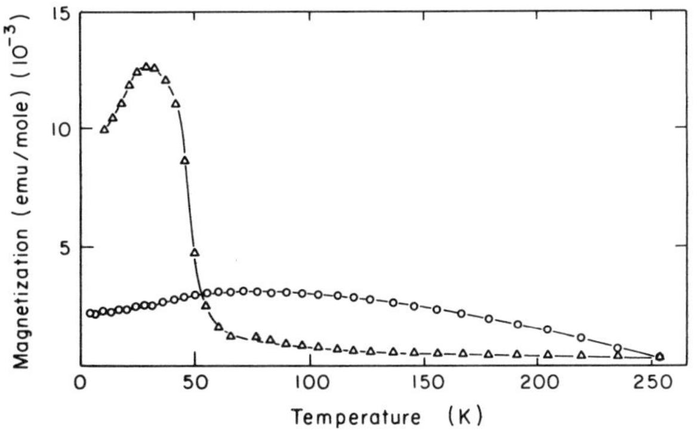

Fig. 7.10. Magnetization-temperature curves for virgin (△) and degassed (○) $PrCo_2$, determined at 6.1 kOe [7.65]. The decrease in magnetization of the virgin material for $T<30$ K does not occur at higher applied fields [7.66]; the decrease is probably due to increasing magnetic hardness at low temperatures. The uncontaminated material does not magnetically order at low temperatures whereas H-contaminated $PrCo_2$ does for $T<40$ K

influence of the hydrogen contaminant is very surprising, it is possible to rationalize the effect. The trend of solubilities observed by *Van Vucht* et al. [7.61] shows a decreasing tendency to absorb hydrogen with decrease in the lattice dimensions of the compound. Thus $CeNi_5$, whose unit cell is considerably smaller than that of $LaNi_5$, is a poor hydrogen solvent. The solubility decreases in the sequence $LaNi_5$, $PrNi_5$, $NdNi_5$, and $SmNi_5$. These trends can readily be understood in terms of anionic hydrogen residing in interstitial positions which become progressively smaller, and hence less energetically favorable, as the atomic number of the rare earth increases. These trends are not understood, or at best can be understood only with great difficulty, in terms of protonic hydrogen. Thus it appears that hydrogen is an electron absorber in the compounds just as it is in the elemental rare earths. Consequently, the electron concentration (e.c.) changes as the concentration of hydrogen in the metal is altered. It is well known that the interactions in rare earth systems are e.c. dependent [7.68]. Evidently, removal of hydrogen alters the interactions sufficiently to suppress substantially the magnetic interactions, resulting in a considerable lowering of the ordering temperatures.

Viewed in this context, the behavior of $PrCo_2$ is not altogether surprising. In the elaboration of the ideas adumbrated in the preceding paragraph, it is convenient to consider $PrNi_2$ and PrBi. $PrNi_2$ has been shown [7.69] to become a Van Vleckt paramagnet at low temperatures. The crystal field spectrum for Pr^{+3} in $PrNi_2$ is like that shown to the left in Fig. 7.4. In it, as in PrBi studied by *Tsuchida* and *Wallace* [7.70], interactions are weak, ordering does not occur, and the ions settle upon cooling into the nonmagnetic singlet (Γ_1) ground state. *Cooper* et al. [7.71] and *Mader* et al. [7.72] have shown that even when the ground state is a singlet, magnetic ordering will occur at reduced temperatures, provided the ratio of exchange to the crystal field interaction exceeds some minimal value. In this case, ordered magnetic structures grow out of a singlet state by a process termed the "bootstrap" process by *Cooper*.

Pr in $PrCo_2$ is a moment-carrying species, its moment at low temperatures being developed out of the Γ_1 state by the bootstrap process. Cobalt in this compound is also a moment-carrying species. However, this is not the case with all the RCo_2 compounds [7.73]. Co is nonmagnetic in $CeCo_2$ and in YCo_2 [7.66]. Accordingly, cobalt is found to carry a moment in the Laves phase compounds only when its partner is magnetic. Thus the moments of both components in $PrCo_2$ are dependent upon the RKKY exchange interactions, that of Pr directly and that of Co indirectly as a consequence of the preexistence of a Pr moment.

Now to return to the influence of dissolved hydrogen on $PrCo_2$, it can be seen that if withdrawal of hydrogen by the degassing process alters the electron concentration so as to weaken the RKKY interaction, first the moment of Pr will collapse and subsequently that of Co will disappear, leaving only a Van Vleck paramagnet at low temperatures. It appears that the alteration of the magnetic properties of $PrCo_2$ upon degassing comes about in this way.

The general effect of hydrogen removal should be the same for $NdCo_2$ as for $PrCo_2$ in regard to the magnetic interactions. However, the ground state for Nd^{+3} in the cubic environment prevailing in $NdCo_2$ is the doublet Γ_6 state. The Γ_6 state remains magnetic, irrespective of the strength of the interactions. The results for $NdCo_2$ suggest that the Co moment in this material is sustained by hydrogen, through its effect on the interactions as described above, and that when hydrogen is withdrawn, cobalt becomes a nonmagnetic species. If so, the ordering temperature would fall substantially, perhaps to temperatures below 4 K. The basis of the assertion that the ordering temperatures will sharply decrease if cobalt is rendered nonmagnetic is the intercomparison of the Curie temperatures of the RNi_2, in which Ni is nonmagnetic, and RCo_2 series [7.74]. For example, Curie temperatures of virgin $NdNi_2$ and $NdCo_2$ are 6.5 and 116 K, respectively. The existence of Co as a magnetic species raises T_c by about

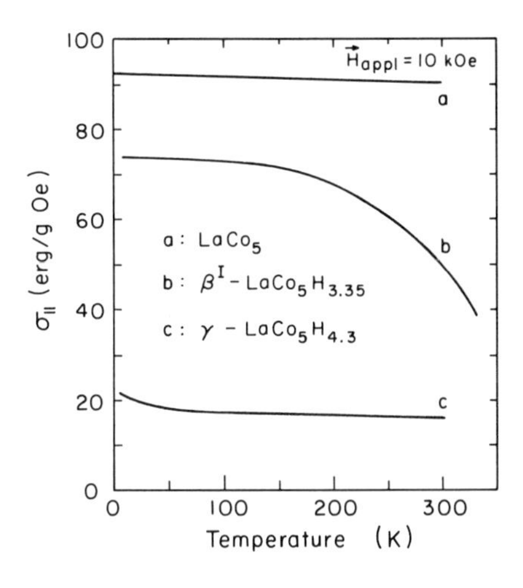

Fig. 7.11. Magnetization versus temperature for $LaCo_5$ (Curve *a*) and two of its hydrides (Curves *b* and *c*). A 10 kOe field is applied along the *c*-axis (the easy axis of magnetization) and the magnetization is measured in the direction of the field [7.75]

100 K in this case. Removal of hydrogen and destruction of the magnetization of cobalt would be expected to lower T_c correspondingly. It is perhaps for this reason that $NdCo_2$ fails to give signs of magnetic ordering at temperatures extending down to 2 K.

7.4.3 RCo_5-H Systems

RCo_5 with R = La, Ce, Pr, and Nd have been studied by *Kuijpers* [7.62, 75]. The reduction of the Co moment noted above for the $PrCo_2$ and $NdCo_2$ systems is also observed for the RCo_5 compounds. The results for $LaCo_5$ (Fig. 7.11), taken from *Kuijpers*' thesis, illustrate the behavior of this family of hydrides. The magnetizations of the β and γ hydrides are about 20 and 80%, respectively, lower than that of $LaCo_5$. The Co moment in $LaCo_5$ is $1.45\mu_B$, whereas that in $LaCo_5D_{3.35}$ is $1.14\mu_B$. Results in Fig. 7.11 show a lowered T_c when $LaCo_5$ is hydrogenated, indicating weakened exchange. Again there is similarity in the behavior of the RCo_5 and RCo_2 systems.

The weakened exchange interaction upon hydrogenation is illustrated in a striking fashion by the neutron diffraction results on $PrCo_5$ and its hydride $PrCo_5D_{3.6}$ obtained by *Kuijpers* and *Loopstra* [1.61]. These are given in Table 7.7. Magnetically, there are two types of cobalt in this material, and these supply the exchange field which serves to align the rare earth moment. From the data in Table 7.7 it is clear that the exchange field acting on Pr^{3+} is reduced in the hydride. This is evident from the fact that the Pr moment has vanished at 300 K, i.e., the Pr moments have become disordered at this temperature. In $PrCo_5$ the Pr moment is sustained to 300 K. The reduction of the Co moments, which generate the exchange field, is also readily apparent.

In summary, Co moments, exchange and Curie temperatures of rare earth-cobalt intermetallics are all reduced when they are hydrogenated. As will be seen in the discussion which follows, this behavior is not always observed in rare earth-iron compounds.

Table 7.7. Magnetic moments in $PrCo_5$ and $PrCo_5D_{3.6}$ [a]

T [K]	μ_{Pr}	$\mu_{Co(1)}$	$\mu_{Co(2)}$
$PrCo_5$			
4.2	1.58	1.30	1.30
300	1.15	1.50	1.50
$PrCo_5D_{3.6}$			
4.2	1.64	0.12	0.79
78	0.50	0.98	1.28
300	0	0.83	0.38

[a] The atomic moments are given in Bohr magnetons.

Table 7.8. Magnetic characteristics of RFe_2 and hydrogenated RFe_2 compounds

R	RFe_2			Fe Moment (μ_B)	RFe_2H_4 [a]			Fe Moment (μ_B)
	T_c [K]	T_{comp} [K]	Moment (μ_B/formula unit)		T_c [K]	T_{comp} [K]	Moment (μ_B/formula unit)	
Y [7.77] [b]	545	— [c]	2.90	1.45	308	—	3.66 [d]	1.83
Ce [7.78]	230	—	2.60	1.30	358	—	4.14 [d]	2.07 (?) [e]
Gd [7.77]	785	—	2.80	2.10	388	—	4.10 [f]	1.45 (?) [e]
Ho [7.76]	614	—	6.1	1.95	287	60	2.35 [g]	
Er [7.76]	596	480	4.75	2.13	280	42	5.60 [g]	1.7 (?) [e]
Tm [7.76]	610	225	2.52	2.24	270	18	6.45 [h]	0.28 (?) [e]

[a] Nominal compositions. The actual compositions used by *Gualtieri* et al. were $HoFe_2H_{4.47}$, $ErFe_2H_{3.9}$, and $TmFe_2H_{4.3}$.
[b] The numbers in brackets indicate the reference.
[c] No entry indicates that there is no compensation point (T_{comp}).
[d] Moment at saturation at 4.2 K.
[f] Moment at 4.2 K and 18 kOe.
[e] This quantity is uncertain (see text).
[g] Moment at 4.2 K and 21 kOe.
[h] Moment at 4.2 K and 120 kOe.

7.4.4 Rare Earth-Iron Intermetallics

RFe_2-H Systems

The effect of dissolved hydrogen on the magnetic properties of RFe_2 compounds has been investigated by *Gualtieri* et al. [7.76] (R = Ho, Er, and Tm) and by *Buschow* and *Van Diepen* [7.77, 79] (R = Y, Ce, and Gd). Results obtained are summarized in Table 7.8. In commenting on those results, it is appropriate to note that hydrogenation of $ErFe_2$ and $TmFe_2$ very substantially

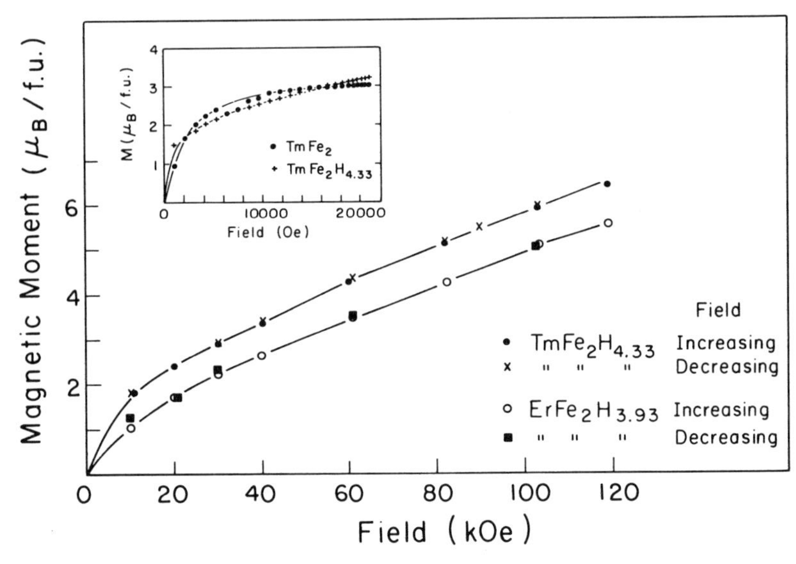

Fig. 7.12. Moment versus applied field for hydrides of $HoFe_2$, $ErFe_2$, and $TmFe_2$ [7.76]

increases the difficulty in determining the saturation magnetization. Data obtained using applied fields up to 120 kOe (Fig. 7.12) show that saturation is not achieved at the highest fields for hydrided $ErFe_2$ and $TmFe_2$. There is no such difficulty in saturating the hydrogen-free RFe_2, and *Buschow* and *Van Diepen* [7.77] indicate that YFe_2H_4 saturated at applied fields "above a few kOe". Clearly, hydrogenation of the Ho, Er, or Tm compound has produced some profound magnetic changes in the system. This is evident from the data in Fig. 7.12. It is equally evident when one uses the measured moment per formula unit (μ_B/f.u.) to estimate the moment per iron atom (μ_{Fe}) in these materials (*vide infra*).

Consider the establishment of μ_{Fe} from μ_B/f.u. In the YFe_2 system this is straightforward since Fe is the only magnetic species present. In this case $\mu_{Fe}=(\mu_B/\text{f.u.})/2$. If Ce in $CeFe_2$ and $CeFe_2H_4$ is quadripositive, it is non-magnetic and μ_{Fe} is obtained similarly. Iron moments calculated in this way are given in Table 7.8. When R is magnetic, obtaining μ_{Fe} is more complicated. It can be calculated assuming the free R^{3+} moment, viz., gJ, and also assuming the coupling systematics observed for RFe_2 systems—ferromagnetic for light rare earths and antiferromagnetic for heavy rare earths, including Gd. Under this assumption, μ_{Fe} in $GdFe_2$ is $1.45\mu_B$. Calculation of μ_{Fe} in $HoFe_2H_4$ in a similar fashion gives $3.82\mu_B$, which is unrealistically large. The large μ_{Fe} is obtained because $HoFe_2H_4$ is not saturated at 21 kOe. The μ_{Fe} values calculated for the Er and Tm ternaries are also suspect because, clearly, saturation is not achieved at field strengths of 120 kOe. The results for $GdFe_2H_4$ are equally suspect for similar reasons.

Hydrogenation produces a decrease in T_c for all RFe_2 compounds except $CeFe_2$, indicating that exchange is weakened by the entry of hydrogen in the lattice. The aberrant behavior of $CeFe_2$ suggests the possibility of a different valence state for Ce in $CeFe_2$ and $CeFe_2H_4$. In the ternary, Ce may be partially or totally tripositive, enhancing exchange and producing a rise in T_c. If Ce is magnetic in this hydride, μ_{Fe} computed assuming quadripositive Ce would be erroneous.

Substantial variation in the compensation temperature (T_{comp}) is produced when $ErFe_2$ and $TmFe_2$ are hydrided. This is a consequence of weakened exchange. From the coupling systematics referred to above, one observed that the R and Fe moments in $ErFe_2$ and $TmFe_2$ couple antiparallel at very low temperatures. As temperature is increased, the Er and Tm moments disorder more rapidly than the Fe moments and, since the former are larger, magnetization passes through a minimum at some temperature T_{comp}. Above T_{comp} the magnetization originates entirely with the Fe sublattice. The weakened exchange accompanying hydrogenation leads to an even more rapid disordering of the R sublattice as temperature is increased with an accompanying depression of T_{comp}, as noted for hydrogenated $ErFe_2$ and $TmFe_2$ in Table 7.8. In $HoFe_2$ the exchange field acting on the Ho sublattice is sufficiently strong that no compensation is observed. In $HoFe_2H_4$, exchange has been weakened and a compensation point at 60 K is observed.

These results make it clear that hydrogenation weakens exchange (as was the case for the RCo_5 systems), except for $CeFe_2$. An increase in Fe moment occurs when YFe_2 is hydrogenated, as evidenced both by a rise in magnetization and an increase in the nuclear hyperfine field. Mössbauer measurements [7.77] indicate a rise in hyperfine field from 163 to 200 kOe when YFe_2 is hydrogenated. In this respect YFe_2 is unlike the RCo_5 systems.

The striking increase in moment when $TmFe_2$ is hydrogenated is truly extraordinary. The increase is so great as to suggest that the antiferromagnetic coupling of the Tm and Fe sublattices may have been modified in the hydride. If this is confirmed, it would be of very great interest, pointing to a way to produce ferromagnetic coupling in heavy rare earth-transition metal intermetallics.

Neutron diffraction results have been obtained recently for $ErFe_2H_4$ [7.78]. It is found that hydrogenation does not alter the Fe moment, but it significantly reduces exchange. These results make it appear that the lack of saturation in the hydrided RFe_2 compounds is a consequence of such strongly reduced exchange that the rare earth ions are behaving as an assemblage of paramagnetic ions.

RFe_3-H Systems

Buschow has reported results on the YFe_3–H system [7.79], and *Malik* et al. [7.80] have given information for the RFe_3–H systems with R = Gd, Dy, and Ho. Results for the RFe_3 systems are summarized in Table 7.9. The results

Table 7.9. Influence of hydrogenation on the magnetic characteristics of RFe_3

	RFe_3			RFe_3H_3 [a]		
R	T_c [K]	T_{comp} [K]	Moment (μ_B/formula unit)	T_c [K]	T_{comp} [K]	Moment (μ_B/formula unit)
Y [7.78] [b]	549	—	5.01	545	—	5.70
Gd [7.81]	725	615	1.87	— [c]	170	1.39 [d]
Dy [7.81]	605	545	4.25	— [c]	175	2.2 [e]
Ho [7.81]	575	395	4.42	— [c]	112	2.53 [d]

[a] Nominal compositions. Actual compositions were YFe_3H_5, $GdFe_3H_{3.1}$, $DyFe_3H_{3.0}$, and $HoFe_3H_{3.6}$.
[b] Numbers in brackets indicate the reference.
[c] T_c could not be observed due to loss of hydrogen.
[d] Measurements made for field strengths up to 21 kOe and extrapolated to $H=\infty$.
[e] Moment at 21 kOe. This compound, unlike $GdFe_2H_3$ and $HoFe_2H_3$, was far from saturation at 21 kOe.

indicate a weakening of exchange[2] in all cases except YFe_3 and seem to indicate a rise in Fe moment with hydrogenation. In this respect they behave like YFe_2 described above. The predominant behavior of Fe-containing rare earth intermetallics is for the Fe moment to increase upon hydrogenation. This contrasts sharply with the behavior described above for the RCo_5 series. Why there is this difference is not clear at this time. A proper interpretation of these differences must await band structure calculations for these hydrides similar to those of *Switendick* on the simple hydrides.

Y_6Mn_{23}-H and Th_6Mn_{23}-H Systems

Recent studies of hydrogenated Y_6Mn_{23} and Th_6Mn_{23} show [7.81] some very interesting, and puzzling, results, The results presented above suggest that the Co–Co or Fe–Fe exchange in R–Co and R–Fe compounds, respectively, is reduced on hydrogenation, although generally μ_{Co} decreased and μ_{Fe} increased. Recent work by *Malik* and *Wallace* [7.83] on $GdNi_2$ showed that the Gd–Gd interaction is also weakened by hydrogenation. The R–Mn compounds show rather different results. For instance, Y_6Mn_{23} is a ferromagnet with $T_c=500$ K and a moment of $0.5\,\mu_B$/Mn. On hydrogen absorption it forms $Y_6Mn_{23}H_{\sim 25}$ in which Mn moment is completely quenched and the compound behaves like a paramagnet. In contrast, Th_6Mn_{23}, which is isostructural with Y_6Mn_{23}, is a paramagnet but $Th_6Mn_{23}H_{\sim 30}$ (see Figs. 7.13 and 7.14) exhibits ferromagnetic ordering with $T_c \sim 329$ K and a moment of $0.8\,\mu_B$/Mn. This contrasting behavior of isostructural Y_6Mn_{23} and Th_6Mn_{23} on hydrogen absorption seems to be unique so far and presumably arises because of the differing band

[2] Recent work [7.82] in this laboratory on RCo_3–H systems with R=Gd, Dy, Ho, and Er shows similar behavior. From the saturation magnetization of $GdCo_3$–H, a reduction of Co moment (to $1.3\,\mu_B$/atom) is observed. In this latter respect the RCo_3 systems behave like the RCo_5–H rather than the RFe_3–H systems.

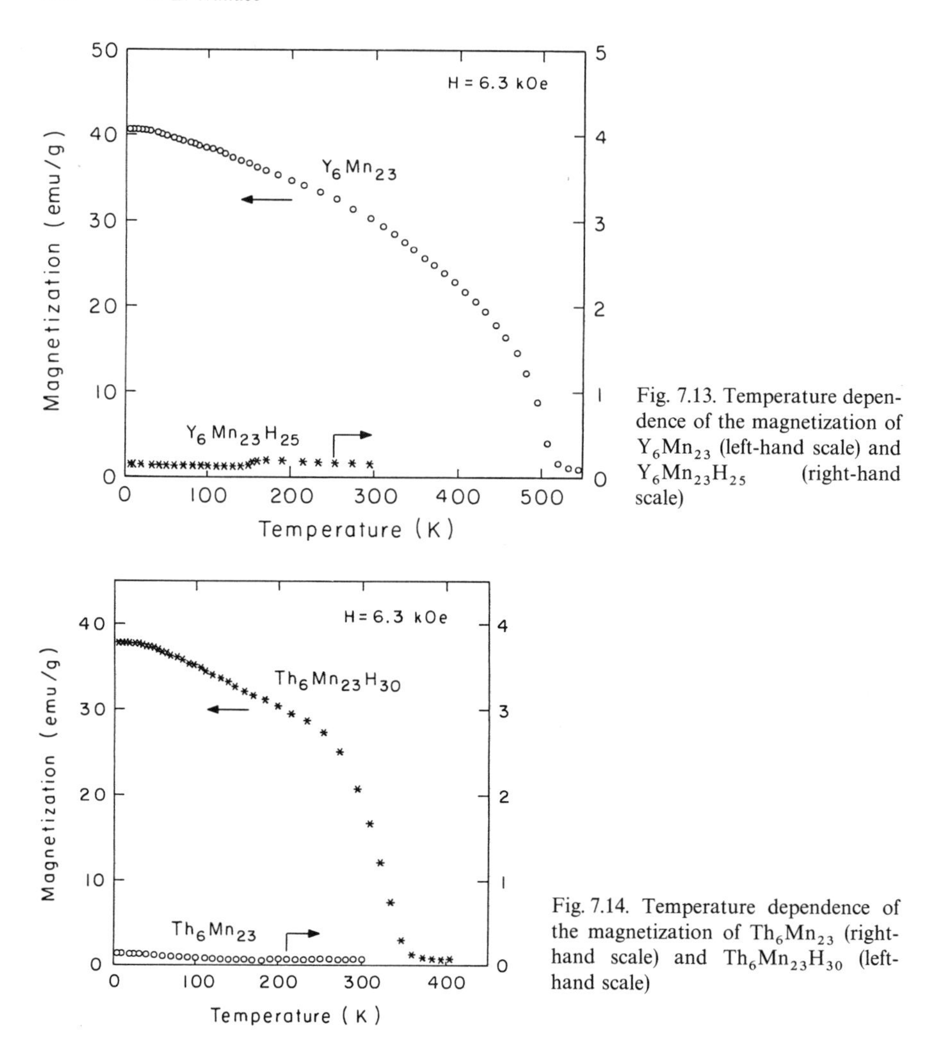

Fig. 7.13. Temperature dependence of the magnetization of Y_6Mn_{23} (left-hand scale) and $Y_6Mn_{23}H_{25}$ (right-hand scale)

Fig. 7.14. Temperature dependence of the magnetization of Th_6Mn_{23} (right-hand scale) and $Th_6Mn_{23}H_{30}$ (left-hand scale)

structure of the two materials. Since Y is trivalent and Th normally quadrivalent, the extra electron from Th fills the spin split *d*-band of Mn and is probably responsible for the fact that Y_6Mn_{23} is a ferromagnet while Th_6Mn_{23} is not. Introduction of hydrogen changes the situation. Band structure calculations by *Switendick* on simple metal hydrides (see above) suggest that the hydrogen creates additional states below the metal *d*-band as well as lowers some of the existing states. Since hydrogen contributes one electron but two spin states, there is a repopulation of levels which presumably results in the quenching of the Mn moment in Y_6Mn_{23}, but it appears as if the extra electron from Th fills the additional state contributed by hydrogen leaving the Mn moment in $Th_6Mn_{23}H_{30}$ unquenched as in Y_6Mn_{23}. Further discussion of the results must await band structure calculations on these compounds.

Hydrogen-Induced Magnetic Ordering in the Superconductor Th_7Fe_3

Another striking observation has been on Th_7Fe_3. According to *Matthias* et al. [7.84], this compound is a superconductor with $T_c \sim 2$ K. However, it has been observed that on hydrogen absorption this compound also exhibits magnetic ordering [7.85]. This is again consistent with the observation that in compounds with iron, μ_{Fe} increases on hydrogen absorption.

7.5 Hydrogen Uptake in Relation to the Photodecomposition of H_2O

The source of the hydrogen contaminant in $PrCo_2$, $NdCo_2$, and other rare earth intermetallics discussed above is not clear. One possibility is that it comes from the photolysis of adsorbed H_2O by the ambient light flux in the laboratory. Possibly a photon strikes adsorbed H_2O, breaking the O–H bond, with the released hydrogen atom being drawn into the lattice by the chemical affinity of the intermetallic. There is no other ready explanation for the dissolved hydrogen. The surfaces of the metal remain bright and shiny with no evidence of a buildup of oxide layer expected if this were merely the reaction of an active metal with H_2O.

H_2O vapor does not absorb at wavelengths in the visible light region and hence visible radiation cannot split H_2O in the vapor state. The situation may be different with adsorbed H_2O. The surfaces of these materials have recently been shown to be catalytically active [7.86, 87]. If future work confirms that the rare earth intermetallic split H_2O with the help of visible radiation, it will be of major technological significance, perhaps leading to a useful way to capture and store solar energy.

Acknowledgements. The author wishes to acknowledge the assistance of Miss Marcia L. Wallace and Dr. A. Elattar in the survey of the literature. He also wishes to acknowledge the support of the Army Research Office in the general program of research dealing with rare earth systems.

References

7.1 H. Biggs: Phil. Mag. **32**, 131 (1916)
7.2 J. Aharoni, F. Simon: Z. Physik. Chem. B**4**, 175 (1929)
7.3 B. Svensson: Ann. Phys. Leipzig **14**, 699 (1932); **18**, 299 (1933)
7.4 J. Wucher: Ann. Phys. Paris **7**, 317 (1952)
7.5 H.C. Jamieson, F.D. Manchester: J. Phys. F. **2**, 323 (1972)
7.6 N.F. Mott, H. Jones: *The Theory of the Properties of Metals and Alloys* (Dover Publications, New York 1958) p. 200
7.7 M.A. Ruderman, C. Kittel: Phys. Rev. **96**, 99 (1954)
7.8 T. Kasuya: Progr. Theor. Phys., Kyoto **16**, 45 (1956)
7.9 K. Yoshida: Phys. Rev. **106**, 893 (1957)
7.10 H. Zijlstra, F.F. Westendorp: Solid State Commun. **7**, 857 (1959)

7.11 G. Alefeld, J. Völkl (eds.): *Hydrogen in Metals. II, Application-Oriented Properties*, in Topics in Applied Physics, Vol. 29 (Springer, Berlin, Heidelberg, New York 1978)
7.12 K. A. Gschneider, Jr.: In *Scandium: Its Occurrence, Chemistry, Physics, Metallurgy, Biology and Technology*, ed. by C. T. Horovitz (Academic Press, New York 1975) pp. 84—86
7.13 W. Trzebiatowski, B. Stalinski: Bull. Acad. Polon. Sci. Cl III, **1**, 131 (1953)
7.14 R. L. Zanowick, W. E. Wallace: J. Chem. Phys. **36**, 2059 (1962)
7.15 H. R. Khan, A. Knödler, Ch. J. Raub, A. C. Lawson: Mat. Res. Bull. **9**, 1191 (1974)
7.16 D. P. Schumacher, W. E. Wallace: Inorg. Chem. **9**, 1563 (1966)
7.17 W. E. Gardner, J. Penfold: Phys. Lett. **26** A, 204 (1968)
7.18 C. D. Parks, W. G. Bos: J. Solid State Chem. **2**, 61 (1970)
7.19 R. A. Andrievskii, E. B. Boiko, R. B. Ioffe: Izv. Akad. Nauk SSSR, Neorg. Mater. **3**, 1591 (1967)
7.20 J. Fitzwilliam, A. Kaufman, C. Squire: J. Chem. Phys. **9**, 678 (1941)
7.21 P. Bickel, T. Berlincourt: Phys. Rev. **119**, 1603 (1960)
7.22 W. Trzebiatowski, B. Stalinski: Bull. Acad. Polon. Sci. Cl III, **1**, 317 (1953)
7.23 B. Stalinski: Bull. Acad. Polon. Sci. Cl III, **5**, 997 (1957)
7.24 B. Stalinski: Bull. Acad. Polon. Sci. Cl III, **2**, 245 (1954)
7.25 B. G. Childs, W. E. Gardner, J. Penfold: Phil. Mag. **5**, 1267 (1960)
7.26 N. F. Mott: Advan. Phys. **13**, 325 (1972)
7.27 H. L. Yakel, Jr.: Acta Cryst. **11**, 46 (1958)
7.28 R. Mehlman, H. Huseman, H. Brodowsky: Ber. Bunsenges. Phys. Chem. **77**, 36 (1973)
7.29 J. Vuillemin, M. Priestley: Phys. Rev. Lett. **14**, 307 (1965)
J. Vuillemin: Phys. Rev. **144**, 396 (1966)
7.30 W. Mueller, J. P. Blackledge, G. G. Libowitz: *Metal Hydrides* (Academic Press, New York 1968)
7.31 A. C. Switendick: Solid State Commun. **8**, 1463 (1970); Intern. J. Quantum Chem. **5**, 459 (1971)
7.32 H. Nagel, H. Goretzki: J. Phys. Chem. Sol. **36**, 431 (1975)
7.33 Y. Kubota, W. E. Wallace: J. Appl. Phys. Suppl. **33**, 1348 (1962)
7.34 Y. Kubota, W. E. Wallace: J. Appl. Phys. **34**, 1348 (1963)
7.35 Y. Kubota, W. E. Wallace: J. Chem. Phys. **39**, 1285 (1963)
7.36 W. E. Wallace, K. H. Mader: J. Chem. Phys. **48**, 84 (1968)
7.37 W. E. Wallace, Y. Kubota, R. L. Zanowick: Advan. Chem. Ser. **39**, 122 (1963)
7.38 R. L. Zanowick, W. E. Wallace: Phys. Rev. **126**, 537 (1962)
7.39 R. N. R. Mulford, C. E. Holley: J. Phys. Chem. **59**, 1222 (1955)
7.40 A. Pebler, W. E. Wallace: J. Phys. Chem. **148**, 66 (1962)
7.41 M. Mannsman, W. E. Wallace: J. Phys. (Paris) **25**, 454 (1964)
7.42 R. Heckman: J. Chem. Phys. **40**, 2958 (1964); **46**, 2158 (1967)
7.43 B. Stalinski: Bull. Acad. Polon. Sci. **5**, 1001 (1957); **7**, 269 (1959)
7.44 F. Trombe: Compt. Rend. **219**, 182 (1944)
7.45 See, for example, T. Kasuya: In *Magnetism*, Vol. II B, ed. by G. T. Rado, H. Suhl (Academic Press, New York 1966)
7.46 For a discussion of the influence of the crystal field interaction on the magnetic properties of rare earth ions see W. E. Wallace: *Rare Earth Intermetallics* (Academic Press, New York 1973) Chaps. 2, 3
7.47 D. P. Schumacher, C. A. Hollingsworth: J. Phys. Chem. Sol. **27**, 749 (1966)
7.48 W. E. Wallace, C. Deenadas, A. W. Thompson, R. S. Craig: J. Phys. Chem. Sol. **32**, 805 (1971)
7.49 A. M. Van Diepen, R. S. Craig, W. E. Wallace: J. Phys. Chem. Sol. **32**, 1853 (1971)
7.50 Z. Bieganski, B. Stalinski: Phys. Stat. Sol. **2**, K 161 (1970)
7.51 K. Knorr, B. E. F. Fender: In *Crystal Field Effects in Metals and Alloys*, ed. by A. Furrer (Plenum Press, New York 1977)
7.52 J. A. Mydosh: Phys. Rev. Lett. **33**, 1562 (1974)
7.53 W. Trzebiatowski, A. Sliwa, B. Stalinski: Roczniki Chemii **26**, 110 (1952) and **28**, 12 (1954)
7.54 A. Sliwa, W. Trzebiatowski: Bull. Acad. Polon. Sci. **5**, 217 (1962)
7.55 D. M. Gruen: J. Chem. Phys. **23**, 1708 (1955)
7.56 W. Spalthoff: Z. Phys. Chem., N.F. **29**, 258 (1961)
7.57 W. E. Henry: Phys. Rev. **109**, 1976 (1958)

7.58 See, for example, W.J.Nellis, M.B.Brodsky: In *The Actinides: Electronic Structure and Related Properties*, Vol. II, ed. by A.J.Freeman, J.B.Darby, Jr. (Academic Press, New York 1974) p. 265
7.59 G.Cinader, D.Zamir, U.El-Hanany, U.Hadari, G.Degani: Solid State Commun. **8**, 1703 (1970)
7.60 H.Neumann: Ph.D. Thesis. Technische Hochschule Darmstadt (1969)
7.61 J.H.N.Van Vucht, F.A.Kuijpers, H.C.A.M.Bruning: Philips Res. Repts. **25**, 133 (1970)
7.62 F.A.Kuijpers, B.A.Loopstra: J. Phys. (Paris) Suppl. **32**, C 1—657 (1971)
7.63 H.H.Neumann, S.Nasu, R.S.Craig, N.Marzouk, W.E.Wallace: J. Phys. Chem. Sol. **32**, 2788 (1971)
7.64 S.Nasu, H.H.Neumann, N.Marzouk, R.S.Craig, W.E.Wallace: J. Phys. Chem. Sol. **32**, 2779 (1971)
7.65 C.Titcomb, R.S.Craig, W.E.Wallace, V.U.S.Rao: Phys. Lett. **39** A, 157 (1972)
7.66 J.J.Farrell, W.E.Wallace: Inorg. Chem. **5**, 105 (1966)
7.67 C.Deenadas, R.S.Craig, N.Marzouk, W.E.Wallace: J. Solid State Chem. **4**, 1 (1972)
7.68 See, for example, D.Mattis, W.E.Donath: Phys. Rev. **128**, 1618 (1962)
K.H.Mader, W.E.Wallace: J. Chem. Phys. **49**, 1521 (1968)
W.M.Swift, W.E.Wallace: J. Solid State Chem. **3**, 180 (1971)
7.69 W.E.Wallace, K.H.Mader: Inorg. Chem. **7**, 1627 (1968)
7.70 T.Tsuchida, W.E.Wallace: J. Chem. Phys. **43**, 2087, 2885 (1965)
7.71 B.R.Cooper, Y.L.Wang: J. Appl. Phys. **40**, 1344 (1969); Phys. Rev. **163**, 444 (1967); **172**, 539 (1968); **185**, 696 (1969)
7.72 K.H.Mader, E.Segal, W.E.Wallace: J. Phys. Chem. Sol. **30**, 1 (1969)
7.73 B.Bleaney: "Rare Earth Research", in *Proceedings of the Third Rare Earth Research Conference*, Vol. 2, ed. by K.S.Vorres (Gordon and Breach, New York 1964) p. 499
7.74 W.E.Wallace: *Rare Earth Intermetallics* (Academic Press, New York 1973) pp. 113, 147
7.75 F.A.Kuijpers: J. Less-Common Metals **27**, 27 (1972); Ph.D. Thesis, University of Delft (1973)
7.76 D.M.Gualtieri, K.S.V.L.Narasimhan, W.E.Wallace: AIP Conf. Proc. **34**, 219 (1976)
7.77 K.H.J.Buschow, A.M.Van Diepen: Solid State Commun. **19**, 79 (1976)
7.78 J.J.Rhyne, S.G.Sankar, W.E.Wallace: *Proc. of 13th Rare Earth Research Conference* (Plenum Press New York, to be published)
7.79 K.H.J.Buschow: Solid State Commun. **19**, 421 (1976)
7.80 S.K.Malik, T.Takeshita, W.E.Wallace: Magnetism Lett. **1**, 33 (1976)
7.81 S.K.Malik, T.Takeshita, W.E.Wallace: Solid State Commun. **23**, 599 (1977)
7.82 S.K.Malik, T.Takeshita, W.E.Wallace: To be published
7.83 S.K.Malik, W.E.Wallace: Solid State Commun. **24**, 283 (1977)
7.84 B.T.Matthias, V.B.Compton, E.J.Corenzwit: J. Phys. Chem. Sol. **19**, 130 (1961)
7.85 W.E.Wallace, S.K.Malik, T.Takeshita, S.G.Sankar, D.M.Gualtieri: J. Appl. Phys. (to be published)
7.86 T.Takeshita, W.E.Wallace, R.S.Craig: J. Catalysis **44**, 236 (1976)
7.87 V.T.Coon, T.Takeshita, W.E.Wallace, R.S.Craig: J. Phys. Chem. **80**, 1878 (1976)

8. Theory of the Diffusion of Hydrogen in Metals

K. W. Kehr

With 9 Figures

8.1 Overview

Hydrogen in metals has a large mobility. At room temperature and below, its mobility is many orders of magnitude larger than that of other interstitially dissolved atoms, as has been pointed out in Chapter 12. The question arises of why the mobility of hydrogen is so large. Is the interaction between hydrogen and the host lattice weak compared to other interstitials, such that the potential barriers between equilibrium sites of the hydrogen atoms are low? Are other mechanisms than thermally activated jumps over the potential barriers the cause of the large mobility? Not much is known about the detailed interaction of hydrogen with the host metal atoms. Here we shall concentrate on the discussion of the possible mechanisms for diffusion of hydrogen atoms. Since hydrogen atoms have a small mass compared to other interstitials, quantum effects in the diffusion are likely to be observed for hydrogen, if they occur at all. The three isotopes of hydrogen have large mass ratios, and in addition the positive muon can be considered as a light isotope of the proton (cf. Chap. 13). Hence isotope effects can be studied over a wide range, which is very important for determining a distinction between different possible diffusion mechanisms.

First a survey of the different possible diffusion mechanisms of a light interstitial in a host lattice will be given. Figure 8.1 gives a rough subdivision of the different possible regimes as a function of temperature. At the lowest temperatures, the interstitial must be delocalized in the form of a band state, unless it is trapped by lattice defects. The propagation in the band state is limited by the scattering on termal phonons or lattice defects. At some higher temperature the interstitial will be localized at or about a specific interstitial

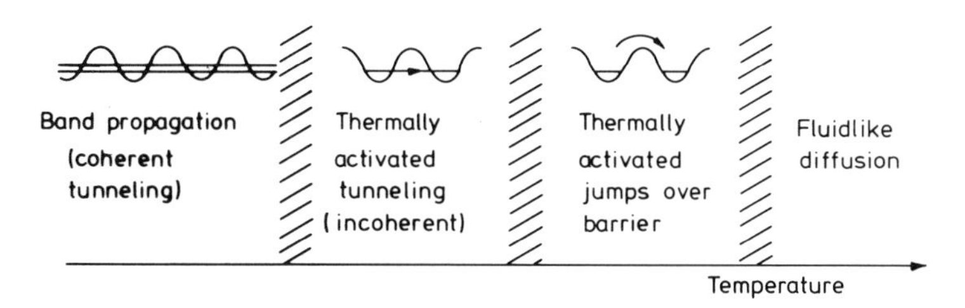

Fig. 8.1. Diffusion processes of a light interstitial at different temperatures

site. This process will be discussed in detail in Section 8.2.3. The elementary step of the diffusion process is now a thermally activated jump from one to another interstitial site. The particle might execute the jump by tunneling from one to the other interstitial site or by hopping over the potential barrier. In the first case, thermal activation is necessary to bring the energy levels of both sites to the same height. In the second case, a higher activation energy is required to overcome the barrier. Hence this process contributes at higher temperatures. Finally, at the highest temperatures the interstitial will be mainly in states above the potential barriers. It should perform a diffusion somewhat similar to a dense gas or a liquid, where many collisions occur, provided in our case by the thermally fluctuating host lattice. This regime will henceforth be called "fluidlike diffusion". The boundaries separating the different regimes are not sharply defined and one can consider subdivisions and modifications within one regime. Also it is not certain that all regimes really occur in a given system. It will be one of the aims of this chapter to assess the applicability of the different mechanisms to the diffusion of hydrogen in metals.

In the regime of thermally activated tunneling transitions and over-barrier jumps, a natural separation between the individual jump processes and the diffusion in the lattice of interstitial sites appears. This separation is well justified as long as the individual jump processes have a short duration compared to the mean time of stay at an interstitial site, and are uncorrelated. Under these conditions the contribution of the diffusive jumps to the lattice diffusion can be evaluated by using master equations (cf. Chap. 10). We shall direct our attention to the statistical-mechanical problem of the individual jump processes.

It is instructive to compare the problem of interstitial diffusion with other related mobility problems. The regime of band diffusion is encountered in the electronic conductivity. In some metal oxides "small polarons" occur, i.e., electrons which are self-trapped within a local lattice deformation. They have a mobility which is determined by the thermally activated tunneling processes corresponding to the second regime. The theory of these processes was first derived for small polarons (see *Holstein* [8.1, 2]). Paraelectric centers in crystals can reorient themselves; molecules or molecular groups in molecular crystals can perform rotational motions. The crystal potentials in which the paraelectric centers or molecular groups move differ from system to system. Thus these systems offer the opportunity of studying the different regimes, their limits, and transitions between them induced by the variation of parameters such as the temperature. All regimes including tunneling at lower temperatures and fluidlike diffusion at higher temperatures are found.

So far, the diffusion processes of one hydrogen atom in a host lattice have been considered. The diffusion of many dissolved hydrogen atoms is important because of the large solubility of hydrogen in some transition metals. Not much theoretical work has been done on the diffusion at higher hydrogen concentration, with the exception of critical phenomena. The hydrogen-metal systems exhibit a critical slowing down of the hydrogen diffusion (*Alefeld* et al.

[8.3]) at the critical point of the phase transition between the α and α' phases. The connection between critical slowing down and the elastic interaction leading to the phase transition has been discussed by *Janssen* [8.4]. Almost no theoretical work has been done on the diffusion in the dense phases of hydrogen in metals such as the β phase of Nb–H.

The next section contains a discussion of the interaction between hydrogen and host metal atoms, and a discussion of the localization problem of hydrogen atoms in metals. The following section describes the quantum mechanical rate theory and discusses its applicability to experiments. In the fourth section the classical rate theory is presented, together with its quantum mechanical modification and the experimental observations. The limits of the description by individual jump processes are discussed in Section 8.5.

The theory of the diffusion of light interstitials has been reviewed by *Sussmann* [8.5], *Stoneham* [8.6], and *Maksimov* and *Pankratov* [8.7]. Additional summary comments have been given by *Stoneham* [8.8, 9].

8.2 Interaction of Hydrogen with Host Metal Atoms

8.2.1 Phenomenological Interaction Parameters

A knowledge of the interaction between the dissolved hydrogen atoms and the host metal atoms is essential in order to discuss the diffusive processes of these interstitials. However, the microscopic derivation of this interaction starting from the interacting system of host metal ions, protons, and electrons seems to be too difficult for the transition metals of interest. On the other hand, the hydrogen-metal interaction produces observable effects which allow the determination of interaction parameters. Since only a few parameters are determined in this way, the theory assumes a model character. The use of model interactions is quite reasonable for the discussion of the basic diffusion mechanisms, which is intended in this chapter. Of course, a derivation of quantities such as the activation energy for diffusion in specific metals would require an explicit investigation of the hydrogen-metal interaction.

In a microscopic derivation of the interaction of hydrogen with the host metal, the screening of a proton at an interstitial site has to be considered (*Friedel* [8.10]). *Friedel* treats the screening of a proton in a free electron gas and obtains a screened Coulomb potential in the linearized Thomas-Fermi limit, and superimposed oscillations ("Friedel oscillations") in the linearized self-consistent Hartree approximation. *Popovic* and *Stott* [8.11] treat the nonlinear screening of a proton in the electron gas and derive activation energies for diffusion in Al and Mg. They have shown that nonlinear screening is essential for protons in metals. The situation in transition metals with narrow *d*-bands such as Pd is even more complicated. A qualitative discussion has been given by *Friedel* [8.10], but no quantitative derivations have appeared in the literature.

Hence it is necessary to use models for the interaction between hydrogen and metal atoms. One possibility is to introduce explicit models such as that of *Boes* and *Züchner* [8.12], who consider the hydrogen and the neighboring metal atoms as spheres interacting with a repulsive Born-Meyer potential. The metal atoms themselves are embedded in an elastic continuum. In this way the energies of the equilibrium positions and saddle-point configurations are estimated. Models of this kind make specific assumptions about the form of the interaction between hydrogen and the metal atoms, which can influence more complicated properties in an uncontrolled way.

The other possibility is to consider models which use, as far as possible, parameters which can be determined experimentally. This approach will be followed in this chapter. In order to introduce phenomenological coupling parameters, one considers the potential energy U of the metal-hydrogen system, which is a function of all hydrogen coordinates Y and metal atom coordinates X (*Wagner* and *Horner* [8.13]; see also Chap. 2),

$$U(X, Y) = \Phi(X) + \Psi(X, Y). \tag{8.1}$$

Φ is the potential energy of the metal lattice without hydrogen. For low hydrogen concentration, the following approximation is made which neglects direct hydrogen-hydrogen interactions

$$U(X, Y) = \Phi(X) + \sum_a \tau_a^a \psi(X, Y^a), \tag{8.2}$$

where τ_a is an occupation number (0, 1) of the hydrogen interstitial site a. We assume a regular lattice of metal atoms and interstitial sites. The energy U is expanded about the equilibrium positions R of the unloaded lattice and the equilibrium positions S of the hydrogen atoms,

$$\begin{aligned} X_\mu^m &= R_\mu^m + u_\mu^m \\ Y_\alpha^a &= S_\alpha^a + v_\alpha^a . \end{aligned} \tag{8.3}$$

m denotes the metal atoms and a the interstitial sites; μ, α denote the Cartesian components. We have

$$\begin{aligned} U = \Phi(R, S) &+ \tfrac{1}{2}\Phi_{\mu\nu}^{mn} u_\mu^m u_\nu^n + \dots \\ &+ \sum_a \tau_a [{}^a\psi(R, S) + {}^a\psi_\mu^m u_\mu^m + \dots]. \end{aligned} \tag{8.4}$$

A summation convention over repeated indices is adopted. First-order terms which do not appear in (8.4) are zero because of equilibrium conditions. The notation for derivatives is, e.g.,

$${}^a\psi_\mu^m = \left.\frac{\partial\, {}^a\psi(X, Y)}{\partial X_\mu^m}\right|_{R_\mu^m, S_\alpha^a} .$$

The basic experimental observation about the hydrogen-lattice interaction is the appearance of lattice strains, resulting in a volume dilatation upon loading the crystal with hydrogen (cf. Chap. 3). As explained by Peisl, the average lattice parameter change is determined by the trace of the double-force tensor $\underset{\sim}{P}$, which is given by

$$^{a}P_{\mu\nu} = -{}^{a}\psi^{m}_{\mu} R^{m}_{\nu} \,. \tag{8.5}$$

The forces $^{a}\psi^{m}_{\mu}$ were introduced by *Kanzaki* [8.14]. When higher order terms in the expansion (8.4) are taken into account, then the $^{a}\psi^{m}_{\mu}$ determined from $\underset{\sim}{P}$ are effective Kanzaki forces which produce the correct volume dilatation in the harmonic approximation for the unloaded lattice. The symmetry of the double-force tensor is determined by the symmetry of the interstitial sites. For octahedral sites in an fcc lattice,

$$P_{\mu\nu} = P\delta_{\mu\nu} \,, \tag{8.6}$$

whereas for an "x" tetrahedral site in a bcc lattice, $\underset{\sim}{P}$ has the form

$$\underset{\sim}{P} = \begin{pmatrix} A & 0 & 0 \\ 0 & B & 0 \\ 0 & 0 & B \end{pmatrix} . \tag{8.7}$$

Results for the trace $A+2B$ for Nb and Ta are given by *Peisl*. It should be emphasized that the trace of the double-force tensor does not show an isotope dependence for Nb–H, but a weak isotope dependence is observed for Ta–H, D, such that $A+2B$ is about 8% smaller for D (*Pfeiffer* and *Peisl* [8.15]). The double-force tensor for H in fcc metals can be deduced from *Baranowski* et al. [8.16], who found the same absolute volume dilatation ΔV per H atom for different fcc metals and alloys (see also Chap. 3). The values for P are of the same magnitude as in bcc metals, e.g., $P = 3.5$ eV for Pd–H. Nothing definite can be stated about an isotope effect of P for fcc metals.

The difference $A-B$ of the double-force tensor in bcc metals can be determined from the "Snoek effect", that is, the reorientation of elastic dipoles under the influence of external strain fields. Since the Snoek effect is discussed in the Chapter 3, only a few important points are mentioned here. *Buchholz* et al. [8.17] and *Magerl* et al. [8.18] have observed no Snoek effect and have given an upper limit to $A-B$. One possibility is to set $A=B$; then force constants to the nearest *and* the next-nearest neighbors can be fitted using the measured value of $A+2B$. Following *Pick* and *Bausch* [8.19], we assume that the absence of the Snoek effect in bcc metals is caused by this interplay of first- and second-nearest neighbor forces. Also the existence of tunneling states has been proposed in order to explain the absence of the Snoek effect [8.17]. We shall come back to this proposition in Section 8.2.3. More detailed information on the coupling constants $^{a}\psi^{m}_{\mu}$ can be obtained in principle from "diffuse scattering" experiments. Again we refer to Chapter 3. In summary, we assume the linear forces ${}^a\psi^m_\mu$ to some near neighbors to be known.

Let us now consider the second-order term in the expansion of ${}^a\psi(X, Y)$. We have

$$\tfrac{1}{2}{}^a\psi^{mn}_{\mu\nu}u^m_\mu u^n_\nu + {}^a\psi^{ma}_{\mu\alpha}u^m_\mu v^a_\alpha + \tfrac{1}{2}{}^a\psi^{aa}_{\alpha\beta}v^a_\alpha v^a_\beta \,. \tag{8.8}$$

The mixed term can be eliminated by using the adiabatic approximation for the proton, in which the proton follows the motion of the host lattice. By decomposing $\boldsymbol{v}=\hat{\boldsymbol{v}}+\boldsymbol{v}'$ and determining $\hat{\boldsymbol{v}}$ from the adiabatic condition, one obtains an additional contribution to the first term (cf. Chap. 2). This term leads to a change in the elastic constants of the host lattice, due to the hydrogen loading. The last term determines the frequencies of the localized vibrations of the light hydrogen atoms, in the harmonic approximation.

One has, in the case of tetrahedral sites in bcc metals, e.g., for an x site,

$${}^a\psi^{aa}_{\alpha\beta} = \begin{pmatrix} m\omega_1^2 & 0 & 0 \\ 0 & m\omega_2^2 & 0 \\ 0 & 0 & m\omega_2^2 \end{pmatrix} . \tag{8.9}$$

Two frequencies have been determined by inelastic neutron scattering on V–H, Nb–H, and Ta–H, with $\hbar\omega_1 \approx 120$ meV and $\hbar\omega_2 \approx 170$ meV in all three metals. For references and more detailed results see Chapter 4. The information available on the local mode frequencies of the isotope deuterium indicates a $1:\sqrt{2}$ reduction compared to hydrogen, which indicates isotope-independent coupling parameters ${}^a\psi^{aa}_{\alpha\beta}$ within the limits of experimental accuracy. In the case of octahedral sites in fcc metals, one expects one frequency in (8.9). This frequency, together with its first harmonic, has been observed in Pd–H by several authors. *Drexel* et al. [8.20] found for low concentration $\hbar\omega_{\mathrm{H}} \approx 68$ meV and $\hbar\omega_{\mathrm{D}} \approx 48$ meV, again consistent with isotope-independent second-order coupling parameters. The localized mode frequencies will be used in the next section to estimate tunneling matrix elements.

Other experimental information is provided by the measured activation energies of diffusion. However, it is not possible to extract information on the potential energy of a hydrogen atom in the lattice as long as the diffusion mechanism is not known. For example, the activation energy for thermally excited tunneling processes is completely different from the height of the potential barrier between two hydrogen sites.

8.2.2 Tunneling Matrix Elements

Given the potential energy of the last subsection, a Hamiltonian for the metal-hydrogen system can be written down immediately. However, the assumption has been made that the protons are localized at definite sites. In order to include quantum effects in the motion of the protons, delocalization of protons over several or all sites has to be allowed for. This will be done by introducing tunneling matrix elements.

Let us consider a host metal lattice fixed at its equilibrium positions. Formally, one could consider hydrogen atoms in a periodic potential and calculate band states. The resulting bands would be narrow, i.e., a tight-binding picture is appropriate with strongly localized wave functions. Hence, one only need estimate tunneling matrix elements of a hydrogen atom to its neighboring sites.

An order-of-magnitude estimate of the tunneling matrix element is obtained by putting a hydrogen atom in a periodic one-dimensional potential and calculating the energy splitting between the lowest (symmetric) and the first excited (antisymmetric) state. The potential shall be given by

$$\Psi(v)=\frac{\Delta\Psi}{2}\cos\left(2\pi\frac{v}{d}\right), \tag{8.10}$$

where d is the distance between two minima. The frequency of a harmonic vibration in one of the potential minima is given by

$$\omega^2=(2\pi)^2\Delta\Psi/(2md^2). \tag{8.11}$$

In the case of a deep potential $\Delta\Psi \gg \hbar\omega$, the energy splitting and thus the tunneling matrix element is obtained from an asymptotic formula for eigenvalues of the Mathieu equation given by *Abramowitz* and *Stegun* [8.21],

$$J=8\frac{\hbar^2}{2md^2}\left(\frac{2md^2\Delta\Psi}{\hbar^2}\right)^{3/4}\exp\left[-\frac{2}{\pi}\left(\frac{2md^2\Delta\Psi}{\hbar^2}\right)^{1/2}\right]. \tag{8.12}$$

There are two possibilities to estimate J. Either one uses as potential heights $\Delta\Psi$ the experimentally determined activation energies for diffusion or one calculates $\Delta\Psi$ from the frequencies of the localized modes. The estimates for J for some metals obtained in both ways are found in Table 8.1. We have taken an

Table 8.1. Estimates for the tunneling matrix elements J according to (8.12). Either a given value of $\hbar\omega_H$ or a given value of $\Delta\Psi$ is considered; the derived quantity $\Delta\Psi$ or $\hbar\omega_H$ is put in parentheses. $\Delta\Psi$ (exp) for V was not used since the resulting value of $\hbar\omega_H$ would be larger than $\Delta\Psi$

Metal	d [Å]	$\hbar\omega_H$ [meV]	$\Delta\Psi$ [meV]	J_H [meV]	J_D [meV]	J_μ [meV]
V	1.07	154	(332)	0.13	3.1×10^{-3}	69
Nb	1.17	156	(404)	2.5×10^{-2}	2.9×10^{-4}	43
		(80)	106	1.45	0.14	85
Pd	2.74	68.5	(431)	6.5×10^{-9}	1.6×10^{-13}	0.20
		(50)	226	4.1×10^{-6}	1.8×10^{-9}	1.3

average frequency ω_H for the bcc metals and assumed that $\Delta\Psi$ is isotope independent. The difference between both procedures is partly due to anharmonic effects. We have included the positive muon with $m_\mu = 0.1126 m_H$ in our table. It is clear that the given values for J are rough estimates. Also the condition $\Delta\Psi \gg \hbar\omega$ is not fulfilled for bcc metals. An accurate estimate requires the knowledge of the interaction potential of a hydrogen atom with the host lattice, beyond the harmonic approximation. One should notice, however, the strong isotope dependence of J, irrespective of the estimate used. It is seen from (8.12) that the tunneling matrix element J depends strongly on d. In the following derivations this dependence will be neglected, except in the case of Section 8.3.3.

8.2.3 Localization; Tunneling States

In the last subsection the tunneling matrix element J of a proton between neighboring interstitial sites was introduced. As a consequence, protons in a fixed ideal lattice at $T=0$ would spread out over the whole crystal and form a band. The exact eigenstates would fulfill Bloch's theorem. The diffusion process at low temperatures would be band propagation similar to electronic conduction. A finite diffusion constant results from interaction with thermal phonons, and static defects. Band models for the diffusion of protons in crystals have been considered, e.g., by *Kley* et al. [8.22, 23], *Blaesser* and *Peretti* [8.24], and *Lepski* [8.25]. In these papers the coupling of the protons to the lattice has been either disregarded or considered as a weak perturbation. The band propagation model of *Sussmann* and *Weissman* [8.26] will be discussed separately in Section 8.3.4. As will be shown below, the coupling to the lattice is strong compared to the tunneling matrix elements. The propagation in band states in the presence of a strong coupling of the protons to the lattice has been considered by *Kagan* and *Klinger* [8.27]. It has been termed "coherent tunneling" by these authors. An earlier, related work is that of *Holstein* [8.2] on the propagation of small polarons. No detailed discussion of propagation in the band regime will be given here. We only mention that the diffusion constant decreases with increasing temperature, due to an increasing number of thermal phonons which can interact with the propagating proton. *Kagan* and *Klinger* found a T^{-9}-law for the temperature dependence of the diffusion constant. No indications of a coherent-tunneling regime for hydrogen diffusion have been observed up to now. Thus we direct the discussion to the expected limits of the coherent-tunneling regime. This question has been considered in detail by *Holstein* [8.2] and *Kagan* and *Klinger* [8.27]. A related question is the possible existence of tunneling states of hydrogen atoms. In a tunneling state the proton would occupy several (e.g., six) interstitial sites at the same time. Also, the existence of tunneling states will be limited to a certain temperature regime.

The other possibility is that of protons localized at given interstitial sites. Let us consider in more detail this case, which corresponds to the usual picture

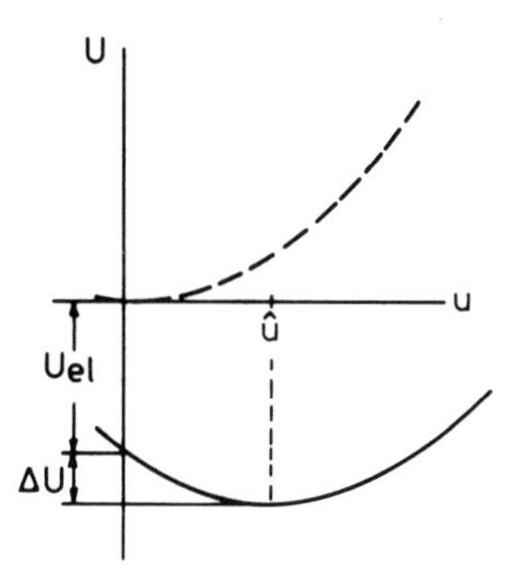

Fig. 8.2. Potential energy of a metal-hydrogen system as a function of one representative lattice coordinate u. – – – potential energy without hydrogen in harmonic approximation. U is a quadratic function of u. —— potential energy after inserting one hydrogen atom. U_{el} is an electronic contribution to the binding energy of the hydrogen atom. The parabola is shifted such that its slope at $u=0$ gives the force $-\Psi$ acting on a representative lattice atom. The lattice relaxes to a new equilibrium configuration $\hat{u}$, lowering thereby the energy by ΔU

of interstitials in a lattice. The neighboring metal atoms of a proton at site a experience the Kanzaki forces ${}^a\psi_\mu^m$ introduced in (8.5), and will be shifted to new equilibrium positions. This process is depicted schematically in Fig. 8.2.

The new equilibrium positions are given by

$$ {}^a\hat{u}_\mu^m = D_{\mu\nu}^{mn}\,{}^a\psi_\nu^n , \tag{8.13}$$

where $D_{\mu\nu}^{mn}$ is the static phonon Green's function in the harmonic approximation,

$$ D_{\mu\nu}^{mn} = \frac{1}{N}\sum_{k \neq 0,\lambda} \frac{e_\mu^\lambda(\boldsymbol{k})\, e_\nu^\lambda(\boldsymbol{k})}{M\omega_{\boldsymbol{k}\lambda}^2}\, \mathrm{e}^{\mathrm{i}\boldsymbol{k}\cdot(\boldsymbol{Q}^m-\boldsymbol{Q}^n)} , \tag{8.14}$$

M is the mass of the metal ions, $\omega_{\boldsymbol{k}\lambda}$ are the harmonic phonon frequencies with wave vector $\boldsymbol{k}$ and polarization index λ, and $\boldsymbol{e}^\lambda(\boldsymbol{k})$ the polarization vector for the phonon mode λ. According to (8.13), short-ranged Kanzaki forces produce large displacements in the vicinity of the proton, but also a long-range displacement field ($\propto r^{-2}$) resulting in a volume dilatation ΔV. The relative volume dilatation $\Delta V/V$ is negligible for one or a few protons. As is suggested in Fig. 8.2, there is an energy gain ΔU by relaxation of the lattice to the new equilibrium positions. ΔU is given by (cf. *Wagner* and *Horner* [8.13])

$$ \Delta U = \tfrac{1}{2}\,{}^a\psi_\mu^m D_{\mu\nu}^{mn}\,{}^a\psi_\nu^n . \tag{8.15}$$

ΔU is estimated by considering an isotropic elastic continuum, and an isotropic double-force tensor,

$$ \Delta U = -\frac{N}{2V}\frac{P^2}{C_{11}} , \tag{8.16}$$

where N is the number of metal atoms, V the volume of the crystal, and C_{11} an elastic constant. ΔU has been calculated by *Horner* and *Wagner* [8.28] for Nb and Ta using realistic phonon Green's functions. Their result for Nb corrected

Table 8.2. Energy gain by relaxation of the lattice for Nb and Pd

Metal	P [eV]	ΔU [meV], Eq. (8.15)	ΔU [meV], Eq. (8.16)
Nb	3.33	−180	−198
Pd	3.5		−296

for the newest value of P (cf. Chap. 3) is found in Table 8.2, together with the estimates for ΔU.

The energy gain ΔU is rather large. Even if there are uncertainties in the tunneling matrix elements, the inequality $\Delta U \gg J$ holds. Since ΔU is a measure of the coupling of a proton to the lattice, the hydrogen-metal system is a system with strong proton-lattice coupling, compared to the tunneling matrix elements. As has been stressed repeatedly (cf. e.g., [8.29]), the effects of J can be considered in perturbation theory, but the coupling to the lattice must be treated in a better approximation.

There is no isotope dependence of ΔU in the approximation considered when the forces ${}^a\psi^m_\mu$ are isotope independent. The experimental results on the isotope dependence of the double-force tensor have been mentioned earlier (cf. also Chap. 3). *Matthew* [8.30] has estimated the isotope dependence of $\Delta V/V$ using a model hydrogen-metal interaction and considering contributions of zero-point motion. His estimate is an isotope effect of less than 3% for $\Delta V/V$, i.e., an isotope effect of less than 6% for ΔU.

The local relaxation of the lattice around a proton implies that the proton is "self-trapped". When the relaxed lattice configuration is fixed, the neighboring hydrogen sites have higher energy, as shown in Fig. 8.3. The energy difference from a self-trapped state to the neighboring sites is given by

$$\delta U = -\tfrac{1}{2}({}^a\psi^m_\mu - {}^b\psi^m_\mu) D^{mn}_{\mu\nu} ({}^a\psi^n_\nu - {}^b\psi^n_\nu)\,. \tag{8.17}$$

Also, the values for δU in Nb and Ta can be obtained from *Horner* and *Wagner* [8.28]. For Nb, $\delta U = 109$ meV.

However, the condition $\Delta U \gg J$ does not give an argument that self-trapping really occurs. Imagine that the proton is quantum mechanically distributed over several sites. Then it is easy to see that, *within the approxima-*

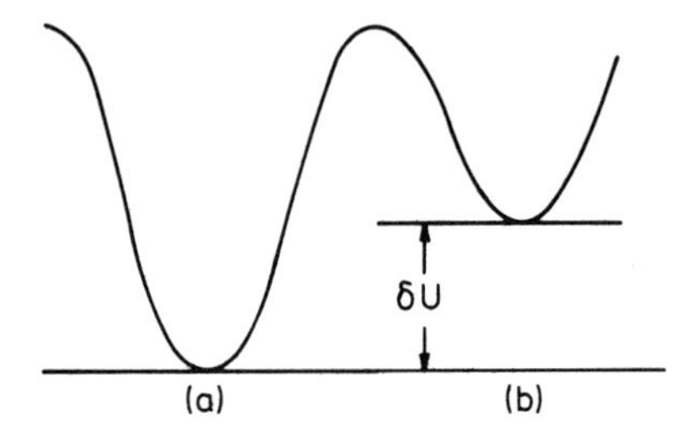

Fig. 8.3. Schematic picture of the potential for a proton when the lattice is kept fixed in the relaxed configuration about site *a*. In the approximation of (8.4), the potential would be given by the horizontal lines

tion of (8.4), the same energy gain ΔU is obtained by relaxation of the lattice about the spread-out configuration. Higher order terms in the interactions have to be taken into account before energetic considerations can favor localized or delocalized configurations. The independence of ΔU on localization or delocalization is implicit in the work of *Holstein* [8.2].

Two additional effects have to be considered before the question of localization or delocalization can be decided. First, the tunneling matrix element should include effects of the local relaxation of the lattice. In the last section, J was derived for a fixed, unrelaxed lattice configuration. The overlap of the lattice wave function for the relaxed configuration about site a and the wave function for the relaxed configuration about site b has to be included. Formally, one either uses the appropriate wave functions (e.g., *Holstein* [8.2]), or one uses a unitary transformation which creates the relaxed lattice configuration (see, e.g., *Sander* and *Shore* [8.29]). The result is an effective tunneling matrix element

$$J_{\text{eff}} = J \exp[-S(T)] . \tag{8.18}$$

$S(T)$ is given by

$$S(T) = \sum_{\boldsymbol{k}\lambda} d_{\boldsymbol{k}\lambda} d^*_{\boldsymbol{k}\lambda} (\bar{n}_{\boldsymbol{k}\lambda} + 1/2) , \tag{8.19}$$

where $\bar{n}_{\boldsymbol{k}\lambda}$ are thermal phonon occupation numbers,

$$\bar{n}_{\boldsymbol{k}\lambda} = [\exp(\hbar\omega_{\boldsymbol{k}\lambda}/k_{\text{B}}T) - 1]^{-1} \tag{8.20}$$

and

$$d_{\boldsymbol{k}\lambda} = \sum_{m,\mu} (2MN\hbar\omega^3_{\boldsymbol{k}\lambda})^{1/2} ({}^b\psi^m_\mu - {}^a\psi^m_\mu) e^\lambda_\mu(\boldsymbol{k}) \, \mathrm{e}^{-i\boldsymbol{k}\cdot\boldsymbol{Q}^m} . \tag{8.21}$$

The factor $\exp[-S(T)]$ is formally analogous to a Debye-Waller factor. Even at $T=0$, there is a reduction of J, which increases with increasing temperature.

No calculations of $S(T)$ exist, and it is difficult to estimate $S(T)$. A very rough estimate has been given in [8.31] for Nb,

$$\exp[-S(T=0)] \sim 10^{-1}$$

$$\exp[-S(T=200\,^\circ\text{C})] \sim 10^{-2} .$$

This estimate leads to a reduction of J of one order of magnitude at $T=0$ and of two orders of magnitude at room temperature. Since J_{eff} determines the bandwidth, any band of protons would be much smaller than indicated by the already small values of J.

The second quantity which limits the existence of band states is their finite lifetime. *Holstein* [8.2] has elaborated this point in the context of small polaron

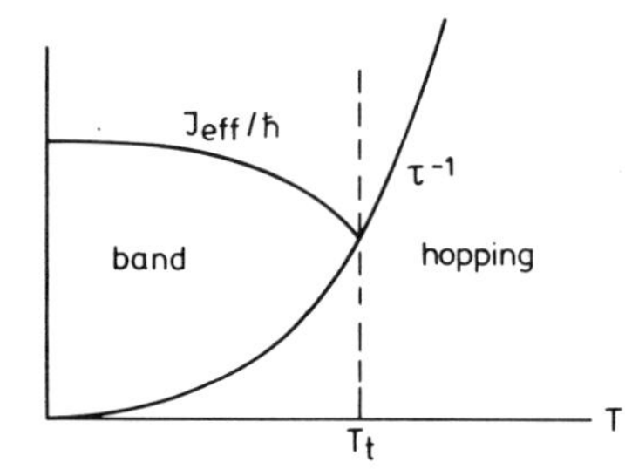

Fig. 8.4. Qualitative behavior of the bandwidth given by J_{eff} and the decay rate τ^{-1} of a band state as a function of temperature

propagation. The essential point of his argument is that the decay rate $\tau^{-1}(T)$ of band states increases with temperature such that above a transition temperature T_t the decay rate τ^{-1} is larger than the effective bandwidth. In this regime, band states are no longer meaningful and the picture of localized particles applies. Hence T_t is given by the condition

$$\hbar\tau^{-1}(T_t) = J_{\mathrm{eff}}(T_t)\,. \tag{8.22}$$

Figure 8.4 gives a schematic picture of the transition between the two regimes.

Holstein [8.2] has investigated τ^{-1} resulting from the interaction with thermal phonons. He showed that τ^{-1} for band states is essentially given by the quantum mechanical hopping rate to be discussed in the next section. Of course, there are other processes possible which lead to a finite lifetime of band states, e.g., transitions to excited states. Such processes were not considered by *Holstein* since he dealt with electrons whose excited states would have a high energy. Another possibility is the destruction of band states by static inhomogeneities or defects. It is clear from these considerations that a reliable estimate of τ^{-1} is difficult at present. In view of the additional uncertainty in $J_{\mathrm{eff}}(T)$, it is impossible to give a reasonable estimate of the transition temperature T_t.

The physical correctness of the picture given above has not yet been confirmed by experiments. Two observations on related systems will be mentioned. First, neutron scattering experiments on tunneling states in molecular crystals show that these states become ill defined above a certain temperature without their widths becoming large. A typical experiment showing these features is that on $(NH_4)_2SnCl_6$ by *Prager* et al. [8.32]. Second, the transition temperature T_t has not yet been found in electronic small-polaron systems (cf. the review of *von Baltz* and *Birkholz* [8.33]). As mentioned previously, a transition to a coherent tunneling regime has never been observed for hydrogen in metals. It will be interesting to see the results of experiments on the diffusion of muons in bcc metals, i.e., whether a coherent tunneling regime is observed for this system.

Kagan and *Klinger* [8.27] invoke another mechanism which limits the regime of coherent tunneling. They consider the "dynamical destruction" of band states through intrawell interaction of protons with phonons, and give an expression for the resulting rate Ω_1, which is given in the limit $T > \Theta_{\text{Debye}}$ by

$$\Omega_1 \sim \omega_{\text{Debye}}^3 (k_B T/\hbar)^2 / \omega_H^4 . \tag{6.23}$$

$\hbar\Omega_1$ is about 0.1 meV at $T = 200\,°\text{C}$, i.e., larger than the estimates of J_{eff}. However, (8.23) does not contain the coupling parameters of the protons to the phonons, as should be the case for this theory.

Finally, the question of the existence of tunneling states will be considered. All that has been said above about the existence of band states can be transferred to a discussion of the existence and limitations of tunneling states. However, there is one additional argument to consider, namely that the dependence of J on the lattice displacements u_μ^m might be important for tunneling states. Physically, some neighboring metal atoms might move in such a way as to allow for large tunneling matrix elements between a group of interstitial sites (*Horner* [8.34]). In view of the absence of knowledge on the interaction potentials, nothing definite can be said about this proposition. The same remarks as before can be made about estimations of the transition temperature T_t for tunneling states.

At present the only way to decide on the existence of tunneling states is through experiments. The absence of the Snoek effect, however, does not give an argument in favor of tunneling states, unless the tunneling matrix elements are unreasonably large. In fact, the Snoek effect has been observed for tunneling states, e.g., in KCl:Li, by *Byer* and *Sack* [8.35]. Neutron scattering experiments can provide information on the transitions between different tunneling states, and on the form factor of the proton. Results will be discussed in Chapter 10. No convincing evidence for tunneling states has been found in neutron scattering experiments up to now. The reported anomalies of the form factor increase with increasing temperature. Hence, it is plausible that they are related to the limitations of the model of jump diffusion at high jump rates. There are specific-heat results in Nb–H and Nb–D which have been interpreted as evidence for tunneling states (*Birnbaum* and *Flynn* [8.36]). However, there are indications [8.37, 38] that the anomalies in the specific heat are connected with the presence of other impurities such as O, N. Thus the possibility remains that there exist tunneling states of the interstitial complexes O–H or N–H at low temperatures. Future experiments will hopefully clarify this subject.

8.3 Quantum Mechanical Rate Theory

8.3.1 Small-Polaron Hopping Theory

In the regime where the proton is localized at a specific interstitial site, a diffusive process to another site requires thermal activation. The localized proton is self-trapped by the relaxation of the lattice about it, and the

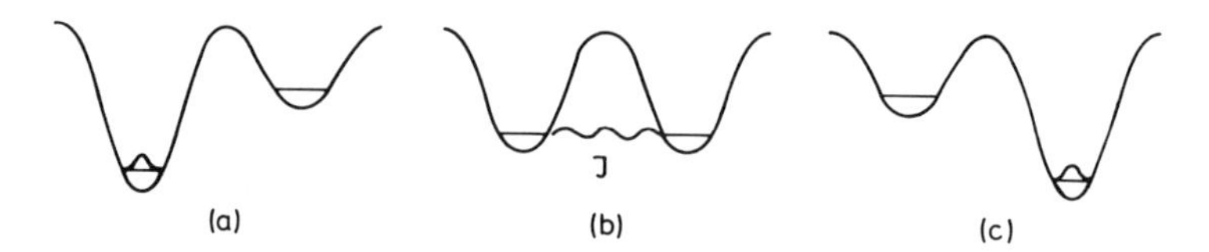

Fig. 8.5a–c. Thermally activated tunneling transition of a proton. (a) The lattice is relaxed about the proton at the left site, and the proton has a lower energy (cf. also Fig. 8.3.). (b) Thermal fluctuations have brought both levels to the same height. A tunneling process is possible. (c) The lattice is relaxed about the new equilibrium position

neighboring sites have a higher potential energy in this configuration (cf. Fig. 8.3). Thermal activation is necessary to overcome the potential barrier between two sites. Also, a tunneling process from the occupied site to a neighboring site requires thermal activation, as indicated in Fig. 8.5. The energy which creates the configuration b of Fig. 8.5 is provided by the thermal fluctuations of the metal atoms. It is easy to guess the transition rate Γ to the neighboring interstitial site by referring to the "Golden rule". One expects

$$\Gamma \sim J^2 \exp(-E_a/k_B T). \tag{8.24}$$

The second factor represents the thermal activation of the configuration Fig. 8.5b, which requires the energy E_a. The first factor is the square of the tunneling matrix element.

A quantitative derivation of the thermally activated tunneling process has been first given for small polarons. We refer to the work of *Holstein* [8.1, 2], who first derived the rate in its correct form and carefully discussed the limits of applicability of the theory. The importance of polaron effects for the mobility of hydrogen in metals has been recognized by *Schaumann* et al. [8.39]. *Flynn* and *Stoneham* [8.40] applied the concepts of the theory of small-polaron hopping to the diffusion of hydrogen in metals and extended the theory. Methods analogous to small-polaron hopping theory are also employed for the re-orientation of defects in crystals (cf. *Sander* and *Shore* [8.29]).

The starting point of small-polaron hopping theory is to consider the tunneling matrix elements as a perturbation in the total Hamiltonian. Without J, the localized states of an electron or proton are eigenstates of the Hamiltonian. J introduces transitions between the localized eigenstates, which are treated in time-dependent perturbation theory. The central point is the inclusion of the self-trapping distortion in the derivation. The amplitude to be considered is

$$a(t) = \langle b, t | a \rangle, \tag{8.25}$$

where $|a\rangle$ is the state with one proton at site a and the lattice relaxed about it, and $|b, t\rangle$ the state at time t with one proton at site b and the lattice relaxed about it. The calculations is straightforward but tedious and can be found in several places (e.g., [8.2, 29, 40]). After thermally averaging, the transition rate is found in the form

$$\Gamma_{b\leftarrow a}(T)=(J^2/\hbar)\exp[-2S(T)]$$
$$\cdot\lim_{t\to\infty}\int_{-t}^{t}dt'\left(\exp\left\{\sum_{\boldsymbol{k}\lambda}d_{\boldsymbol{k}\lambda}d^*_{\boldsymbol{k}\lambda}[\bar{n}_{\boldsymbol{k}\lambda}e^{i\omega_{\boldsymbol{k}\lambda}t'}+(\bar{n}_{\boldsymbol{k}\lambda}+1)e^{-i\omega_{\boldsymbol{k}\lambda}t'}]\right\}-1\right). \quad (8.26)$$

$S(T)$ is defined in (8.19) and $d_{\boldsymbol{k}\lambda}$ in (8.21). The integral in (8.26) is evaluated differently in the high- and in the low-temperature regime (see *Holstein* [8.2] and *Flynn* and *Stoneham* [8.40]).

In the high-temperature region $T>\Theta_{\text{Debye}}$, phonon processes of all orders contribute to the integral in (8.26). One obtains

$$\Gamma(T)=\sqrt{2\pi}\,(J/\hbar)^2\left[\sum_{\boldsymbol{k}\lambda}d_{\boldsymbol{k}\lambda}d^*_{\boldsymbol{k}\lambda}\frac{\omega^2_{\boldsymbol{k}\lambda}}{\sinh(\hbar\omega_{\boldsymbol{k}\lambda}/2k_{\text{B}}T)}\right]^{1/2}$$
$$\cdot\exp\left[-\sum_{\boldsymbol{k}\lambda}d_{\boldsymbol{k}\lambda}d^*_{\boldsymbol{k}\lambda}\tanh(\hbar\omega_{\boldsymbol{k}\lambda}/4k_{\text{B}}T)\right]. \quad (8.27)$$

The high-temperature limit of this formula is

$$\Gamma(T)=\frac{J^2}{\hbar}\left(\frac{\pi}{4E_{\text{a}}k_{\text{B}}T}\right)^{1/2}\exp(-E_{\text{a}}/k_{\text{B}}T), \quad (8.28)$$

where the activation energy is given by

$$E_{\text{a}}=\tfrac{1}{4}\sum_{\boldsymbol{k}\lambda}\hbar\omega_{\boldsymbol{k}\lambda}d_{\boldsymbol{k}\lambda}d^*_{\boldsymbol{k}\lambda}. \quad (8.29)$$

Thus the high-temperature limit of the transition rate gives the expected result of (8.24). The activation energy E_{a} is determined by the coupling between the proton and the lattice, and by the force constants of the host lattice, as is seen from (8.29) and (8.20). It is readily seen that $E_{\text{a}}=\delta U/4$. E_{a} is the increase in relaxation energy when the proton is *classically* half at site a and half at site b. Hence E_{a} is the saddle-point energy for transferring the proton from site a to site b.

In the low-temperature region $T\ll\Theta_{\text{Debye}}$, the contributions to the integral (8.26) can be decomposed in $1, 2, \ldots$-phonon processes. The lowest nonvanishing contribution is a 2-phonon process. The expression "2-phonon process" is somewhat misleading, since a rearrangement of the lattice distortion takes place during the jump process. An explicit result has been derived by *Flynn* and *Stoneham* [8.40] using an isotropic Debye model

$$\Gamma(T)\approx 57600\pi\omega_{\text{D}}$$
$$\cdot\frac{J^2E_{\text{a}}}{(\hbar\omega_{\text{D}})^4}\left(\frac{T}{\Theta_{\text{D}}}\right)^7\exp\left(-\frac{5E_{\text{a}}}{\hbar\omega_{\text{D}}}\right). \quad (8.30)$$

Stoneham [8.41] has also given a representation of the jump rate for an isotropic Debye model as an integral which can be evaluated numerically. One of the results of this representation was the limits of validity of the asymptotic expressions. The parameter governing the limits is $E_a/\hbar\omega_D$. For actual values of this ratio, only small deviations from the Arrhenius law (8.28) were seen for $T \gtrsim \Theta_D$. The T^7-law (8.30) was only found for temperatures of the order of a few percent of Θ_{Debye}.

8.3.2 Comparison with Experiment

Can the small-polaron hopping theory developed so far describe experiments on the diffusion of hydrogen in metals? This question will be discussed in this subsection, before extensions of the theory are reviewed in the next subsection.

The following predictions amenable to experiments are obtained from the expressions of the preceding subsection. The jump rate should follow an Arrhenius law for $T \gtrsim \Theta_{Debye}$ and a T^7-law for $T \ll \Theta_{Debye}$. The prefactor being proportional to J^2 should be strongly isotope dependent, since tunneling matrix elements should be isotope dependent (cf. also [8.40]). The activation energy E_a should be isotope independent, if the Kanzaki forces ${}^a\psi_\mu^m$ are isotope independent.

It is useful to discuss bcc and fcc metals separately. The activation energies for diffusion are smaller in bcc metals than in fcc metals. This is a hint to smaller barriers between the interstitial sites in bcc metals, which also have a smaller distance than in fcc metals. Both data lead to larger tunneling matrix elements in bcc than in fcc metals. Hence, the bcc metals are the primary candidates for an application of the small-polaron hopping theory of hydrogen diffusion.

The experiments on bcc metals (cf. Chap. 12) show Arrhenius laws for the diffusion coefficients or jump rates at temperatures above 250 K, which is of the order of the Debye temperature of the metals considered. An Arrhenius law with a smaller activation energy (68 meV) and prefactor has been found in Nb–H below 250 K by *Schaumann* et al. [8.39] and confirmed by *Wipf* [8.42] and *Richter* et al. [8.43]. There exist some quenching experiments at low temperature [8.44–47]. The interpretation of these experiments is difficult. No T^7-law has been observed up to now. *Engelhard* [8.47] found in Nb–H between 30 and 60 K an activation energy of 65 ± 10 meV, in agreement with that of the Arrhenius law mentioned above. The results above 250 K seem to be reliable now. In this region, no significant isotope effect has been found in the prefactors of the Arrhenius law. This result is at variance with a prefactor simply determined by the square of a tunneling matrix element J. An isotope effect has been observed in the activation energies above 250 K, i.e., $E_a^D > E_a^H$. No isotope effect is expected for Nb–H, Nb–D where the double-force tensors are equal [8.15], and a reversed isotope effect of E_a of about 16% is expected in Ta–H, Ta–D according to the newest measurements of P in tantalum [8.15]. Thus the small-polaron theory in its simplest form does not describe the experiments on hydrogen diffusion in bcc metals above 250 K. One possible exception will be discussed below.

One might speculate that the "flat branch" of the Arrhenius law in Nb–H below 250 K is explained by the small-polaron hopping theory [8.31, 43], since the flat branch has not been observed for Nb–D in the same temperature region. If Nb–D would show a crossover to a flat branch at some lower temperature (e.g., 110 K) with an activation energy of about 60 meV and a prefactor of about 10^{-2} of that of Nb–H, then the interpretation of the flat branch as small-polaron hopping would be much less speculative. One might further speculate that the diffusion of protons in vanadium is described by the small-polaron hopping theory in the whole temperature region observed so far. This speculation is stimulated by the low activation energy for H of only 40 meV, and the large isotope effect in the activation energy. V has a smaller lattice constant than Nb or Ta, hence the tunneling matrix element could be larger than in Nb or Ta. Again, experiments on V–D at lower temperatures can clarify the interpretation.

The experiments on the diffusion of low concentrations of hydrogen in fcc metals have given Arrhenius laws with isotope-dependent prefactors and activation energies (cf. Chap. 12). Nothing is known about the low-temperature behavior of the hopping rate. The isotope dependence of the prefactors is approximately described by the square root of the mass ratios (cf., e.g., [8.48–51]). One exception is Pd–T (*Sicking* and *Buchold* [8.52]). In view of the small tunneling matrix elements J which one obtains in estimates for fcc metals (cf. Table 8.1), it is very unlikely that the small-polaron hopping theory applies to the fcc metals, at least in its form discussed so far. For muons, however, the estimated tunneling matrix elements are much larger (cf. Table 8.1). The diffusion of muons in copper has been observed in the temperature regime from 77 to 300 K by *Gurevich* et al. [8.53], who found an activation energy of 46 meV and a prefactor $\Gamma_0 = 2.6 \times 10^6\,\mathrm{s}^{-1}$. The activation energy is much less than that of hydrogen in Cu (403 meV). It is persuasive to assume that the muon hopping process is described by the small-polaron hopping theory. Although δU has not yet been calculated for hydrogen in fcc metals, $E_\mathrm{a} = 45$ meV is of the right order (cf. $\delta U/4 = 26$ meV for Nb–H). The prefactor obtained from the estimate $J = 0.47$ meV of (8.12) applied to Cu is $1.1 \times 10^{10}\,\mathrm{s}^{-1}$, i.e., several orders too large, but there are uncertainties in the experimental analysis and in the estimate of J.

The quantum mechanical jump rate theory should also apply, under appropriate conditions, to the reorientation of defects. *Chen* and *Birnbaum* [8.54] have investigated the anelastic relaxation of H–O and H–N complexes in Nb. They can describe their results together with older results over a wide temperature range by the formula of *Stoneham* [8.41] for small-polaron jumps in an isotropic Debye solid. J is found to be as large as 53 meV. It would be very interesting to have the results for the reorientation of D–O or D–N complexes, in order to study the isotope effect of the reorientation rate.

The main conclusion is that the small-polaron hopping theory in its simplest form generally cannot be applied to the diffusion of hydrogen in metals above 250 K. The same conclusion is implicit in the paper of *Matthew* [8.30]. The diffusive processes of protons in niobium below 250 K, of protons in vanadium at all temperatures, and of muons in copper are candidates for the small-polaron jump process. The reorientation of defects composed of hydrogen and other interstitials might also be governed by the quantum mechanical jump theory, although further investigations are needed.

8.3.3 Extensions of the Small-Polaron Theory

In the last two subsections, thermally activated tunneling transitions from one ground state level to an adjacent ground state level have been considered. In addition, the approximation of constant tunneling matrix elements has been made, i.e., J was assumed independent of the host lattice coordinates. There are certainly other, more complicated, processes by which a hydrogen atom can perform its diffusive motion. Two of them will be considered in this section.

First, there is the possibility of "lattice activated" processes which were introduced and investigated by *Flynn* and *Stoneham* [8.40]. *Flynn* and *Stoneham* noted that only "antisymmetric" displacement modes contribute to the activation energy (8.29). E_a can be written in the following form by introducing displacements instead of forces:

$$E_a = 1/8 \delta u_\mu^m \Phi_{\mu\nu}^{mn} \delta u_\nu^n ,$$

where (8.31)

$$\delta u_\mu^m = {}^a\hat{u}_\mu^m - {}^b\hat{u}_\mu^m .$$

${}^a\hat{\boldsymbol{u}}$, ${}^b\hat{\boldsymbol{u}}$ has been introduced in (8.13). Only the "antisymmetric" parts of the displacements, where ${}^b\hat{\boldsymbol{u}} = -{}^a\hat{\boldsymbol{u}}$, contribute to E_a. The symmetric parts of $\hat{\boldsymbol{u}}$, which do not contribute to E_a, can have a strong influence on the tunneling matrix element J. They may enlarge J by "opening the door" for a large overlap of the wave functions. *Flynn* and *Stoneham* considered jumps between the octahedral sites in fcc and in bcc lattices. In fcc metals there are two host metal atoms in a plane perpendicular to the jump path between octahedral sites, which can strongly affect the overlap; in bcc metals there are no such atoms. It is now known with more certainty that hydrogen occupies the tetrahedral sites in bcc lattices. Lattice activation appears possible for jumps between them (cf. Fig. 8.6). *Flynn* and *Stoneham* introduce normal coordinates $q_{k\lambda}$ in order to have independent symmetric and antisymmetric modes. They assume $J(q_s)$ to be small for $q_s < q_{s,c}$ and equal J_{lim} for $q_s > q_{s,c}$ (cf. Fig. 8.7). They perform the thermal average with $J(q_s)$ and obtain the following formula for the jump rate:

$$\Gamma = \frac{J_{\text{lim}}^2}{4\hbar (E_a E_s)^{1/2}} \exp[-(E_a + E_s)/k_B T] . \tag{8.32}$$

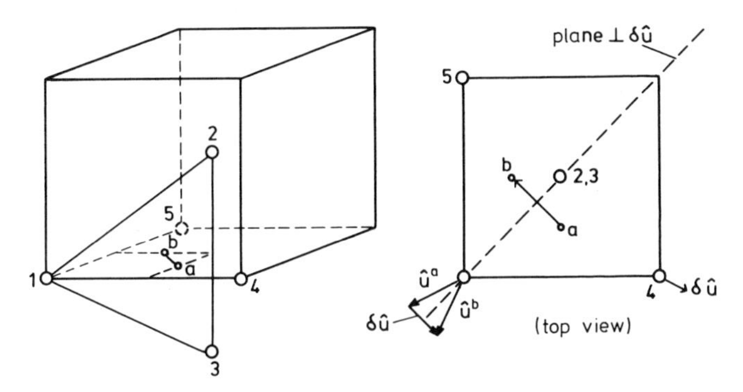

Fig. 8.6. Lattice activation for jumps between two tetrahedral sites a, b in a bcc lattice. Displacements of the atoms 1, 2, 3 in the plane defined by them do not influence E_a (symmetric mode). These displacements can influence the tunneling matrix element between a and b

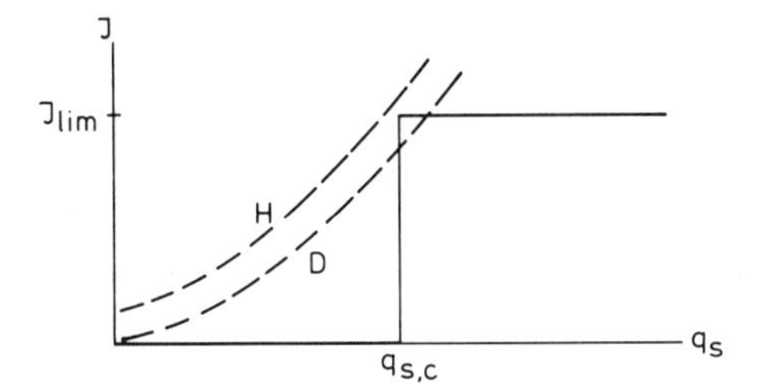

Fig. 8.7. Schematic picture of the tunneling matrix element J as a function of a symmetric mode q_s. —— assumption of Flynn and Stoneham; – – – likely behavior of J for H, D

E_s is the energy necessary to create the lattice configuration with normal coordinates $q_{s,c}$. After introduction of the corresponding displacements $u_{s,c}$, one has

$$E_s = 1/2 u^m_{s,c\mu} \Phi^{mn}_{\mu\nu} u^n_{s,c\nu} \,. \tag{8.33}$$

J probably varies more smoothly with q_s, as indicated qualitatively in Fig. 8.7. Thus one expects an isotope dependence of $q_{s,c}$ where $J = J_{\text{lim}}$, and a resulting isotope dependence of E_s. E_s should be larger for the heavier isotopes. On the other hand, the prefactor should weakly decrease with increasing mass. It will be less isotope dependent than in the simple expression (8.28).

It is difficult to decide whether lattice activated processes occur, before realistic calculations on E_s have been made. These calculations would require a better knowledge of the hydrogen-host metal interaction. It is plausible that lattice activation is important in bcc metals, since the initial J is already appreciable, and only small displacements q_s are necessary to increase the overlap sufficiently. The derivation of *Flynn* and *Stoneham* was made for fcc metals, where the initial J is small, and considerable excursions q_s are necessary to reach J_{lim}. Then the competition of other processes, e.g., the contribution of excited states, has to be considered.

The importance of the contribution of the excited states of the localized hydrogen vibrations to the thermally activated tunneling process has been recognized by several authors. *Flynn* and *Stoneham* [8.9, 40] have expressed the view, which has been shared by this author [8.31], that explicit calculations seem not useful, since the tunneling matrix elements and other parameters are not well enough known. *Kagan* and *Klinger* [8.27] take into account contributions of the excited states in a general way without explicitly specifying the matrix elements. They consider only direct transitions from an excited state to the same one in the adjacent potential well. Hence they can write the jump rate as a sum of all individual contributions. In the paper of *Gorham-Bergeron* [8.55], direct as well as indirect transitions are taken into account. In this work, an algebraically simple potential is used, which allows a summation of all processes yielding a closed formula. The result for the diffusion constant D is given in the limit $(\hbar\omega)^2 \ll 2\Delta U k_B T$ where ω is the frequency of the localized vibrations and ΔU the self-trapping energy.

$$D = \frac{\hbar}{m}\left(\frac{\pi k_B T}{\Delta U + \hbar\omega \bar{b}^2/2}\right)^{1/2} \exp\left[-\frac{\Delta U + \hbar\omega\bar{b}^2/2}{2k_B T} - \frac{\hbar^2\omega^2}{2k_B T(\Delta U + 2\hbar\omega\bar{b}^2)}\right], \quad (8.34)$$

where $\bar{b}^2 = b^2 m\omega/2\hbar$ and b the distance of the two harmonic oscillator wells used. The limit used is appropriate for fcc metals. It should be noted that the jump rate Γ contains a factor $1/b^2$. According to (8.34), the activation energy for diffusion is one-half the self-trapping energy, and a correction which decreases with increasing isotope mass. This result is in qualitative accord with the observations in fcc metals. The physical origin of this correction, however, is not well understood.

8.3.4 Other Theoretical Developments

We first comment on a simple theory of tunneling processes for protons, which was put forward in the papers by *Heller* [8.56] and *Arons* et al. [8.57, 58] (cf. also *Blaesser* and *Peretti* [8.24]). These authors consider tunneling of protons from interstitial to interstitial site. The coupling of the protons to the lattice with its many degrees of freedom is neglected. If this coupling were weak, then the protons would be delocalized in the form of band states. Since the coupling is strong, it has to be taken into account and leads to the incoherent tunneling processes in the manner discussed in Section 8.3.1.

The second group of theories which will be discussed in this section can be described as "propagation of wave packets". *Weiner* and co-workers have investigated diffusion problems with the aim of deriving rate theory from the Schrödinger equation using this approach. Three papers will be mentioned; references to earlier work are found in them. *Weiner* and *Partom* [8.59], have studied the motion of wave packets near saddle points in an N-dimensional configuration space. *Weiner* and *Forman* [8.60] calculated numerically the motion of wave packets in a one-dimensional periodic potential using a quantum mechanical Langevin equation. These papers cover important aspects of the quantum mechanical diffusion problem, but a coherent picture has not been reached. *Weiner* [8.61] also treats the anomalous isotope effect of the activation energy E_a in fcc lattices by the motion of wave packets out of a potential well. In this paper, the coupling of the interstitial to the lattice and the frictional forces resulting from it are disregarded.

Sussmann and *Weissman* [8.26] have suggested the propagation of wave packets from another point of view. They consider a proton in its ground state to be localized due to interaction with the lattice, i.e., by self-trapping. They then assert that there are bands of the excited states which, however, are broadened due to the interaction with the lattice. The diffusive process is described as the formation of a wave packet of these band states, which can propagate and decay at the neighboring interstitial site. It is not recognized that band states no longer exist when their decay rate is larger than the bandwidth. Since, in this theory, the propagation of protons occurs in bands of excited states, the energies of excited states and the activation energies for diffusion should coincide. There is a fortuitous coincidence for Nb–H above 250 K; $E_a > \hbar\omega_1$ for Ta–H while $E_a < \hbar\omega_1$ for V–H, where $\hbar\omega_1$ is the lower energy of the localized vibration. No excited state has been found in Nb–H at 68 meV corresponding to the activation energy for diffusion below 250 K. While there are some appealing elements in the theory of *Sussmann* and *Weissman*, it does not give a description of hydrogen diffusion which is free of contradictions.

8.4 Classical Rate Theory

8.4.1 Derivation of Classical Rate Theory

At higher temperatures, the over-barrier jumps of hydrogen atoms from one equilibrium site to another should become important. Although they require higher activation energies than the thermally activated tunneling processes, they should have larger prefactors and there should be more states available. We shall call the regime of over-barrier jumps "classical", although the discreteness of the states might bring important modifications. We want to study the classical rate theory in this section, in order to see how far it already describes the results on hydrogen diffusion. The classical rate theory should be derived from a quantum mechanical rate theory by letting $\hbar \to 0$. Quantum theories which include over-barrier diffusion were considered in the last section. The expressions given by *Kagan* and *Klinger* [8.27] or by *Gorham-Bergeron* [8.55], however, do not reduce to the classical rate theory in the limit $\hbar \to 0$. In view of the lack of a quantum mechanical derivation, the classical rate theory must be derived from classical considerations.

There is a potential barrier against jumps of a proton to a neighboring site, resulting from the self-trapping distortion, and from additional interactions

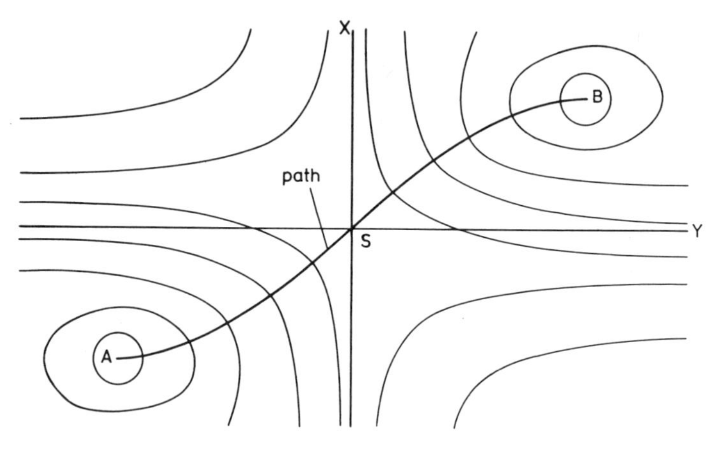

Fig. 8.8. Two-dimensional picture of the configuration space, where X is one lattice coordinate and Y is one hydrogen coordinate. A and B are minima and S is a saddle point

with the host metal atoms (cf. Fig. 8.3). When the particle has performed a jump to a neighboring site, the lattice has relaxed to a new equilibrium configuration. Thus the diffusive motion can be considered classically as the motion of a system point in a $3N+3$-dimensional configuration space. Figure 8.8 gives a two-dimensional picture of the motion. It is useful to introduce new coordinates such that one coordinate, called ξ, runs along the diffusion path between A and B. v is the corresponding velocity. The jump rate Γ can be found by considering the probability $W(\xi, v)d\xi dv$ that the system point has the coordinate ξ between ξ and $\xi+d\xi$ and the velocity between v and $v+dv$. Γ is obtained from the current of system points over the saddle point ξ_S in the positive direction (*Zener* [8.62]),

$$\Gamma=(1/W_A)\int_0^\infty dv v W(\xi_S, v) \tag{8.35}$$

W_A is the probability of finding the system point near the minimum A. The probabilities in (8.35) can be expressed by partition functions, and the velocity integration carried out. The result is that of the "absolute rate theory" of *Eyring* [8.63] (cf. also *Wert* and *Zener* [8.64]),

$$\Gamma=\frac{k_B T}{2\pi\hbar}\,\frac{Z^*(\xi_S)}{Z_A}. \tag{8.36}$$

$Z^*(\xi_S)$ is a partition function where ξ is kept fixed at ξ_S and v does not appear, and Z_A is the partition function restricted to a region around the minimum A. Thermodynamic potentials are introduced in the usual way, e.g., the free enthalpy $H^*(\xi)$ where the coordinate ξ is kept fixed. In terms of the free enthalpy $H^*(\xi)$ and entropy $S^*(\xi)$, the rate is given by

$$\Gamma=\Gamma_0 \exp(-\Delta H/k_B T) \tag{8.37}$$

where

$$\Gamma_0=\nu_{A,\xi} \exp(\Delta S/k_B),$$
$$\Delta H=H^*(\xi_S)-H^*(\xi_A),$$
$$\Delta S=S^*(\xi_S)-S^*(\xi_A).$$

$\nu_{A,\xi}$ is the "attempt frequency", i.e., the frequency of harmonic vibrations in the minimum A in the ξ-direction. It is renormalized to Γ_0 by the entropy factor. A comparison of the experimentally determined prefactors of the jump rates of hydrogen and the frequencies of the localized vibrations shows that the entropy factor would be less than one for hydrogen in metals, or that ΔS is negative. However, the theory cannot be applied to hydrogen in metals in its present form, as will be discussed in the next subsection.

ΔH and ΔS have an obvious equilibrium interpretation, e.g., $H^*(\xi_S)$ is the free enthalpy where the ξ-coordinate of the system point is kept fixed in the saddle-point configuration. Another formulation of the classical rate theory is obtained by integrating out all momenta in (8.36) (*Vineyard* [8.65]). One finds

$$\Gamma=\left(\frac{k_B T}{2\pi m^*}\right)^{1/2} \frac{P^*(\xi_S)}{P_A}. \tag{8.38}$$

Here the P's are the configuration parts of the partition functions. The meaning of $P^*(\xi_S)$ and P_A is analogous to that of $Z^*(\xi_S)$ and Z_A in (8.36). m^* is the mass associated with motions in the ξ-direction. It is easy to derive the well-known frequency-product formula of *Vineyard* from (8.38). Here only the isotope effect resulting from (8.38) will be considered. As long as the interaction potentials are indepedent of the isotope of the diffusing particle, one has for two isotopes α, β

$$\Gamma_\alpha/\Gamma_\beta=(m_\alpha^*/m_\beta^*)^{-1/2}. \tag{8.39}$$

The task is to relate m^* to the mass m of the diffusing particle, and the mass M of the lattice atoms. This can be done formally by introducing the ΔK-factor (*Mullen* [8.66] and *Le Claire* [8.67]). *Glyde* [8.68] has argued that for $m \ll M$ the isotope effect should be determined by m itself. *Katz* et al. [8.48] have made an estimate of m^* for hydrogen in Cu and Ni and found that $m^* \approx m$ within parts of a percent. Thus the prefactor of the hydrogen jump rate should show the "classical" isotope dependence $\propto (m)^{-1/2}$ if the classical theory applies. On the other hand, no isotope effect is expected in the activation energy of the classical theory, as long as the interactions are isotope independent.

8.4.2 Quantum Mechanical Modification

It is assumed in the derivation of the classical rate theory that the system point can have any energy; in other words, that the possible energy levels of the system form a continuum. The energy levels of a hydrogen atom, however, are discrete, and widely spaced. The energy of the localized vibration of a proton in Pd fulfills $\hbar\omega > k_B T$ at room temperature, the localized modes in bcc metals fulfill $\hbar\omega \gg k_B T$ at room temperature. Hence the discreteness of the excitation energies has to be taken into account in the derivation of rate theory. A modification of the classical theory which includes this aspect is obtained by replacing the classical partition functions in (8.36) by quantum mechanical partition functions. It is not clear that quantum mechanical effects are adequately taken into account by this replacement, since classical concepts have been used in an essential way to obtain (8.36) (cf. the remarks of *Wert* and *Zener* [8.64]). The quantum mechanical modification has been carried out by *Le Claire* [8.67] and by *Ebisuzaki* et al. [8.69, 70]. *Alefeld* [8.71] has investigated the quantum mechanical modification using the "dynamical" approach to classical rate theory.

The following identifications are made

$$Z_A = \sum_\nu \exp(-E_\nu^A/k_B T),$$
$$Z^*(\xi_S) = \sum_\nu \exp(-E_\nu^{*S}/k_B T). \tag{8.40}$$

E_ν^A are the energy eigenvalues for states near the minimum of A, and E_ν^{*S} eigenvalues for states near the saddle-point configuration. The star indicates that the degree of freedom corresponding to ξ has been omitted. E_ν^{*S} contains a term $U_S - U_A$, the difference in potential energy between saddle-point and minimum configuration. Z_A and $Z(\xi_S)$ is easily evaluated in the harmonic approximation for E_ν^A, E_ν^{*S}. The result is (cf. *Katz* et al. [8.48])

$$\Gamma = \frac{\prod_{i=1}^{3N+3} \nu_{A,i} \prod_{j=1}^{3N+3} f(2\pi\hbar\nu_{A,j}/k_B T)}{\prod_{i=1}^{3N+2} \nu_{S,i} \prod_{j=1}^{3N+2} f(2\pi\hbar\nu_{S,j}/k_B T)} \exp[-(U_S - U_A)/k_B T], \tag{8.41}$$

where $f(x) = \sinh(x)/x$ and $\nu = E/2\pi\hbar$. In the limit $\hbar\omega \ll k_B T$ where $f(x) \to 1$, the frequency-product formula of *Vineyard* [8.65] is obtained. Also the corrections to the classical result can be given. The following assumptions will be made for simplification:

a) There are three equal localized vibration frequencies ω_A in the minimum configuration,

b) there are two equal localized vibration frequencies ω_S in the saddle-point configuration,

c) the lattice vibrations are decoupled from the localized vibrations and are equal in the minimum and saddle-point configuration.

Then (cf. *Katz* et al. [8.48]),

$$\Gamma = \frac{v_A^3 f^3(2\pi\hbar v_A/k_B T)}{v_S^2 f^2(2\pi\hbar v_S/k_B T)} \exp[-(U_S - U_A)/k_B T]. \tag{8.42}$$

In the limit $2\pi\hbar v_{A,S} \gg k_B T$ which is relevant to bcc metals, the result is found

$$\Gamma = \frac{k_B T}{2\pi\hbar} \exp\left[-\frac{U_S - U_A - (3/2)\hbar\omega_A + \hbar\omega_S}{k_B T}\right]. \tag{8.43}$$

Thus there appears a general prefactor, independent of isotope mass or host lattice properties. *Ebisuzaki* et al. [8.70] first stated the existence of a general prefactor in the limit considered, without explicitly displaying it. The activation energy which appears in (8.43) has a simple physical interpretation. It is the difference of the potential energy of the saddle-point and minimum configuration, together with the difference in zero-point energies, where the minimum configuration contains one more degree of freedom.

8.4.3 Comparison with Experiment

Does the modified classical rate theory adequately describe the experimental results on the jump rates of hydrogen in metals at higher temperature? Again this question has to be examined for fcc and bcc metals separately.

In fcc metals, $\hbar\omega$ is expected to be of the order of $k_B T$ somewhat above room temperature. If the simplifying assumptions of the last subsection are made, (8.42) should apply. In the limit $T \to \infty$, an isotope effect $\propto m^{-1/2}$ is obtained, since the localized modes are proportional to this factor. An approximate $m^{-1/2}$-dependence of the prefactor has been found for H, D in Pd by *Bohmholdt* and *Wicke* [8.49], *Holleck* and *Wicke* [8.50], and *Völkl* et al. [8.51], and for all three isotopes in Ni and Cu by *Katz* et al. [8.48]. The data on T in Pd (cf. Chap. 12) are an exception. The main problem is the explanation of the observed reversed isotope effect in the activation energy. *Jost* and *Widmann* [8.72, 73] pointed out that a reversed isotope effect can be explained by the assumption of higher frequencies of the (transverse) localized vibrations ω_S in the saddle-point configuration, compared to the frequencies ω_A in the minimum configuration. Then the energy difference of the corresponding ground state levels is larger for the lighter isotope (Fig. 8.9). The results at intermediate temperatures for all three hydrogen isotopes in Cu and Ni can be fitted by choosing appropriate parameters ω_A and ω_S. *Katz* et al. used for Ni, $\hbar\omega_A = 115$ meV, $\hbar\omega_S = 200$ meV and for Cu, $\hbar\omega_A = 330$ meV, $\hbar\omega_S = 470$ meV—all values for protons. The reversed isotope effect follows from $\omega_S > \omega_A$. Although no measurements of the local mode frequencies of H in Cu and Ni exist, the

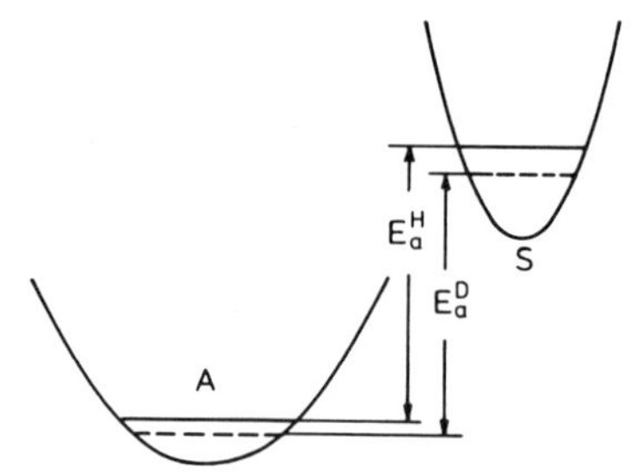

Fig. 8.9. Schematic explanation of how a reversed isotope effect of the activation energy can arise from $\omega_S > \omega_A$. —— are the lowest oscillator states of the proton in the minimum and saddle point respectively; --- are those of the deuteron

values obtained by the fit are rather large, especially for Cu, and violate the condition $\hbar\omega \approx k_B T$. *Katz* et al. [8.48] have included anharmonic corrections into (8.42) and obtained more reasonable values for ω_A, ω_S in Cu. No fit of the experiments on H, D, T in Pd has been made yet.

In the bcc metals, the condition $\hbar\omega \gg k_B T$ is fulfilled at room temperature and in a range above. According to (8.43), the prefactor should be isotope and metal independent, in fair agreement with the experimental findings above 250 K (cf. Chap. 12). The absolute value of the prefactor is also reasonable. Using the experimentally determined prefactor, one obtains agreement with (8.43) at a temperature of about 250 K. *Schaumann* et al. [8.39] have pointed out that the low-temperature limit of the modified classical rate theory gives an isotope effect of the activation energy according to (8.43) and that the experiments above 250 K agree fairly well with the following estimate. If one puts $\omega_A = \omega_S = \omega$,

$$(1/2)\hbar\omega^H - (1/2)\hbar\omega^D = (1/2)\hbar\omega^H(1 - 1/\sqrt{2}). \tag{8.44}$$

Using the average value $\hbar\omega^H = 155$ meV, one estimates

$$E_a^D - E_a^H \approx 23\,\text{meV}.$$

The good agreement with two systems (Nb, Ta) is probably fortuitous in view of the many simplifications which have been made.

In the opinion of this author, the modified classical theory of over-barrier jumps gives a plausible description of the hydrogen diffusion data in fcc metals, and in bcc metals above 250 K. The task remains of deriving the modified classical rate theory starting from a quantum mechanical rate theory.

8.5 Anomalies at Higher Temperatures

Up to now, two regimes of interstitial diffusion of hydrogen have been discussed in some detail: the regime of thermally activated tunneling transitions, and the regime of thermally activated over-barrier jumps. The primary task was the derivation of the jump rate of a hydrogen atom from a given site to a neighboring interstitial site. It was tacitly assumed that the jump processes are rare events, such that the total diffusion in the metal lattice could be obtained by a combination of uncorrelated jumps. This is the usual picture of interstitial diffusion where the duration of a jump process is short compared to the average time between two consecutive jumps. At higher temperatures, hydrogen atoms perform so many jumps per second, especially in bcc metals, that the average time between two consecutive jumps is not long compared to the duration of a jump. Then the subdivision into individual jumps and lattice diffusion loses its meaning. Expressed differently, a transition to the regime of "fluidlike diffusion" takes place, where states above the barriers contribute increasingly to the diffusion process.

There is experimental evidence of anomalies at higher temperatures, although the picture is not yet clear at the moment. The anomalies were found in quasielastic neutron scattering experiments in which the Fourier transform of the self-correlation function is determined (cf. Chap. 10). The self-correlation function characteristic for jump diffusion of protons is derived from the *Chudley-Elliott* model [8.74] in which the proton either performs localized vibrations about an equilibrium site, or executes sudden, uncorrelated jumps to neighboring sites. The result for the scattering law is a quasielastic line whose width is determined by the jump rate and jump geometry and whose intensity is given by the Debye-Waller factor of the proton. The anomalies which have been observed are described in Chapter 10. Due to difficulties in the analysis of the measured quasielastic lines, the correctness and consistency of the results obtained so far have to be carefully verified.

On the theoretical side, the general problem of diffusion in a multiwell potential formed by the oscillating host lattice atoms should be treated, including the hydrogen-host lattice interaction. Of course this is a formidable task which cannot be accomplished without simplifying assumptions. *Kagan* and *Klinger* [8.27] introduce a density matrix formalism and derive expressions for the diffusion constant at high temperature. However, the self-correlation function which contains detailed information on the diffusion mechanism is not derived. *Fulde* et al. [8.75] treat the problem of diffusion of a particle in a periodic potential under the influence of a stochastic force. The self-correlation function for this model has been calculated by *Dieterich* et al. [8.76]. The application to the diffusion of hydrogen in metals requires an extension to the case of discrete oscillator levels in the potential minima. A finite jump time has been included in a phenomenological model for hydrogen diffusion at higher temperatures by *Gissler* and *Stump* [8.77]. A two-state model analogous to that of *Singwi* and *Sjölander* [8.78] is used for the calculation of the self-correlation function, including rest periods and flight periods with finite lifetimes. The detailed calculations of these authors have not been applied to experiments. The simplified version of the model used by *Gissler* et al. [8.79] to describe experiments on Nb–H cannot be correct. A still more phenomenological model has been considered by *Wakabayashi* et al. [8.80] to fit experiments in Nb–H at different temperatures. Here the density distribution of a proton, averaged over the mean time of stay at a site, has a localized and a delocalized component. However, there are uncertainties in the experimental analysis of this experiment.

In summary, the deviations from simple jump diffusion at high temperature are not yet fully explored both experimentally and theoretically. It will be one of the regimes which will attract interest in the near future.

8.6 Outlook

The picture of hydrogen diffusion in metals which has emerged from the experimental findings and the theoretical interpretations is by no means clear. The question of the origin of the high mobility of hydrogen in metals has not yet been satisfactorily answered. The elementary diffusion process seems to be at most a marginal case with respect to tunneling effects. Instead of repeating the detailed conclusions of the chapter, a list will be given of those experiments and theoretical derivations which should be done in the future in order to achieve a better understanding of the problems.

Most desirable are experiments of hydrogen diffusion at lower temperatures and for all of its isotopes. Only experiments at low temperature, performed with all isotopes, can clarify the possible quantum nature of the individual jump processes. Experiments on muons, which represent an extremely light isotope of hydrogen, should be performed for different fcc and bcc metals over as wide a temperature range as possible. Investigation of hydrogen diffusion in alloys seems to be a promising field as well. One intriguing question is the possible existence of tunneling states of protons, either centered around other defects, or possibly in a pure host metal. There are indications of tunneling states around other interstitials (cf. Sect. 8.2.3). The existence of those tunneling states will probably be clarified by inelastic neutron scattering experiments. Another topic of interest is the very rapid diffusion of hydrogen at higher temperatures, where the usual picture of jump diffusion breaks down. Quasielastic neutron scattering experiments to study this problem have been mode recently [8.81].

On the theoretical side, a derivation of the interaction of a proton with the host metal atoms is urgently needed. This requires the treatment of the screening of a proton in the transition metals. In the statistical-mechanical treatment of the diffusion problem, the derivation of the modified classical rate theory starting from a quantum mechanical basis seems to be the most important unsolved problem. The theory of the fluidlike diffusion at high temperatures, including the quantum nature of the excitations, presents another interesting problem.

In summary, the diffusion of hydrogen in metals presents many questions which, on the one hand, are scientifically appealing and, on the other hand, bear a close relation to technologically important problems.

References

8.1 T.Holstein: Ann. Phys. (N.Y.) **8**, 325 (1959)
8.2 T.Holstein: Ann. Phys. (N.Y.) **8**, 343 (1959)
8.3 G.Alefeld, G.Schaumann, J.Tretkowski, J.Völkl: Phys. Rev. Lett. **22**, 297 (1969)
8.4 K.-H.Janssen: Z. Physik B**23**, 245 (1976)
8.5 J.A.Sussmann: Ann. Phys. (Paris) **6**, 135 (1971)
8.6 A.M.Stoneham: Ber. Bunsenges. Physik. Chem. **76**, 816 (1972)
8.7 E.G.Maksimov, O.A.Pankratov: Sov. Phys.—Uspekhi **18**, 481 (1976)
8.8 A.M.Stoneham: Collective Phenomena **2**, 9 (1975)
8.9 A.M.Stoneham: In *Proceedings of the International Conference on Properties of Atomic Defects on Metals*, Argonne Ill. (1976); J. Nucl. Mater. **69/70**, 109 (1978)
8.10 J.Friedel: Ber. Bunsenges. Physik. Chem. **76**, 828 (1972)
8.11 Z.D.Popovic, M.J.Stott: Phys. Rev. Lett. **33**, 1164 (1974)
8.12 N.Boes, H.Züchner: Z. Naturforsch. **31**a, 760 (1976)
8.13 H.Wagner, H.Horner: Advan. Phys. **23**, 587 (1974)
8.14 H.Kanzaki: J. Phys. Chem. Sol. **2**, 24 (1957)
8.15 J.Pfeiffer, H.Peisl: Phys. Lett. **60** A, 363 (1977)
8.16 B.Baranowski, S.Majchrzak, T.B.Flanagan: J. Phys. F **1**, 258 (1971)
8.17 J.Buchholz, J.Völkl, G.Alefeld: Phys. Rev. Lett. **30**, 318 (1973)
8.18 A.Magerl, B.Berre, G.Alefeld: Phys. Stat. Sol. **36**, 161 (1976)
8.19 M.A.Pick, R.Bausch: J. Phys. F **6**, 1751 (1976)
8.20 W.Drexel, A.Murani, D.Tocchetti, W.Kley, I.Sosnowska, D.K.Ross: J. Phys. Chem. Sol. **37**, 1135 (1976)
8.21 M.Abramowitz, I.A.Stegun: *Handbook of Mathematical Functions* (National Bureau of Standards, Washington D.C. 1964) p. 727
8.22 W.Kley, J.Peretti, R.Rubin, G.Verdan: BNL Conf., Report Brookhaven National Laboratory 940 (C54) p. 105 (1966)
8.23 W.Kley, J.Peretti, R.Rubin, G.Verdan: J. Phys. (Paris) **28**, C1–26 (1967)
8.24 G.Blaesser, J.Peretti: JÜL-Bericht JÜL-Conf-2 (Kernforschungsanlage Jülich 1968) Vol. II, p. 886
8.25 D.Lepski: Phys. Stat. Sol. **35**, 697 (1969)
8.26 J.A.Sussmann, Y.Weissman: Phys. Stat. Sol. **53**, 419 (1972)
8.27 Yu.Kagan, M.J.Klinger: J. Phys. C **7**, 2791 (1974)
8.28 H.Horner, H.Wagner: J. Phys. C **7**, 3305 (1974)
8.29 L.M.Sander, H.B.Shore: Phys. Rev. B **3**, 1472 (1971)
8.30 J.A.D.Matthew: Phys. Stat. Sol. **42**, 841 (1970)
8.31 K.W.Kehr: "Diffusion von Wasserstoff in Metallen", JÜL-Bericht JÜL-1211 (Kernforschungsanlage Jülich 1975)
8.32 M.Prager, W.Press, B. Alefeld, A.Hüller: To be published
8.33 R.von Baltz, U.Birkholz: "Polaronen", in *Advances in Solid State Physics*, Vol. 12, ed. by O.Madelung (Pergamon-Vieweg, Oxford-Braunschweig 1972) p. 233
8.34 H.Horner: Private communication
8.35 N.E.Byer, H.S.Sack: J. Phys. Chem. Sol. **29**, 677 (1968)
8.36 H.K.Birnbaum, C.P.Flynn: Phys. Rev. Lett. **37**, 25 (1976)
8.37 G.Alefeld, H.Wipf: Fizika Nizkikh Temperatur **1**, 600 (1975); Sov. J. Low Temp. Phys. **1**, 317 (1975)
8.38 Ch. Morkel: Diploma Thesis, Technische Universität München (1977)
8.39 G.Schaumann, J.Völkl, G.Alefeld: Phys. Rev. Lett. **21**, 891 (1968)
8.40 C.P.Flynn, A.M.Stoneham: Phys. Rev. B**1**, 3966 (1970)
8.41 A.M.Stoneham: J. Phys. F**2**, 417 (1972)
8.42 H.Wipf: Dissertation Technische Hochschule München (1972) and JÜL-Bericht JÜL-876-FF (Kernforschungsanlage Jülich 1972)
8.43 D.Richter, B.Alefeld, A.Heidemann, N.Wakabayashi: J. Phys. F**7**, 569 (1977)

8.44 K.Faber, H.Schultz: Scripta Metall. **6**, 1065 (1972)
8.45 R.Hanada, T.Suganuma, H.Kimura: Scripta Metall. **6**, 483 (1972)
8.46 R.Hanada: Scripta Metall. **7**, 681 (1973)
8.47 J.Engelhard: Verhandl. DPG (VI) **12**, 296 (1977)
8.48 L.Katz, M.Guinan, R.J.Borg: Phys. Rev. B**4**, 330 (1971)
8.49 G.Bohmholdt, E.Wicke: Z. Physik. Chem. NF **56**, 133 (1967)
8.50 G.Holleck, E.Wicke: Z. Physik. Chem. NF **56**, 155 (1967)
8.51 J.Völkl, G.Wollenweber, K.-H.Klatt, G. Alefeld: Z. Naturforsch. **26**a, 922 (1971)
8.52 G.Sicking, H. Buchold: Z. Naturforsch. **26**a, 1973 (1971)
8.53 I.I.Gurevich, E.A.Meleshko, I.A.Muratova, B.A.Nikolsky, V.S.Roganov, V.I.Selivanov, B.V.Sokolov: Phys. Lett. **40**A, 143 (1972)
8.54 C.G.Chen, H.K.Birnbaum: Phys. Stat. Sol. (a) **36**, 687 (1976)
8.55 E.Gorham-Bergeron: Phys. Rev. Lett. **37**, 146 (1976)
8.56 W.R.Heller: Acta Metall. **9**, 600 (1961)
8.57 R.R.Arons: JÜL-Bericht JÜL-Conf-2 (Kernforschungsanlage Jülich 1968) Vol. II, p. 901
8.58 R.R.Arons, Y.Tamminga, G.de Vries: Phys. Stat. Sol. **40**, 107 (1970)
8.59 J.H.Weiner, Y.Partom: Phys. Rev. B**1**, 1533 (1970)
8.60 J.H.Weiner, R.E.Forman: Phys. Rev. B**10**, 325 (1974)
8.61 J.H.Weiner: Phys. Rev. B**14**, 4741 (1976)
8.62 C.Zener: In *Imperfections in Nearly Perfect Crystals*, ed. by W.Shockley (Wiley, New York 1952) p. 289
8.63 H.Eyring: J. Chem. Phys. **3**, 107 (1932)
8.64 C.Wert, C.Zener: Phys. Rev. **76**, 1169 (1949)
8.65 G.H.Vineyard: J. Phys. Chem. Sol. **3**, 121 (1957)
8.66 J.G.Mullen: Phys. Rev. **121**, 1649 (1961)
8.67 A.D.Le Claire: Phil. Mag. **14**, 1271 (1966)
8.68 H.R.Glyde: Phys. Rev. **180**, 722 (1969)
8.69 Y.Ebisuzaki, W.J.Kass, J.O'Keeffe: J. Chem. Phys. **46**, 1373 (1967)
8.70 Y.Ebisuzaki, W.J.Kass, J.O'Keeffe: Phil. Mag. **15**, 1071 (1967)
8.71 G.Alefeld: Phys. Rev. Lett. **12**, 372 (1964)
8.72 W.Jost, A.Widmann: Z. Physik. Chem. B**29**, 247 (1935)
8.73 W.Jost, A.Widmann: Z. Physik. Chem. B**45**, 285 (1940)
8.74 C.T.Chudley, R.J.Elliott: Proc. Phys. Soc. **77**, 353 (1961)
8.75 P.Fulde, L.Pietronero, W.R.Schneider, S.Strässler: Phys. Rev. Lett. **35**, 1776 (1975)
8.76 W.Dieterich, I.Peschel, W.R.Schneider: Z. Physik B **27**, 177 (1977)
8.77 W.Gissler, N.Stump: Physica **65**, 109 (1973)
8.78 K.S.Singwi, A.Sjölander: Phys. Rev. **167**, 152 (1968)
8.79 W.Gissler, B.Jay, R.Rubin, L.A.Vinhas: Phys. Lett. **43**A, 279 (1973)
8.80 N.Wakabayashi, B.Alefeld, K.W.Kehr, T.Springer: Solid State Commun. **15**, 503 (1974)
8.81 V.Lottner, A.Heim, K.W.Kehr, T.Springer: In *Proceedings of the IAEA Symposium on Neutron Inelastic Scattering*, Vienna (1977) to be published

9. Nuclear Magnetic Resonance on Metal-Hydrogen Systems

R. M. Cotts

With 13 Figures

In the first twenty-five years after the discovery of nuclear magnetic resonance (NMR) by *Bloch* et al. [9.1] and *Purcell* et al. [9.2], much of the research in the subject was on the understanding of the response of the nuclear spin system to magnetic fields, both those generated by the experimenter and by the matter containing the spin system. With an abundance of induced dynamic spin processes described in the literature, NMR has matured as a technique so that now the emphasis is on its application to the understanding of physical and chemical processes in condensed matter.

One of its earliest applications, reported in 1952 by *Norberg* [9.3], was on a metal-hydrogen (M–H) system, PdH_x. The breadth of application of NMR has grown with the development of the subject. In 1972, *Cotts* [9.4] reviewed its application to the study of atomic motion and defined the basic NMR experiments. In 1965, *Pedersen* [9.5] reviewed all aspects of NMR relative to M–H systems.

The purpose of this review is to examine the use of NMR as a tool in the investigation of M–H systems. Emphasis is in those areas where it has already seen application, without attempting to make an inclusive listing of the work in the literature. Some examples of its application are cited.

NMR does not usually stand alone as a research tool in materials science. In many instances it probes internal fields microscopically on an atomic scale so that its measurements can be very sensitive to small changes in the M–H system under observation. This sensitivity can indeed be an advantage of the technique. Information from NMR supplements other techniques. Without basic knowledge of host-metal structure and phase diagrams as general guidelines, interpretation of NMR data could be difficult. For example, the high degree of sensitivity can be used to locate phase boundaries with improved precision, to determine existence of hysteresis at phase transitions, or to measure the temperature width of a transition.

NMR is sensitive to electronic structure, and through measurements of relaxation rates and Knight shifts new insight can be acquired into changes of electron structure of the metal with addition of hydrogen. Usually electron specific heat and magnetic susceptibility data are needed for full utilization of NMR data.

One area where NMR is substantially independent is in diffusion measurements. The onset of diffusion of hydrogen has a clear effect on NMR observations. Diffusion constants of hydrogen can be measured directly using

NMR with only a very limited knowledge of other properties of the M–H system. Furthermore, with a knowledge of host metal structure, diffusion mechanisms can be identified.

Application of NMR is usually limited to paramagnetic systems. Internal fields in ferromagnetic systems can overwhelm the nuclear spin systems. A substantial number of nuclear spins are needed for adequate signal-to-noise ratio. Hydrogen concentrations must exceed 0.01 as the H–M ratio for most experiments.

In the first section of this review, the use of NMR on structure studies is discussed. Theory and experiments for determination of hydrogen diffusion parameters are presented in the second section. The third deals with the effect of electronic structure on NMR.

9.1 Structural Information

9.1.1 Nuclear Dipole-Dipole Interaction

The interaction common to all nuclear spins in a solid is the magnetic dipole-dipole interaction. Nuclear dipolar fields are the order of magnitude of a few gauss. In this section, we assume the absence of nuclear electric quadrupole interactions and large internal fields due to sample ferromagnetism or strong paramagnetism. The dipolar fields are the principal source of resonance linewidth.

The form of the dipole-dipole interaction between two spins is

$$\mathscr{H}_{12}=\frac{\hbar^2\gamma_1\gamma_2}{r_{12}^3}\left[\boldsymbol{I}^{(1)}\cdot\boldsymbol{I}^{(2)}-\frac{3(\boldsymbol{I}^{(1)}\cdot\boldsymbol{r}_{12})(\boldsymbol{I}^{(2)}\cdot\boldsymbol{r}_{12})}{r_{12}^2}\right], \tag{9.1}$$

where γ_1 and γ_2 are the gyromagnetic ratios of the nuclei, $\boldsymbol{I}^{(1)}$ and $\boldsymbol{I}^{(2)}$ are spin operators, and $\boldsymbol{r}_{12}$ is the displacement from spin 1 to spin 2. Because of its large anisotropy and relatively strong dependence upon $\boldsymbol{r}$, the dipole-dipole interaction can be used to obtain structural information about solids over very large ranges of absolute temperature.

At low temperatures with diffusive motion frozen out, the dipole-dipole interaction determines the NMR linewidth and shape. As diffusive motion of hydrogen increases at higher temperatures, the NMR line is observed to become narrower in width because the dipolar fields are averaged toward zero by random atomic motion. In this high temperature regime, the time-dependent dipolar fields can control the value of the spin-lattice relaxation time of the nuclear spin system. Even under motional narrowing, an understanding of the dipolar contribution to relaxation processes can yield improved knowledge of structure.

Second Moment of the NMR Absorption Spectrum

In this subsection, we consider the low-temperature (rigid lattice) regime of the dipole-dipole interaction. In a metal-hydrogen (M–H) system, there is interaction between H spins, between H and M, and between M and M. In what follows let the I spin system be hydrogen and the S spin system be the host metal nuclei. Interactions between H nuclei and conduction electron spins are narrowed to a negligible amount. For two spin systems, the full nuclear Hamiltonian is

$$\mathscr{H} = -\hbar\gamma_I H_0 \sum_j I_z^j - \hbar\gamma_S H_0 \sum_k S_z^k + (\mathscr{H}_{\mathrm{d}}^{II} + \mathscr{H}_{\mathrm{d}}^{IS} + \mathscr{H}_{\mathrm{s}}^{SS}).$$

The first two terms are the Zeeman energies of the spin systems in the constant, homogeneous magnetic field H_0. $\mathscr{H}_{\mathrm{d}}^{IS}$ is the dipolar interaction between "unlike" spins I and S, while $\mathscr{H}_{\mathrm{d}}^{II}$ and $\mathscr{H}_{\mathrm{d}}^{SS}$ are the interactions between "like" spins within the I or S spin systems, respectively. The Zeeman energy of the jth spin is $-\gamma_1 \hbar H_0 m_j$, where m_j is the magnetic quantum number.

A useful form for $\mathscr{H}_{\mathrm{d}}$ is suggested by *Van Vleck* [9.6]. As an example, consider $\mathscr{H}_{\mathrm{d}}^{IS}$,

$$\mathscr{H}_{\mathrm{d}}^{IS} = \sum_{j<k} \frac{\hbar^2\gamma_I\gamma_S}{r_{jk}^3}(A^{jk} + B^{jk} + C^{jk} + D^{jk} + E^{jk} + F^{jk}), \tag{9.2}$$

where

$$A^{jk} = (1 - 3\cos^2\theta)\, I_z^j S_z^k$$

$$B^{jk} = -\tfrac{1}{4}(1 - 3\cos^2\theta)(I_+^j S_-^k + I_-^j S_+^k)$$

$$C^{jk} = -\tfrac{3}{2}\sin\theta\cos\theta\exp(-\mathrm{i}\phi)(I_z^j S_+^k + S_z^k I_+^j)$$

$$D^{jk} = -\tfrac{3}{2}\sin\theta\cos\theta\exp(+\mathrm{i}\phi)(I_z^j S_-^k + S_z^k I_-^j)$$

$$E^{jk} = -\tfrac{3}{4}\sin^2\theta\exp(-\mathrm{i}2\phi)\, I_+^j S_+^k$$

$$F^{jk} = -\tfrac{3}{4}\sin^2\theta\exp(+\mathrm{i}2\phi)\, I_-^j S_-^k\,,$$

and θ and ϕ are the polar and azimuthal angles of $\boldsymbol{r}_{jk}$ relative to the z-axis of quantization parallel to H_0. The usual notation $I_+ = (I_x + \mathrm{i}I_y)$ and $I_- = (I_x - \mathrm{i}I_y)$ has been used.

Note that the A terms connect states of the same magnetic quantum number for both spins I and S. The B terms connect states $\Delta m_I = \pm 1$ and $\Delta m_S = \mp 1$. Terms C and D connect states $\Delta m_I = 0, \Delta m_S = \pm 1$ and states $\Delta m_I = \pm 1$, $\Delta m_S = 0$. Finally E and F connect $\Delta m_I = +1$, $\Delta m_S = +1$ and states $\Delta m_I = -1$, $\Delta m_S = -1$.

For $\mathscr{H}_{\mathrm{d}}^{II}$ and $\mathscr{H}_{\mathrm{d}}^{SS}$ the A and B terms connect states which do not change the total m for each spin system. That is, in $\mathscr{H}_{\mathrm{d}}^{II}$ and $\mathscr{H}_{\mathrm{d}}^{SS}$, A and B conserve Zeeman energy and angular momentum. In $\mathscr{H}_{\mathrm{d}}^{IS}$, A and B conserve total angular momentum but only A conserves Zeeman energy.

Because each spin interacts with many other spins, the resonance line is split into many lines in what is essentially a continuum. Only the A terms (and for like spins the A and the B terms) contribute to this broadening. In principle the exact shape of the resonance line contains much structural information, but, because of the many-body nature of the system, it is not practical to calculate the exact shape.

Instead the second moment of the resonance is used. By using a sum rule of *Van Vleck* [9.6], it can be calculated exactly for each assumed crystal structure, with or without random site occupancy. Since most NMR experiments are done on polycrystalline samples, only the equations for second moments of resonance lines in powder samples will be given here.

For a powder sample the mean squared width measured from the center of the symmetric dipolar broadened resonance, the second moment, is [9.6]

$$\begin{aligned}\langle \Delta\omega_{\mathrm{d}}^2 \rangle &= \tfrac{3}{5}\gamma_I^4 \hbar^2 I(I+1) \sum_n (r_{mn}^{-6}) + \tfrac{4}{15}\gamma_I^2 \gamma_S^2 \hbar^2 S(S+1) \sum_{n'} (r_{mn'}^{-6}) \\ &= \langle \Delta\omega_{II}^2 \rangle + \langle \Delta\omega_{IS}^2 \rangle ,\end{aligned} \tag{9.3}$$

where the sum $\sum_n (r_{mn}^{-6})$ is over all hydrogen sites n from the typical site m, $\sum_{n'} (r_{mn'}^{-6})$ is over all metal sites n' having nuclear spins S, and ω is in angular frequency units.

For a resonance line having a shape function $g(\omega)d\omega$,

$$\langle \Delta\omega^2 \rangle = \frac{\int_0^\infty (\omega-\omega_0)^2 g(\omega) d\omega}{\int_0^\infty g(\omega) d\omega}, \tag{9.4}$$

where ω_0 is the central resonant frequency, $\omega_0 = \gamma H_0$.

The usefulness of the second moment in obtaining structural information in the M–H systems depends upon the relative size of the two terms in (9.3). If the second term is large, due to large values of S and γ_S and the fact that M atoms are closer to H atoms than are other H atoms, then the total value of $\langle \Delta\omega_{\mathrm{d}}^2 \rangle$ might not be sensitive to changes in models for H site occupancy.

Valuable information has been obtained by second moment analysis in a number of M–H systems. For example, for $1.6 < x < 2.0$ in γ phase TiH_x, *Stalinski* et al. [9.7], measured the hydrogen second moment using a continuous wave (cw) NMR spectrometer. They tested their data against an assumption that a fraction of the ^1H atoms occupied the tetrahedral interstices of the fcc Ti host lattice and that the remaining fraction occupied the octahedral sites. They also assumed that the occupied sites were randomly distributed in the lattice. The data fit a model in which hydrogens occupy the tetrahedral sites randomly. The upper limit to the fraction of H atoms occupying octahedral sites was found to be 0.0125. Second moments were determined for a lattice temperature of 77 K to assure rigid lattice conditions. Because of the very small

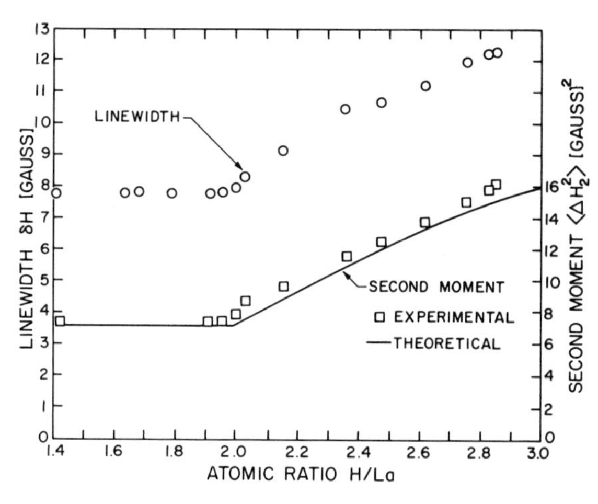

Fig. 9.1. Rigid lattice linewidth and second moment of ^{1}H NMR in LaH_x as a function of x. The theoretical curve for the second moment is based upon the assumption that for $x<2$, H atoms occupy tetrahedral sites in the LaH_2 phase and for $x>2$, octahedral sites are occupied randomly while tetrahedral sites are full [9.8]

nuclear magnetism of the titanium spin system, the second term of (9.3) was negligible and omitted in their analysis.

In the lanthanum-hydrogen system LaH_x, for $x<2$, coexistence of two phases, lanthanum and the dihydride, was demonstrated by the second moment measurements of *Schreiber* and *Cotts* [9.8] as shown in Fig. 9.1. (The possibility that the "lanthanum" phase was actually α-LaH_x with a finite maximum hydrogen solubility was not considered). In Fig. 9.1, the independence of linewidth (plotted as full width at maximum slope) and the second moment upon hydrogen concentration for $x<2$ shows that most of the hydrogen is in the dihydride phase. In an analysis similar to that of *Stalinski* et al. [9.7] in TiH_x, it was suggested that for $x>1.95$, hydrogen atoms occupy the octahedral sites randomly in the fcc metal host, while the tetrahedral sites are effectively full at $x=1.95$ with a defect fraction of 0.05. The contribution of the nuclear magnetic moments of the host La spins was included in this analysis.

If the second term of (9.3) is considerably larger than the first, the total value of $\langle \Delta\omega_d^2 \rangle$ might not be sensitive to different models of H atom location. This problem can be especially acute when trying to determine a particular ordering of H atoms. The value of $\langle \Delta\omega_{II}^2 \rangle$, which is sensitive to various models of H atom ordered structures, can be separated from $\langle \Delta\omega_d^2 \rangle$ by a technique suggested by *Engelsberg* and *Norberg* [9.9], following the development of *Mansfield* [9.10]. When $\langle \Delta\omega_{IS}^2 \rangle \gg \langle \Delta\omega_{II}^2 \rangle$, a 90°-$\tau$-180° spin echo sequence produces a "solid echo" for τ the order of $\tau \approx \langle \Delta\omega_{II}^2 \rangle^{-1/2}$. They show that the ratio of the echo peak for a given τ to the extrapolated maximum at $\tau=0$ is

$$R(\tau)=1-\langle \Delta\omega_{II}^2 \rangle \frac{(2\tau)^2}{2!}+\varepsilon_4 \frac{(2\tau)^4}{4!}-\dots, \tag{9.5}$$

where ε_4 is a predictable "error" term. *Bowman* [9.11] has recently used the 90°-τ-180° echo experiment to measure the H–H second moment term in the VH_x system in which $\langle \Delta\omega_{HV}^2 \rangle \gg \langle \Delta\omega_{HH}^2 \rangle$.

For the models tested by *Bowman*, the measured values of $\langle \Delta\omega_{HH}^2 \rangle$ fell below the theoretical values with the exception that for one concentration, $VH_{0.5}$, there was agreement with experiment for a model of H atom ordering which has maximum separation between H atoms using the octahedral sites of the bct V host lattice. A rather puzzling and as yet unexplained result of these experiments is that the experimental values of $\langle \Delta\omega_{HH}^2 \rangle$ fell *below* the calculated theoretical values associated with the site models, including random occupancy. Data were taken at 100 K, where diffusive motion was frozen out. More tests of this experimental technique in other systems might suggest the reason for the discrepancy.

Measurement of $\langle \Delta\omega^2 \rangle$: Methods

The second moment is usually defined, as in (9.4), in terms of the cw resonance absorption shape function. Application of this definition to recorded NMR data from a cw spectrometer is straightforward. In wide-line NMR spectroscopy, the spectrometer output is usually recorded as the first derivative of the resonance line shape function $g(\omega)$ taken with respect to the applied magnetic field. Usually the magnetic field is modulated to produce the signal for the phase-sensitive detector. Although signal might be recorded as a function of field as the variable, the conversion to frequency (or vice versa) is obtained through the relationship $\omega = \gamma H$. If the modulation amplitude is kept small compared to the linewidth, the observed line shape data accurately represent the resonance line first derivative, so it can be integrated directly to obtain the second moment. To correct for a small error caused by finite sinusoidal modulation field of amplitude H_m, the quantity $\gamma_I^2 H_m^2/4$ must be subtracted from the experimentally determined second moment.

It is also possible to obtain the second moment from the free induction decay (FID) signal of the pulsed NMR experiment. Since the FID $f(t)$ is the cosine Fourier transform of the cw absorption line shape, it follows that [9.12]

$$\frac{f(t)}{f(0)} = 1 - \langle \Delta\omega^2 \rangle \frac{t^2}{2!} + \langle \Delta\omega^4 \rangle \frac{t^4}{4!} - \dots \tag{9.6}$$

so that

$$\langle \Delta\omega^2 \rangle = \frac{-1}{f(0)} \left[\left. \frac{\partial^2 f(t)}{\partial t^2} \right|_{t=0} \right]. \tag{9.7}$$

Usually the FID is recorded and $[f(t)/f(0)]$ is plotted against t^2. From the limiting value of the slope of $[f(t)/f(0)]$ vs t^2, the second moment is determined.

Both methods require especially good signal-to-noise (S/N) ratio for accurate measurement of $\langle \Delta\omega^2 \rangle$. In cw, the signal from the wings of the resonance line are given more weight by the factor $(\omega - \omega_0)^2$. Without good S/N, normal noise fluctuations in the wings can cause large errors in the second moment. In the pulse experiment one must have good data for times $t \ll T_2^*$

where T_2^* is the lifetime of the FID. The receiver of the pulse apparatus must have excellent recovery time to amplify linearly the weak NMR signal in a time t after being "overloaded" by the transmitter pulse used to drive the spin system. Typically $T_2^* \approx 1/\gamma_I \Delta H$ where ΔH, the half-width of the NMR cw line, is approximately equal to $(\langle \Delta\omega_d^2 \rangle^{1/2}/\gamma_I)$ and is the order of a few gauss. It is therefore required that $t \ll 10$–$15\,\mu$s. The requirement for fast recovery can be somewhat alleviated by use of the solid echo experiment [9.10]. Again, good S/N will be required because the bandwidth of the electronics must be large to avoid distortion of the rapidly changing FID. Increased bandwidth of course means increased noise. The first requirement for fast receiver recovery is helped by performing the experiment at a high Larmor frequency. This improves S/N and also helps in the second problem. A high Larmor frequency also improves S/N in the cw experiment and thus makes it more accurate.

Accuracy of better than 5% in second moment measurements requires considerable care. At relatively low temperatures needed to assure rigid lattice conditions, the spin lattice relaxation time T_1 is long. A long T_1 inhibits the rf level in the cw experiment to maintain the stringent no-saturation condition [9.13] required in line shape studies, $(\gamma_I H_1)^2 \ll \langle \Delta\omega_d^2 \rangle^{1/2}/T_1$, where H_1 is the amplitude of the rf field. Since signal is proportional to H_1, signal is necessarily limited.

The long T_1 means that the repetition period for performing the pulse experiment must be increased to allow the spin system to recover its change in magnetization. In both the cw and FID experiments, signal averaging is needed and the effect of the above inhibitions on signal strength is to increase averaging times for the required S/N.

Use of the NMR second moment for structural information starts where x-ray diffraction leaves off. X-ray diffraction determines the metal host lattice structure but usually provides limited information on H atom location. Neutron diffraction experiments must be done on the ^{2}H isotope because of large incoherent scattering from ^{1}H. Furthermore, in some systems, for example the vanadium hydrogen system, there is a considerable difference between ^{1}H and ^{2}H phase diagrams (see [Ref. 9.14, Chap. 2]). Therefore, structural information on the ^{1}H system itself is needed and can be obtained through a study of the hydrogen second moment and its dependence upon H-concentration. The limitations of the cw experiment that can yield only the full value of $\langle \Delta\omega_d^2 \rangle$ have already been discussed. Perhaps the pulse experiments which effectively partition the equation for the second moments into its M and H contributions [9.11], will receive further test and development and so improve the second moment technique for obtaining structural information. For both cw and pulsed experiments, prior knowledge of the metal host lattice structure is essential in setting the scale of distance and in limiting the number of interstitial site models to be tested.

Where there are large internal magnetic fields associated with sample ferromagnetism or unusually high paramagnetism, the inhomogeneous magnetization can mask the information available from nuclear dipole interactions.

The hydrogen nucleus, having spin 1/2, has zero nuclear electric quadrupole moment so that purely dipolar interactions with it are assured. The deuteron, though, has a quadrupole moment and a relatively small magnetic dipole moment. Electric quadrupole interactions for the deutron can, and usually do, exceed its dipole interaction strength by an order of magnitude, thus rendering the dipole-dipole second moment of no value with ^{2}H in metals. Nevertheless, as will be shown in the next section, the deuterium quadrupole interaction has proved to be a second valuable NMR probe for structure study.

The host metal nuclear quadrupole interaction can be quite large and comparable to the metal nuclear Zeeman interaction with the applied magnetic field H_0. In this instance, an implicit assumption that these spins are quantized along H_0 would be erroneous, and the second term of (9.3) could be in error (*Pedersen* et al. [9.15]).

9.1.2 Nuclear Electric Quadrupole Interaction

Nuclei with spins greater than $I=1/2$ can possess a nuclear electric quadrupole moment that couples to the electric field gradient tensor (EFG) produced by the lattice at the site of the nucleus. Deuterons with spin $I=1$ have a quadrupole moment, as do many of the host metal nuclei in M–H systems. The strength of the quadrupole interaction for most nuclei of interest in M–H systems is small enough that it can be treated as a perturbation of the Zeeman interaction. If all nuclei of one spin system experience the same quadrupole interaction, the NMR cw line is split into a multiplet. Analysis of the multiplet gives structural and phase diagram information.

Electric Field Gradient Tensor

Before presentation of some experimental examples, the form of the quadrupole interaction will be outlined. A complete description can be found in *Cohen* and *Reif* [9.16].

The EFG tensor is most simply expressed relative to its principal axis coordinate system (x', y', z'), which is determined by the crystalline environment around the nucleus. Since the EFG tensor is traceless, it can be expressed with two parameters, the field gradient $(eq)=\partial^2 V/\partial z'^2$ and the axial asymmetry parameter $\eta=(\partial^2 V/\partial x'^2-\partial^2 V/\partial y'^2)/eq$, where V is the electric potential in the vicinity of the nucleus. For points in a cubic lattice, $eq=0$, but interstitial sites in a cubic lattice can have low enough symmetry that $eq\neq 0$. By definition, $0\leqq\eta\leqq 1$.

The quadrupolar Hamiltonian is

$$\mathscr{H}_Q=\frac{e^2Qq}{4I(2I-1)}[3I_{z'}^2-I(I+1)+\eta(I_{x'}^2-I_{y'}^2)], \tag{9.8}$$

where Q is the nuclear electric quadrupole moment and e is the electronic charge. The value of Q is a nuclear property and it varies in magnitude by over a factor of 10^3 from deuterons to tantalum (181) nuclei. We shall assume $\mathscr{H}_Q \ll \mathscr{H}_z$, where $\mathscr{H}_z$ is the Zeeman interaction, so that first-order perturbation theory applies. After transforming the principal axes of the EFG tensor to the coordinate system defined by the Zeeman field, the quadrupolar shift of the mth Zeeman level is

$$\Delta E_m = \frac{h\nu_Q}{12}[3m^2 - I(I+1)][3\cos^2\theta - 1 + \eta\cos 2\psi(\cos^2\theta - 1)] \tag{9.9}$$

$$= \frac{h\nu_Q}{12}[3m^2 - I(I+1)]K(\eta, \theta, \psi), \tag{9.10}$$

where θ and ψ, the Euler angles in the transformation, are measured relative to the magnetic field and $\nu_Q = 3e^2qQ/h2I(2I-1)$. Each magnetic resonance transition $\Delta m = 1$ is then split off from the Larmor frequency by an amount

$$(\nu_{m \leftrightarrow m-1} - \nu_0) = \frac{\nu_Q}{4}(2m-1)K(\eta, \theta, \psi). \tag{9.11}$$

If every nucleus in a spin system experiences the same quadrupole interaction and if the sample is a single crystal, a well-resolved multiplet is produced. Observations of the multiplet splittings for several sample orientations can be analyzed to obtain values of (e^2qQ/h), η, and the orientation of (x', y', z') in the lattice. The opportunity to analyze a homogeneous, single-crystal quadrupole pattern is rare in M–H systems. The occupation of a fraction of interstitial sites means that there will be a number of nonequivalent types of crystal environments for host metal nuclei, whether or not the hydrogen atoms are ordered on interstitial sites. Similar considerations apply to the deuterons.

In powder samples the intensity of each $(\Delta m) = 1$ transition of the cw NMR absorption signal is distributed over a finite range of frequencies and is describable by a distribution function $P(x, \eta)$, where $x = (\nu - \nu_0)/(1/2\nu_Q)$. $P(x, \eta)$ retains features which can be analyzed for (e^2qQ/h) and η, providing the quadrupole interaction is sufficiently homogeneous. Since many of the experiments done have utilized the deuteron quadrupole interaction, we consider spin $I = 1$. The shape of $P(x, \eta)$ is an even function of x centered on ν_0 and is produced by two satellite line shapes symmetrically situated about ν_0 [9.17]. Each satellite line is associated with one $\Delta m = 1$ transition. The shapes of one of the satellite lines for three values of η are shown in Fig. 9.2. This idealized shape function neglects broadening due to dipolar interactions or due to inhomogeneous quadrupole interactions. The satellite lines are characterized by a singularity at $(\nu_1 - \nu_0) = -(\nu_Q/4)(1-\eta)$, a "shoulder" at $(\nu_2 - \nu_0) = -(\nu_Q/4)(1+\eta)$, and a "step" at $(\nu_3 - \nu_0) = (\nu_Q/2)$. For $\eta = 0$, $\nu_1 = \nu_2$, and for $\eta = 1$, $\nu_1 = \nu_0$ and the step and shoulder are symmetrically displaced about ν_0.

From the separations between pairs of singularities $\Delta\nu_1 = (\nu_Q/2)(1-\eta)$, between shoulders and between steps, $\Delta\nu_2 = (\nu_Q/2)(1+\eta)$, and $\Delta\nu_3 = \nu_Q$, the values of η and ν_Q are readily determined.

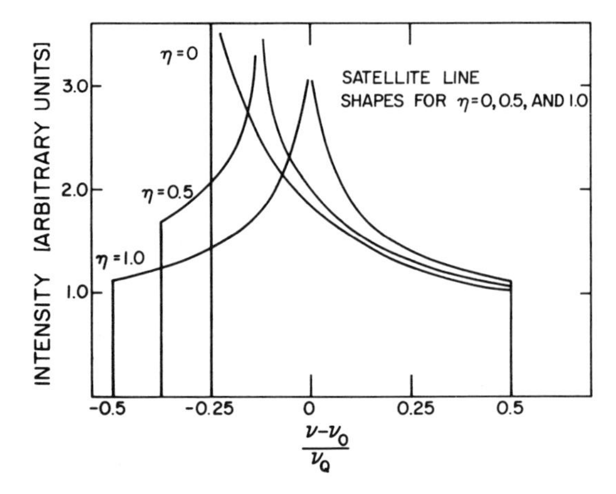

Fig. 9.2. Plots of $P(x,\eta)$, for the low-frequency satellite intensity for $\eta=0.0$, 0.5, and 1.0. For deuterons with $I=1$, the complete quadrupole powder pattern for a given value of η is the superposition of the line shape shown and the high frequency satellite found by reflecting the line shape in the figure across the $x=0$ line [9.17]

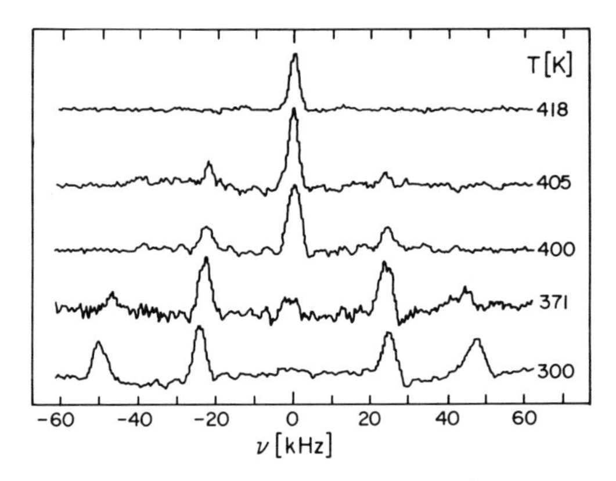

Fig. 9.3. Temperature dependence of the quadrupole spectra of the ^{2}H NMR in $V^2H_{0.42}$ in foil samples. The outer pair of satellites is for those portions of the sample having $H_0 \| c$ axis and the inner pair is for $H_0 \perp c$ axis. As the transition is made from the β phase to the α' phase, the central line of the α' phase grows in intensity [9.19]

Applications Utilizing Quadrupole Interactions

A considerable amount of experimental work has been done on the deuteron quadrupole interaction in V, Nb, and Ta.

In annealed foils of Nb [9.18] and V [9.19], some preferential alignment of crystal axes occurred in the Nb foils with the rolling plane being (110) and the direction of rolling (001). Nearly single crystals of V appeared in similarly treated foils, the plane of the foils being (100), with a nearly random distribution of cube edges in the plane. In [9.19] the z' principal axis direction was along a crystal axis, so it was possible to obtain spectra appropriate to the applied field parallel and perpendicular to z' in $V^2H_{0.42}$ as shown in Fig. 9.3. All of the other work was done in powders.

It can be seen in Fig. 9.3 that as the temperature increases, signal intensity of α' deuterons (central line) grows while intensity of β phase doublets declines. The data for Fig. 9.3 were determined from the Fourier transform of the recorded average of about one thousand free induction decays. By following temperature dependence of quadrupole splittings, *Arons* et al. [9.19] deduced a tentative phase diagram for the deuterated VH_x system.

In all instances, knowledge of the host metal structure in the various phases was known to the NMR experimenters, and in some instances, knowledge of deuteron site occupancy and ordering existed from neutron diffraction work. Values of eq and η were determined with fair to good agreement between different laboratories on similar samples. Deuteron quadrupole interactions were measured as a function of deuterium concentration and temperature.

These group Vb metals have a number of features in common. In their α and α' phases, rapid diffusion of ^2H atoms among the bcc interstitial sites averages out the quadrupolar splitting and a single NMR line is observed. For this motional narrowing to occur, each ^2H atom must sample with equal probability the interstitial sites with z' along each of the unit cell cube edges. The model of disordered site occupancy is thus confirmed.

In the lower temperature, high concentration phases, a quadrupolar spectrum is observed even though the ^2H atoms are known to be diffusing fast enough to motionally narrow the spectrum. The lack of motional narrowing means that ^2H atoms move among a subset of interstitial sites which, for example, in the δ orthorhombic phase of VH_x, have their z' axis along the c edge of the unit cell. In all three metal systems, ordered site occupancy has thus been confirmed by the quadrupole splittings in β phases. The magnitudes of the electric field gradients are similar for tetrahedral sites in the δ phase for V and the β phases for Nb and Ta and the asymmetry parameters small in these phases. Values of (e^2qQ/h) vary from 32–37 kHz with a weak dependence upon temperature and concentration of deuterons. Values of η vary between about 0.1 and 0.2. In TaH_x, η decreases as x increases in the β phase. The electric field gradient at the ^2H atom site is, to a good approximation, determined by the positions of the nearest neighbor metal atoms and is thus sensitive to phase transitions.

The coexistence of α' and β phases is evident in the quadrupole spectrum as a superposition of patterns of both phases in Fig. 9.4. The temperature width of the phase transition has been measured by monitoring the two quadrupole spectra [9.20]. In obtaining data for Fig. 9.4, *Barnes* et al. [9.20] used a cw NMR spectrometer with phase-sensitive detection which records the first derivative of the absorption signal with respect to applied magnetic field. At $T = 396$ K, the separation between steps is shown as 45.8 kHz. Shoulders are not resolved from the singularities in Fig. 9.4.

Host metal nuclear quadrupole splittings have been observed for V. *Arons* et al. [9.19] note that since the V quadrupole splittings are essentially the same for V^1H_x as V^2H_x, ^1H and ^2H must occupy the same octahedral sites in the tetragonal β phase. The larger quadrupole moments of Nb and Ta make them more sensitive to lack of homogeneity in the quadrupole coupling. As a result, the quadrupole effects in $Nb\ ^2H_x$ are observable only as a composite of shifted central $m = 1/2$ to $m = -1/2$ transitions for this $I = 9/2$ nucleus. The ^{181}Ta NMR has not been observed in high hydrogen concentrations due to large

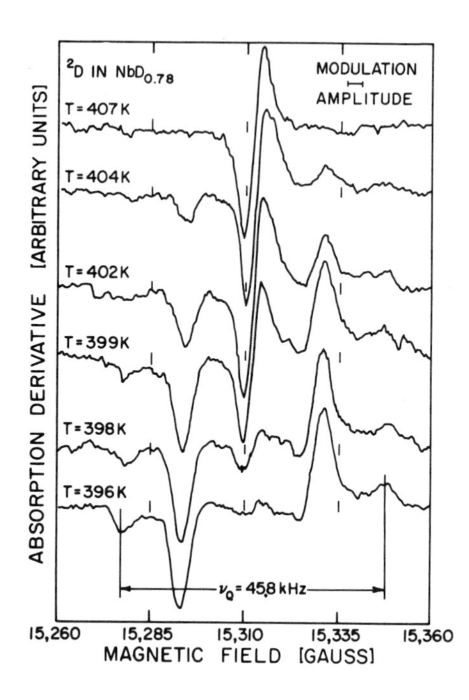

Fig. 9.4. First derivative cw absorption curves of the ^{2}H NMR in a powder sample of $Nb^2H_{0.78}$. The central line is characteristic of the α' phase, while the satellites are from the β phase. As temperature increases the α' phase grows while intensity from the β phase diminishes. See also Fig. 9.6 for the ^{93}Nb NMR in the same sample [9.20]

inhomogeneous quadrupole interactions [9.21]. ^{181}Ta has been observed, using nuclear acoustic resonance techniques in a tantalum single crystal containing small amounts of hydrogen [9.22]. In these experiments, evidence for onset of precipitation of the β phase in Ta is observed for the order of 10 ppm hydrogen. Motional narrowing of the quadrupole broadening (see Sect. 9.2) in the α phase is also observed.

The work done utilizing quadrupole interactions has contributed significantly to detailed understanding of structure effects in M–H systems. The interpretation of quadrupole spectra supplements other experiments and can give evidence of phase transitions, site occupation, ordering, and small changes in symmetry and curvature of the interstitial potential well in which the ^{2}H atom resides. The smallness of the deuteron's quadrupole moment and the electron screening of the neighboring deuterons reduces effects of inhomogeneous quadrupole broadening and makes the deuteron quadrupole interaction an effective tool for these studies. The large quadrupole moments of the host metal nuclei and the greater variety of environments around the metal nuclei, depending upon H atom site occupation, make them more susceptible to inhomogeneous electric field gradients. It can be anticipated that the deuterons will continue to be more effectively utilized than the host metal nuclei in analysis of quadrupole interactions of M–H systems. A well-known problem in working with the ^{2}H NMR is its weak signal strength due to its small nuclear gyromagnetic ratio. Since signal strength is approximately proportional to the square of the applied magnetic field, the largest available fields are usually selected. It should be noted that even with fields of over 20 k gauss, signal averaging times of an hour or more (in one instance [9.20] 26 h) for one spectrum were needed in $Nb\,^2H_{0.82}$. Because of the weak signal from deuterons, experiments are usually done in powders rather than in single crystals where NMR signals are weak due to the shallow penetration of rf fields. The nearly single crystals found in rolled foils [9.18, 19] are the exception. Being thin, the rf field fully penetrates every foil in a stack separated by insulating material. The use of the single crystal would allow measurement of the directions of the principal axes of the EFG tensor and would improve identification of the interstitial site. Even though absolute values of the EFG tensors are not readily calculated in M–H systems, often the direction of the z'-axis and a rough estimate of the axial asymmetry parameter can be estimated for each site and compared with experimental results. As mentioned earlier, the powder experiment yields only the values of eq and η.

9.2 Diffusive Motion of Hydrogen

9.2.1 Mean Residence Times and the cw Spectrum

Observations of the NMR of hydrogen in metals can be analyzed using established theories to determine the mean residence time τ_D for a hydrogen atom on a lattice site. From knowledge of temperature dependence of τ_D, the activation energy for thermal motion can be deduced. Observations of the spectrum using cw spectrometers are, of course, related to the transient FID signal through the Fourier transform.

Characteristic narrowing of the resonance line can occur as temperature is increased beyond the point where $(\tau_D^{-1}) > \Delta\omega$, where $\Delta\omega$ is the width of the low-temperature rigid lattice spectrum. Marked changes in relaxation times of the NMR spin system are also produced by thermal motion. These will be discussed in Section 9.2.2.

Motional Narrowing of Dipolar Interactions

As described in "Second Moment of NMR Absorption Spectrum", only the A and B terms of the dipolar Hamiltonian could contribute to the dipolar broadening of the line. Both of these terms contain $(1-3\cos^2\theta)$ dependence, where θ is the polar angle between the applied magnetic field and the radius vector from one spin to the second. If thermal motion makes θ time dependent, and if the frequency of motion exceeds a measure of the mean dipolar width $\langle \Delta\omega_d^2 \rangle^{1/2}$, then the linewidth will depend only upon the time average of terms A and B. For isotropic motion in a liquid, or over jumps on lattice sites in a three-dimensional lattice, the average value $\langle (1-3\cos^2\theta) \rangle$ is zero. This characteristic line narrowing occurs gradually as τ_D decreases. The cw linewidth ΔH is related to the transverse relaxation time by $\Delta H = (\gamma T_2)^{-1}$. *Kubo* and *Tomita* [9.23] relate the observed linewidth to τ_D in an equation which applies for $\tau_D \lesssim \langle \Delta\omega_d^2 \rangle^{-1/2}$

$$\tau_D^{-1} = \frac{\pi^2(\Delta\nu)}{4\ln 2}\left\{\tan\left[\frac{\pi(\Delta H)^2}{2(\Delta H_0)^2}\right]\right\}^{-1}, \tag{9.12}$$

where $\Delta\nu$ is the observed linewidth (half-width at half maximum) in Hz, ΔH_0 is the low-temperature rigid lattice linewidth, and $\Delta H = 2\pi(\Delta\nu)/\gamma_I$ is the observed linewidth in gauss.

For $(\tau_D^{-1}) \gg \langle \Delta\omega_d^2 \rangle^{1/2}$, the observed linewidth is given, to good approximation, by

$$2\pi(\Delta\nu) = \left(\frac{\tau_D}{2}\right)\langle \Delta\omega_d^2 \rangle. \tag{9.13}$$

In (9.13), $\langle \Delta\omega_d^2 \rangle$, the dipolar second moment, is usually known, or can be measured, so that from a measure of $\Delta\nu$, τ_D is easily calculated when (9.13) applies.

The attractive feature of the line narrowing experiment is that it is convenient experimentally, unless, of course, signals are so weak that extensive signal averaging is required. A disadvantage is the limited range of values of τ_D over which data can be collected. The longest value of τ_D observable is the order of the reciprocal rigid lattice linewidth, $\tau_D \approx \langle \Delta\omega_d^2 \rangle^{1/2} \approx 5 \times 10^{-6}$ s. The shortest value of τ_D corresponds to that point where the narrowed dipolar linewidth becomes just less than the inhomogeneity of the applied magnetic field. Even with a perfectly trimmed magnet, the field in the sample can be inhomogeneous due to particle magnetization of the powder sample. This inhomogeneity can easily exceed 0.1 gauss or about $3 \times 10^3\,\mathrm{s}^{-1}$ for protons. Setting this width equal to the narrowed resonance linewidth gives a typical lowest value of τ_D as $\tau_D \approx 0.15 \times 10^{-6}$ s. The resulting range of τ_D is a factor of about 30–100 which is not competitive with pulse techniques that can determine τ_D over three decades or more. By using thin foils as samples, it would be possible to reduce inhomogeneous broadening by another order of magnitude.

The simple observation of line narrowing is convincing evidence that diffusion is occurring. In many instances, sufficient data for τ_D is obtained to determine the activation energy for diffusive motion.

The dipolar line narrowing of the deuterium NMR is complicated in most systems by quadrupole interactions which are not known. Most line narrowing experiments are done on protons.

Spalthoff [9.24] made a survey of line narrowing in $ThH_{3.5}$, UH_3, $ZrH_{1.40}$, $TaH_{0.66}$, and $PdH_{0.63}$ from $-190\,^\circ$C to $+300\,^\circ$C and obtained activation energies. *Stalinski* et al. [9.7] measured temperature dependence of linewidths of a number of γ phase TiH_x samples over a temperature range from $-196\,^\circ$C to about 200 °C and fit data to Arrhenius relationship. From line narrowing data of Fig. 9.5, *Stalinski* and *Zogal* [9.25] measured activation energies in β-NbH_x for $x > 0.70$. The linewidth in Fig. 9.5 is the full width between points of

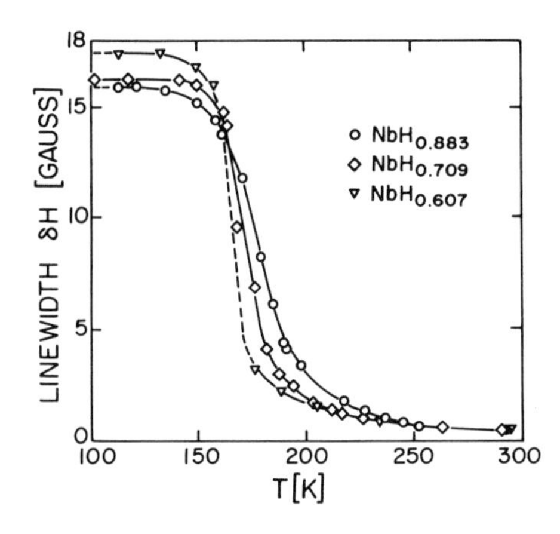

Fig. 9.5. Linewidth of the ^{1}H NMR as a function of temperature in three samples of β NbH_x with $x=0.607$, 0.709, and 0.883. Linewidth narrows as a consequence of self diffusion of H atoms [9.25]

maximum slope on the absorption curve. *Stalinski* and *Zogal* fit their data to (9.12), but the $x=0.607$ curve, which is unusually steep, could not be fit. They report essentially temperature-independent linewidths of about 0.5 gauss near 300 K caused by inhomogeneous magnetization of the powder sample.

Thermal Motion and Quadrupole Splittings

It is also possible to average out the nuclear electric quadrupole splitting of the deuteron NMR spectrum with random thermal motion. The splitting of the line, as shown in (9.9) is proportional to an angular factor $[3\cos^2\theta - 1 + \eta\cos 2\psi(\cos^2\theta - 1)]$. The three-dimensional average of the angular factor is zero. For most sets of interstitial sites for hydrogen in metals, the mean value of the angular factor averaged over all possible sites is also zero. It can therefore be expected that thermal motion will reduce or average out nuclear electric quadrupole spectra for the deuteron and for host metal nuclei which have nuclear quadrupole moments.

This averaging indeed has been observed to occur in the α and α' phases of group Vb metals. The collapse of the quadrupole splitting has been observed to occur discontinuously at the transition to the disordered α' state from the ordered, lower temperature β phase. As discussed in "Applications Utilizing Quadrupole Interactions", the quadrupole splitting in the ordered β phase, say in NbH_x, for example, remains unchanged as temperature is increased even though the jump frequency exceeds $10^8\,s^{-1}$ and the quadrupole splitting is less than 10^5 Hz. Because hydrogen atoms move only on a subset of interstitial sites, the angular factor does not become modulated, and the splitting remains essentially constant. This observation is a strong indication of ordering of hydrogen in these lower temperature phases of group Vb metals [9.18–20].

Little information appears to have been published on the linewidth of the ^{2}H single resonance line in the disordered α' phases of V, Nb, or Ta. It is only noted that the linewidth is less than 10^3 Hz [9.20]. In some recent exploratory measurements in Nb and Pd deuterides with which this author was associated,

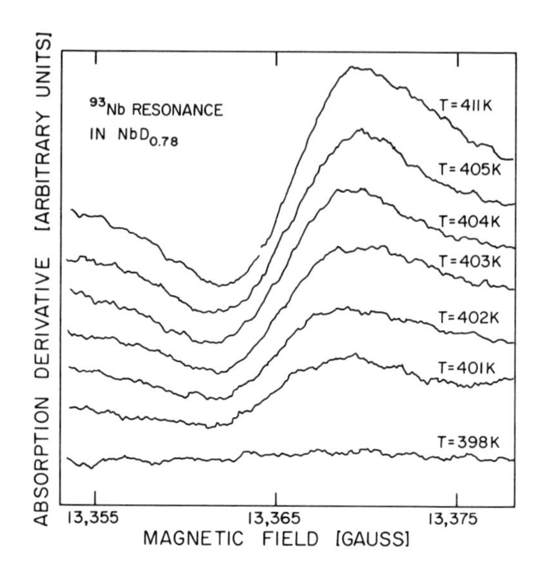

Fig. 9.6. First derivative cw absorption curve of the ^{93}Nb NMR in a powder sample of Nb$^2H_{0.78}$. In the disordered α' phase, quadrupolar broadening of the ^{93}Nb NMR is averaged to a relatively low value allowing observation of the central (1/2, $-1/2$) transition of the resonance. At the lowest temperature, in the β phase, quadrupole broadening wipes out even the central transition of the ^{93}Nb NMR (see also Fig. 9.4) [9.20]

the spin echo decay times (T_2) are found to be less than 5×10^{-3} s in the α' phase where the motional narrowing formula would indicate that T_2 should exceed 1 s. Since the magnetic dipole interactions are weaker for deuterons than for protons, they can be eliminated as a cause of the additional breadth. It is more likely that strain fields in the sample produce small long-range field gradients which are not completely averaged by the rapid diffusion in the α' phase.

In a few systems (VH$_x$ [9.19], NbH$_x$ [9.20], and LaH$_x$ [9.8]), the motional narrowing of quadrupole broadening of the host metal nuclei has been observed. In the NbH$_x$ system, rapid ^{1}H diffusion on disordered α' phase lattice sites averages out quadrupole splittings of the ^{93}Nb NMR in the β phase (*Zamir* and *Cotts* [9.26]). *Barnes* et al. [9.20] concluded from temperature dependence of the ^{93}Nb spectrum that the boundary between ($\alpha'+\beta$) and α' phases in Nb^2H_x falls at $0.78<x<0.82$ at 300 K. They also measured the temperature width of this transition to be 7 K as shown in Fig. 9.6. The single Nb NMR line seen in the high-temperature α' phase is believed to be from the central (1/2, $-1/2$) transition only. A distribution of quadrupolar interactions due to strain and lattice imperfections prevents observation of other $\Delta m=1$ transitions. If the sample were to consist of perfect α' crystallites, all of the transitions would be superimposed at the Larmor resonant applied field, $H_0=\omega_0\gamma^{-1}$. In Fig. 9.6, the loss of signal in the lower temperature β phase is due to the β phase quadrupolar coupling that broadens and shifts the (1/2, $-1/2$) transition in the powder sample.

In a related experiments the motional narrowing of quadrupole interactions and isomer shifts of the Mössbauer resonance of a host metal nucleus can be used to measure mean residence times of H atoms. Details of the Mössbauer technique are contained in Chapter 6.

9.2.2 Mean Residence Times and Relaxation Rates

Much of the study of atomic motion using NMR is through measurements of the spin lattice relaxation time T_1, the spin-spin, or transverse, relaxation time T_2, and the rotating frame spin lattice relaxation time $T_{1\varrho}$.

Of these relaxation times, measurement of T_1 is preferred over T_2 as being more reliably representative of effects of dipole-dipole interactions. T_1 measures relaxation of the component of magnetization parallel to the field and, as indicated in (9.14), requires Fourier components at the Larmor frequency for relaxation. T_2 measures relaxation of transverse components of the magnetization which are very sensitive to extraneous interactions at or near zero frequency. Any interaction which affects the phase of a precessing spin can contribute to the transverse relaxation rate, even in a spin echo experiment that averages out effects of static inhomogeneities in H_0. $T_{1\varrho}$ is especially useful for obtaining information on low-temperature motion. Most of the discussion that follows relates to calculation and measurement of T_1.

The theory presented is for nuclear dipole-dipole interactions only. The same formalism applies to the nuclear electric quadrupole interaction which for deuterons is stronger than dipolar interactions. The dipolar interaction can be completely specified for any given set of H sites but the quadrupole interactions cannot, since the electric field gradient tensor is not, in general, completely known. Consequently observations of T_1 due to quadrupole effects for diffusing deuterons give less accurate measures of mean residence times of the diffusing atom than can similar measurements on protons. The random jumps of H atoms modulate the nuclear dipole-dipole interactions between hydrogen nuclei and between hydrogen and metal nuclei.

Relaxation Rate Theory

Relaxation rates are expressed in terms of power spectra of randomly varying dipolar dields $J^{(q)}(\omega)$, which are, for H–H magnetic dipolar interactions only [9.27] [Ref. 9.12, Chap. VIII],

$$\frac{1}{T_1}=\frac{3}{2}\gamma_I^4\hbar^2 I(I+1)[J^{(1)}(\omega_0)+J^{(2)}(2\omega_0)]\,, \tag{9.14}$$

$$\frac{1}{T_{1\varrho}}=\frac{3}{8}\gamma_I^4\hbar^2 I(I+1)[J^{(0)}(2\omega_1)+10J^{(1)}(\omega_0)+J^{(2)}(2\omega_0)]\,, \tag{9.15}$$

$$\frac{1}{T_2}=\frac{3}{8}\gamma_I^4\hbar^2 I(I+1)[J^{(0)}(0)+10J^{(1)}(\omega_0)+J^{(2)}(2\omega_0)]\,, \tag{9.16}$$

where

$$J^{(q)}(\omega) = \int_{-\infty}^{\infty} G^{(q)}(t)\,\mathrm{e}^{-\mathrm{i}\omega t}\,dt\,, \tag{9.17}$$

$$G^{(q)}(t) = \sum_{j} \langle F_{ij}^{(q)}(t')\,F_{ij}^{(q)*}(t'+t)\rangle\,, \tag{9.18}$$

$$F_{ij}^{(q)}(t) = \frac{d_q Y_{2q}(\Omega_{ij})}{r_{ij}^3}\,, \tag{9.19}$$

$$d_0^2 = \frac{16\pi}{5}\,, \quad d_1^2 = \frac{8\pi}{15}\,, \quad d_2^2 = \frac{32\pi}{15}\,. \tag{9.20}$$

In the above equations $Y_{2,q}$ are normalized spherical harmonics with coordinates $(r,\theta,\phi)_{jk}$ made time dependent by diffusion of hydrogen. M–H nuclear spin interactions can be included by extensions of the equations (Ref. [9.12], Chap. VIII). Equations (9.14)–(9.20) are valid in the motionally narrowed regime where $\langle \Delta\omega_{\mathrm{d}}^2\rangle^{1/2}\tau_{\mathrm{D}} \ll 1$. The key to application of these equations to relaxation time measurements is in the evaluation of correlation functions $G^{(q)}(t)$.

The simplest form of $G^{(q)}(t)$ was suggested by *Bloembergen*, *Purcell*, and *Pound* (BPP) [9.28]. It is assumed that dipolar field correlations decay as $\exp(-t/\tau_{\mathrm{c}})$, where τ_{c} is the correlation time. Then the power spectrum of the dipolar interaction at angular frequency ω becomes

$$J^{(q)}(\omega) = G^{(q)}(0)\frac{2\tau_{\mathrm{c}}}{(1+\omega^2\tau_{\mathrm{c}}^2)} \tag{9.21}$$

and typically for H–H interactions, $\tau_{\mathrm{c}} = (\tau_{\mathrm{D}}/2)$.

This model for $G^{(q)}(t)$ is independent of the existence of a discrete lattice and has been widely applied to analyze atomic motion in fluids. It has also been applied to polycrystalline solids including M–H systems [9.4]. From temperature dependence of T_1, values of activation energies for the hydrogen jumping rate have been determined to be within about 10% of those found by other techniques. However, absolute values of τ_{c} found from the BPP model can be in error by as much as about 50%. Typical dependences of T_1, T_2, and $T_{1\varrho}$ upon τ_{c} are shown in Fig. 9.7. The BPP theory was used to calculate the relaxation rates, expressed in arbitrary units, in Fig. 9.7. It has been assumed that only H–H spin interactions exist so that $\tau_{\mathrm{c}} = (\tau_{\mathrm{D}}/2)$. For all theories, the maximum in $(1/T_1)$ occurs near $\omega_0 \approx (1/\tau_{\mathrm{D}})$ while the maximum in $(1/T_{1\varrho})$ occurs near $\omega_1 = (1/\tau_{\mathrm{D}})$. For Fig. 9.7, $\omega_0 = 500\omega_1$. In M–H systems, typical maximum values of $(1/T_1)$ are 10^2 to $10^3\,\mathrm{s}^{-1}$.

Some association with actual lattice sites for H is made when applying the theory to solid M–H systems by making the summation of (9.18) over all lattice sites. This summation can be calculated for either ordered or disordered arrangements of H atoms.

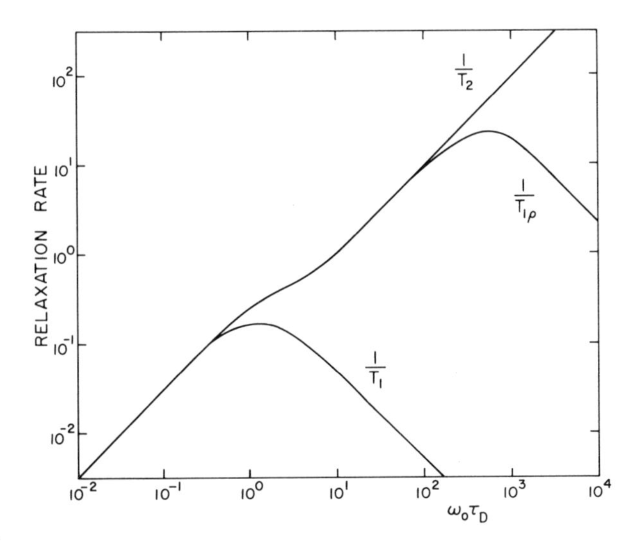

Fig. 9.7. Relaxation rate, in arbitrary units, as a function of $\omega_0\tau_D$ for $(1/T_1)$, $(1/T_{1\varrho})$, and $(1/T_2)$ as calculated for an exponential correlation function and for $\omega_0 = 500\,\omega_1$

Torrey [9.29] and *Resing* and *Torrey* [9.30] developed a way to calculate $G^{(q)}(t)$ for specific lattices. The Torrey model for a polycrystalline sample assumes that isotropic diffusion is possible from a lattice site to all points on a spherical surface with steps of equal length. A consequence of this assumption is that the value of $G^{(q)}(0)$ does not fit known sum rules for the dipolar interaction and must be normalized. *Torrey* introduced a normalization parameter k calculated for each type of lattice considered. Values of k for fcc and bcc lattices appear in [9.30].

It can be expected that the assumption of isotropic diffusion would be quite good for a lattice with a large number of possible nearest neighbor jumps of equal distance, as in the fcc lattice, while in bcc or sc lattices with smaller coordination numbers the approximation might not be as good.

In Torrey's theory the motion of the spins is uncorrelated. He shows that, insofar as the dipolar interactions between like spins are concerned, the relative displacement of a pair of spins after a time t is the same as that of one spin fixed and the other moving for a time $2t$. An equivalent statement in the BPP model is that the correlation time $\tau_c = (\tau_D/2)$, assuming that one H atom jump changes nearest neighbor H–H dipole interactions by a fraction the order of one. Insofar as H–M interactions are concerned, the M atoms are fixed in place, $\tau_c = \tau_D$ in the BPP model.

The Torrey theory has been applied, for example, to T_1 measurements in $ScH_{1.7}$, $Ti^1H_{1.55}$, $Ti^3H_{1.5}$ by *Weaver* and *Van Dyke* [9.31]. Discussion of these experiments will be postponed until after presentation of the models of *Sholl* [9.32] and of *Wolf* [9.33], who have calculated the theory of nuclear spin relaxation due to H atom diffusion in a more rigorous manner.

Sholl's theory differs from Torrey's by replacing the assumption of isotropic diffusion with calculation of $G^{(q)}(t)$ explicitly summed over all lattice sites before the orientational averaging is done for polycrystalline samples. He also

Table 9.1. Spin lattice relaxation rates ($1/T_1$) in polycrystalline samples for $\omega_0\tau_D \ll 1$ and $\omega_0\tau_D \gg 1$

$\left(\frac{1}{T_1}\right)^{a}$ Theory	bcc lattice		fcc lattice	
	$\omega_0\tau_D \ll 1$	$\omega_0\tau_D \gg 1$	$\omega_0\tau_D \ll 1$	$\omega_0\tau_D \gg 1$
Torrey	$0.578\,\alpha\tau_D$	$0.408\,\frac{\alpha}{\omega_0^2\tau_D}$	$2.22\,\alpha\tau$	$1.651\,\frac{\alpha}{\omega_0^2\tau_D}$
Sholl	$0.625\,\alpha\tau_D$	$0.395\,\frac{\alpha}{\omega_0^2\tau_D}$	$2.27\,\alpha\tau$	$1.63\,\frac{\alpha}{\omega_0^2\tau_D}$
Wolf	$0.508\,\alpha\tau_D$	$0.400\,\frac{\alpha}{\omega_0^2\tau_D}$	$1.917\,\alpha\tau$	$1.63\,\frac{\alpha}{\omega_0^2\tau_D}$

[a] $\alpha = 96\,\frac{\gamma^4\hbar^2 I(I+1)c}{b^6}$,

where c = fraction of sites occupied and b = length of one side of unit cell.

evaluated explicitly by numerical integration the equation for the probability of finding an H atom at a given lattice site near the site of origin at a time $2t$ after it was at the site of origin. *Sholl* has tabulated values of T_1 against a wide range of values of $(\frac{1}{2}\omega_0\tau_D)$ for fcc [9.32] and in a subsequent paper [9.34] for bcc and sc lattice assuming nearest neighbor jumps.

Wolf, as did *Sholl*, starts with the Torrey formulation of the random walk problem. The isotropic diffusion model is also replaced but by expanding the correlation function $G^{(q)}(t)$ into a product of the simple exponential decay function and a series in increasing powers of time. From the spectral densities calculated for a single crystal, the average over all crystallographic orientations is obtained for polycrystalline samples.

Since T_1 in the single crystal has a small anisotropy (except for $\omega\tau_D \ll 1$ where T_1 is isotropic in all models), the relaxation recovery of the nuclear magnetization in the polycrystal will, in principle, be nonexponential. However, experimental measurements usually fit a single exponential sufficiently well in polycrystalline samples that in many systems the deviation from a single exponential must be small. *Wolf* approximates the relaxation of the polycrystal by one relaxation rate equal to the orientational average of the single-crystal relaxation rate. This approximation appears then to be equivalent to *Sholl*'s taking the orientation average of the correlation functions before the spectral densities are calculated.

The anisotropies in T_2 and in $T_{1\varrho}$ calculated by *Wolf* are considerably larger than in T_1 in the range $\omega_0\tau_D \gg 1$ for T_2 and $T_{1\varrho}$ and $\omega_1\tau_D \gg 1$ for $T_{1\varrho}$, where $\omega_1 = \gamma_I H_1$. Both Sholl's and Wolf's methods give the detailed shape of the dependence of T_1 upon $y = (\omega\tau_D/2)$.

The calculations of *Torrey*, *Wolf*, and *Sholl* are done numerically and the reader is referred to their papers for details. A brief comparison of their predictions can be made in the extreme cases, $\omega_0\tau_D \ll 1$ and $\omega_0\tau_D \gg 1$ for bcc and fcc lattices, and is shown in Table 9.1. (It is recognized that only the fcc lattice applies to H interstitials, as in PdH_x, and that specific numerical calculations would need to be done for each H atom occupancy model.)

It is seen in Table 9.1 that for $\omega_0\tau_D \gg 1$, on the low-temperature side of the T_1 minimum, the relaxation rates predicted by the improved theories of *Sholl* and *Wolf* are less than the Torrey value but only by amounts up to 3%. For $\omega_0\tau_D \ll 1$, Sholl's values of $(1/T_1)$ exceed Torrey's by 2.2% in the fcc lattice and by 8.1% in the bcc lattice, while Wolf's values of $(1/T_1)$ are less than Torrey's by 13.6% for the fcc lattice and by 12% in the bcc lattice.

If data from both sides of the T_1 minimum are used to calculate τ_D as a function of temperature at constant ω_0, as is usually done experimentally, the resulting values of activation energies for the diffusion process would not vary by more than 1–2% from model to model. Values of E_a from Sholl's theory would exceed Torrey's by less than 1% while Wolf's would be less than Torrey's by about 1.5%.

Applications of Theory

In the experimental comparison of the Torrey theory with the BPP model made by *Weaver* and *Van Dyke* [9.31] for three hydrides having the CaF_2 structure, $ScH_{1.7}$, $Ti^1H_{1.55}$, and $Ti^3H_{1.5}$, no significant difference in activation energies between models was obtained. The values of τ_D obtained through use of the BPP theory were about half the values obtained using the Torrey theory.

The Torrey theory has also been applied to determine the temperature dependence of τ_D in a series of TiH_x samples [9.35] in a series of TiH_x, ZrH_x, and HfH_x samples [9.36], and in $PdH_{0.7}$ [9.37]. The Sholl theory has been applied to some PdH_x samples [9.38] in conjunction with NMR measurements of diffusion coefficients of ^1H in the same samples, to be discussed in Section 9.2.3.

Arons et al. [9.39] used $T_{1\varrho}$ measurements to extend determinations of values of τ_D in $PdH_{0.73}$ down to 170 K. An advantage of using $T_{1\varrho}$ at lower temperatures is that its relaxation rate due to diffusion is so large that the relaxation rate due to conduction electrons, to be discussed in Section 9.3, is negligible.

Figure 9.8 shows temperature dependence of T_1 from *Lütgemeier* et al. [9.40] in the NbH_x system. Increased diffusivity of hydrogen in the α' phase above 75–100 °C causes the steep rise of T_1 for $x=0.3$, 0.58, and 0.78. The curvature of the line for the $x=0.03$ sample is due principally to the conduction electron mechanism.

In summary, the more precise calculations of *Sholl* and *Wolf* do not appear to contain any surprises insofar as determination of activation energies E_a or mean residence times τ_D are concerned. As expected, the deviations of the Torrey theory, though small, are greater for a lattice with low coordination number such as scc than one with a high number such as the fcc lattice. There in some concern that for $\omega_0\tau_D \ll 1$, the Sholl and Wolf theories differ by about 15–20% in the predicted values of $(1/T_1)$, with the Torrey value *in between*. But it is

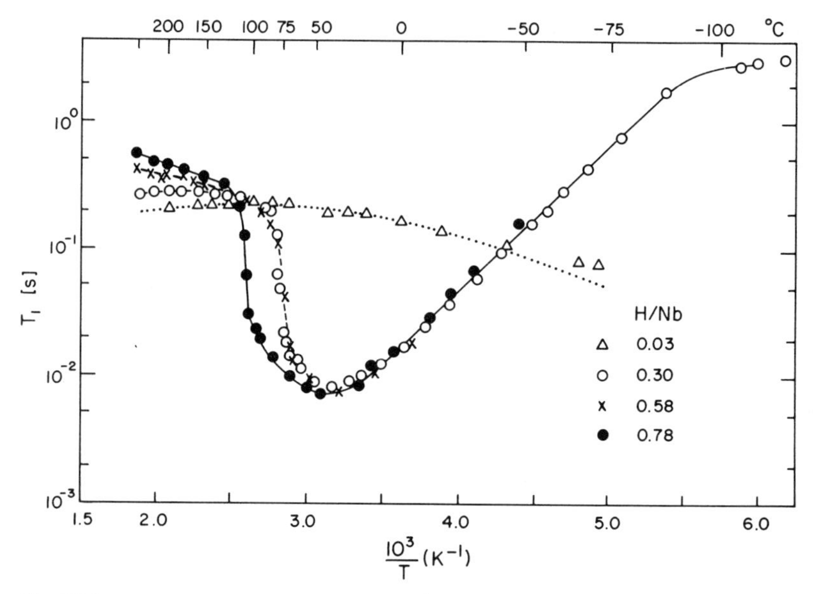

Fig. 9.8. Temperature dependence of T_1 of the ^{1}H spin system in Nb^{1}H$_x$ for $x=0.03$, 0.30, 0.58, and 0.78. The shapes of the T_1 minima shown are asymmetric because of the sharp increase in diffusivity of hydrogen as the transition to the α' phase occurs. The flattening of these same T_1 curves at high temperatures is due to effects of T_{1e}. Since diffusion of ^{1}H is so fast in α-Nb$_{0.03}$ over most of the temperature range shown, the effects of T_{1e} dominate its shape for $(10^3/T)<3.5$ [9.40]

in this regime that the numerical calculations appear to be most difficult and slowly converging.

An attractive feature of the Sholl and the Wolf calculations is that they provide detailed calculations of the shape and width of the peak in $(1/T_1)$ vs $\omega_0\tau_D$. It has been suggested by *Wolf* that the shape and widths of the peak can indicate which diffusion mechanism is in effect. The anisotropy of T_1 and $T_{1\varrho}$ in M–H single crystals could also help identify the diffusion mechanism. Wolf's theory for these anisotropies has been applied to a recent study of self-diffusion in BaF_2 single crystals [9.41]. Application to single-crystal metals will be limited to systems with optimum signal-to-noise ratio and probably to T_1 measurements only, since accurate T_2 and $T_{1\varrho}$ measurements require good homogeneity of the rf field. The rf field is, of course, very inhomogeneous in samples thicker than the rf skin depth.

On the other hand, the nearly single crystals obtained [9.19] in annealing rolled foils of Nb and V offer the opportunity for limited observation of anisotropy in $T_{1\varrho}$ under good signal conditions.

NMR and Quasielastic Neutron Scattering

The theory of quasielastic neutron scattering by hydrogen has a form similar to spin relaxation theory in NMR. According to *Van Hove* [9.42], the differential scattering cross section per unit solid angle and angular frequency interval is proportional to

$$\frac{d^2\sigma}{d\Omega d\omega} \propto C_{\mathrm{H}}\left(\frac{k}{k_0}\right)\sigma_{\mathrm{b}} S(\boldsymbol{K},\omega), \tag{9.22}$$

where C_{H} is the hydrogen concentration, $\hbar\omega = (E - E_0)$ the energy transfer, $\hbar\boldsymbol{K} = \hbar(\boldsymbol{k} - \boldsymbol{k}_0)$ the momentum transfer, and σ_{b} is the incoherent bound scattering cross section of the proton.

The scattering function is

$$S(\boldsymbol{K},\omega) = \int_{-\infty}^{\infty} I(\boldsymbol{K},t)\,\mathrm{e}^{-\mathrm{i}\omega t}\,dt \tag{9.23}$$

where

$$I(\boldsymbol{K},t) = \langle \mathrm{e}^{\mathrm{i}\boldsymbol{K}\cdot\boldsymbol{r}(t')}\,\mathrm{e}^{-\mathrm{i}\boldsymbol{K}\cdot\boldsymbol{r}(t'+t)} \rangle . \tag{9.24}$$

The ensemble average denoted by the angle brackets in (9.24) can be expressed in terms of a probability function $P_{\mathrm{s}}(\boldsymbol{r},t)$

$$I(\boldsymbol{K},t) = \frac{1}{2\pi}\int \mathrm{e}^{\mathrm{i}\boldsymbol{K}\cdot\boldsymbol{r}}\,P_{\mathrm{s}}(\boldsymbol{r},t)\,d\boldsymbol{r} . \tag{9.25}$$

$P_{\mathrm{s}}(\boldsymbol{r},t)d\boldsymbol{r}\,dt$ is the probability that the proton at $\boldsymbol{r}=0$, $t=0$, will be at a later time, t, at $\boldsymbol{r}$. A similar representation of the ensemble average in the definition of the dipolar field correlation function of (9.18) and (9.19) is, [9.32]

$$G^{(q)}(t) = C d_q^2 \sum_{j,k} \frac{Y_{2q}(\Omega_j)}{r_j^3}\,\frac{Y_{2q}^*(\Omega_k)}{r_k^3}\,P(\boldsymbol{r}_k - \boldsymbol{r}_j, 2t) \tag{9.26}$$

where $P(\boldsymbol{r}_k - \boldsymbol{r}_j, 2t)$ is the probability of a spin being displaced by $(\boldsymbol{r}_k - \boldsymbol{r}_j)$ in a time $2t$. Equation (9.26) has been expressed as a lattice sum, and when the evaluation of $I(\boldsymbol{K},t)$ is applied to a specific lattice its integral would necessarily be replaced by the appropriate summation.

Both experiments, measurement of a relaxation rate in NMR and observations of the quasielastic peak in neutron scattering, are measures of Fourier transforms of appropriate self-correlation functions associated with diffusive motion in the lattice. The solutions of the problems lie in finding accurate representations of the probability functions. The effects of finite jumps of atoms on the form of $P_{\mathrm{s}}(\boldsymbol{r},t)$ were considered by *Chudley* and *Elliot* [9.43].

One expects that the numerical methods of *Sholl* and of *Wolf* used to calculate $J^{(q)}(\omega)$ in NMR could be usefully applied to the neutron scattering problem to account for finite jump effects in specific lattices.

In the limit of small momentum transfer, the scattering is determined by the form of $P_{\mathrm{s}}(\boldsymbol{r},t)$ after many jumps have occurred. The application of the solution of the macroscopic diffusion equation then applies, and *Vineyard* [9.44] has calculated the shape of $S(\boldsymbol{K},\omega)$ in this case. He finds

$$S(\boldsymbol{K},\omega) = \frac{2DK^2}{(DK^2)^2 + \omega^2} . \tag{9.27}$$

For large momentum transfer, the theory of *Chudley* and *Elliot* [9.43] with improvements by *Gissler* and *Stump* [9.45] must be applied. Small momentum transfer requires that $K^2 D\tau_D \ll 1$, which is equivalent to the requirement that $K \ll (2\pi/l)$ where l is the step length in the diffusion process. Large momentum transfer corresponds to having $K \gtrsim (2\pi/l)$.

It follows then that at small momentum transfer the form of $S(\boldsymbol{K}, \omega)$ is independent of the details of the jump process and, of course, the width, $\Delta\omega_{1/2}$, of the peak yields directly the value of D. This direct determination of D is to be compared to D measurements in NMR from spin echo attentuation in an applied magnetic field gradient to be discussed in Section 9.2.3. Both techniques yield values of D independent of assumptions of the diffusion mechanism.

At large momentum transfer, the dependence of $\Delta\omega_{1/2}$ upon K is sensitive to the diffusion mechanism and can, in principle, identify the mechanism, providing an accurate theory exists for $\Delta\omega_{1/2}$ for each specific lattice. Quasielastic neutron scattering in the large K regime has been used to identify the diffusion mechanism in PdH_x [9.46], but in NbH_x and VH_x the interpretation has been less clear. Experiments in single crystals are essential. These experiments are similar to NMR experiments in the regime $\omega_0\tau_D \gg 1$ in which the relaxation rates are anisotropic and sensitive to the diffusion mechanism (*Wolf* [9.33]). Unfortunately, in NMR the fast T_2 in the regime $\omega_0\tau_D \gg 1$ makes observation of signals in single crystals of M–H systems difficult because of the rf skin depth problem. In insulating solids the two techniques would be comparable in their sensitivity to diffusion mechanisms, but in single-crystal metals the neutron scattering has the advantage over measurements of relaxation rates in NMR. It will be seen in Section 9.2.3, however, that by combining NMR and relaxation rate data and diffusion data the diffusion mechanism can also be identified in NMR.

9.2.3 Direct Measurement of Hydrogen Diffusion Coefficients

Two-Pulse Spin Echo Technique

The spin echo experiments in NMR provide the means for making direct measurement of D, the self-diffusion coefficient. In its simplest form a standard 90°–180° spin echo is observed with a uniform magnetic field gradient $G = (\partial H_z/\partial z)$ applied to the sample. The echo amplitude at time 2τ is

$$M(2\tau) = M_0 \exp[-(2\tau/T_2) - (2/3)\gamma_I^2 G^2 D\tau^3]. \tag{9.28}$$

By fitting data for $M(2\tau)$ as a function of G and τ, the value of D is found. This is a well-established technique and has been exploited in insulating liquids and gases.

In every direct measurement of D there must be some label for the diffusing atoms such as its radioactivity or its mass. In the NMR experiment the spin of the nucleus is its label. As spins move and change z coordinates in the sample, their Larmor precession frequencies change as $\omega(z) = \gamma_I[H_0 + z(\partial H_z/\partial z)]$. Due to the randomness in diffusive motion, disorder develops in the phase angles of the spins. Predictable echo attenuation, according to (9.28), is the effect of this disorder. Typical values of 2τ in an experiment are $2\tau \approx 10$–100 ms.

In solids with smaller values of D, G must be increased. For a given size sample, increasing G broadens the frequency bandwidth since $\Delta\omega \approx \gamma_I GR$, where R is the sample radius. To observe the broader resonance, the bandwidth of the electronics must be increased, more noise appears on the signal, and signal-to-noise ratio is reduced.

Instrumentation bandwidth does not need to be increased if the gradient is turned off during the appearance of the echo. Such is the case in the pulsed gradient experiment. In the basic form of this experiment, G is applied for a time δ at constant amplitude between the 90° and 180° pulses and between the 180° pulse and formation of the echo. To a good approximation, the usual magnet inhomogeneity can be represented as a constant gradient G_0. Recognizing its presence, the attenuation of the echo for the pulsed gradient is [9.47]

$$\begin{aligned} M(2\tau) = M_0' \exp - \gamma_I^2 D \{ (2/3)\tau^3 G_0^2 + \delta^2 (3\Delta - \delta) G^2/3 \\ - \delta [(t_1^2 + t_2^2) + \delta(t_1 + t_2) + (2\delta^2)/3 - 2\tau^2] \boldsymbol{G} \cdot \boldsymbol{G}_0 \}, \end{aligned} \tag{9.29}$$

where Δ is the time between gradient pulses, $t_2 = 2\tau - (t_1 + \Delta + \delta)$, and M_0' includes the T_2 term.

Assume $t_1 = 0$ and $\Delta = \tau$ and that the experiment is done at fixed τ. The G_0^2 term in the exponent can be combined with M_0' as $M_0'' = M_0' \exp - [\gamma^2 D \tau^3 G_0^2 (2/3)]$. Then the echo attenuation is

$$M(2\tau)/M_0'' = \exp - \gamma_I^2 D \{ \delta^2 [\tau - (\delta)/3] G^2 + \delta [\tau^2 + \delta(3\tau - 2\delta)/3] \boldsymbol{G} \cdot \boldsymbol{G}_0 \}. \tag{9.30}$$

In many experiments the G^2 term is made much larger than the $(\boldsymbol{G} \cdot \boldsymbol{G}_0)$ term. The $(\boldsymbol{G} \cdot \boldsymbol{G}_0)$ term can be neglected if, to good approximation, the inequality

$$\frac{|\boldsymbol{G}|}{|\boldsymbol{G}_0|} \gg \frac{\tau}{\delta}$$

is met. Values of G up to $200\,\mathrm{G\,cm^{-1}}$ can be generated without too much difficulty and thus the inequality can usually be satisfied.

A special problem is encountered in application of (9.29) in M–H systems. In powdered or granular samples the sample magnetization can be quite inhomogeneous. Internal demagnetization fields shown by *Drain* [9.48] to be the order of $3\chi_0 H_0$ change by a large fraction over distances the order of R, the radius of the particle size. These demagnetization fields produce a distribution of "background gradients", G_0, that can exceed $20\,\mathrm{G\,cm^{-1}}$. Under these circumstances the inequality above might be barely satisfied. Correction for the presence of the $(\boldsymbol{G} \cdot \boldsymbol{G}_0)$ term cannot easily be made since there is a distribution of values of G_0 across the sample and the form of G_0 is not predictable except possibly for the impractical assumption that particles are smooth spheres.

To minimize the effect of G_0, H_0 is reduced to as low a value as practical (since signal-to-noise ratio is proportional to H_0^2), and R is increased to as large a value as allowed by the rf skin depth for good rf field penetration of the particles. Usually the mean value of R is chosen to approximately equal the skin depth. Typical values of R equal 100 μ.

An estimate of the mean squared value G_0^2 can be made from spin echo experiments in the absence of the pulsed gradient. *Murday* and *Cotts* [9.49] have described a series expansion of the $(\boldsymbol{G}\cdot\boldsymbol{G}_0)$ term that allows correction for the effects of G_0^2 in the determination of D. The series expansion is valid only for $G_0 < G$. It might seem that the effect of the $(\boldsymbol{G}\cdot\boldsymbol{G}_0)$ term in (9.30) should be separable from the G^2 term because the latter is quadratic in G. The problem is that there is a distribution of values of G_0 that is probably symmetric about $G_0 = 0$. In a series expansion of the $(\boldsymbol{G}\cdot\boldsymbol{G}_0)$ term of (9.30),

$$\exp - \gamma^2 D\delta[\tau^2 + \delta(3\tau - 2\delta)/3]\,\boldsymbol{G}\cdot\boldsymbol{G}_0 \approx 1 - A(\boldsymbol{G}\cdot\boldsymbol{G}_0) + \frac{A^2(\boldsymbol{G}\cdot\boldsymbol{G}_0)^2}{2} - \ldots$$

the effect of the term linear in $(\boldsymbol{G}\cdot\boldsymbol{G}_0)$ vanishes for a symmetric distribution of G_0. It is the quadratic term that introduces the small systematic error in D.

A second possible correction is for effects of restricted diffusion if the particle size is too small. The derivation of (9.29) assumes that the average distance a nuclear spin diffuses in time τ is much less than R, that is, $(2D\tau)^{1/2} \ll R$. For large values of D, say $10^{-4}\,\mathrm{cm^2\,s^{-1}}$ and τ as small as 10 ms, $(2D\tau)^{1/2} \approx 14$ μ. *Murday* and *Cotts* [9.49] have also described corrections in the value of D due to restricted diffusion and based upon calculations of *Neuman* [9.50] and *Robertson* [9.51].

One way to avoid both correction factors is to use foil samples with the z-axis in the plane of the foils. If the foils are metallographically homogeneous, the internal fields will be quite uniform and, since only diffusion along the z-axis is effective, the restricted diffusion corrections can be negligible. Inhomogeneous samples of metal hydride foils might result from thermal cycling through phase transitions [9.52] and lead to unacceptable values of G_0.

Even with these restrictions, values of D for ^{1}H ranging from just below $10^{-7}\,\mathrm{cm^2\,s^{-1}}$ to above $10^{-4}\,\mathrm{cm^2\,s^{-1}}$ have been measured to about 10% accuracy in M–H systems. All samples used have been in the α or α' phases. Powdered samples of $VH_{0.3}$ [9.53] and NbH_x [9.54], foil samples of PdH_x [9.55] and $Pd_yAg_{1-y}H_x$ [9.38], and a single crystal of $NbH_{0.6}$ [9.56] have been used. Both temperature and concentration dependence have been observed. Data have been fit to the Arrhenius relation $D = D_0 \exp(-E_a/kT)$, and values of D_0 and E_a have been determined using NMR.

Multiple Pulse Technique

The background gradient problem can become a serious limitation in the application of NMR to measure D of hydrogen in metals. By use of multiple pulse techniques the problem can be overcome. The application of a pattern of many rf (usually 180°) pulses during the time between application of the two magnetic gradient pulses can eliminate the effect of the constant background gradients.

In the definitive *Carr* and *Purcell* [9.57] paper it is shown that a train of 180° rf pulses controls the amount by which a particular echo is attenuated by diffusion in a uniform, constant gradient. A train of 180° pulses is applied a time 2τ apart, with the first pulse a time τ following the 90° pulse. Spin echoes occur at times midway between the 180° pulses, with one final echo occurring a time τ after the last rf pulse. When done in a constant field gradient G_0, the peak amplitude of the echoes is [9.57],

$$M(t_n)=M_0\exp-\left(\frac{t_n}{T_2}+\frac{D\gamma_I^2G_0^2t_n^3}{12n^2}\right), \tag{9.31}$$

where t_n is the time of the nth echo. By increasing the number of pulses for a given echo time (equivalent to decreasing τ), echo attenuation due to G_0 is reduced.

Williams [9.54] has developed a modified *Carr-Purcell-Meiboom-Gill* (CPMG) [9.58] pulse technique to eliminate most of the error due to the $(\boldsymbol{G}\cdot\boldsymbol{G}_0)$ term in measuring D. In this technique two gradient pulses G are applied for time δ and separated by time Δ in the CPMG pulse train. The nth echo is observed where $n=(4L+1)$ and L is an integer which is not changed in the experiment. *Williams* shows that the echo amplitude is

$$\begin{aligned}M(t_n)=M_0\exp(-t_n/T_2)\exp-\gamma_I^2D\Bigg\{&\frac{G_0^2t_n^3}{12n^2}+\delta^2G^2\left(\Delta-\frac{\delta}{3}\right)\\&-(\boldsymbol{G}\cdot\boldsymbol{G}_0)\,\delta\left[(t_1^2+t_3^2)+\delta(t_1+t_3)+\frac{2\delta^2}{3}-2\tau^2\right]\Bigg\},\end{aligned} \tag{9.32}$$

where $t_3=(\tau-\delta-t_1)$ and 2τ is the 180° pulse spacing. The first gradient pulse is applied starting at time t_1 after the 90° pulse. The second pulse is applied starting at time t_1 after the $(4N+1)$th 180° pulse where $N\leqq L$. Thus, Δ can take on values τ, 9τ, 17τ,... as shown in Fig. 9.9. In Fig. 9.9, $\Delta=9\tau$, and the echo at $t=18\tau$ (normally the ninth echo) is observed. If G, t_n, n, t_1, δ, and τ are held constant, then

$$M(t_n)=M_0'''(t_n,T_2,G_0,n,\delta,\tau,t_1)\exp\left[-\gamma_I^2DG^2\delta^2\left(\Delta-\frac{\delta}{3}\right)\right]. \tag{9.33}$$

From the dependence of $M(t_n)$ upon Δ in (9.33) the value of D can be determined independent of G_0. The technique assumes that G_0 is constant in time and uniform over the region covered by a diffusing spin during the total duration of the experiment which is $2n\tau$. With largest possible values of particle radius R, G_0 will be as small as possible.

Williams [9.54] has tested this technique in low and high G_0 samples in the NbH_x system. The technique works well but has some limitations. The echo observed is always the $(4L+1)$th echo and it can be significantly attenuated by

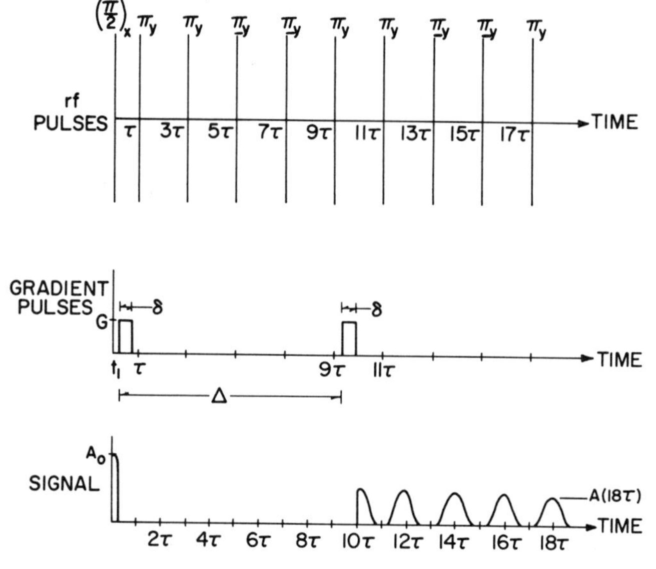

Fig. 9.9. Time sequence of rf pulses and gradient pulses in the multiple pulse spin echo experiment designed to eliminate effects of background gradients in the determination of the diffusion coefficient. The echo amplitude at $t = 18\tau$ is observed as a function of the time of the second gradient pulse which can occur after the first, fifth, or ninth $\pi(180°)$ rf pulse. The subscripts x, y, $-y$ give the direction of the rf field in the coordinate system rotating about H_0 (z-axis) at the Larmor frequency [9.54]

G_0 in very magnetically inhomogeneous samples. The technique sacrifices in some samples a considerable fraction of the signal to gain further immunity from G_0. In samples with G_0, a very short value of τ is needed to maintain signal amplitude, but τ cannot be less than the grad pulse width δ.

The reason that the above spacing of gradient pulses is changed by every four 180° pulses rather than every two 180° pulses is that the phase of the rf during the 180° pulses must be shifted by π radius every second pair of pulses to compensate for inhomogeneities in the rf field.

The accuracy of the NMR technique for measurement of D depends upon the signal-to-noise ratio from the sample and upon the degree to which errors from the background gradient can be eliminated. If the multiple pulse technique is applicable, the errors due to G_0 can be eliminated. Of course, signal-to-noise ratio improvement, if necessary, requires long averaging times in data taking.

Because of the signal-to-noise limitations it is unlikely that any NMR pulsed gradient method will be practical for samples having less than about 1% hydrogen. The minimum hydrogen concentration depends upon the temperature range over which D is to be measured. Results on PdH_x [9.55] and NbH_x [9.54, 56] are in good agreement with results from Gorsky effect, quasielastic neutron scattering, and electrochemical methods [9.59].

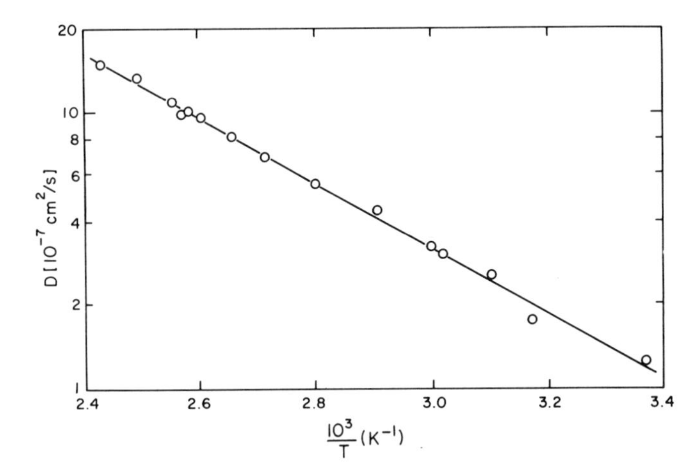

Fig. 9.10. Temperature dependence of the diffusion coefficient of ^1H in $PdH_{0.7}$ measured with the NMR two-pulse spin echo pulsed gradient technique. The straight line is for $D = 9.0 \times 10^{-4} \exp(-0.228 \pm 0.006\,\mathrm{eV}/k_B T)\,\mathrm{cm}^2\,\mathrm{s}^{-1}$ [9.55]

Temperature dependence of D for $PdH_{0.70}$ measured by the two-pulse spin echo pulsed gradient technique is shown in Fig. 9.10. The values of D and the activation energy for the data of Fig. 9.10 are in good agreement with data from other methods for measuring D. Combination of D and T_1 data in [9.55] and [9.37] confirms octahedral site occupancy for the hydrogen [9.38]. Figure 9.11 shows a compilation of data for PdH_x including NMR data [9.38]. Results are seen to be consistent with a blocking factor $(1-x)$ in that when the data for $D/(1-x)$ are plotted against temperature, as in Fig. 9.11, the results fall on a single line. In Fig. 9.11, NMR results for direct measurement of D and values of D calculated from T_1 data assuming octahedral site occupancy are compared with results from permeation studies and quasielastic neutron scattering showing good, general consistency. The similar effect of a blocking factor, $(2-x)$, as the dihydride is approached in TiH_x was observed by *Korn* and *Zamir* [9.64] and by *Schmolz* and *Noack* [9.35].

A compilation of values of E_a of the NbH_x system including multiple pulse NMR results [9.54] is shown in Fig. 9.12. In Fig. 9.12, values of E_a for the α and α' phases are nearly independent of hydrogen concentration of $H/Nb \gtrsim 0.4$.

Application to ^2H

The pulsed gradient spin echo technique can also be used to study diffusion of deuterons in M–H systems. At a given frequency of operation of the NMR spectrometer, the ratio of deuteron signal to proton signal amplitude is $(S_2/S_1) = 0.409$ for equal numbers of isotopes. This ratio assumes that any quadrupolar splitting of the deuteron resonance is completely averaged by motional narrowing.

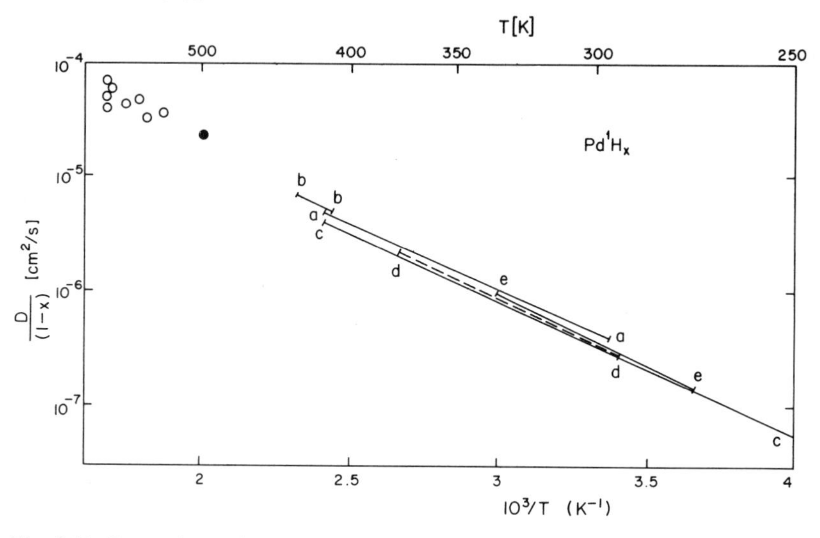

Fig. 9.11. Comparison of measured diffusion coefficients, D, for $\alpha' - PdH_x$, corrected for site-blocking factor $(1-x)$. *aa*, NMR spin echo $x=0.7$ [9.37]; *bb*, NMR spin echo, $x=0.54$ [9.38]; *cc*, NMR T_1, $x=0.69$ [9.38]; *dd*, permeation, $\bar{x}=0.74$ [9.60]; *ee*, permeation $\bar{x}=0.69$ [9.61]; ○, neutron scattering $x=0.2$, 0.4, 0.48 [9.62]; ●, neutron scattering, $x=0.55$ [9.63] (from [9.38])

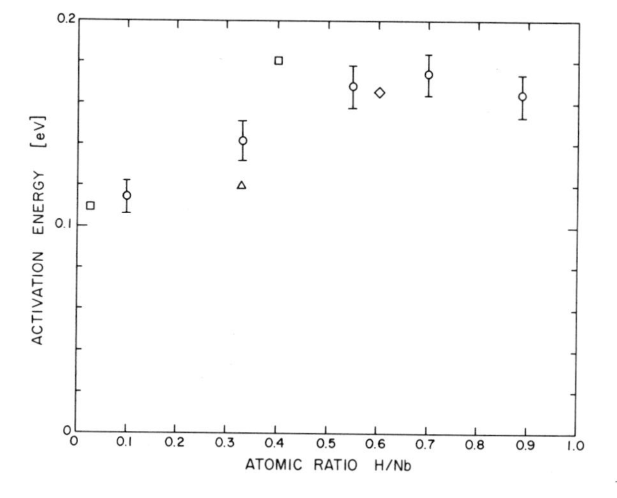

Fig. 9.12. Compilation of values of the activation energy for diffusion of ^1H in $\alpha' - NbH_x$. The methods of measurement are: ○-multiple pulse NMR spin echo pulsed field gradient [9.54]; □-Gorsky effect [9.59]; △-quasielastic neutron scattering [9.59]; ◇-NMR two pulse spin echo pulsed gradient [9.56]

Because of the lower value of the gyromagnetic ratio for ^2H, $(\gamma_2/\gamma_1)=0.153$, the value of G must be increased over the value used for ^1H to measurably attenuate the echo. Only a longer value of T_2 for ^2H that would allow larger values of δ and Δ could reduce the values of G needed for ^2H. As noted above in "Thermal Motion and Quadrupole Splittings", some initial surveys of ^2H NMR in the α and α' phases of bcc transition metal hydrides indicate that T_2 is surprisingly fast. It could be expected that for T_2 the order of 10 ms or longer, good measurements of D should be obtainable for ^2H in M–H systems, especially through use of the stimulated echo [9.47]. The only known diffusion isotope effect observed by NMR in M–H systems is in the Ti^3H_x system [9.31] and is discussed briefly in Section 9.4.

Identification of the Step Length

Interpretation of the temperature and Larmor frequency dependence of T_1 and $T_{1\varrho}$ of ^1H can give reliable values of τ_D. The model for site occupancy must be known or assumed, and the absolute magnitudes of T_1 or $T_{1\varrho}$ at the minimum in the τ_D dependence of either must be consistent with the model to verify its validity. But the step length for diffusion does not come from the T_1 data in a well-defined way.

For example second moment data and T_1 data for ^1H in γ phase TiH_x demonstrate that hydrogen atoms are on tetrahedral sites of the Ti fcc lattice. The mean residence time for H atoms on these sites has been determined, [9.31, 64]. The question is, do the H atoms move on their sc lattice by jumps to nearest neighbor tetrahedral (T–nnT) sites or by jumps across the vacant octahedral site to their third nearest neighbor site (T–3nnT)? The ratio of step lengths is $\sqrt{3}$.

The BPP model contains no information about step length and so cannot answer such a question.

In the Torrey theory the spectral functions $J(\omega)$ of (9.14) are only weakly dependent upon the step length l. For example, *Seymour* [9.65] has estimated from Torrey theory that the minimum values of T_1 differ by only 15% between (T-nnT) and (T-3nnT) steps. The two T_1 values from the Sholl theory for the minimum in T_1 differ by considerably less [9.65].

It is possible to determine the value of l with independent knowledge of the value of D, the self-diffusion coefficient, through the relationship,

$$D=\frac{1}{6}\frac{\langle l^2\rangle}{\tau_D}, \tag{9.34}$$

where $\langle l^2\rangle$ is the mean squared step length. By applying the Sholl results for the sc lattice to published T_1 data to obtain τ_D, *Seymour* [9.65] has calculated the values of the diffusion coefficients for the two models. He finds that is, for example at $T=777$ K, $D_{\text{T-3nnT}}=(4.75\pm0.22)\times10^{-7}\ \text{cm}^2\,\text{s}^{-1}$ and $D_{\text{T-nnT}}=(2.20\pm0.10)\times10^{-7}\ \text{cm}^2\,\text{s}^{-1}$. Since D can be measured to 10–20% using NMR, the measurement of D can clearly settle the question of the step length in γ titanium hydride.

A similar combination of T_1 and D measurements was made for $NbH_{0.33}$ and $NbH_{0.9}$ in the ordered β phase by *Alefeld* et al. [9.66], except that their values of D were determined by quasielastic neutron scattering. They found $\langle l^2\rangle^{1/2}=4.8\pm0.8$ Å. In the α' phase of $NbH_{0.6}$, *Zogal* and *Cotts* [9.56], using NMR for both T_1 and D measurements and only the BPP theory, concluded that $1.7\ \text{Å}<\langle l^2\rangle^{1/2}<2.7\ \text{Å}$.

9.3 Electronic Structure and NMR

9.3.1 Host Metal Knight Shift and T_{1e}

The nuclear spin systems in metal hydrides are affected by the electronic structure of the M–H alloys through the hyperfine fields produced at the site of the nucleus. The time-average component of the hyperfine field parallel to H_0 causes a shift in the Larmor frequency known as the Knight shift [9.67]. The time dependence of the transverse components of the hyperfine fields provides a relaxation mechanism and contributes a conduction electron component $(1/T_{1e})$ to the spin lattice relaxation rate. In metals, $(1/T_1)_e \propto T$, where T is absolute temperature. The Knight shift in most metals is toward higher frequencies. Since hyperfine fields are involved, these interactions are weakest for hydrogen and strongest for heavy elements. For hydrogen the magnitude of these hyperfine fields is comparable to normal demagnetization fields associated with the bulk susceptibility of the sample. For the transition and rare earth host metal nuclei of M–H systems, the hyperfine fields are much greater than demagnetization fields. Measurements of Knight shift and T_{1e} in M–H systems reflect changes in electronic structure.

The Knight shift is defined as $K=(\nu_m-\nu_s)/\nu_s$ where ν_m and ν_s are the Larmor precession frequencies in the metal and salt (reference) for a given applied magnetic field. Theories for K and $(1/T_{1e})$ will be presented in terms of a simplified model frequently used in data analysis. *Narath* [9.68] presents a more detailed model.

The partitioned models for K and $(1/T_{1e})$ are due to *Obata* [9.69], *Yafet* and *Jaccarino* [9.70], and *Korringa* [9.71].

$$\begin{aligned} K &= K_s + K_d + K_{\mathrm{orb}} \\ &= 1.79 \times 10^{-4}[H_{\mathrm{hf}}(s)\chi_s + H_{\mathrm{hf}}(d)\chi_d + H_{\mathrm{hf}}(\mathrm{orb})\chi_{\mathrm{vv}}]\,, \end{aligned} \quad (9.35)$$

$$\begin{aligned} \frac{1}{T_{1e}T} &= 4\pi\hbar\gamma^2 k_{\mathrm{B}}\Big\{[H_{\mathrm{hf}}(s)N_s]^2 + [H_{\mathrm{hf}}(d)N_d]^2[(f^2/3)+(1-f)^2/2] \\ &\quad + \frac{26}{25}[H_{\mathrm{hf}}(\mathrm{orb})N_d]^2[2f(6-5f)/9]\Big\}\,, \end{aligned} \quad (9.36)$$

where $H_{\mathrm{hf}}(s)$, $H_{\mathrm{hf}}(d)$, and $H_{\mathrm{hf}}(\mathrm{orb})$ are s electron contact, core polarization, and orbital hyperfine fields; χ_s, χ_d, and χ_{vv} are s electron spin, d electron spin, and Van Vleck second-order paramagnetic susceptibilities, respectively; N_s and N_d are s-band and d-band electron density of states per atom, and f is the fraction of Γ_5 symmetry d orbitals at the Fermi surface. The hyperfine fields are given by $H_{\mathrm{hf}}(s)=(8/3)\pi\beta\langle\psi_s(o)^2\rangle$, $H_{\mathrm{hf}}(d)=(8/3)\pi\beta\langle\psi_{\mathrm{cp}}(o)^2\rangle$ and $H_{\mathrm{hf}}(\mathrm{orb})=2\beta\langle r^{-3}\rangle$ where β is the Bohr magneton, and the averages are at the Fermi surface for one unpaired s electron, one unpaired d electron, and one unit of orbital angular momentum, respectively [9.70].

For host metal nuclei in M–H systems, the core polarization and orbital terms dominate K_M and T_{1e}. In most transition metals $H_{hf}(d)$ is negative and partially conceals the orbital term. Only $T_{1e}T$ and K are measured in an NMR experiment. A combination of theoretical calculations, data from susceptibility, specific heat, and electron spin resonance experiments is needed to identify enough parameters in (9.35) and (9.36) to draw conclusions about the electron structure.

Observation of the host metal NMR is hindered in some M–H systems by the electric quadrupole broadening of the host NMR due to asymmetric distributions of occupied H atom sites. Nevertheless, the K_M and $(1/T_{1e})$ of host metal nuclei were among the first to be made including the LaH_x [9.8] and NbH_x [9.26] systems.

Schreiber and *Cotts* [9.8] found that for LaH_x the Knight shift of ^{139}La was strongly dependent upon hydrogen concentration for $x>2.0$. The effects of diffusing H atoms were also observed on the temperature dependence of the La NMR.

Schreiber [9.72] and *Zamir* [9.73] reviewed the data on host metal NMR, and they suggested that K_M and T_{1e} could be interpreted by utilizing the rigid band model for the *d*-band electrons in group Vb M–H systems. In this model the H atom gives its electron to the conduction band of the host affecting its electron density of states in the same way in which an alloying element of the next higher atomic number would do. Support for this viewpoint was provided by the experiments of *Rohy* and *Cotts* [9.74], who measured electron specific heats K_M and T_{1e} in the $V_{1-y}Cr_yH_x$ system and showed to good approximation that these quantities depended only upon the sum $(x+y)$ and not upon the relative amounts of Cr or H.

9.3.2 K and T_{1e} for Hydrogen

The effects of the $(1/T_{1e})$ term on relaxation of the ^1H spin system was first identified by *Korn* and *Zamir* [9.64] in γ phase TiH_x. They found $T_{1e}T$ ranged from 190 to 62 s K as x varied from 1.55 to 1.90. *Schoep* et al. [9.75] found that the value of $T_{1e}T=70\pm3$ s K applied to PdH_x for $x=0.73$, 0.76, and 0.80, and in $PdH_{0.70}$, *Cornell* and *Seymour* [9.37] measured $T_{1e}T=75\pm3$ s K.

If the temperature dependence $(1/T_{1e})\propto T$ were not enough evidence for the conduction electron contribution to the ^1H relaxation rate, *Lütgemeier* et al. [9.40] added further proof in their observation that $(T_{1e}T)_H/(T_{1e}T)_{Nb}=400$ for all hydrogen concentrations in the α' phase where $(T_{1e}T)_{Nb}$ changes by a factor of five with changing hydrogen concentration. Since $(T_{1e}T)_{Nb}$ is principally determined by the *d*-band density of states, then $(T_{1e}T)_H$ must be also. Knowledge of T_{1e} is important also in hydrogen diffusion studies using NMR since the relaxation rate can contain a large contribution from $(1/T_{1e})$ at temperatures well above and well below the temperature of the T_1 minimum.

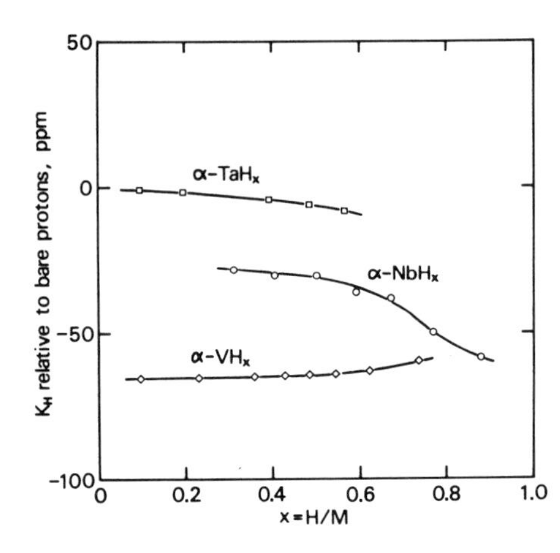

Fig. 9.13. Knight shifts for ^{1}H relative to the bare proton in $\alpha-TaH_x$, $\alpha-NbH_x$, and $\alpha-VH_x$. Measurements are made in a single foil of each sample. Corrections for demagnetizing field effects are determined with the single foil sample geometry [9.76]

Because of the small hyperfine fields in hydrogen it can be expected that K will be small and possibly masked by demagnetization fields. *Norberg* [9.3] and *Schreiber* and *Graham* [9.76] have described this problem and suggested ways in which the shift due to K can be separated from the shift due to the demagnetizing fields.

Recently *Kazama* and *Fukai* [9.77, 78] have used a high resolution NMR spectrometer to observe the NMR of ^{1}H in α' phase of VH_x, NbH_x, and TaH_x using a single sheet of foil as a specimen. By measuring the shift of the resonance as a function of orientation of the foil in the magnetic field, they determined values of K_H and χ_v, the volume magnetic susceptibility, in a single experiment. They also measured $(T_{1e}T)_H$ for VH_x. In the single foil the effects of the demagnetizing fields are predictable. K_H for all samples was negative relative to bare protons, and concentration dependences are shown in Fig. 9.13. The dependence of K_H upon H–M in Fig. 9.13 is not identical to the dependence of the susceptibility upon H–M, indicating that the rigid band model is not applicable for these alloys. To account for the negative values of K_H, *Kazama* and *Fukai* suggest that K_H is principally determined by direct contact interaction with the spin density of d-electrons of the host metal. It is known from experiments using polarized neutrons that the spin density distribution in transition metals is peaked at metal sites and can be slightly negative over the interstitial regions.

They also consider the possibility that K_H is determined by bonding states formed from hydrogen 1s and host metal 3d orbitals, but a lack of detailed knowledge of the nature of these H bonding states inhibits further analysis at this time.

9.3.3 Interpretations

There is no question that K and T_{1e} in M–H systems are determined by conduction electrons, but uncertainties exist in the detailed theory of the relationship between K and T_{1e} and the electronic structure of the system.

The demonstration of qualitative success of the rigid band model in explaining H concentration dependence of K_M and T_{1e} in group Vb metals has attracted experimenters, including this author, to use this model with (9.35) and (9.36) to analyze K and (T_{1e}) data for host nuclei. However, a series of band structure calculations for M–H systems by *Switendick* [9.79] has questioned the application of the rigid band model here. His APW calculations show a bonding state splitting off from the d-band in stoichiometric metal hydrides. Evidence for the existence of this bonding state appears in recent photoemission spectroscopy experiments on PdH_x [9.80] and soft x-ray emission spectroscopy in NbH_x [9.81] and VH_x [9.82]. Addition of hydrogen does affect the density of states at the Fermi surface but perhaps not by filling available states in a rigid band.

Thus a gap exists between the data and its interpretation in terms of changes in electronic structure. The problem is not limited to NMR data. It is to be hoped that theorists will be attracted to this problem and that through mutual interaction between experimentalists and theorists the gradual understanding of electronic structure in metal-hydrogen systems will evolve.

9.4 Summary

The ways in which parameters of cw and pulsed NMR depend upon structure, H atom diffusion, and changes in electronic structure in metal-hydrogen systems have been described. NMR is a microscopic probe sensitive to static and dynamic changes in the lattice over distances of a few interatomic spacings. Observations of the resonance second moment have been used to confirm or reject proposed structures of hydrogen site occupancy. Small structure changes can be monitored through the electric quadrupolar splitting of the deuteron NMR, in some systems, the host metal NMR. There is interest in identifying the strength and asymmetry parameter of the quadrupole interaction to compare these values with what would be expected for specific site occupancy models. In three systems cited, the group Vb metal-hydrogen systems, just the observation of the existence of a temperature independent quadrupolar splitting in the β phase proved to quite informative itself. Knowing that diffusion was fast enough (from NMR T_1 measurements) to average out the quadrupolar splittings, but did not, is enough to say that the H atom motion must have been confined to a subset of interstitial sites, consistent with an ordering of H atoms in the β phase. The theory of relaxation rates of ^1H due to diffusive motion has reached a high level of development. By measure of $(T_1)^{-1}$, for example, and through comparison between measured and predicted values of $(T_1)^{-1}$ in polycrystalline samples, models for H atom site occupancy can be confirmed or rejected with about the same degree of certainty as through measurement of the second moment. From $(T_1)^{-1}$ or from $(T_{1\varrho})^{-1}$, values of τ_D and its temperature dependence can be found. Values of E_a can then be established with accuracy limited only by experimental uncertainty. While $(T_1)^{-1}$ theory for polycrystalline samples is sensitive to H atom site occupancy, the values of $(T_1)^{-1}$ are not very sensitive to the step length or the mechanism for diffusion.

To establish the value of the step length, independent measure of the self-diffusion coefficient is necessary. D can be measured using one of a number of varied spin echo pulsed gradient methods. For systems with a fast T_2 and a slow T_1 the stimulated echo method is effective [9.47]. For systems with large "background" gradients, a multiple pulse technique can produce a series of echoes whose change in amplitude upon changing gradient pulse timing is independent of background gradients [9.54]. Values of D ranging upward from 10^{-8} cm^2 s^{-1} can be measured with these techniques. If D exceeds 10^{-4} cm^2 s^{-1}, there can be excessive errors due to restricted diffusion effects [9.50] in small particle samples.

Measurement of D for deuterons will probably be limited to those phases of M–H systems in which the quadrupole splitting has been averaged to essentially zero. But even in these systems, special care in sample preparation and in hydriding will be necessary to avoid introduction of microstructure which can produce inhomogeneous field gradients caused by internal strains. With the combination of diffusion and strain fields, usual quadrupolar echo techniques used in NMR [9.12] will probably not be effective.

All of the NMR experiments applicable to hydrogen can in principle be applied to tritium which also has spin 1/2 and therefore has zero quadrupole moment. In comparing values of τ_D obtained from T_1 data for $Ti^1H_{1.55}$ and $Ti^3H_{1.5}$, *Weaver* and *Van Dyke* [9.31] found that tritium and hydrogen had essentially equal activation energies but the pre-exponential factor D_0 of hydrogen exceeds that of tritium by a factor of 1.3 instead of the classical factor $\sqrt{3}$. Because of the complications due to quadrupole interactions with deuterons and the radioactivity of tritons, there will be limited observation of diffusion isotope effects using NMR.

Diffusion constants and activation energies measured by NMR are generally consistent with values obtained by other means. The NMR technique for measuring D cannot be applied to systems that are so strongly paramagnetic or which break up into such a fine powder that background gradients become so excessive that echo attenuation by the background gradient limits T_2 to less than the order of 1 ms.

The experimenter can be tantalized by the prospect of contributing to the understanding of electronic structure in M–H systems. With the single foil experiment [9.77], the very small Knight shift of ^{1}H can be separated from effects of demagnetizing fields. From measurements of $T_{1e}T$ for ^{1}H, the Korringa relation, which applies only to s contact hyperfine interaction and to effectively free electron metals, can be tested. This relation is, in a generalized form

$$K^2 T_{1e} T = \frac{\hbar}{4\pi k_B} \left(\frac{\gamma_e}{\gamma_I}\right)^2 f\,,$$

where γ_e is the electron gyromagnetic ratio and f is an enhancement factor which corrects for electron-electron correlation effects. In alkali metals $f=1.5$–1.8. For ^1H in $VH_{0.042}$, *Kazama* and *Fukai* [9.77] find $f=3$. In view of the limited available theoretical predictions for the interstitial hydrogen hyperfine interaction, they postpone detailed discussion of the meaning of this particular value of f. These measurements of K_H and $T_{1e}T$ should prove useful as one test of wave function calculations for the interstitial H atom, when they appear.

With cw NMR, the measurement of K_H in a foil sample is limited to motionally narrowed ^1H resonances, i.e., to relatively high temperatures. By use of multiple pulse techniques designed to average dipolar fields to zero [9.83], values of K_H could be measured at low temperatures as well. For the interested reader, a new textbook provides an excellent development of the theory and physical understanding of spin processes in magnetic resonance [9.84].

Acknowledgements. Extensive and instructive discussions with Professor E. F. W. Seymour have been much appreciated. Undoubtedly this review would have benefited considerably from his comments if limitations of time and distance had not prevented their solicitation.

Mr. L. D. Bustard made T_2 measurements on ^2H ("Motional Narrowing of Dipolar Interactions") and assisted in analysis of effects of background gradients on D measurements.

References

9.1 F. Bloch, W. W. Hansen, M. Packard: Phys. Rev. **70**, 474 (1946)
9.2 E. M. Purcell, H. C. Torrey, R. V. Pound: Phys. Rev. **69**, 37 (1946)
9.3 R. E. Norberg: Phys. Rev. **86**, 745 (1952)
9.4 R. M. Cotts: Ber. Bunsenges. Physik. Chem. **76**, 760 (1972)
9.5 B. Pedersen: Tidsskr. Kem. Berg. Met. **3**, 63 (1965)
9.6 J. H. Van Vleck: Phys. Rev. **74**, 1168 (1948)
9.7 B. Stalinski, C. K. Coogan, H. S. Gutowsky: J. Chem. Phys. **34**, 1191 (1961)
9.8 D. S. Schreiber, R. M. Cotts: Phys. Rev. **131**, 1118 (1963)
9.9 M. Engelsberg, R. E. Norberg: Phys. Rev. B **5**, 3395 (1972)
9.10 P. Mansfield: Phys. Rev. **137**, A961 (1965)
9.11 R. C. Bowman: Congress Ampere, 19th, Heidelberg, W. Germany, 1976 (to be published)
9.12 A. Abragam: *The Principles of Nuclear Magnetism* (Oxford University Press, London 1961)
9.13 M. Goldman: *Spin Temperature and Nuclear Magnetic Resonance in Solids* (Oxford University Press, London 1970) pp. 110—112
9.14 G. Alefeld, J. Völkl (eds.): *Hydrogen in Metals II, Application-Oriented Properties*, Topics in Applied Physics, Vol. 29 (Springer, Berlin, Heidelberg, New York 1978) to be published
9.15 B. Pedersen, T. Krogdahl, O. E. Stokkeland: J. Chem. Phys. **42**, 72 (1965)
9.16 M. H. Cohen, F. Reif: *Solid State Physics*, Vol. 5, ed. by F. Seitz, D. Turnbull (Academic Press, New York 1959) p. 311
9.17 R. G. Barnes, J. W. Bloom: J. Chem. Phys. **57**, 3082 (1972)
9.18 H. Lütgemeier, H. G. Bohn, R. R. Arons: J. Magn. Resonance **8**, 80 (1972)
9.19 R. R. Arons, H. G. Bohn, H. Lütgemeier: J. Phys. Chem. Sol. **35**, 207 (1974)
9.20 R. G. Barnes, K. P. Roenker, H. R. Brooker: Ber. Bunsenges. Physik. Chem. **80**, 875 (1976)
9.21 K. P. Roenker, R. G. Barnes, H. R. Brooker: Ber. Bunsenges. Physik. Chem. **80**, 470 (1976)
9.22 D. G. Westlake, T. H. Wang, R. K. Sundfors, P. A. Fedders, D. I. Bolef: Phys. Stat. Sol. (a) **25**, K 35 (1974)
B. Ströbel, K. Läuger, H. E. Bömmel: Appl. Phys. **9**, 39 (1976)
9.23 R. Kubo, K. Tomita: J. Phys. Soc. Japan **9**, 888 (1954)
9.24 W. Spalthoff: Z. Physik. Chem. **29**, 258 (1961)

9.25 B.Stalinski, O.J.Zogal: Bull. Acad. Polon. Sci. **13**, 397 (1965)
9.26 D.Zamir, R.M.Cotts: Phys. Rev. **134**, A 666 (1964)
9.27 D.C.Look, I.J.Lowe: J. Chem. Phys. **44**, 2995 (1966)
9.28 N.Bloembergen, E.M.Purcell, R.V.Pound: Phys. Rev. **73**, 679 (1948)
9.29 H.C.Torrey: Phys. Rev. **92**, 962 (1953); **96**, 690 (1954)
9.30 H.A.Resing, H.C.Torrey: Phys. Rev. **131**, 1102 (1963)
9.31 H.T.Weaver, J.P.Van Dyke: Phys. Rev. B **6**, 694 (1972)
9.32 C.A.Sholl: J. Phys. C. Solid State Phys. **7**, 3378 (1974)
9.33 D.Wolf: J. Magn. Resonance **17**, 1 (1975)
9.34 C.A.Sholl: J. Phys. C.: Solid State Phys. **8**, 1737 (1976)
C.A.Sholl, W.A.Barton: J. Phys. C.: Solid State Phys. **9**, 4315 (1976)
9.35 A.Schmolz, F.Noack: Ber. Bunsenges. Physik. Chem. **78**, 339 (1974)
9.36 E.F.Khodosov, N.A.Shepilov: Phys. Stat. Sol. (b) **47**, 693 (1971)
9.37 D.A.Cornell, E.F.W.Seymour: J. Less-Common Metals **39**, 43 (1975)
9.38 P.P.Davis, E.F.W.Seymour, D.Zamir, W.D.Williams, R.M.Cotts: J. Less-Common Metals **49**, 159 (1976)
9.39 R.R.Arons, H.G.Bohn, H.Lütgemeier: Solid State Commun. **14**, 1203 (1974)
9.40 H.Lütgemeier, R.R.Arons, H.G.Bohn: J. Magn. Resonance **8**, 74 (1972)
9.41 S.Figueroa, J.H.Strange, D.Wolf: To be published
9.42 L.Van Hove: Phys. Rev. **95**, 249 (1954)
9.43 C.T.Chudley, R.J.Elliot: Proc. Phys. Soc. **77**, 353 (1960)
9.44 G.H.Vineyard: Phys. Rev. **110**, 999 (1958)
9.45 W.Gissler, N.Stump: Physica **65**, 109 (1973)
9.46 W.Sköld, G.Nelin: J. Phys. Chem. Sol. **28**, 2369 (1967)
9.47 E.O.Stejskal, J.E.Tanner: J. Chem. Phys. **42**, 288 (1965)
J.E.Tanner: J. Chem. Phys. **52**, 2523 (1970)
9.48 L.E.Drain: Proc. Phys. Soc. (London) **80**, 1380 (1962)
9.49 J.S.Murday, R.M.Cotts: Z. Naturforsch. **26**a, 85 (1971)
9.50 C.H.Neuman: J. Chem. Phys. **60**, 4508 (1974)
9.51 B.Robertson: Phys. Rev. **151**, 273 (1966)
9.52 H.Zabel, G.Alefeld, H.Peisl: Scripta Met. **9**, 1345 (1975)
9.53 G.J.Krüger, R.Weiss: *Magnetic Resonance and Related Phenomena Proceedings of the 18th Congress Ampere Nottingham*, Vol. 2, ed. by P.S.Allen, E.R.Andrew, C.A.Bates (North-Holland, Amsterdam 1974) pp. 339—340
9.54 W.D.Williams: Ph.D. Thesis (1976), Cornell University, Ithaca, NY (unpublished)
9.55 E.F.W.Seymour, R.M.Cotts, W.D.Williams: Phys. Rev. Lett. **35**, 165 (1975)
9.56 O.J.Zogal, R.M.Cotts: Phys. Rev. B **11**, 2443 (1975)
9.57 H.Y.Carr, E.M.Purcell: Phys. Rev. **94**, 630 (1954)
9.58 S.Meiboom, D.Gill: Rev. Sci. Instr. **29**, 688 (1958)
9.59 J.Völkl, G.Alefeld: *Diffusion in Solids: Recent Developments*, ed. by A.S.Nowick, J.J.Burton (Academic Press, New York 1975) pp. 231—302
9.60 G.Bohmholdt, E.Wicke: Z. Phys. Chem. (N.F.) **56**, 133 (1967)
9.61 G.Holleck, E.Wicke: Z. Phys. Chem. (N.F.) **56**, 155 (1967)
9.62 G.Nelin, K.Sköld: J. Phys. Chem. Sol. **36**, 1175 (1975)
9.63 D.K.Ross: Private communication
9.64 C.Korn, D.Zamir: J. Phys. Chem. Sol. **31**, 489 (1970)
9.65 E.F.W.Seymour: Private communications
9.66 B.Alefeld, H.G.Bohn, N.Stump: Ber. Bunsenges. Physik. Chem. **76**, 781 (1972)
9.67 W.D.Knight: Phys. Rev. **76**, 1259 (1949)
9.68 A.Narath: *Hyperfine Interactions*, ed. by A.J.Freeman, R.B.Frankel (Academic Press, New York 1967) pp. 287—363
9.69 Y.Obata: J. Phys. Soc. Japan **18**, 1020 (1963)
9.70 Y.Yafet, V.Jaccarino: Phys. Rev. **133**, A 1630 (1964)
9.71 J.Korringa: Physica **16**, 601 (1950)
9.72 D.S.Schreiber: Phys. Rev. **137**, A 860 (1965)

9.73 D. Zamir: Phys. Rev. **140**, A 271 (1965)
9.74 D. A. Rohy, R. M. Cotts: Phys. Rev. B **1**, 2070 (1975); B **1**, 2484 (1970)
9.75 G. K. Schoep, N. J. Poulis, R. R. Arons: Physica **75**, 297 (1974)
9.76 D. S. Schreiber, L. D. Graham: J. Chem. Phys. **43**, 2573 (1965)
9.77 S. Kazama, Y. Fukai: J. Phys. Soc. Japan **42**, 119 (1977)
9.78 S. Kazama, Y. Fukai: J. Less-Common Metals **53**, 25 (1977)
9.79 A. C. Switendick: Ber. Bunsenges. Physik. Chem. **76**, 535 (1972)
9.80 D. E. Eastman, J. K. Cashion, A. C. Switendick: Phys. Rev. Lett. **27**, 35 (1971)
9.81 E. Gilberg: Phys. Stat. Sol. (b) **69**, 477 (1975)
9.82 Y. Fukai, S. Kazama, K. Tanaka, M. Matsumoto: Solid State Commun. **19**, 507 (1976)
9.83 W. K. Rhim, D. D. Elleman, R. W. Vaughan: J. Chem. Phys. **59**, 3740 (1973)
9.84 C. P. Slichter: *Principles of Magnetic Resonance*, Springer Series in Solid-State Sciences, Vol. 1, 2nd ed. (Springer, Berlin, Heidelberg, New York 1978)

10. Quasielastic Neutron Scattering Studies of Metal Hydrides*

K. Sköld

With 11 Figures

10.1 Overview

The high mobility of hydrogen in certain metals is of considerable interest both from a scientific and from a technological point of view, and this property of metal hydrides has been studied extensively for many years and by a variety of techniques [10.1]. Information about the macroscopic rate of diffusion is obtained from permeation experiments, from electrical resistivity measurements, and also from measurements of the anelastic relaxation of a sample under initial stress (Gorsky effect). Microscopic information about the diffusion process is obtained from NMR measurements and, more recently, also from quasielastic neutron scattering studies and from the motional narrowing of Mössbauer lines [10.2].

If we want to study the diffusion of an atom in a solid in all its detail, we need a probe which is sensitive to the motion of the atom over distances of a few Angstrom and on a time scale of the order of the inverse of the jump frequency of the atom. The technique of quasielastic scattering of thermal neutrons offers such a probe and is, in fact, the only method which allows the simultaneous measurement of both the time and the space development of the elementary step in the diffusion process. Information in this regard is obtained from a study of the detailed shape of the neutron scattering function at small energy transfers and over a range of wave vector transfers. The width of the peak centered at zero energy transfer (quasielastic peak) is proportional to the jump rate. A major limitation of the technique is that the energy resolution of standard neutron spectrometers ($\Delta E \approx 10^{-4}$ eV) limits the range of relaxation times that can be studied to $\tau \lesssim 10^{-11}$ s. For this reason the majority of experiments reported so far have been for systems where the diffusion rate is unusually large (PdH_x, NbH_x, VH_x, and TaH_x) and for temperature of several hundred degrees centigrade. However, recent development of high resolution backscattering spectrometers [10.3] ($\Delta E \approx 10^{-6}$ eV) has extended the accessible range of relaxation times to $\tau \lesssim 10^{-9}$ s. The backscattering technique has been used to study the diffusion of H in niobium down to 165 K [10.4] and has also been used to study the diffusion of H in niobium with nitrogen impurities [10.5, 6] which act as traps for the hydrogen and therefore slow down the effective rate of diffusion.

---

* Work supported by the U.S. Energy Research and Development Administration and by the Swedish Science Research Council.

Information about the geometry of the interstitial sites over which the diffusing atoms moves is obtained from the variation of the width of the quasielastic peak with wave vector transfer. This information is often unique, as standard diffraction methods are difficult to apply in the case of samples with low concentration of hydrogen. By measuring the variation of the intensity of the quasielastic scattering with wave vector transfer, information is obtained about the spreading out of a proton due to the thermal motion at the equilibrium site. Recent results indicate an abnormal behavior of the intensity which may imply that the density distribution of the particle along the jump path is also contributing to the intensity form factor in the case of very high jump rates.

In this chapter we shall review results obtained so far from neutron quasielastic scattering studies of hydrides. Our objective is not only to survey the results obtained to date but also to indicate areas where more work is needed and where major achievements can be expected in the future. The material which is presented in more detail was selected with this in mind, and the review is not necessarily comprehensive in all areas.

10.2 Theoretical Background

The theory of the diffusion of hydrogen in metals is treated by *Kehr* in Chapter 8; in the present chapter we shall discuss only those aspects of the theory which are of immediate interest for the analysis of the neutron scattering results. In developing the appropriate theory we shall assume that the motion of the proton can be treated in the classical limit.

The correlation function of interest in the present case is the Van Hove self-correlation function $G_s(\boldsymbol{r},t)$ which, in a classical interpretation, measures the probability that a particle which is at the origin at $t=0$ has moved to position $\boldsymbol{r}$ during the time t. $G_s(\boldsymbol{r},t)$ is related to the neutron incoherent scattering function $S_{\text{inc}}(\boldsymbol{q},\omega)$ and the intermediate scattering function $I_s(\boldsymbol{q},t)$ via Fourier transforms [10.7]

$$\begin{aligned}S_{\text{inc}}(\boldsymbol{q},\omega) &= \frac{1}{2\pi}\int\int G_s(\boldsymbol{r},t)\exp[\mathrm{i}(\boldsymbol{q}\boldsymbol{r}-\omega t)]\,d\boldsymbol{r}\,dt\\ &= \frac{1}{2\pi}\int I_s(\boldsymbol{q},t)\exp(-\mathrm{i}\omega t)\,dt\,,\end{aligned} \tag{10.1}$$

where $\hbar\omega$ and $\hbar\boldsymbol{q}$ are the energy and the momentum transfered in the scattering process.

Apart from trivial factors, $S_{\text{inc}}(\boldsymbol{q},\omega)$ is observed directly in the neutron scattering experiment. The strength of this particular component to the scattering is determined by the incoherent scattering cross section of the diffusing particle, and it is of particular interest in the present context that the incoherent scattering cross section of the proton is 1 to 2 orders of magnitude larger than the scattering cross section of most other nuclei. Also, in most cases the elastic scattering from the metal atoms is predominantly coherent and therefore concentrated to the reciprocal lattice points; it is therefore possible to isolate the quasielastic scattering from the protons even at hydrogen concentrations as low as a few tenths of an atomic percent in favorable cases.

As seen from (10.1), a complete mapping of $S_{\mathrm{inc}}(\boldsymbol{q},\omega)$ would in principle allow a direct determination of $G_{\mathrm{s}}(\boldsymbol{r},t)$ by double Fourier inversion of the experimental scattering function. Such a procedure has not been attempted to date, however, and recourse has instead been taken to the reverse procedure, namely, to calculate the scattering function from a theoretical model for $G_{\mathrm{s}}(\boldsymbol{r},t)$ and compare the result to the experimental scattering function. In cases where the theoretical model adequately describes the data, the parameters of the model can be obtained by fitting to the experimental curves.

A particularly simple model for the process of interstitial diffusion in a lattice is the jump-diffusion model of *Chudley* and *Elliott* [10.8]. Despite its simplicity, this model brings out many of the important features of the problem and, furthermore, clearly shows how these features are related to the neutron scattering function.

In the Chudley-Elliott model, $G_{\mathrm{s}}(\boldsymbol{r},t)$ is calculated with the following assumptions:

1) The diffusing particle moves from one site to a nearest neighbor site by random instantaneous jumps.

2) The sites over which the particle can jump form a simple Bravais lattice, i.e., the distribution of nearest neighbor sites is the same for all sites.

3) The diffusive jumps are not correlated to the thermal vibration of the particle at the equilibrium site.

4) All sites are equally available to the diffusing particle, e.g., interaction between diffusing particles and correlation effects are not important.

With these assumptions, the motion of the particle over the interstitial lattice can be calculated from a simple rate equation

$$\frac{\partial P(\boldsymbol{r},t)}{\partial t}=\frac{1}{n\tau}\sum_{\boldsymbol{l}}[P(\boldsymbol{r}+\boldsymbol{l},t)-P(\boldsymbol{r},t)]\,, \tag{10.2}$$

where $P(\boldsymbol{r},t)$ is the probability of finding the particle at $\boldsymbol{r}$ at time t, $\boldsymbol{l}$ is the set of vectors which connects the site at $\boldsymbol{r}$ with the nearest neighbor sites, n is the number of nearest neighbor sites, and τ is the "rest" time at a particular site. With the boundary condition $P(\boldsymbol{r},o)=\delta(\boldsymbol{r})$, the function $P(\boldsymbol{r},t)$ is equivalent to that part of the Van Hove self-correlation function which describes the diffusion.

The assumption under 3) implies that we may write

$$I_{\mathrm{s}}(\boldsymbol{q},t)=I^{\mathrm{D}}(\boldsymbol{q},t)\,I^{\mathrm{V}}(\boldsymbol{q},t)\,, \tag{10.3}$$

where $I^{\mathrm{D}}(\boldsymbol{q},t)$ describes the diffusion and $I^{\mathrm{V}}(\boldsymbol{q},t)$ describes the vibrational motion of the diffusing particle. If we Fourier transform both sides of (10.2), we obtain

$$\frac{\partial}{\partial t}\int e^{i\boldsymbol{qr}}P(\boldsymbol{r},t)d\boldsymbol{r}=\frac{1}{n\tau}\sum_{\boldsymbol{l}}\left[\int e^{i\boldsymbol{qr}}P(\boldsymbol{r}+\boldsymbol{l},t)d\boldsymbol{r}-\int e^{i\boldsymbol{qr}}P(\boldsymbol{r},t)d\boldsymbol{r}\right] \tag{10.4}$$

which, according to (10.1) and using the fact that $P(\boldsymbol{r},t)=G_{\mathrm{s}}^{\mathrm{D}}(\boldsymbol{r},t)$, is equivalent to

$$\frac{\partial I^{\mathrm{D}}(\boldsymbol{q},t)}{\partial t}=\frac{1}{n\tau}\sum_{\boldsymbol{l}}\left[e^{-i\boldsymbol{ql}}I^{\mathrm{D}}(\boldsymbol{p},t)-I^{\mathrm{D}}(\boldsymbol{p},t)\right] \tag{10.5}$$

with the boundary condition $I^{\mathrm{D}}(\boldsymbol{q},o)=1$. The solution of (10.5) is

$$I^{\mathrm{D}}(\boldsymbol{q},t)=e^{-f(\boldsymbol{q})\cdot t/\tau} \tag{10.6}$$

with

$$f(\boldsymbol{q})=\frac{1}{n}\sum_{\boldsymbol{l}}(1-e^{-i\boldsymbol{ql}}). \tag{10.7}$$

The scattering function, finally, is obtained by Fourier transformation of $I^{\mathrm{D}}(\boldsymbol{q},t)$ and is

$$S_{\mathrm{inc}}^{\mathrm{D}}(\boldsymbol{q},\omega)=\frac{1}{\pi}\frac{f(\boldsymbol{q})/\tau}{\omega^2+|f(\boldsymbol{q})/\tau|^2}. \tag{10.8}$$

If inelastic contributions to $S_{\mathrm{inc}}^{\mathrm{V}}(\boldsymbol{q},\omega)$ are small for energies of interest for $S_{\mathrm{inc}}^{\mathrm{D}}(\boldsymbol{q},\omega)$, the quasielastic scattering can be written

$$S_{\mathrm{inc}}^{\mathrm{q.e.}}(\boldsymbol{q},\omega)=\frac{1}{\pi}e^{-q^2\langle u^2\rangle}\frac{f(\boldsymbol{q})/\tau}{\omega^2+|f(\boldsymbol{q})/\tau|^2}, \tag{10.9}$$

where the Debye-Waller factor $\exp(-q^2\langle u^2\rangle)$ has been included to account for the form factor due to vibrational motion of the particle at the equilibrium site; the Debye-Waller factor will be discussed in more detail below. Typical width curves for octahedral-octahedral (o-o) and tetrahedral-tetrahedral (t-t) nearest neighbors jumps in a fcc structure are shown in Fig. 10.1. For a polycrystalline sample the scattering function must be averaged over the direction of $\boldsymbol{q}$ with respect to the lattice.

It is worth noting that for sufficiently small values of q the scattering function is a Lorentzian of width $2\hbar Dq^2$. Thus, in this limit the neutron scattering results make a connection to the results obtained by macroscopic techniques.

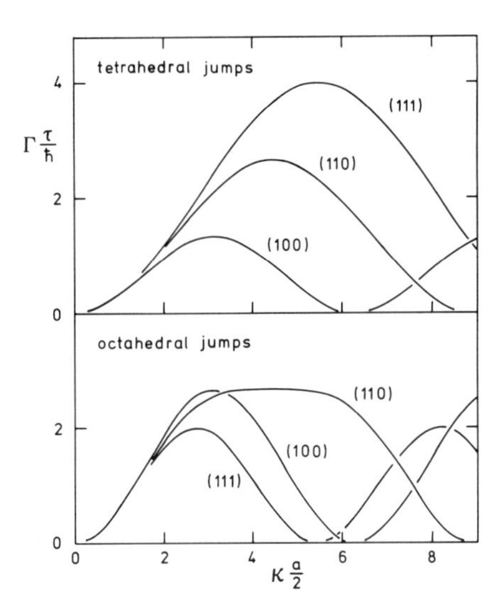

Fig. 10.1. Width of quasielastic peak for random instantaneous jumps between octahedral and tetrahedral sites, respectively, for selected directions in an fcc lattice

The information which is obtained from $S^{\mathrm{D}}_{\mathrm{inc}}(\boldsymbol{q},\omega)$ is thus contained in the "width function" $f(\boldsymbol{q})/\tau$ which is determined by the geometry of the sites and by the residence time at a given site. As we shall see below, the only case for which the theory in this simple form is able to account for the experimental results is that of fcc PdH_x in the α and in the β phase. Other systems investigated to date are all bcc metals, and in all cases considerable deviation from the prediction of the Chudley-Elliott model are observed. In order to understand this situation we must examine the assumptions underlying the derivation of (10.9). The assumption under 1) will be discussed later. Assumption 2) is easily relaxed, and the results for the bcc latice, in which the sites do not form a simple Bravais lattice, were first obtained by *Blaesser* and *Peretti* [10.9]. Similar results have also been obtained by *Gissler* and *Rother* [10.10] using a random walk method and by *Rowe* et al. [10.11] using an extension of the Chudley-Elliott method. We shall review briefly the derivation of $S^{\mathrm{D}}_{\mathrm{inc}}(\boldsymbol{q},\omega)$ using the approach of *Rowe* et al. [10.11].

In the case of a non-Bravais interstitial lattice we must replace the single rate equation [(10.2)] by a set of coupled differential equations. [Strictly speaking (10.2) already implies an infinite number of differential equations which are solved by the Fourier transform technique.]

If there are m inequivalent sites per unit cell, we have

$$\frac{\partial P_i(\boldsymbol{r},t)}{\partial t}=\frac{1}{n\tau}\sum_{j,k}[P_j(\boldsymbol{r}+\boldsymbol{l}_{ijk},t)-P_i(\boldsymbol{r},t)]\,, \tag{10.10}$$

where $i=1,\ldots,m$, $\boldsymbol{l}_{ijk}$ connects site i with site k of local symmetry j, and the summation is over all nearest neighbors of the site at $\boldsymbol{r}$. The corresponding

Van Hove self-correlation function is

$$G_s^D(\boldsymbol{r},t)=\frac{1}{m}\sum_j P^j(\boldsymbol{r},t)$$

with (10.11)

$$P^j(\boldsymbol{r},t)=\sum_i P_i(\boldsymbol{r},t)$$

subject to the boundary condition

$$P_i(\boldsymbol{r},t)=\delta(\boldsymbol{r})\delta_{ij}, \tag{10.12}$$

where δ_{ij} is the Kronecker delta. Proceeding as above, we obtain from (10.10)

$$\frac{\partial I_i^D(\boldsymbol{q},t)}{\partial t}=\frac{1}{n\tau}\left[\sum_j I_j^D(\boldsymbol{q},t)\sum_k \mathrm{e}^{-\mathrm{i}\boldsymbol{q}\boldsymbol{l}_{ijk}}-\mathrm{I}_i^D(\boldsymbol{q},t)\right] \tag{10.13}$$

which may be written

$$|A|\boldsymbol{I}=\frac{\partial}{\partial(t/\tau)}\boldsymbol{I}, \tag{10.14}$$

where $|A|$ is an $m\times m$ matrix with elements $A_{ij}=\frac{1}{n}\sum_k \mathrm{e}^{-\mathrm{i}\boldsymbol{q}\boldsymbol{l}_{ijk}}-\delta_{ij}$ and $\boldsymbol{I}$ is a column vector with m elements $I_i(\boldsymbol{q},t)$. Equation (10.14) is solved using standard eigenvalue methods, and we refer to the paper by *Rowe* et al. [10.11] for mathematical detail. The resulting scattering function is

$$S_{\mathrm{inc}}^D(\boldsymbol{q},\omega)=-\frac{1}{m\pi}\sum_{j=1}^{m}\left|\sum_i \alpha_j^i\right|^2\frac{M_j(\boldsymbol{q})/\tau}{[M_j(\boldsymbol{q})/\tau]^2+\omega^2}. \tag{10.15}$$

In this case the scattering function is a sum of m Lorentzians with half-widths which depend on the relaxation time τ and also on the interstitial lattice through the functions $M_i(\boldsymbol{q})$. Using this formalism it is easy to incorporate jumps to sites other than nearest neighbor sites and also to allow for differences in the jump probabilities for different sets of neighbors. Typical width curves for o-o and t-t nearest neighbor jumps in a bcc structure are shown in Fig. 10.2.

In deriving the results above, we have assumed that the vibrational motion of the proton is not correlated to the jump motion. If, furthermore, the jumps are instantaneous and the vibrations are harmonic, the contribution from $I^V(\boldsymbol{q},t)$ in the quasielastic region is properly accounted for through the Debye-Waller factor as given in (10.9), where $\langle u^2\rangle$ is the mean square amplitude of vibration of the proton at the interstitial site. However, experimental evidence

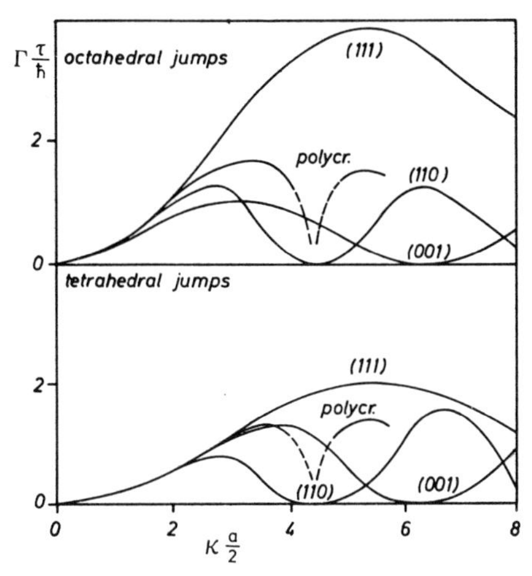

Fig. 10.2. Width of quasielastic peak for random instantaneous jumps between octahedral and tetrahedral sites, respectively, for selected directions in a bcc lattice. Also shown is the result for a polycrystal (*Gissler* and *Rother* [10.10])

indicates that in some cases $\langle u^2 \rangle$ is much larger than would be expected from the frequency of the proton modes and that, furthermore, the integrated intensity does not obey the simple q-dependence given by (10.9) but rather shows a fast initial decay followed by a much slower decay at large q. This behavior is in all cases more pronounced at elevated temperatures, and it was conjectured by *de Graaf* et al. [10.12] that the anomaly is related to the high diffusion rate. A phenomenological theory to explain these results was formulated by *Wakabayashi* et al. [10.13]. They assumed that the intensity form factor of the proton is due to the thermal motion in the ground state plus a contribution from a excited state which is characterized by a much more extended wave function; diffusion is assumed to occur by decay of this excited state to the ground state of an adjacent site. If the decay time of correlations of motions other than diffusion is small compared to the rest time of the particle at a given site, the average of the density correlation over a period of the order of the rest time may be written as

$$\int d\boldsymbol{r}' \langle \overline{\varrho(\boldsymbol{r}'-\boldsymbol{r},o)\varrho(\boldsymbol{r}',t)} \rangle \approx \int d\boldsymbol{r}' \bar{\varrho}(\boldsymbol{r}-\boldsymbol{r}')\bar{\varrho}(\boldsymbol{r}') \tag{10.16}$$

and the quasielastic intensity, which is obtained by integration of the scattering function over an energy interval of the order of the inverse of the rest time, is then given by

$$I_{\text{q.e.}}(\boldsymbol{q}) \approx |f(\boldsymbol{q})|^2 \tag{10.17}$$

with

$$f(\boldsymbol{q}) = \int d\boldsymbol{r}\, \mathrm{e}^{\mathrm{i}\boldsymbol{q}\boldsymbol{r}} \bar{\varrho}(\boldsymbol{r}) \,. \tag{10.18}$$

The authors [10.13] made the following ansatz for the average density distribution:

$$\bar{\varrho}(\boldsymbol{r}) = (1-\alpha)\varrho_0(\boldsymbol{r}) + \alpha\varrho_1(\boldsymbol{r}), \tag{10.19}$$

where ϱ_0 and ϱ_1 correspond to the ground state and the excited state, respectively, and are gaussians of width u_0 and u_1; α is the weight of the extended part. Using this ansatz, the quasielastic intensity is given by

$$I_{\mathrm{q.e.}}(\boldsymbol{q}) = (1-\alpha)^2 \exp(-u_0^2 q^2) + 2\alpha(1-\alpha)\exp\left(-\frac{u_0^2+u_1^2}{2} q^2\right) + \alpha^2 \exp(-u_1^2 q^2) \tag{10.20}$$

which, if $u_0^2 \ll u_1^2$, has the same characteristics as the experimental results at high temperatures.

A rather different approach to this problem was taken by *Gissler* and *Stump* [10.14]. They assumed that the self-correlation functions for the residence phase and the jump phase can be superimposed and that each phase is characterized by its own Debye-Waller factor. It is further assumed that the motion in the jump phase is that of a free particle; the precise meaning of the Debye-Waller factor is therefore somewhat obscure in this case. We refer to the original paper for details of this model.

It is well known that atoms such as oxygen, nitrogen, and carbon can exist as interstitial impurities in transition metals and act as traps for dissolved hydrogen [10.15]. This technologically very important aspect of the transport of hydrogen in metals has recently been studied by the Jülich group both experimentally and theoretically. Theoretical results were derived for a phenomenological random-walk model by *Richter* et al. [10.16] and later by *Kehr* and *Richter* [10.17] using the average *T*-matrix approximation in the limit of dilute concentration of traps. We shall review briefly the results obtained in [10.16] and refer to the original paper for the work of *Kehr* and *Richter*.

In the approach taken by *Richter* et al. [10.16] it is assumed that the traps are randomly distributed in the metal lattice. The proton diffuses through the undisturbed parts of the lattice for an average time τ_1 and is then trapped at an impurity site for an average time τ_0. The scattering function is obtained using a mathematical procedure which is similar to that employed by *Singwi* and *Sjölander* [10.18] for the derivation of jump diffusion of molecules in liquids. The result is

$$S_{\mathrm{inc}}^{\mathrm{D}}(\boldsymbol{q},\omega) = R_1 \frac{\omega_1/\pi}{\omega^2+\omega_1^2} + R_2 \frac{\omega_2/\pi}{\omega^2+\omega_2^2} \tag{10.21}$$

with

$$\omega_{1,2} = \tfrac{1}{2}\{\tau_0^{-1} + \tau_1^{-1} + \Lambda(\boldsymbol{q}) \pm \sqrt{[(\tau_0^{-1}+\tau_1^{-1}+\Lambda(\boldsymbol{q})]^2 - 4\Lambda(\boldsymbol{q})/\tau_0}\}\,. \tag{10.22}$$

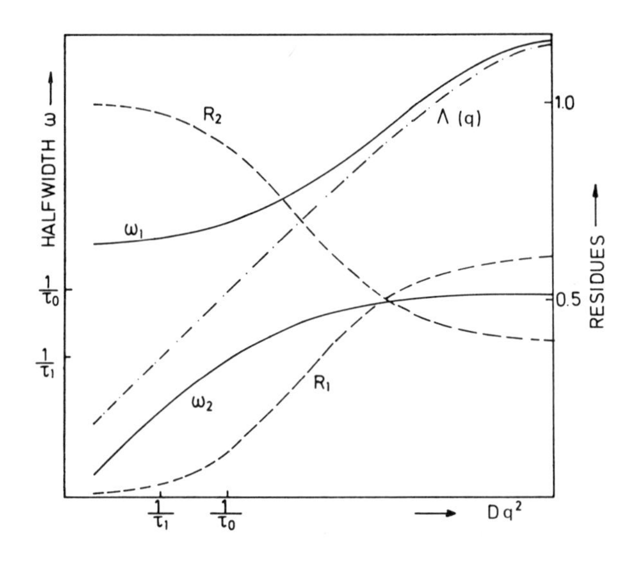

Fig. 10.3. Weights and widths of the two Lorentzians in (10.21) as function of Dq^2. The meaning of the symbols is given in the text (*Kehr* and *Richter* [10.17])

The weights of the two Lorentzians are

$$R_1 = \frac{1}{2} + \frac{1}{2}\left[\Lambda(\boldsymbol{q})\frac{\tau_1-\tau_0}{\tau_1+\tau_0} - \tau_1^{-1} - \tau_0^{-1}\right] \Big/ \sqrt{[\tau_0^{-1}+\tau_1^{-1}+\Lambda(\boldsymbol{q})]^2 - 4\Lambda(\boldsymbol{q})/\tau_0} \tag{10.23}$$

$$R_2 = 1 - R_1 \,. \tag{10.24}$$

$\Lambda(\boldsymbol{q})$ is the width function for the case of a particle diffusing in the undisturbed lattice. The resulting widths and weights of the Lorentzian are shown in Fig. 10.3 as functions of Dq^2. The results given in (10.21–24) have been verified by the calculations of *Kehr* and *Richter* referred to above.

10.3 Experimental Results

In this section we shall review the experimental results which have been obtained from neutron quasielastic scattering studies of metal hydrides. For information regarding the experimental technique we refer to the article by *Gissler* [10.2], in which the various types of neutron spectrometers which are commonly used for studies of this kind are discussed.

10.3.1 Quasielastic Width for PdH_x

The first experiment in which the quasielastic scattering from a metal hydride was interpreted in terms of the diffusion of the proton was that by *Sköld* and *Nelin* [10.19] for fcc PdH_x in the α phase. In this case the half-width was

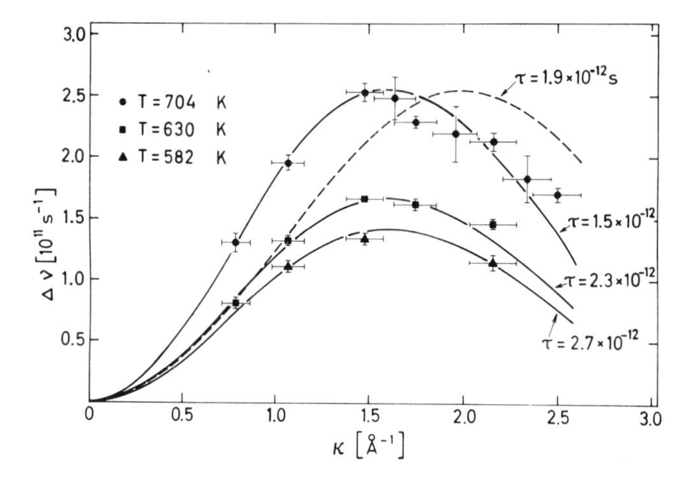

Fig. 10.4. Width of the quasielastic peak for polycrystalline PdH_x in the α phase, together with the corresponding theoretical curves for o-o jumps (solid lines) and t-t jumps (dashed line), respectively (*Sköld* and *Nelin* [10.19])

measured for a polycrystalline sample and was found to be in good agreement with the results of the Chudley-Elliott model as outlined above. The results are shown in Fig. 10.4, together with the theoretical predictions for nearest neighbor o-o and t-t jumps, respectively. The width curves are in excellent agreement with the assumption that the protons diffuse by rapid jumps between nearest neighbor octahedral sites. This conclusion was later verified by the results obtained by *Rowe* et al. [10.20] for a single-crystal sample. In a recent experiment by *Kley* et al. [10.21], the quasielastic scattering was measured for a single crystal of $PdH_{0.006}$ at 400 °C. The width curves were compared to the predictions of the Chudley-Elliott model after normalization of the theoretical curve to the experimental results at the two smallest values of q measured ($q \approx 0.5\,\text{Å}^{-1}$). Using this procedure, they found that the theoretical curve shows deviations from the experimental results which are well outside the experimental error, even if o-o site jumps are assumed. Also, the diffusion constant which corresponds to the normalized theoretical result is $\approx 60\%$ larger than the average value from a large number of other measurements. The authors conjecture that this may be due to the fact that they are using a sample of lower concentration than that used in most other experiments. However, the samples used in the neutron studies reported above were $PdH_{0.02-0.04}$ [10.19] and $PdH_{0.03}$ [10.20], respectively, and it seems unlikely [10.22] that differences in the concentration can explain the discrepancy between these studies and the results of *Kley* et al. [10.21]. It may be worth noting that, if the theoretical curves are normalized such that the diffusion constant obtained in other experiments is reproduced, the o-o model is in much better agreement with the results of *Kley* et al.

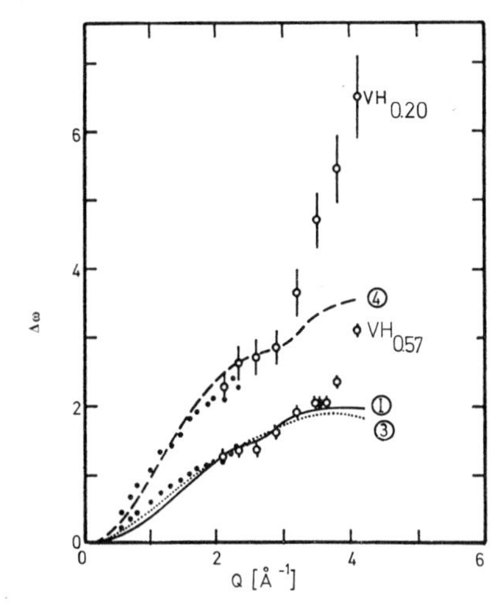

Fig. 10.5. Width of the quasielastic peak for α phase VH_x at 485 K. Solid points are from [10.11] and open circles are from [10.12]. Dashed line is the theoretical curve corresponding to mixed jumps between octahedral and tetrahedral sites. For further details regarding the theoretical curves we refer to the original paper (*de Graaf* et al. [10.12])

Neutron scattering results have also been obtained for PdH_x in the β phase by *Beg* and *Ross* [10.23] and by *Nelin* and *Sköld* [10.24]. From the results obtained by *Beg* and *Ross* it was concluded that a complex picture involving jumps between octahedral and tetrahedral sites as well as jumps between octahedral sites was indicated. This is in contradiction to the results obtained by *Nelin* and *Sköld* which are well explained by o-o jumps only. Recent NMR studies by *Davis* et al. [10.22] are consistent with diffusion by a random walk on octahedral sites. Also, previous neutron diffraction results clearly indicate octahedral site occupancy [10.25–27]. The discrepancy between the two sets of neutron scattering results can possibly be explained by differences in the experimental arrangements in the two cases. In particular, the energy resolution used by *Nelin* and *Sköld* was significantly better than that used in the experiment of *Beg* and *Ross*. The former results would, therefore, seem to be more reliable, at least in this regard, and are also, as seen above, in agreement with the NMR results and the neutron diffraction results.

In the experiment by *Nelin* and *Sköld*, the results were also compared to the model including finite jump times by *Gissler* and *Stump* [10.14]. However, the data were not accurate enough to allow a determination of the significance of this modification to the instantaneous jump model.

10.3.2 Quasielastic Width for VH_x, TaH_x, NbH_x

In the case of bcc metals, neutron scattering results have been obtained for VH_x, TaH_x, and NbH_x. Although considerable effort has been devoted to the calculation of the width curves for complex jump patterns, in neither case has it been possible to account for all the features of the experimental results.

In Fig. 10.5 we show the results obtained by *de Graaf* et al. [10.12] for $VH_{0.20}$ and $VH_{0.57}$ at 485 K; both samples are in the α phase. We observe a very strong concentration dependence of the width of the quasielastic peak which reflects the slowing down of the rate of diffusion at higher concentration. It is worth noting that the neutron scattering method does not require a macroscopic concentration gradient and that the concentration dependence of the diffusion process can therefore be studied under equilibrium conditions.

As far as the detailed shapes of the width curves are concerned, several models for the jump pattern were considered, including multisite occupation (compare Fig. 10.5). Although no model gave complete agreement with the experimental results, it was concluded that the diffusion occurs predominantly between tetrahedral sites. The monotonic increase of the width for large q is clearly inconsistent with the theoretical pictures as developed above. The authors conjecture that the major deficiency of the theory is that it does not properly treat the jump phase of the motion; this could be important in the case of very high jump rates, where the assumption of instantaneous jumps is less valid, i.e., the particle spends a larger fraction of the time in the jump phase which therefore cannot be neglected in the evaluation of the self-correlation function. However, in a later publication by *Rush* et al. [10.28], it is pointed out that a corresponding increase of the linewidth is not observed for $NbH_{0.07}$ at 256 °C, although the relaxation time is similar to that of $VH_{0.2}$ at 212 °C. Also, in a recent study by *Sköld* [10.29] the quasielastic scattering for $VH_{0.12}$ at 210 °C was measured. A single crystal was used and the width function was determined in the three symmetry directions (100), (110), and (111) for $1.8\,\text{Å}^{-1} \leqq q \leqq 6.2\,\text{Å}^{-1}$. In neither direction is the dramatic increase in the linewidth for $q > 3\,\text{Å}^{-1}$ reported by *de Graaf* et al. observed, and the data are on the whole rather consistent with diffusion on the tetrahedral interstitial lattice. The experimental situation for α phase VH_x is thus somewhat confusing and there is an obvious need for further careful studies, preferably using a single-crystal sample. A particular difficulty in this case is the correction that must be applied for the strong incoherent elastic scattering from the vanadium, and it is conceivable that this is the source of the discrepancies between the two experiments.

Quasielastic neutron scattering results were obtained for α phase $TaH_{0.15}$ and for Ta_2H in the α and β_1 phases for polycrystalline samples by *Rush* et al. [10.28] and for a single-crystal sample of $TaH_{0.02}$ by *Rowe* et al. [10.30]. The polycrystal results do not show the marked increase of the width at large q observed for VH_x, but even in this case it was not possible to explain the data satisfactorily using a simple jump diffusion model. The single-crystal results are shown in Fig. 10.6 for two values of $|\boldsymbol{q}|$ as function of the direction of $\boldsymbol{q}$. The experimental results show much less anisotropy than expected from the models; the overall behavior at large q is quite consistent with simple t-t jumps, however. Recent results for $TaH_{0.13}$ at 410 °C by *Lottner* et al. [10.31] have been analyzed in terms of nearest neighbor jumps, direct jumps between nearest and second nearest neighbor sites, and also for a model in which the particle jumps to a second nearest neighbor site via a nearest neighbor site. These authors conclude that the first two models do not reproduce the directional

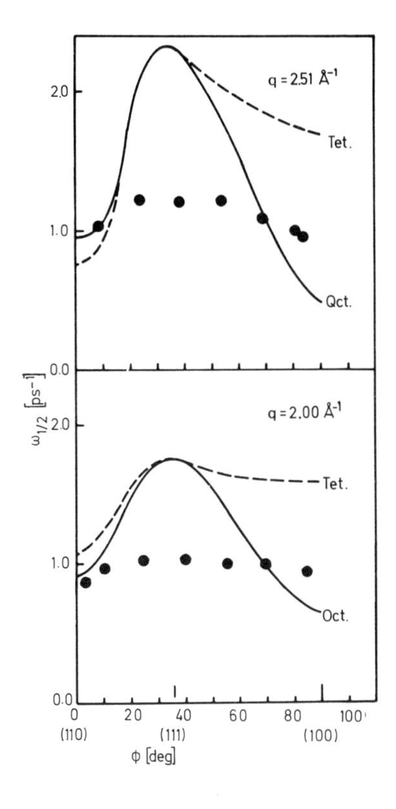

Fig. 10.6. Width of the quasielastic peak for $TaH_{0.02}$ at 584 K at two values of $|\boldsymbol{q}|$ compared to the theoretical predictions for o-o and for t-t jumps, respectively. The widths are shown as function of the angle between $\boldsymbol{q}$ and the (110) direction in the single-crystal sample (*Rowe* et al. [10.30])

dependence of the widths but that the latter model is in good agreement with the experimental results for $0.5\,\text{Å}^{-1} \leqq q \leqq 2.5\,\text{Å}^{-1}$.

The quasielastic scattering for NbH_x has also been studied for both polycrystals [10.32–35] and for single-crystal samples [10.31, 36, 37]. The measurements reported in [10.35] were for a sample in the heterogeneous $(\alpha+\beta)$ phase; the other measurements on NbH_x are all for α phase samples. Extensive results for a single-crystal sample have been reported by *Stump* et al. [10.37] and are shown in Fig. 10.7, together with model calculations for tetrahedral sites. Although detailed agreement is not observed, it is interesting to note that the general behavior of the width curves is consistent with the model calculations and that the experimental results indicate predominantly tetrahedral sites for this bcc system also.

From the work reported in [10.31] it is concluded that a nearest neighbor jump model is in agreement with the width curves obtained for $NbH_{0.02}$ at 20 °C. At higher temperatures the data can be explained only by assuming a multistep model as in the case of $TaH_{0.13}$ at 410 °C discussed above. This suggests that the difficulties encountered in explaining the experimental results for bcc systems may be related to the high jump rates in most of the cases studied.

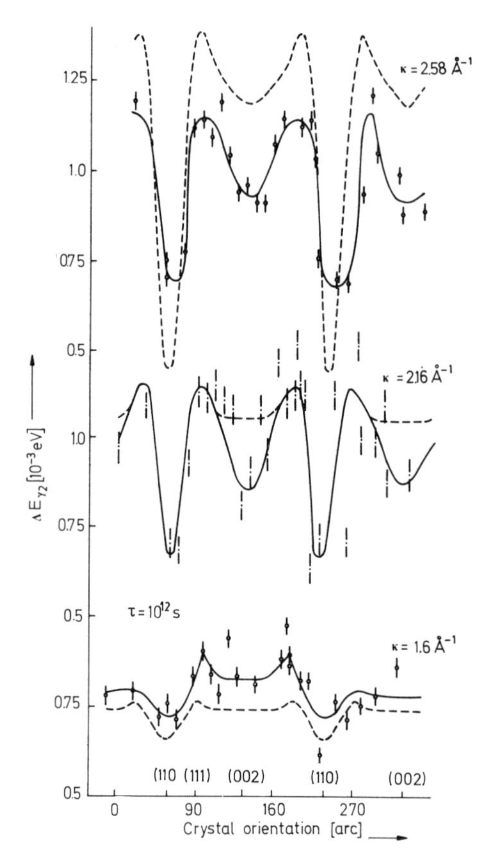

Fig. 10.7. Width of the quasielastic peak for $NbH_{0.07}$ at 256 °C for three values of $|\boldsymbol{q}|$ compared to the theoretical prediction for t-t jumps. The theoretical curves (dashed lines) are folded with the instrumental resolution function. The solid lines are the authors' estimate of the experimental points. The results are shown as function of the direction of $\boldsymbol{q}$ in the single-crystal sample (*Stump* et al. [10.37])

The recent study of $NbH_{0.0012}$ for temperatures in the range $165\,K \leqq T \leqq 300\,K$ by *Richter* et al. [10.4] is a good demonstration of the present capability of the neutron scattering method, both with regard to the temperature range covered and with regard to the low concentration of hydrogen. The measurements were made in order to investigate the behavior of the diffusion coefficient around 250 K where previous macroscopic measurements [10.38] indicate a change in the activation energy for diffusion; the low concentration of hydrogen was necessary in order to avoid precipitation at the lower temperatures. The diffusion coefficient was obtained by fitting the data at small q with a Lorentzian of width $2\hbar Dq^2$ [compare discussion following (10.9)] and are shown in Fig. 10.8, together with the result obtained by *Schaumann* et al. [10.38] from Gorsky effect measurements. We note that the neutron results are in good agreement with the macroscopic measurements and thus confirm the low value of the activation energy below 250 K.

As should be obvious from the discussion above, the neutron scattering studies have not so far yielded conclusive information about the diffusion process in any of the bcc systems studied. One possible reason for this is that for

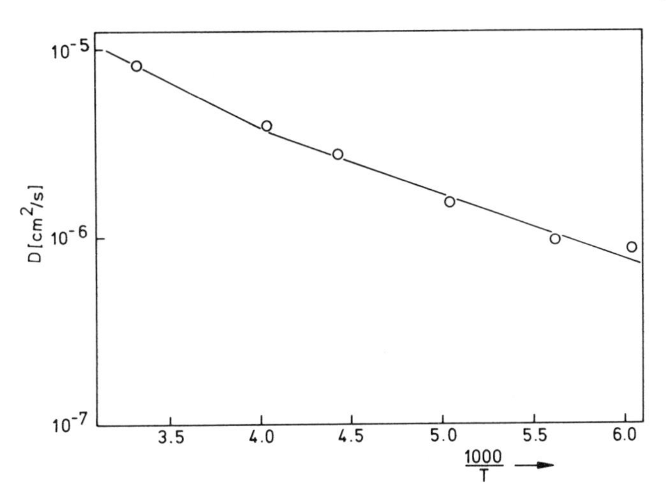

Fig. 10.8. Diffusion coefficient of H in $NbH_{0.0012}$ as function of temperature. The solid line shows the result obtained from Gorsky effect measurements (*Richter* et al. [10.4])

a given diffusion constant, the jump rates are higher in the bcc case than in the fcc case and that, therefore, a proper treatment of the jump phase is more important for bcc systems [10.12]. Another possibility is that the actual jump pattern is more complex than that assumed for any of the models calculated so far. For example, from the recent work of *Birnbaum* and *Flynn* [10.39], one may conjecture that a proton performs rapid diffusion over the four tetrahedral sites surrounding an octahedral site and that the actual transport occurs by infrequent transfers from one such four-membered ring to another. It is also possible that the triangular sites [10.40] in the t-t path serve as quasiequilibrium sites and must be included in the calculation of $G_s^D(\boldsymbol{r}, t)$. The experimental results are also contradictory in several cases, and it is important that further measurements under improved experimental conditions be made. There are reasons to believe that sample purity and the degree of annealing can sometimes be important; we shall return to this question later.

10.3.3 Quasielastic Intensity

The integrated quasielastic intensity has been measured for several hydrides. In the case of VH_x in the α phase, two separate experiments covering different ranges of wave vector transfer have been reported. In the experiment by *Rowe* et al. [10.11], the intensity was determined for $q \lesssim 2.2\,Å^{-1}$ for several concentrations and temperatures, and in the experiment by *de Graaf* et al. [10.12], results were obtained for $q \gtrsim 2.0\,Å^{-1}$ for $VH_{0.198}$ and $VH_{0.570}$ at 485 K. In both cases the mean square amplitude was determined from the slope of $\ln I_{\mathrm{q.e.}}$ versus q^2. From the measurements at small q, the values $0.182\,Å^{-2}$ and $0.124\,Å^{-2}$ are obtained for $VH_{0.198}$ and $VH_{0.570}$, respectively, at 483 K; from

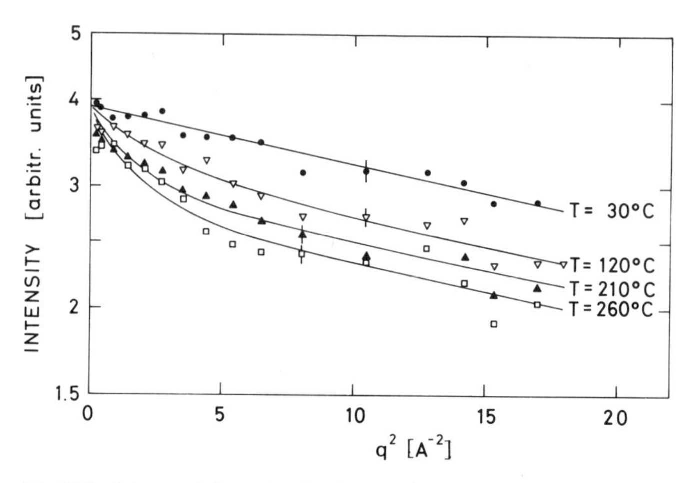

Fig. 10.9. Integrated intensity in the quasielastic peak for poly-crystalline $NbH_{0.16}$ at four temperatures is shown as function of q^2. The solid curves show the best fit of (10.20) to the experimental points as explained in the text (*Wakabayashi* et al. [10.13])

Table 10.1. Parameters obtained by fitting (10.20) to the experimental points shown in Fig. 10.9 (from [10.13])

T[°C]	u_0^2 [Å^2]	u_1^2 [Å^2]	α
30	0.019	—	0
120	0.019	0.69	0.086
210	0.019	1.16	0.121
260	0.019	1.05	0.153

the results at large q, the corresponding values are 0.082 Å^2 and 0.056 Å^2. The mean square displacements are unusually large and, furthermore, depend on the range of q over which they are determined.

For PdH_x in the β phase [10.24] and for NbH_x [10.13, 41] in the α phase, the results again indicate a complex behavior at elevated temperature; the intensity decreases rapidly at small q, followed by a much slower decay at large q. Rather than discussing all these experiments in detail, we shall use the data obtained by *Wakabayashi* et al. [10.13] for $NbH_{0.16}$ at several temperatures to illustrate the kind of results that are obtained. From the curves shown in Fig. 10.9 we notice that the intensity shows a simple Debye-Waller behavior at $T=30$ °C but that at higher temperatures the intensity decreases much more rapidly for small q and then assumes the low-temperature slope for $q \gtrsim 2$ Å^{-1}. The solid lines show the best fit of the function given in (10.20) to the experimental points, assuming that u_0^2 does not depend on temperature; the corresponding parameters are given in Table 10.1. The large value of u_1^2 and the temperature dependence of α are both consistent with the philosophy underlying the derivation of (10.20).

However, a recent series of experiment by *Lottner* et al. [10.42] has shown that for $0.3\,\text{Å}^{-1} < q < 2.5\,\text{Å}^{-1}$ the intensity of the quasielastic line for H in Nb follows a simple Debye-Waller behavior with a mean square amplitude of the order of the value expected for harmonic vibration of the H atom. The authors [10.42] suggest that the discrepancies between these results and those obtained earlier [10.13, 41] are due to differences in the methods used to evaluate the area of the quasielastic peak and, in particular, the accuracy by which the wings of the peak are included in the integration.

As we have seen above, the intensity of the incoherent quasielastic scattering is governed by the mean square displacement of the particle due to the thermal motion at the equilibrium site. In the case of coherent Bragg scattering, a further contribution to the intensity form factor arises from the local relative displacement of atoms from their regular lattice sites. Therefore, by combining information obtained from incoherent scattering from the hydrogen-containing metal with information obtained from the intensity in the Bragg peaks from the deuterium-containing metal, it is possible to obtain a measure of the local distortion around the interstitial. Interesting preliminary results in this regard have been obtained for $PdH_{0.63}$ and $PdD_{0.63}$ [10.43]; this possibility has yet to be fully explored, however.

10.3.4 Diffusion with Impurities

In recent papers by the Jülich group [10.5, 6], neutron quasielastic scattering studies of the diffusion of hydrogen in niobium in the presence of nitrogen impurities have been reported. This is a topic of considerable theoretical interest which also has important consequences in many practical applications. The results were obtained under carefully controlled conditions with regard to sample purity and are therefore also of value in the evaluation of the potential importance of unintentional impurities in the kinds of studies discussed above.

Richter et al. [10.5, 6] reported measurements on two samples of niobium containing nitrogen impurities. In one case [10.5] $NbN_{0.007}H_{0.004}$ was measured for $0.15\,\text{Å}^{-1} \leqq q \leqq 1.9\,\text{Å}^{-1}$ and for temperatures between 180 and 373 K, and in the other case [10.6] $NbN_{0.004}H_{0.003}$ was measured for $0.1\,\text{Å}^{-1} \leqq q \leqq 1.6\,\text{Å}^{-1}$ and for temperatures between 198 and 303 K; we shall refer to these as sample 1 and sample 2, respectively. The measurements were made at the backscattering spectrometer at the Grenoble high flux reactor. This instrument has an energy resolution of (0.7–1.5) μeV and scans over an energy range of ±5.3 μeV. It is only due to the recent development of this high resolution technique that measurements of kind are now feasible.

Figure 10.10 shows the width of the quasielastic peak as function of q^2 for sample 1 at several temperatures. Also shown is the result for pure Nb, and we notice the dramatic decrease in the width due to the impurity. The scattering function as given in (10.21)–(10.24) was fitted to the experimental results for $q < 1\,\text{Å}^{-1}$ for both samples with τ_0 and τ_1 as fitting parameters. The resulting values are shown in Fig. 10.11 where we notice that τ_0^{-1}, the rate of escape from

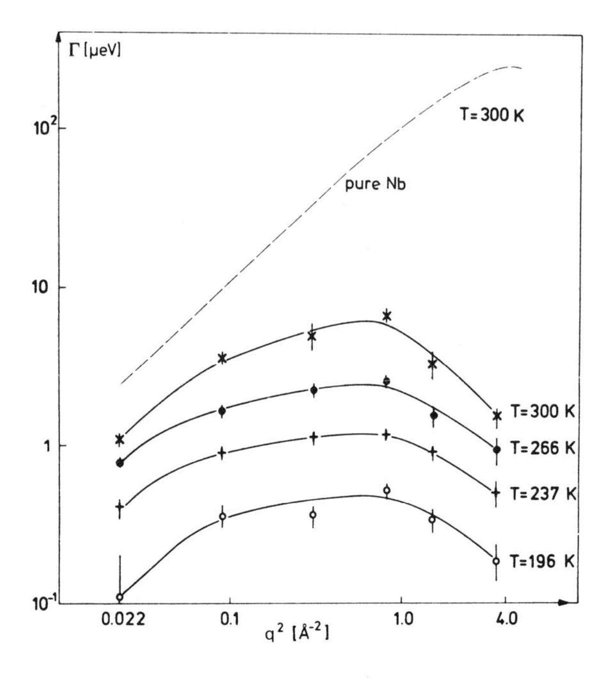

Fig. 10.10. Width of the quasielastic peak for $NbN_{0.007}H_{0.004}$ at four temperatures and the corresponding result for pure niobium [10.33] (dashed line) as function of q^2 (*Richter* et al. [10.5])

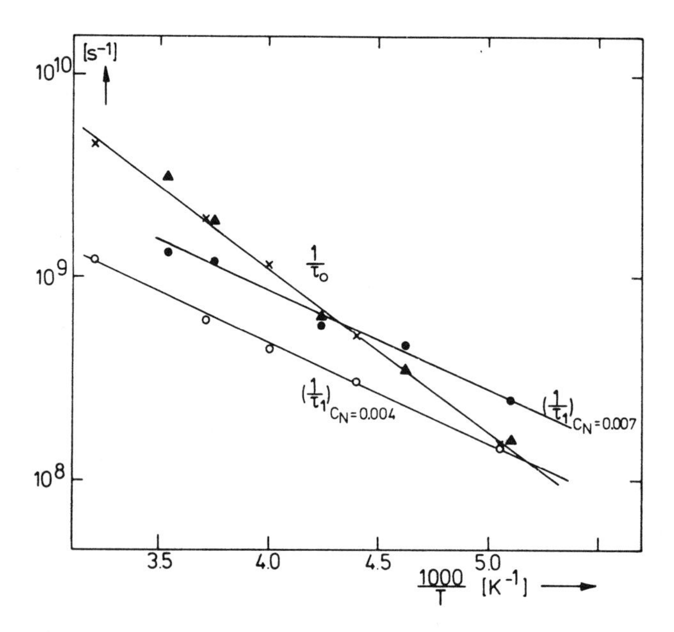

Fig. 10.11. The rate of escape ($1/\tau_0$) and the rate of trapping ($1/\tau_1$) of the proton at the impurity site for $NbN_{0.004}H_{0.003}$ and $NbN_{0.007}H_{0.004}$ as function of inverse temperature (*Richter* et al. [10.6])

a trap, is concentration independent while τ_1^{-1}, the trapping rate, is concentration dependent and, within the errors of the determination, is proportional to the concentration of traps as expected. Both rates follow an Arrhenius behavior with the activation energies: τ_0^{-1} (166 meV), τ_1^{-1} (sample 1) (96 meV), and τ_1^{-1} (sample 2) (94 meV). The activation energy for τ_0^{-1} is in good agreement with the expected value, namely the sum of the binding energy to the impurity (100 meV) and the activation energy for diffusion in the undisturbed lattice (70 meV). The activation energy for τ_1^{-1} is expected to be similar to the activation energy for diffusion; the authors suggest that the observed discrepancy can be explained by saturation effects of the traps at lower temperatures.

It is clear that studies of this kind can yield very detailed information about the process of diffusion in the presence of traps, and neutron scattering results can be expected to play an important role in the future advancement of our understanding of these phenomena. For further details regarding this work, we refer to the articles cited above and also to a recent paper by *Kehr* et al. [10.44]. For an extensive treatment of trapping of hydrogen in metals we refer to the discussion in [Ref. 10.45, Chap. 8].

10.4 Discussion and Conclusions

We hope to have demonstrated that the neutron scattering technique is a very powerful method for the study of the microscopic details of the process of interstitial diffusion of hydrogen in metals. In the limit of small wave vector transfers, the neutron results connect to the results obtained by macroscopic techniques. An advantage of the neutron method is that it does not require a macroscopic concentration gradient; the diffusion process can therefore be studied under equilibrium conditions and at specified concentration of the diffusing particle. The method has not yet been utilized for systematic studies of the concentration dependence, but the capabilities in this regard are evident from the results for VH_x [10.11, 12] and TaH_x [10.28] described above.

It is somewhat discouraging that the neutron experiments have not yet led to a good understanding of the diffusion process in bcc systems despite considerable efforts in this regard. The main reason for this is that we do not yet have a firm theoretical basis for the analysis of the neutron results. Molecular dynamics simulation of the diffusion of hydrogen in tantalum hydride has recently been reported by *Jeulink* et al. [10.46]. It is conceivable, that molecular dynamics results will be as useful as guides to the interpretation of the neutron results in this case as they have been in the case of neutron scattering studies of liquids.

It is also somewhat disturbing that the experimental results in several cases are inconsistent. From the measurements on NbN_xH_x [10.5, 6] discussed above we know that impurities can play an important role in the determination of the quasielastic width. It has also been suggested that clustering of hydrogen may occur in PdH_x at concentrations as low as 0.2 atomic percent, unless the sample is properly annealed [10.21]. This was found to have a large influence on the proton vibration spectrum and could also be important for the diffusion process. It is possible that discrepancies between experimental results for the quasielastic scattering in some cases are due to badly characterized samples.

Recent development of high-resolution neutron spectrometers has expanded the range of relaxation times which can be studied to $\tau \lesssim 10^{-9}$ s. This technique has already yielded very interesting results in studies of diffusion of hydrogen in niobium with nitrogen impurities [10.5, 6] and in studies of diffusion in niobium at low temperatures [10.4].

The dependence of the quasielastic intensity on the wave vector transfer has been studied in several cases and has been interpreted in terms of phenomenological models for the density distribution of the proton [10.13]. We anticipate that the neutron scattering results will be even more important once a more basic theoretical picture is available for the interpretation of these results.

References

10.1 For recent summaries of the field we refer to: Ber. Bunsenges. Phys. Chem. **76** (1972) (Proceedings of an International Conference on Hydrogen in Metals); *Diffusion in Solids: Recent Developments*, ed. by A.S.Nowick and J.J.Burton (Academic Press, London 1975) and J. Less-Common Metals **49** (1976) (Proceedings of an International Conference on Hydrogen in Metals)

10.2 For recent reviews of quasielastic neutron scattering studies of metal hydrides we refer to W.Gissler: Ber. Bunsenges. Phys. Chem. **76**, 770 (1972)
J.M.Rowe: *Proceedings of the Conference on Neutron Scattering*, Gatlinburg, Tennessee (CONF-760601-P1) pp. 491–506 (1976)
T.Springer: "Diffusion and Rotational Motions", in *Topics in Current Physics*, Vol. 3, ed. by S.Lovesey and T.Springer (Springer, Berlin, Heidelberg, New York 1977). The utilization of the Mössbauer technique for the study of diffusion of H in metals has recently been described by
A.Heidemann, G.Kaindl, D.Salomon, H.Wipf, G.Wortmann: Phys. Rev. Lett. **36**, 213 (1976)

10.3 B.Alefeld, M.Birr, A.Heidemann: Naturwissenschaften **56**, 410 (1969)

10.4 D.Richter, B.Alefeld, A.Heidemann, N.Wakabayashi: J. Phys. F **7**, 569 (1977)

10.5 D.Richter, J.Töpler, T.Springer: J. Phys. F **6**, L93 (1976)

10.6 D.Richter, T.Springer, K.W.Kehr: Presented at the *Second International Congress on Hydrogen in Metals*, Paris, June 6–11, 1977

10.7 L.Van Hove: Phys. Rev. **95**, 249 (1954)

10.8 C.T.Chudley, R.J.Elliott: Proc. Phys. Soc. (London) **77**, 353 (1961)

10.9 G.Blaesser, J.Peretti: *Proc. Int. Conf. Vacancies and Interstitials in Metals*, Jül-Conf. 2, Vol. 2 (KFA Jülich 1968) p. 886

10.10 W.Gissler, H.Rother: Physica **50**, 380 (1970)

10.11 J.M.Rowe, K.Sköld, H.E.Flotow, J.J.Rush: J. Phys. Chem. Sol. **32**, 41 (1971)

10.12 L.A.de Graaf, J.J.Rush, H.E.Flotow, J.M.Rowe: J. Chem. Phys. **56**, 4574 (1972)

10.13 N.Wakabayashi, B.Alefeld, K.W.Kehr, T.Springer: Solid State Commun. **15**, 503 (1974)

10.14 W.Gissler, N.Stump: Physics **65**, 109 (1973)

10.15 See Ref. 10.5 for references in this regard

10.16 D. Richter, K. W. Kehr, T. Springer: *Proceedings of the Conference on Neutron Scattering*, Gatlinburg, Tennessee (CONF-760601-P1) pp. 568–574 (1976)
10.17 K. W. Kehr, D. Richter: Solid State Commun. **20**, 477 (1976)
10.18 K. S. Singwi, A. Sjölander: Phys. Rev. **119**, 863 (1960)
10.19 K. Sköld, G. Nelin: J. Phys. Chem. Sol. **28**, 2369 (1967)
10.20 J. M. Rowe, J. J. Rush, L. A. de Graaf, G. A. Ferguson: Phys. Rev. Lett. **29**, 1250 (1972)
10.21 W. Kley, W. Drexel, A. Murani, D. Tocchetti, I. Sosnowska, D. K. Ross: *Proceedings of the Conference on Neutron Scattering*, Gatlinburg, Tennessee (CONF-760601-P1) pp. 558–567 (1976)
W. Drexel, A. Murnai, D. Tocchetti, W. Kley, I. Sosnowska, D. K. Ross: J. Phys. Chem. Sol. **37**, 1135 (1976)
10.22 P. P. Davis, E. F. W. Seymour, D. Zamir, W. David Williams, R. M. Cotts: J. Less-Common Metals **49**, 159 (1976)
10.23 M. M. Beg, D. K. Ross: J. Phys. C: Solid State Phys. **3**, 2487 (1970)
10.24 G. Nelin, K. Sköld: J. Phys. Chem. Sol. **36**, 1175 (1975)
10.25 J. E. Worsham Jr., M. K. Wilkinson, C. G. Shull: J. Phys. Chem. Sol. **3**, 303 (1957)
10.26 J. Bergsma, J. A. Goedkoop: Physica **26**, 744 (1960)
10.27 G. Nelin: Phys. Stat. Sol. (b) **45**, 527 (1971)
10.28 J. J. Rush, R. C. Livingston, L. A. de Graaf, H. E. Flotow, J. M. Rowe: J. Chem. Phys. **59**, 6570 (1973)
10.29 K. Sköld: Unpublished
10.30 J. M. Rowe, J. J. Rush, H. E. Flotow: Phys. Rev. B **9**, 5039 (1974)
10.31 V. Lottner, A. Heim, T. Springer: To be published
10.32 G. Verdan, R. Rubin, W. Kley: *Inelastic Scattering of Neutrons in Solids and Liquids*, Proc. Symp. IAEA, Vienna, 223 (1968)
10.33 W. Gissler, G. Alefeld, T. Springer: J. Phys. Chem. Sol. **31**, 2361 (1970)
10.34 J. H. L. Birchall, D. K. Ross: *Proc. International Meeting on Hydrogen in Metals*. Jül-Conf. 6, Vol. 1 (KFA Jülich 1972) p. 313
10.35 B. Alefeld, H. G. Bohn, N. Stump: *Proc. International Meeting on Hydrogen in Metals*, Jül-Conf. 6, Vol. 1 (KFA Jülich 1971) p. 286
10.36 G. Kistner, R. Rubin, I. Sosnowska: Phys. Rev. Lett. **27**, 1576 (1971)
10.37 N. Stump, W. Gissler, R. Rubin: Phys. Stat. Sol. (b) **54**, 295 (1972)
10.38 G. Schaumann, J. Völkl, G. Alefeld: Phys. Stat. Sol. **42**, 401 (1970)
10.39 H. K. Birnbaum, C. P. Flynn: Phys. Rev. Lett. **37**, 25 (1976)
10.40 S. A. Steward: Solid State Commun. **17**, 75 (1975)
10.41 W. Gissler, B. Jay, R. Rubin, L. A. Vinhas: Phys. Lett. **43** A, 279 (1973)
10.42 V. Lottner, A. Heim, K. Kehr, T. Springer: *International Symposium on Neutron Inelastic Scattering* (IAEA, Vienna 1977) (to be published)
10.43 K. Sköld, M. H. Mueller, J. Faber, Jr., C. A. Palizzari: Unpublished
10.44 K. W. Kehr, D. Richter, R. H. Swendsen: To be published
10.45 G. Alefeld, J. Völkl (eds.): *Hydrogen in Metals II, Application-Oriented Properties*, Topics in Applied Physics, Vol. 29 (Springer, Berlin, Heidelberg, New York 1978)
10.46 J. Jeulink, L. A. deGraaf, M. R. Ypma, L. M. Caspers: Delft Progr. Rep., Series A **1**, 115 (1975), Chem. Phys. Engin. **1**, 115 (1975)

11. Magnetic Aftereffects of Hydrogen Isotopes in Ferromagnetic Metals and Alloys

H. Kronmüller

With 28 Figures

11.1 Foreword

This chapter presents a review on the method of magnetic relaxation for the study of hydrogen in ferromagnetic metals and alloys. This technique so far has been applied mainly for the study of the radiation damage in metals after fast particle irradiation. The high sensitivity of magnetic aftereffect measurements, however, predestinates this technique for applications in the field of "hydrogen in metals".

The usual techniques for the investigation of hydrogen are measurements of solubility, diffusivity, and phase transitions. Furthermore, measurements of the electrical resistivity, the mechanical internal friction (Gorsky effect), nuclear magnetic resonance, and of neutron diffraction have revealed many important features of metal-hydrogen systems. The techniques mentioned so far are, in general, applicable for systems with large H-concentrations. For the case of systems with small hydrogen solubilities, however, the application of the above-mentioned methods becomes difficult. Therefore, the application of more sensitive techniques sometimes seems to be desirable. This holds also for investigations concerning the interaction between hydrogen and other impurity atoms or lattice defects.

It is the purpose of this chapter to present experimental results on the properties of hydrogen in ferromagnetic metals and alloys which were obtained by the technique of magnetic relaxation and which cannot be obtained by less sensitive techniques. In the following we shall present results on the formation of hydrogen complexes with impurity atoms on substitutional and interstitial positions. Complexes of this kind give rise to relaxation effects from which we may derive activation energies and the isotope effects for diffusion of H. In particular, such measurements will give information on the so-called Snoek effect of hydrogen which so far could not be detected in the Nb–H or Ta–H systems. It will be shown that in a number of ferromagnetic alloys (**Ni**C, **Fe**Ti, **Ni**Fe), Snoek-type relaxations of hydrogen can be observed. In addition, by measurements of this kind the dependence of the electronic state of the H-atoms in transition metal alloys on the electronic structure of the metal atoms involved can be studied.

For a basic insight into the technique of the magnetic aftereffect it is necessary to consider the interaction energy between the spontaneous magnetization and hydrogen. Within a domain wall (dw) this interaction energy varies with position of the H-atoms, and, therefore, under the influence of thermally activated motions a rearrangement of the H-atoms in the dw takes place. In Section 11.3 we shall give a brief review on the magnetic interaction of hydrogen with a ferromagnetic spin order, and Section 11.4 will be concerned with various relaxation types of H-atoms in ferromagnetic metals and alloys. Section 11.5 presents experimental results and their interpretation. Within the framework of the measured isotope effects we shall discuss the validity of quantum mechanical tunneling models for hydrogen. It will be seen that, in general, the diffusion of H takes place by a combined thermally activated and phonon assisted tunneling process.

11.2 Experimental Background for the Measurement of Relaxation Spectra

The investigation of magnetic relaxation phenomena has been used successfully in the past 20 years for the study of point defects produced by fast particle irradiation [11.1–4] and of light impurity interstitial atoms as H, C, N, and O in fcc and bcc crystals. The magnetic property which can be measured with highest accuracy and maximum sensitivity as a function of time t and of temperature T is the initial susceptibility $\chi(t, T)$. From a theoretical point of view it is more appropriate (see Sect. 11.4) to consider the reciprocal value of χ, the so-called reluctivity

$$r(t, T)=\frac{1}{\chi(t, T)}. \tag{11.1}$$

In general, $r(t, T)$ is measured as a function of time for different temperatures T in a time interval between t_1 and t_2. From these *isothermal relaxation* curves we may construct *isochronal relaxation* curves of the relaxation amplitudes

$$\left.\begin{array}{l} \Delta\chi(t_1, t_2, T)=\chi(t_1, T)-\chi(t_2, T) \\ \text{or} \\ \Delta r(t_1, t_2, T)=\dfrac{1}{\chi(t_2, T)}-\dfrac{1}{\chi(t_1, T)} \end{array}\right\}. \tag{11.2}$$

The relaxation amplitudes $\Delta\chi$ or Δr have turned out to be rather suitable for detecting relaxation processes and for classifying the type of relaxation process (long-range or short-range diffusion) [11.2, 3]. As shown schematically in Fig. 11.1, the shape of the relaxation spectrum observed is characterized by the maximum relaxation amplitude $\Delta r_{max}(t_1, t_2)$, the half-width $\Delta T_H(t_1, t_2)$, and the position $T_{max}(t_1, t_2)$ of the maximum. All these quantities depend on the measuring times t_1 and t_2 in a characteristic manner. In particular, in the case

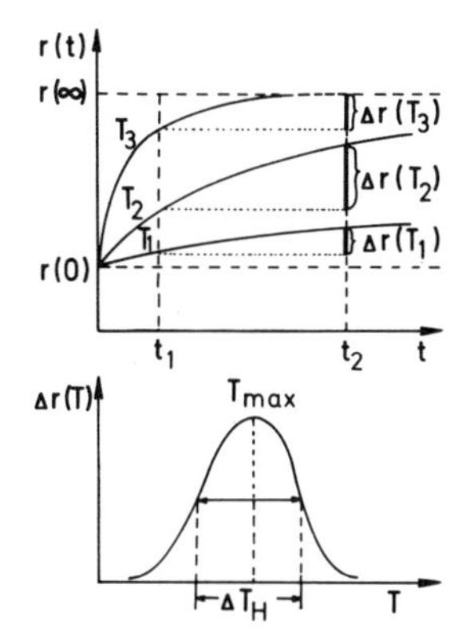

Fig. 11.1. Construction of isochronal relaxation curves from isothermal relaxation curves

of a simple Debye process characterized by one relaxation time $\tau(T)$, T_{max} is implicitly determined by

$$\tau(T_{max}) = \frac{t_2 - t_1}{\ln(t_2/t_1)}. \tag{11.3}$$

In the case of more complex relaxation processes, no simple relations between T_{max} and $t_{1,2}$ can be derived. In these cases the temperature dependence of the relaxation time must be determined by more general techniques described previously in a number of papers [11.5–7].

In the case of a conventional, thermally activated diffusion process, in general a relaxation maximum can be observed with characteristic properties of the shape of the relaxation spectrum. In the case of hydrogen the diffusion mechanism, however, may be influenced by the existence of discrete energy levels of the H-atom as well as by a phonon-assisted tunneling process [11.8–10]. Therefore, the following relaxation spectra may be observed:

1) *A relaxation maximum* is observed if we deal with a classical thermally activated relaxation process in which the relaxation time obeys an Arrhenius law

$$\tau = \tau_0 e^{Q/kT}. \tag{11.4}$$

The pre-exponential factor τ_0 in the case of hydrogen is of the order of magnitude of 10^{-15}–10^{-14} s, i.e., is a factor of 10 smaller in comparison with eigendefects. The temperature, T_{max}, of the maximum depends on t_1, t_2, $\tau(T)$, and the special time dependence of the relaxation process. If we deal with a Debye process (exponential time dependence), $\tau(T_{max})$ is given by (11.3). By varying t_2 and T, the activation parameters τ_0 and Q can be determined from an Arrhenius plot.

2) *A continuous relaxation spectrum* is observed if the diffusion takes place by a quantum mechanical tunneling process from the ground state. The relaxation amplitude remains finite at 0 K and decreases continuously with increasing temperature due to the exhaustion of defects in the ground states (see

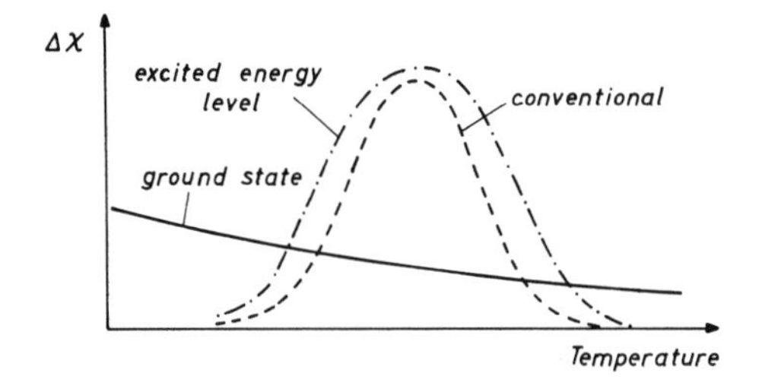

Fig. 11.2. Relaxation spectra for the conventional relaxation process (---); tunneling from the ground state (——); tunneling from excited energy levels (-·-·-)

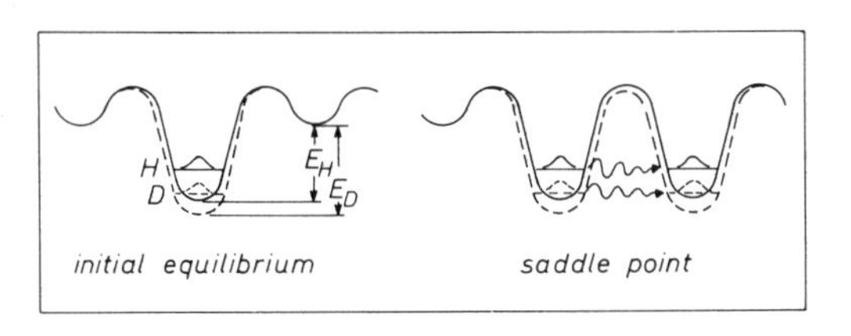

Fig. 11.3. Saddle-point configuration of phonon-assisted tunneling processes from the ground state and excited energy levels

Fig. 11.2). It should be noted that this relaxation spectrum appears only if no self-trapping of H takes place. The relaxation time shows an isotope effect due to the mass-dependent tunneling probability and zero point energy.

3) If the H-atom falls into a self-trapped state, a relaxation peak is observed similar to the conventional thermally activated relaxation process. If the tunneling takes place from the ground state, the activation energy corresponds to the deformation energy required to produce a symmetrical configuration for two neighboring potential wells (see Fig. 11.3). At higher temperatures the tunneling takes place from one of the thermally excited energy levels. The relaxation times show isotope effects in τ_0 as well as Q. The pre-exponential factor τ_0 is expected to be considerably larger than the reciprocal eigenfrequency of the localized phonon mode of the H-atom because the transition frequency τ^{-1} is reduced by the factor of the tunneling probability. Therefore these relaxation peaks have a larger half-width in comparison with the conventional relaxation peak.

11.3 Interaction of H-Atoms with the Spontaneous Magnetization

11.3.1 General Remarks

Magnetic aftereffects of hydrogen are due to the interaction energy of H-atoms with dws. A determination of this interaction energy from first principles seems to be impossible since even a quantum mechanical treatment of ideal ferromagnetic metals is only possible by means of simple models. In the following we discuss the interaction energy on the basis of the theory of micromagnetism using some quantum mechanical aspects for numerical calculations. Following the considerations developed previously [11.3, 11–14], the total interaction energy according to

$$E_{\text{int}} = E_{\text{l.r.}} + E_{\text{s.r.}} \tag{11.5}$$

is given by two terms, where $E_{\text{l.r.}}$ results from long-range and $E_{\text{s.r.}}$ from short-range interactions. Here $E_{\text{l.r.}}$ is due to elastic and magnetic dipolar interactions which decrease according to a $1/r^3$-law, whereas $E_{\text{s.r.}}$ is attributed to local perturbations of the exchange interactions and the spin-orbit coupling energy.

11.3.2 Elastic Long-Range Interactions in Domain Walls

The elastic dipole energy of point defects in dws is due to the interaction of the elastic strain field of the point defect with the magnetostrictive internal stresses $\underset{\sim}{\sigma}^B$ or the magnetostrictive strains $\underset{\sim}{\varepsilon}^B$ of the domain walls [11.3, 15] and is given by [11.16]

$$E_{\text{l.r.}} = -\underset{\sim}{\boldsymbol{Q}} \cdot \underset{\sim}{\sigma}^B = -\underset{\sim}{\boldsymbol{P}} \cdot \underset{\sim}{\varepsilon}^B , \tag{11.6}$$

where $\underset{\sim}{\boldsymbol{Q}}$ corresponds to the so-called dipole displacement tensor which is related to the double force tensor $\underset{\sim}{\boldsymbol{P}}$ by

$$P_{ij} = c_{ijkl} Q_{kl} \tag{11.7}$$

($\underset{\sim}{c}$ tensor of elastic constants). It should be noted that the trace $Q_I = Q_{11} + Q_{22} + Q_{33}$ corresponds to the total volume dilatation produced by the defect. The dipole tensors $\underset{\sim}{\boldsymbol{Q}}$ and $\underset{\sim}{\boldsymbol{P}}$ reveal the symmetry of the interstitial site occupied by the H-atom. Interstitials of cubic symmetry possess an isotropic dipole tensor, i.e., $\underset{\sim}{\boldsymbol{Q}} = Q_1 \delta_{ij}$ (δ_{ij} = Kronecker symbol). H-atoms in bcc crystals or the diatomic complexes in substitutional alloys (FeH or TiH in **Ni**Fe and **Fe**Ti) have tetragonal symmetry and $\underset{\sim}{\boldsymbol{P}}$ is given by

$$\underset{\sim}{\boldsymbol{P}} = \begin{pmatrix} A & & \\ & A & \\ & & B \end{pmatrix}, \tag{11.8}$$

where the components A refer to the twofold axes and B to the fourfold tetragonal axis of the interstitial site. According to the different possible orientations of the tetragonal axis of the H-atoms, these may occupy three different energy levels

$$E_{\text{l.r.}}^{(i)} = -\underset{\sim}{\boldsymbol{Q}}^{(i)} \cdot \underset{\sim}{\sigma}^B = -\underset{\sim}{\boldsymbol{P}}^{(i)} \cdot \underset{\sim}{\varepsilon}^B \tag{11.9}$$

with

$$\underset{\sim}{P}^{(1)} = \begin{pmatrix} B & & \\ & A & \\ & & A \end{pmatrix}; \quad \underset{\sim}{P}^{(2)} = \begin{pmatrix} A & & \\ & B & \\ & & A \end{pmatrix}; \quad \underset{\sim}{P}^{(3)} = \begin{pmatrix} A & & \\ & A & \\ & & B \end{pmatrix}. \tag{11.10}$$

In the case of a (001) 180° domain wall in a bcc crystal where the spontaneous magnetization rotates from the [100] axis into the [$\bar{1}$00] axis, the interaction energies of the three different dipole orientations are given by [11.3]

$$\begin{aligned} E^{(1)}_{\text{l.r.}} &= \tfrac{3}{2}\lambda_{100}(A-B)\sin^2\varphi(z), && [100] \\ E^{(2)}_{\text{l.r.}} &= -\tfrac{3}{2}\lambda_{100}(A-B)\sin^2\varphi(z), && [010] \\ E^{(3)}_{\text{l.r.}} &= 0, && [001]. \end{aligned} \tag{11.11}$$

Here $\varphi(z)$ corresponds to the rotation angle of $\underset{\sim}{M}_s$ with respect to the [100] axis. λ_{100} corresponds to the magnetostriction in a $\langle 100 \rangle$ direction. The following interesting results which are representative also for other dws may be derived from (11.11):

1) Due to the splitting into three energy levels, tetragonal defects give rise to an orientation aftereffect. The maximum splitting energy is obtained for $\varphi = \pi/2$ and is given by $\Delta E = 3\lambda_{100}(A-B)$.

2) The average interaction energy

$$\bar{E}_{\text{l.r.}} = \frac{1}{3}\sum_{i=1}^{3} E^{(i)}_{\text{l.r.}} \tag{11.12}$$

vanishes and therefore the long-range interactions are not the origin of a diffusion aftereffect. The average interaction energy vanishes for all those dws which are characterized by a zero volume magnetostriction, i.e., where the trace $\sum_{i=1}^{3} \varepsilon^B_{ii} = 0$ vanishes.

3) For defects of cubic symmetry ($A = B$), $E^{(i)}_{\text{l.r.}}$ and $\bar{E}_{\text{l.r.}}$ vanish, and therefore in an (100) 180° wall in iron such defects give rise neither to an orientation nor to a diffusion aftereffect.

The (100) 90° wall shows a behavior similar to that of the (100) 180° wall. In Ni the H-atoms occupy cubic interstitial sites. Therefore, in this case hydrogen can give rise only to a diffusion aftereffect. In the (001) 109° and the (001) 71° walls, furthermore, the average interaction energy is zero and therefore these walls are indifferent with respect to magnetic aftereffects. On the other hand the (110) 180° and the (112) 180° walls have a finite interaction energy $\bar{E}_{\text{l.r.}}$ due to their nonvanishing volume magnetostriction and therefore give rise to a magnetic diffusion aftereffect. The energy levels $E^{(i)}_{\text{l.r.}}$ and $\bar{E}_{\text{l.r.}}$ for tetragonal defects in Fe are shown qualitatively in Fig. 11.4.

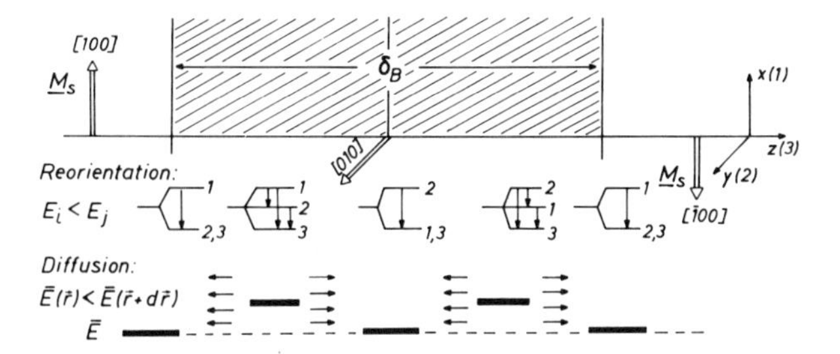

Fig. 11.4. Pictorial representation of the energy spectrum of tetragonal defects in a 180° wall. The upper part represents the dw of width δ_B with $\boldsymbol{M}_s$ rotating from [100] into [$\bar{1}$00]. In the middle part the splitting into three energy levels for $\varphi=0$, $\pi/4$, $\pi/2$, $3\pi/4$ and π is shown. The numbers 1, 2, 3 denote the energy level of the corresponding orientation of the ⟨100⟩ axis as defined on the right side. In the lower portion the full bars represent the dependence of the average interaction energy on the angle φ. The arrows denote the direction of the flux of the long-range diffusion of the defects

11.3.3 Magnetostatic Interactions

In general the additional electron of the H-atom changes the spin moments of the matrix atoms in the neighborhood of the proton. In order to get a qualitative understanding of this type of interactions, we consider the following two characteristic cases:

1) In metals at the end of the transitional series (NiH, PdH), the H-electron is transferred mainly to the d-bands and the positively charged proton is screened mainly by d-electrons. The screening cloud in the neighborhood of the proton reduces the magnetic moment by one Bohr magneton μ_B. This *magnetic hole* has an extension which corresponds to twice the screening radius [11.17, 18]

$$d=[\pi e^2 n(E_F)]^{-1/2} \tag{11.13}$$

[e = electronic charge, $n(E_F)$ = density of electronic states at the Fermi level E_F].

2) Metals of the first half of the transitional series (TiH, ZrH) transfer d-electrons to the H-atom producing an H^--state [11.17–20]. This situation also holds for the case of alloys where the diluted component lies on the left of the periodic table. Examples for such alloys are **Fe**Ti or **Fe**Zr where TiH and ZrH complexes are formed. In contrast to this behavior, in **Fe**Ni and **Fe**Pd we have a case similar to that in pure Ni. Because part of the Ti or Zr electrons are attracted by the H-atom, in this case hydrogen in contrast to NiH acts indirectly as a "ferromagnetic impurity".

If the change of the magnetic moment per H-atom corresponds to a fraction α of a Bohr magneton, the magnetostatic energy within a dw may be written as [11.21]

$$E_{\text{l.r.}}=2\pi\alpha\mu_B M_s\left(1-\frac{1}{24}\frac{d^2}{\delta_0^2}\sin^2\varphi\right), \tag{11.14}$$

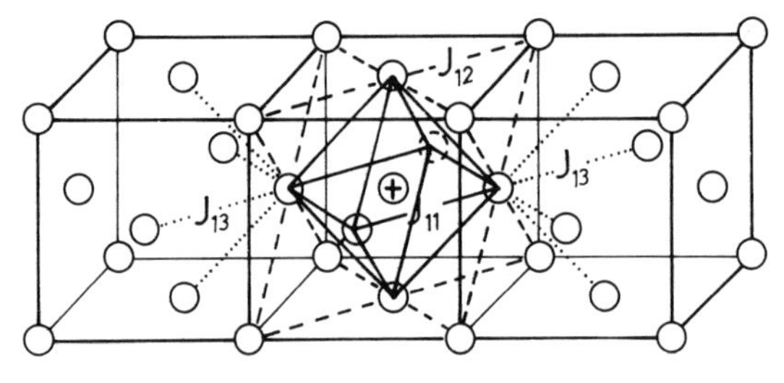

Fig. 11.5. Representation of equivalent exchange integrals in the neighborhood of octahedral H-interstitials in fcc crystals

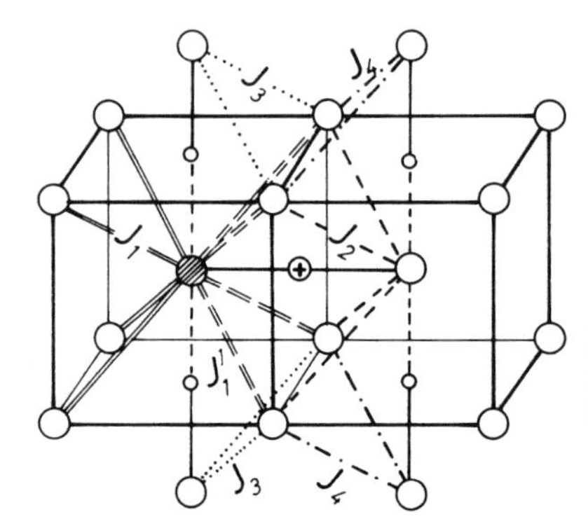

Fig. 11.6. Representation of equivalent exchange integrals in the neighborhood of a TiH complex in α iron

where $\delta_0 = \sqrt{A/K_1}$ (A is the exchange constant and K_1 the crystalline energy constant). Only the second term depends on the position of the H-atom within the dw.

11.3.4 Short-Range Interactions

Due to the electronic rearrangements in the neighborhood of the proton, the spin-dependent part of the total interaction energy is changed. The contribution of the H-atom to the magnetic energy may be neglected if we deal with the electronic configurations $H^+(p)$ or $H^-(1s^2)$. In this case only the magnetic energy terms (exchange and spin-orbit energy) of the nearest or next-nearest matrix atoms are affected.

Figures 11.5 and 11.6 represent schematically the exchange integrals which are suggested to be modified in the case of TiH complexes in an Fe matrix, and in the case of H-interstitials on the octahedral sites of an fcc crystal. In the case of localized electronic states the magnetic perturbation may be described by changes of the exchange integrals $J_{n,n'}$ and of the spin-orbit coupling constant λ_n. Denoting the spin matrix of the transition metal atom of type i on the nth lattice site $\underset{\sim}{\boldsymbol{R}}_n^{(i)}$ by $\underset{\sim}{\boldsymbol{S}}_n^{(i)}$, and the angular moments by $\underset{\sim}{\boldsymbol{L}}_n^{(i)}$, we may write for the short-range contribution to the magnetic energy

$$E_{\text{s.r.}} = -2\sum_{n,n'} J_{nn'}(|\underset{\sim}{\boldsymbol{R}}_{n'}^{(i)} - \underset{\sim}{\boldsymbol{R}}_n^{(j)}|)\underset{\sim}{\boldsymbol{S}}_{n'}^{(i)}(\underset{\sim}{\boldsymbol{R}}_{n'}^{(i)}) \cdot \underset{\sim}{\boldsymbol{S}}_n^{(j)}(\underset{\sim}{\boldsymbol{R}}_n^{(j)}) + \sum_n \lambda_n \underset{\sim}{\boldsymbol{L}}_n^{(i)} \underset{\sim}{\boldsymbol{S}}_n^{(i)}. \tag{11.15}$$

Here the sums extend over the nearest and next-nearest neighbors of the H-atom or the diatomic complex. In order to obtain $E_{\mathrm{s.r.}}$ dependent upon the position of the H-atom in the dw, the spin matrices may be replaced by the vector of spontaneous magnetization $\underset{\sim}{\boldsymbol{M}}_{\mathrm{s}}$ and its direction cosines γ_i, which gives

$$E_{\mathrm{s.r.}} = \sum_{m,n}^{3} C_{mn} \sum_{i=1}^{3} (\nabla_m \gamma_i)(\nabla_n \gamma_i) + W(\gamma_i). \tag{11.16}$$

The tensor C_{mn} corresponds to the local change of the exchange energy due to the H-atom. C_{mn} and the spin-orbit coupling energy $W(\gamma_i)$ reveal the symmetry of the lattice site occupied by the H-atom.

In the case of cubic, tetragonal, or orthorhombic symmetry of the interstitial site of the H-atom we obtain

1) cubic

$$E_{\mathrm{s.r.}} = C_0 \sum_{i=1}^{3} (\nabla \gamma_i)^2 + \varepsilon \sum_{i \neq j} \gamma_i^2 \gamma_j^2 \tag{11.17}$$

2) tetragonal

$$E_{\mathrm{s.r.}} = C_{11} \sum_{i=1}^{3} \left[\left(\frac{\partial \gamma_i}{\partial x_1} \right)^2 + \left(\frac{\partial \gamma_i}{\partial x_2} \right)^2 \right] + C_{33} \left(\frac{\partial \gamma_i}{\partial x_3} \right)^2 + \varepsilon (1 - \gamma_3^2) \tag{11.18}$$

(tetragonal axis parallel to the x_3-direction)

3) orthorhombic

$$E_{\mathrm{s.r.}} = \sum_{i=1}^{3} , \sum_{j=1}^{3} C_{jj} \left(\frac{\partial \gamma_i}{\partial x_j} \right)^2 + \sum_{i=1}^{3} \varepsilon_i \gamma_i^2 \tag{11.19}$$

(coordinates x_i parallel to the three twofold axes of the interstitial or complex configuration). The interaction constants C_{mn} and ε_i are due to the local perturbation of the exchange and the spin-orbit coupling energy, and correspond to the integral deviation from the exchange constant A and the magnetocrystalline energy constant K_1

$$\left. \begin{aligned} C_{mn} &= \int [A - A_{mn}(\boldsymbol{r})] \, d^3 r \\ \varepsilon_i &= \int [K_1 - K_i(\boldsymbol{r})] \, d^3 r \end{aligned} \right\} . \tag{11.20}$$

In (11.20) $A_{mn}(\boldsymbol{r})$ and $K_i(\boldsymbol{r})$ correspond to the r-dependent, local exchange, and magnetocrystalline energy.

For quantitative calculations we must replace $\partial \gamma_i / \partial x_j$ by the corresponding quantities holding for dws. In the case of a (001) 180° wall in α-Fe, the dw equation becomes (neglecting magnetostrictive terms) [11.22, 23]

$$A\left(\frac{d\varphi}{dz}\right)^2 - \frac{1}{4}K_1 \sin^2 2\varphi = 0\,, \tag{11.21}$$

where φ corresponds to the angle of rotation of $\boldsymbol{M}_s$ in the dw plane with respect to the [100] axis (see Fig. 11.4). This gives for $E_{\text{s.r.}}$ of different symmetries with $\gamma_1 = \cos\varphi$, $\gamma_2 = \sin\varphi$, and $\gamma_3 = 0$

1) cubic

$$E_{\text{s.r.}} = C_0 \cdot \frac{K_1}{4A} \sin^2 2\varphi + \frac{\varepsilon}{4} \sin^2 2\varphi\,; \tag{11.22}$$

2) tetragonal

$$\begin{aligned}
E^{(1)}_{\text{s.r.}} &= \frac{K_1}{4A}(C_{33}\sin^2\varphi + C_{11}\cos^2\varphi)\sin^2 2\varphi + \varepsilon \sin^2\varphi \\
E^{(2)}_{\text{s.r.}} &= \frac{K_1}{4A}(C_{33}\cos^2\varphi + C_{11}\sin^2\varphi)\sin^2 2\varphi + \varepsilon \cos^2\varphi \\
E^{(3)}_{\text{s.r.}} &= C_{33}\frac{K_1}{4A}\sin^2 2\varphi + \varepsilon\,.
\end{aligned} \tag{11.23}$$

11.4 Magnetic Aftereffects of Hydrogen

11.4.1 The Stabilization Energy

The magnetic aftereffects of atomic defects may be derived from a calculation of the time dependence of the total interaction energy between the defects and the dws. For a given distribution $c_i(z,t)$ of H-atoms in their different configurations i, the total interaction energy with a dw at position z_0 is given by [11.3]

$$W_s(z_0, t) = \sum_i \int_{-\infty}^{\infty} c_i(z,t)\, E_i(z - z_0)\, d^3r\,. \tag{11.24}$$

The interaction energy W_s is related to the reluctivity change $\Delta r(t)$ by

$$\Delta r(t) = \frac{1}{M_s^2}\,\frac{1}{L_3 p^2}\,\left.\frac{d^2 W_s(z_0,t)}{dz_0^2}\right|_{z_0 \to 0} \tag{11.25}$$

(L_3 = mean distance between dws, p is the numerical factor of the order of magnitude 4). The time dependence of the reluctivity (susceptibility) results from a rearrangement of the H-atoms within the dws. The H-atoms may lower their interaction energy with the dw by two types of rearrangements: 1) long-range diffusion over distances of the dw width δ_0 and 2) rearrangement within the different energy levels if the H-interstitials are in anisotropic configurations (see Fig. 11.4). The first of these rearrangement modes gives rise to the *diffusion aftereffect*, whereas the second process leads to the so-called *orientation aftereffect*.

11.4.2 Diffusion Aftereffect of Hydrogen

The long-range diffusion of H-atoms in dws is governed by the diffusion equation with drift term [11.24]

$$\frac{\partial c}{\partial t}=\frac{\partial}{\partial z}\left(D_z\frac{\partial c}{\partial z}+\frac{c}{kT}\frac{\partial \bar{E}}{\partial z}\right). \tag{11.26}$$

(D_z is the linear diffusion coefficient parallel to dw-normal.) The equilibrium distribution of the concentration is given by

$$c^{\infty}(z)=c_0\mathrm{e}^{-\bar{E}(z)/kT}, \tag{11.27}$$

where c_0 corresponds to the average concentration of H-atoms. At large times, the reluctivity follows a $t^{-3/2}$-law

$$\frac{1}{\chi(t)}=\frac{1}{\chi_\infty}-\frac{c_0\varepsilon_{\mathrm{eff}}^2}{kT}\frac{1}{\delta_{\mathrm{B}}}\frac{1}{(t/\tau_{\mathrm{D}})^{3/2}}+\dots. \tag{11.28}$$

The time dependence is characterized by a relaxation time

$$\tau_{\mathrm{D}}=\frac{1}{\Gamma}\left(\frac{\delta_0}{a}\right)^2\frac{1}{\nu}, \tag{11.29}$$

where the geometrical factor Γ is defined as

$$\Gamma=\frac{1}{\alpha n_E}\sum_{i=1}^{n_E}\sum_{j=1}^{N}\left(\frac{s_{ij,z}}{a}\right)^2. \tag{11.30}$$

Here we have introduced the following quantities: $\varepsilon_{\mathrm{eff}}$ = effective interaction parameter, in general corresponding to a combination of long-range and short-range interactions [11.3, 13, 14]; domain wall width $\delta_{\mathrm{B}}=\pi\sqrt{A/K_1}$; $\delta_0=\delta_{\mathrm{B}}/\pi$; a = lattice constant; n_E = number of energetically different orientations of anisotropic defects; α = number of saddle points available for the transition between configurations i and j; N = number of nearest neighbor positions; $s_{ij,z}=z$ component of the distance between positions i and j.

According to (11.29) the relaxation time of the diffusion aftereffect is a factor of $(\delta_0/a)^2$ larger than the atomic relaxation time ν^{-1} because the number of jumps required to migrate through a dw of width δ_0 is given by $(\delta_0/a)^2$. In fcc crystals, H-atoms occupy octahedral interstitial sites and τ_D is given by

$$\tau_D = \frac{1}{2}\left(\frac{\delta_0}{a}\right)^2 \frac{1}{\nu}. \tag{11.31}$$

H-atoms on tetrahedral or octahedral interstitial sites in bcc lattices have the relaxation times

$$\tau_D^T = 6\left(\frac{\delta_0}{a}\right)^2 \frac{1}{\nu}, \quad \tau_D^O = 3\left(\frac{\delta_0}{a}\right)^2 \frac{1}{\nu}. \tag{11.32}$$

The time dependence represented by (11.28) holds for the case of an unrestricted diffusion path. In many cases, however, the diffusion path is limited due to sinks or the formation of agglomerates of the migrating atoms which leads to

$$\frac{1}{\chi(t)} = \frac{1}{\chi(0)} + \frac{c_0 \varepsilon_{\mathrm{eff}}^2}{kT} \frac{1}{\delta_B} \{b_n(l)[1 - \exp(t/\tau_n)]\} \tag{11.33}$$

with

$$\tau_n = \frac{1}{\Gamma}\left(\frac{l}{n\pi a}\right)^2 \frac{1}{\nu} \quad (n = 1, 2, \ldots). \tag{11.34}$$

Here l corresponds to the free migration path and the b_n's are parameters depending somewhat on l and the special spin structure of the dw. Our results show that (11.33) corresponds to the diffusion of H in a thin sheet, i.e., the limited magnetic diffusion aftereffect corresponds to the mechanical Gorsky effect [11.25, 26].

11.4.3 Orientation Aftereffect of Hydrogen

In bcc crystals the H-atoms occupy lattice sites of tetragonal symmetry [11.27–30], whereas in fcc crystals, interstitial sites of cubic symmetry are occupied. Configurations of lower symmetry than cubic are expected also in bcc and fcc alloys (**Fe**Ti, **Ni**Fe) or in the case where the H-atoms form diatomic complexes with other interstitials (CH in Ni or OH in Nb). In these cases the interaction energy of the H-atoms with the dw decreases by a local rearrangement on the different possible energy levels E_i. This rearrangement is governed by the rate equation

$$\frac{\partial c_i}{\partial t} = -\sum_{i \neq j} (\nu_{ij} c_i - \nu_{ji} c_j), \tag{11.35}$$

where ν_{ij} corresponds to the atomic jump frequency between neighboring positions. In general, ν_{ij} can be replaced by one unique frequency [11.2–4]

$$\nu = \nu_0 \mathrm{e}^{-Q/kT} \tag{11.36}$$

(ν_0 is the attempt frequency and Q the activation energy). The thermodynamic equilibrium distribution on the different energy levels is given by

$$c_i^\infty = \frac{c_0 \mathrm{e}^{-E_i/kT}}{\sum_{i=1}^{n_E} \mathrm{e}^{-E_i/kT}}. \tag{11.37}$$

The time dependence of the reluctivity obeys an exponential law

$$r(t) = \frac{1}{\chi(t)} = \frac{1}{\chi(0)} + \frac{c_0 \varepsilon^2}{kT} \frac{1}{\delta_\mathrm{B}} \sum_i a_i [1 - \exp(-t/\tau_i)] \tag{11.38}$$

(a_i are the amplitudes of different relaxation modes), where the relaxation times τ_i are related to the atomic jump frequency ν by a simple geometrical factor. For example, in the case of H-atoms on octahedral or tetrahedral interstitial sites in bcc crystals, we have the relaxation times

$$\tau^\mathrm{O} = 1/6\nu, \tau^\mathrm{T} = 1/3\nu. \tag{11.39}$$

11.4.4 Magnetic Cold-Work Aftereffect

In a previous paper *Schöck* [11.31] discussed the mechanical relaxation phenomena due to dislocation which are decorated by interstitial atoms. He showed that the so-called cold-work peaks measured in the FeC system [11.32, 33] can be attributed to the creep motion of dislocations. More recently the author showed [11.34, 35] that the conventional mechanical cold-work peak can also be investigated by means of the internal magnetostrictive stresses of dws. Under the influence of the elastic stresses of dws, dislocations are attracted or repelled by the dws (see Fig. 11.7). If the dislocations interact with hydrogen atoms, they approach their equilibrium position by a creep motion.

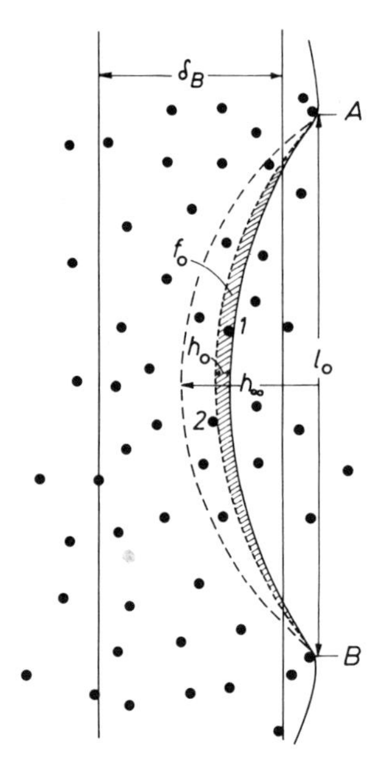

Fig. 11.7. Dislocation segment of length l_0 interacting with a dw and impurity atoms (●). The shaded area f_0 is swept out by passing the impurity atom (1). The amplitude h_0 of this elementary jump is determined by impurity atom (2). In the equilibrium configuration the dislocation segment is elongated to an amplitude h_∞ by the magnetostrictive stress of the dw

This process is connected with a time-dependent reduction of the mobility of the dws. Within the framework of the string model we find

$$\frac{1}{\chi(t)}=\frac{1}{\chi(0)}+\Delta\chi_\infty\left\{1-\frac{2kT}{\bar{\sigma}^B v}\operatorname{arth}\left[\exp(-t/\tau)\operatorname{th}\left(\frac{\bar{\sigma}^B v}{2kT}\right)\right]\right\} \tag{11.40}$$

($\Delta\chi_\infty$ is the amplitude of the relaxation, $\bar{\sigma}^B$ the average internal stress amplitude in the dw, and $v=bdl$ is the activation volume of the thermally activated process). The relaxation time is given by

$$\tau=\frac{kTl_0}{8Gb^3 d\nu_0}\exp(W/kT) \tag{11.41}$$

(l_0 is the mean segment length of dislocations, G the shear modulus, b the modulus of Burgers vector, d the width of the potential well of the H-atom absorbed by the dislocation, and ν_0 the attempt frequency of H-atoms in the dislocation core).

The activation energy $W=Q^{(\mathrm{H})}+\varepsilon_\mathrm{B}$ of the relaxation process is composed of the migration energy $Q^{(\mathrm{H})}$ of hydrogen in the ideal lattice, and the binding energy ε_B of hydrogen to the dislocation. Figures 11.8 and 11.9 present the mechanism for the displacement of a dislocation with a H-atom at the core. The motion of the dislocation according to this model is determined by the thermally activated diffusion of the H-atom.

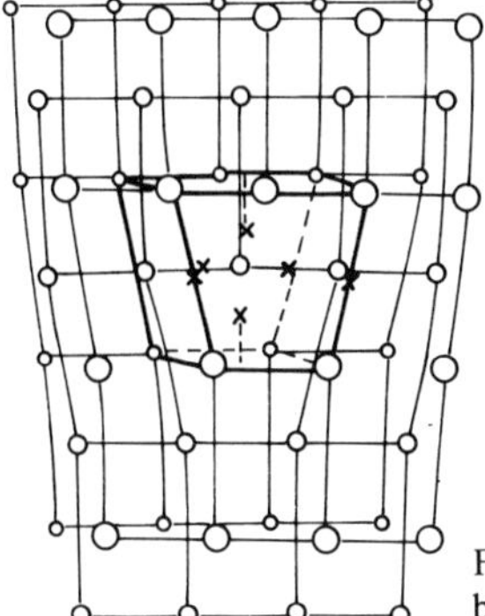

Fig. 11.8. Possible positions of H-atoms in the dislocation core of a bcc crystal (— × —)

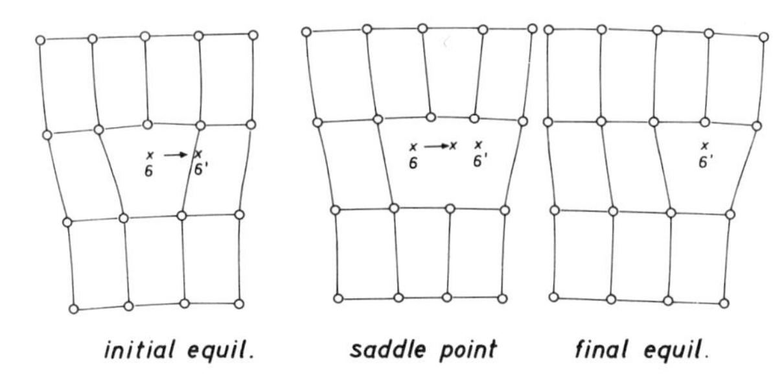

Fig. 11.9. Pictorial representation of the combined motion of dislocations and impurity atom: 1) initial equilibrium; 2) saddle point; 3) nearest neighbor equilibrium position

The string model discussed so far applies if the temperature lies sufficiently above the temperature where dislocation kinks are generated. Below this temperature the dislocations lie as straight segments in the potential wells of the Peierls potential. Segments lying in neighboring Peierls wells are connected by so-called geometrical kinks. A movement of the dislocation is possible by the displacement of geometrical kinks [11.36, 37] or by the formation of double kinks [11.38]. Both processes are thermally activated and are influenced by hydrogen. These types of relaxations are described by (11.41) with an appropriate definition of the activation parameters. For example, W corresponds to the sum of the formation energy of double kinks and the binding energy of hydrogen to the dislocation.

11.5 Discussion of Experimental Results

11.5.1 Nickel

The isochronal magnetic relaxation spectrum of Ni single crystals electrolytically charged with H or D is shown in Figs. 11.10 and 11.11. The specimen contained in the bulk ~ 100 at ppm H. In a surface layer of 10 µm thickness the Ni hydride phase $NiH_{0.6}$ was present. At about 205 K (212 K) a broad

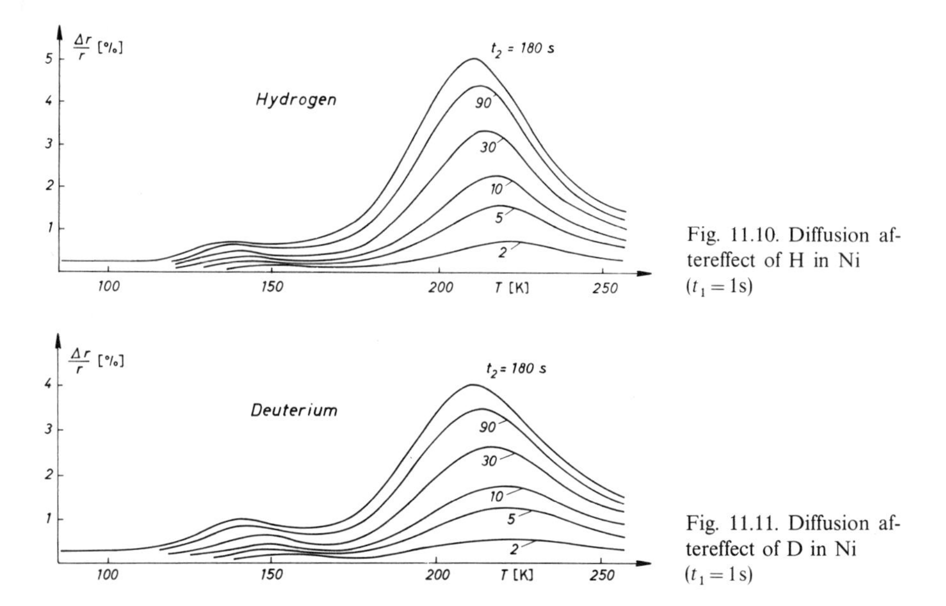

Fig. 11.10. Diffusion aftereffect of H in Ni ($t_1 = 1$s)

Fig. 11.11. Diffusion aftereffect of D in Ni ($t_1 = 1$s)

relaxation maximum (half-width 50 K) is observed, revealing a significant shift of T_{max} with varying t_2. The relaxation curves are well separated on the high- and the low-temperature sides of the maxima. All these properties of the maxima are characteristic features of a diffusion aftereffect. Orientation aftereffects would lead to relaxation maxima with a half-width of 25 K and a fitting on the high-temperature side.

The activation energies for the diffusion of H and D were determined from the temperature dependence of the isothermal relaxation curves with the result: $Q^{(H)} = 0.46 \pm 0.04$ eV; $Q^{(D)} = 0.42 \pm 0.04$ eV. From other measurements [11.39–41], the migration energies for the free diffusion of H or D in Ni were found to be $Q^{(H)} = 0.41$ eV and $Q^{(D)} = 0.40$ eV.

The discrepancy between the present results and the results for the free diffusion are attributed to the fact that we are dealing with large H-concentrations (~3000 at·ppm). Under these conditions the diffusion of hydrogen is influenced by the elastic interactions between H-atoms. If the specimen is annealed at temperatures between 100 °C and 200 °C, the H-concentration decreases and the relaxation peaks shift to lower temperatures (see Fig. 11.12). Due to the reduced elastic interaction, the migration energies now are found to be smaller: $Q^{(H)} = 0.41 \pm 0.04$ eV; $Q^{(D)} = 0.39 \pm 0.04$ eV.

11.5.2 Cobalt

In a previous investigation in rather impure, hydrogen-charged Co a rather structurized relaxation spectrum in the temperature range from 130 K to RT was observed [11.42]. Figure 11.13 presents the relaxation spectrum measured

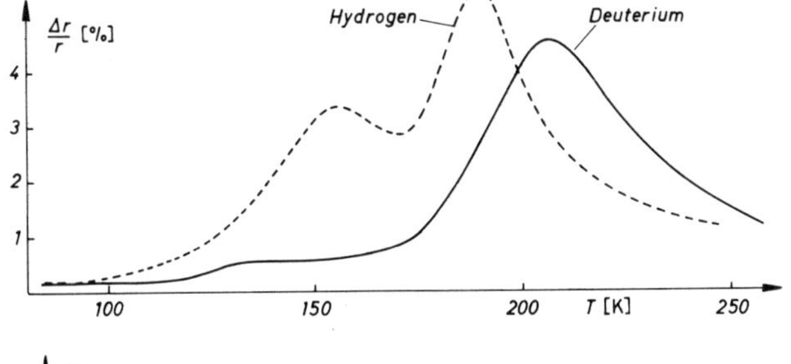

Fig. 11.12. Relaxation spectrum of H and D in Ni after annealing at 300 K for 200 h ($t_1 = 1$ s; $t_2 = 180$s)

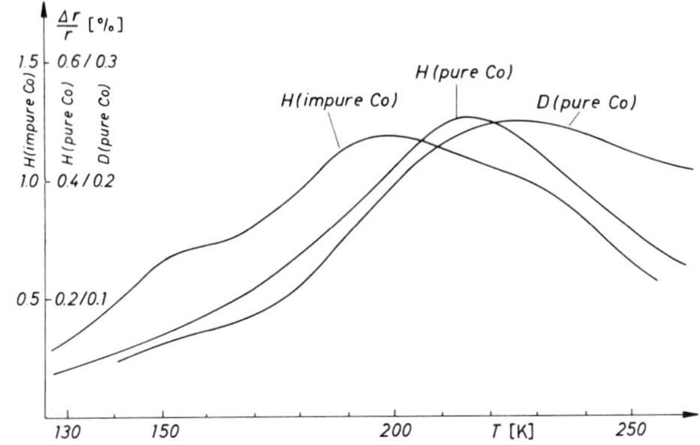

Fig. 11.13. Relaxation spectra of H and D in hcp cobalt crystals [11.43] ($t_1 = 1$s; $t_2 = 180$s)

recently [11.43] for a rather pure Co specimen after hydrogen charging, together with the results for an impure crystal containing ~150 ppm Ni. In both cases a major relaxation maximum with a half-width of 65 K is observed at ~215 K. Satellite maxima are observed in the impure specimen on the low- and the high-temperature side, giving rise to shoulders in the relaxation spectrum. The central relaxation peak at 215 K is attributed to a diffusion aftereffect of hydrogen and the relaxation shoulders in the impure specimen to complexes of Ni and H in the hexagonal-close-packed (hcp) lattice.

Since in hcp Co the dws are oriented perpendicular to the basal plane, any diffusion aftereffect is due to a motion of hydrogen parallel to the basal planes. The existence of a diffusion aftereffect points to the fact that hydrogen occupies interstitial sites of hexagonal or trigonal symmetry. Interstitial positions of this type are the octahedral and the tetrahedral interstitial sites as shown in Fig. 11.14. Measurements of the local magnetic field performed by means of muons [11.44] have shown that light interstitials in Co prefer the octahedral sites. Applying these results to hydrogen, we must assume that the tetrahedral position either corresponds to a metastable configuration or to a saddle point for hydrogen. The relaxation time for the octahedral site is given by

$$\tau_D = 2\left(\frac{\delta_0}{a}\right)^2 \frac{1}{\nu_{||}}, \tag{11.42}$$

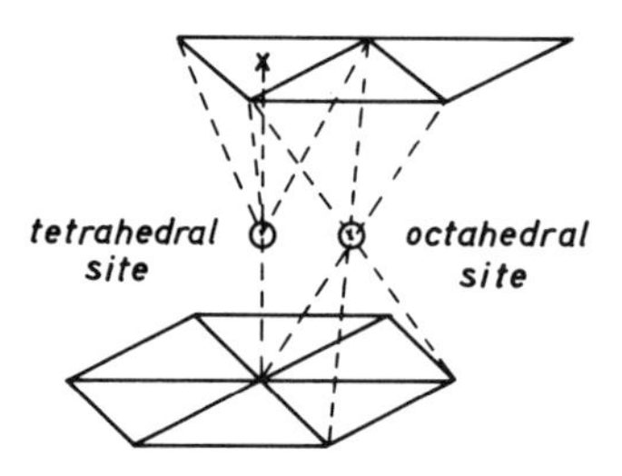

Fig. 11.14. Octahedral and tetrahedral interstitial sites in hcp crystals

where a corresponds to the lattice parameter in the basal plane and $\nu_{||}$ to the jump frequency from one octahedral site to a neighboring one. From (11.42) we may derive the activation energy for the migration of H. With $2(\delta_0/a)^2 = 600$, $\tau_D = 60$ s, $\nu_0 = 10^{13}$ s at $T_{max} = 215$ K, we find $Q^{(H)} = 0.50$ eV. For deuterium, $T_{max} = 223$ K, we find, with the same parameters as for H, $Q^{(D)} = 0.52$ eV. If the relaxation shoulders are analyzed under the assumption that we deal with orientation aftereffects, we obtain $Q = 0.40$ eV and $Q = 0.60$ eV for the low- and the high-temperature shoulders, respectively.

11.5.3 Iron

Many experiments were performed previously in order to investigate the Snoek-type relaxation of hydrogen in a bcc lattice. In a series of papers [11.45–50] the technique of internal friction was applied. Other authors used the technique of magnetic relaxation [11.35, 51–53]. These latter experiments showed that relaxation spectra appear in the temperature ranges 15–40 K and 80–150 K. *Heller* and *Gibala* [11.45, 46] originally suggested that the low-temperature peak is due to the Snoek relaxation of hydrogen. However, there are a number of reasons why the Snoek relaxation of H in α iron is difficult to be detected: 1) The solubility of H in α iron below room temperature lies below 1 ppm [11.54]. 2) The elastic as well as the magnetic interaction energies with dws are rather small (see Sect. 11.3). 3) Furthermore, it should be noted that the diffusion aftereffect of H in α iron is even less probably detectable by the magnetic aftereffect because the average magnetoelastic interactions with the dws vanish in α iron, as discussed in Section 11.3.

In the light of these facts it seems to be self-suggesting to look for other explanations of the low-temperature relaxation. Recent experiments have indeed shown [11.35, 51, 55, 56] that the 15–40 K spectrum as well as the spectrum from 80–150 K appears only in connection with dislocations. This is qualitatively supported by the results represented in Fig. 11.15. In the plastically deformed crystal a relaxation peak at about 18 K is observed. After charging the crystal electrolytically with hydrogen, the 18 K peak is suppressed completely, and a large relaxation peak appears at ~ 130 K. This latter relaxation peak is attributed to the cold-work peak of dislocations and hydrogen (see Sect. 11.4.4). A more detailed analysis shows that the activation

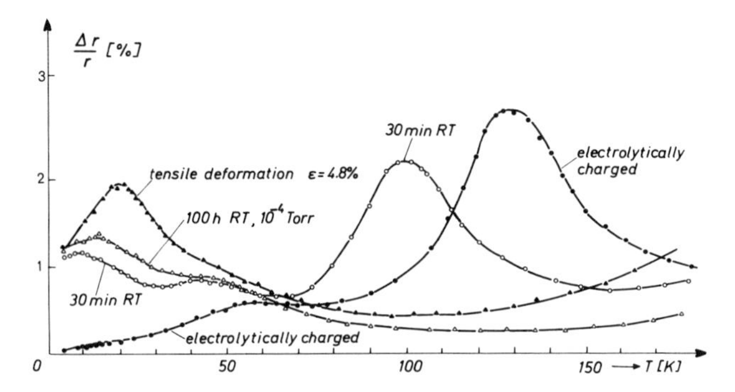

Fig. 11.15. Magnetic relaxation spectra of plastically deformed and hydrogen charged α-Fe single crystals. (▲ tensile deformation, $\varepsilon=4.8\%$; ● electrolytically charged at RT; ○ 30 min annealed at RT; △ 100 h annealed at RT, at 10^{-4} Torr) ($t_1=1$ s; $t_2=180$ s)

energy of this relaxation peak is given by 0.30 eV [11.55]. Assuming an activation energy of 0.10 eV [11.57] for the migration energy of hydrogen, we obtain a binding energy of 0.20 eV for H-atoms and edge dislocations. The pre-exponential factor is found to be $\tau_0=3.2\cdot10^{-10}$ s.

An annealing treatment of 30 min at RT shifts the 130 K peak to 90 K, and the low-temperature peak reappears with smaller amplitude at ~10 K. After further annealing (100 h, RT, 10^{-6} Torr), the cold-work peak disappears, and the amplitude of the low-temperature peak increases further and shifts to 15 K. The relaxation spectra shown in Fig. 11.15 show a shoulder at 40–50 K. This relaxation peak can be obtained separately by quenching in a H atmosphere from high temperatures (1000 K) [11.35]. The 18 K deformation peak has been attributed to the formation of double kinks in nonscrew dislocations (Bordini peak) [11.35, 53]. The activation parameters are $Q=0.048$ eV, $\tau_0=1.6\cdot10^{-11}$ s [11.35].

Whereas the interrelation between the 18 K and the 130 K peak can be conclusively explained by the role of hydrogen, no final model has been developed for the 40–50 K relaxation. Possibly this relaxation process is due to the migration of geometrical kinks in screw dislocations [11.36] which is influenced by the presence of hydrogen [11.35].

The dependence of the 18 K peak on plastic deformation and on hydrogen charging is incompatible with the assumption of a hydrogen Snoek effect. Also, recent investigations of *Au* and *Birnbaum* [11.52], who have studied the effect of impurities on the low-temperature spectrum, give no clear indication for the presence of a hydrogen Snoek effect.

11.5.4 Diatomic CH Complex in Ni

The relaxation spectrum in Ni single crystals containing different amounts of carbon interstitials was measured after electrolytical charging with hydrogen or

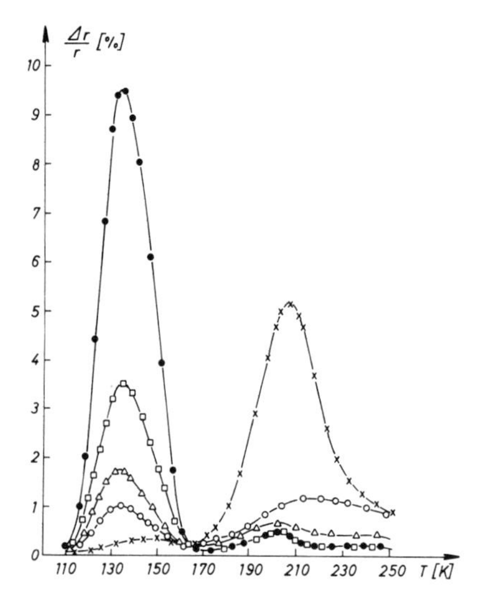

Fig. 11.16. Relaxation spectra of a carbon- and hydrogen-charged Ni single crystal: ● 500 wt.ppm C, 200 wt.ppm C, △ 80 wt.ppm C, ○ 40 wt.ppm C, $x<5$ wt.ppm C and doped 18 h electrolytically by hydrogen ($25\,\mathrm{mA\,cm^{-2}}$) ($t_1=1$ s; $t_2=180$ s)

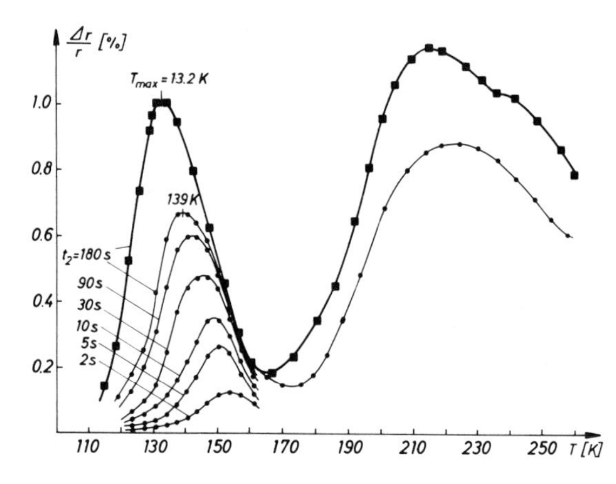

Fig. 11.17. Comparison of the relaxation spectra of the hydrogen and the deuterium doped crystal (40 wt.ppm C); (–·–·– Ni+C+H), (—— Ni+C+D) ($t_1=1$ s)

deuterium [11.35, 58]. Figure 11.16 shows the relaxation spectra of the carbon free crystal and crystals with 40–500 wt·ppm C. In addition to the 205 K peak of the hydrogen diffusion aftereffect, a relaxation peak at 132 K is observed. The diffusion aftereffect vanishes nearly completely in the specimens containing carbon whereas the amplitude of the 132 K peak increases linearly with carbon content. This latter behavior is an indication for the formation of diatomic CH complexes. The concentration of these complexes is given by

$$C_{\mathrm{CH}}=\frac{12C_{\mathrm{H}}C_{\mathrm{C}}}{1-12C_{\mathrm{H}}C_{\mathrm{C}}}\,\mathrm{e}^{\varepsilon_{\mathrm{B}}/kT} \tag{11.43}$$

(C_{H} is the concentration of H-atoms, C_{C} the concentration of C-atoms, and ε_{B} the binding energy between C- and H-atoms). Figure 11.17 shows a series of

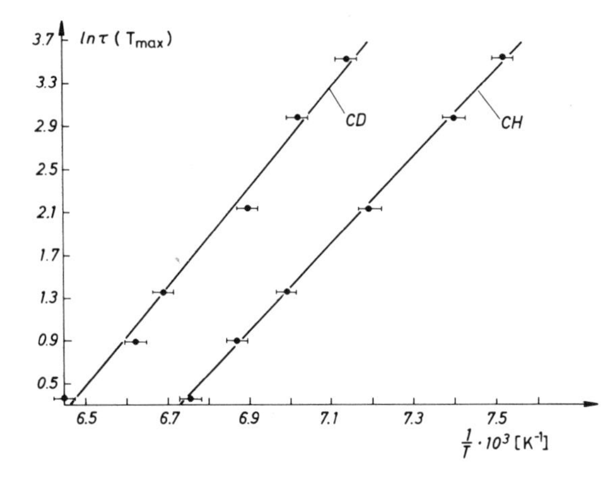

Fig. 11.18. Arrhenius diagrams of the relaxation times of the CH and the CD complexes

isochrones for the CH complex and an isochrone for the CD complex. The temperature dependence of the relaxation time τ could be determined from the temperature shift of the relaxation maximum according to (11.3). As shown in Fig. 11.18, the relaxation times for CH as well as for CD complexes obey Arrhenius relations with the activation parameters $Q^{(H)}=0.34\,eV$, $\tau_0^{(H)}=4 \cdot 10^{-12}\,s$; $Q^{(D)}=0.40\,eV$, $\tau_0^{(D)}=8 \cdot 10^{-14}\,s$. The complexes were found to anneal out by dissociation in the temperature range between 100 and 200 °C [11.58, 59]. The dissociation energies ε_{Di} were obtained by means of a Brinkman-Meechan analysis with the results: $\varepsilon_{Di}^{(H)}=0.65 \pm 0.03\,eV$, $\varepsilon_{Di}^{(D)}=0.57 \pm 0.03\,eV$. Together with the results for the migration energies for H and D (see Sect. 11.5.1), we find for the binding energies: $\varepsilon_B^{(H)}=0.24 \pm 0.03\,eV$; $\varepsilon_B^{(D)}=0.18 \pm 0.03\,eV$.

The experimental results obtained for Q and τ_0 reveal significant isotope effects for the CH and CD complexes. In the classical description of thermally activated processes the activation energies for H and D should be equivalent because their electronic structures are the same. Only the pre-exponential factors should be different by a factor of $\sqrt{2}$. The activation energies measured, however, differ by 20%, and $\tau_0^{(D)}$ is a factor of 50 smaller than $\tau_0^{(H)}$. For an explanation of the isotope effects we describe the diatomic CH" complexes by a quantum mechanical harmonic oscillator model. In this model the transition between neighboring octahedral positions is described by a tunneling process of the H-atoms from their excited energy levels. Since both interstitial atoms are supposed to occupy neighboring octahedral lattice sites, the diatomic defect has orthorhombic symmetry as shown in Fig. 11.19.

The CH complexes possess a preferred ⟨110⟩ axis parallel to the connection line between the two interstitials. Therefore the complexes have six energetically different orientations, similar to the divacancies in fcc crystals [11.60]. A reorientation of the CH complex takes place by a hopping of the H-isotope into one of the neighboring four octahedral positions. Two successive jumps are required for the perpendicular orientation of the CH complex (position F in

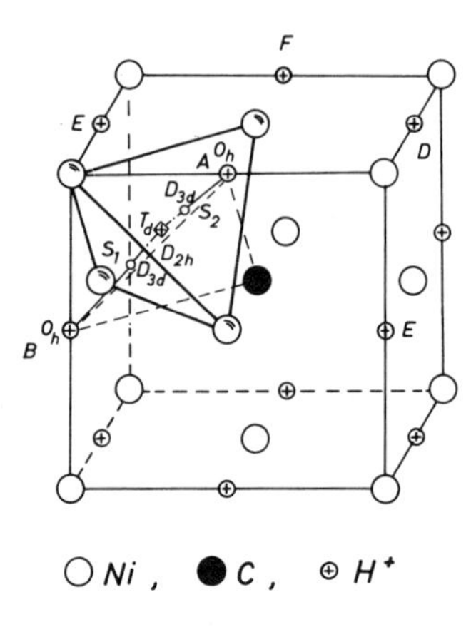

Fig. 11.19. Model of the orthorhombic CH complexes showing octahedral positions of H-atoms ⊕. Saddle points are denoted by $S_{1,2}$

Fig. 11.19). Therefore the relaxation process is described by two relaxation times. A solution of the rate Equation (11.24) gives

$$\frac{1}{\chi(t)}=\frac{1}{\chi(0)}+\frac{C_{\mathrm{CH}}}{kT}\{A_1[1-\exp(-t/\tau_1)]+A_2[1-\exp(-t/\tau_2)]\}, \tag{11.44}$$

with $\tau_1=1/6\nu$ and $\tau_2=1/4\nu$, where the amplitudes A_i are proportional to the squares of the interaction constants defined in Section 11.3.

Two possible transition paths of the hydrogen isotope, as shown in Fig. 11.19, must be considered for the reorientation of the preferred axis of the CH complex. One path leads through an orthorhombic configuration D_{2h}, and the other leads through the tetrahedral site T_d of cubic symmetry. It is supposed that this latter path is energetically favored because in the tetrahedral position the distances of the H-isotopes to the neighboring metal atoms are larger than on all other transition paths. Therefore, it is assumed that in the T_d position a metastable configuration of the H-atom exists. This assumption is also supported by the fact that dilatation centers attract each other in $\langle 111\rangle$ directions [11.61]. According to Fig. 11.19, the H-atoms have to overcome two successive saddle points (S_1 and S_2) of trigonal symmetry D_{3d}. This leads to the potential model for the saddle-point configuration represented in Fig. 11.20 with two rather narrow potential barriers around the D_{3d} positions.

The measured isotope effects are due to the following facts: 1) The hydrogen isotopes occupy discrete energy levels. 2) The hopping of the hydrogen isotopes takes place by a tunneling process through the potential barrier.

The energy levels in a first approximation may be determined from an oscillator model treating the CH complex by an harmonic oscillator with a potential

$$V(x)=V_0\cdot\sin^2\frac{2\pi x}{\lambda}, \tag{11.45}$$

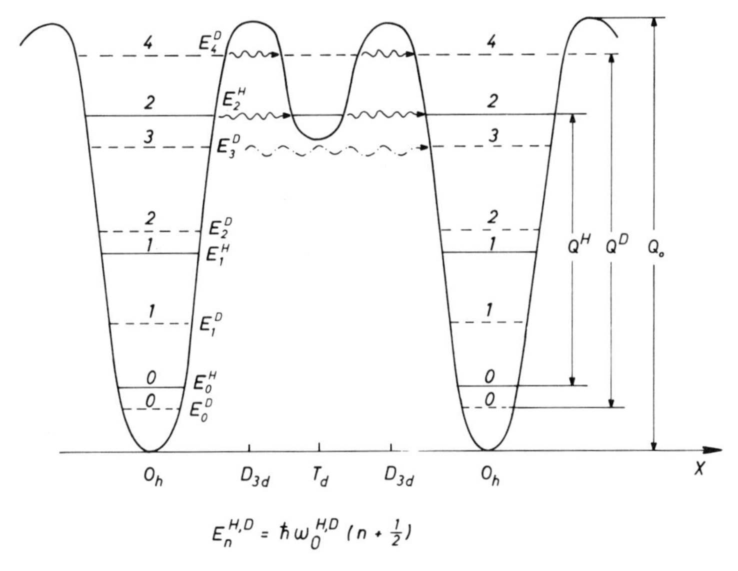

Fig. 11.20. Model for the saddle-point configuration of the potential for the reorientation of diatomic CH complexes

where $\lambda = (a/2)\sqrt{3}$ and the amplitude V_0 is somewhat larger than the measured activation energies. The eigenfrequency of the harmonic oscillators is given by [11.58]

$$\nu_0^{(\mathrm{H,D})} = \frac{1}{3}\,\frac{2\cdot\sqrt{3}}{a}\sqrt{\frac{2V_0}{m_r}} \tag{11.46}$$

$(m_r = m_{\mathrm{H,D}} \cdot m_{\mathrm{C}}/(m_{\mathrm{H,D}} + m_{\mathrm{C}})$, m_{H} is the hydrogen mass, m_{D} the deuterium mass, and m_{C} the carbon mass). With $V_0 = 0.5$ eV, we find $\nu_0^{(\mathrm{H})} = 3.4 \cdot 10^{13}\ \mathrm{s}^{-1}$, giving $\tau_{0,1}^{(\mathrm{H})} = 1/6\nu_0^{(\mathrm{H})} = 5 \cdot 10^{-15}$ s. Similarly we find for deuterium $\tau_{0,1}^{\mathrm{D}} = 6.7 \cdot 10^{-15}$ s. Our experimental results were $\tau_0^{(\mathrm{H})} = 4 \cdot 10^{-12}$ s and $\tau_0^{(\mathrm{D})} = 10^{-13}$ s, i.e., our theoretical results fail by a factor of 10^3 and 15, respectively. These large discrepancies in τ_0, and the isotope effect measured for Q, support the model of tunneling protons from an excited energy level. In this model the jump frequency may be written as

$$\nu^{(\mathrm{H,D})} = \frac{\sum\limits_{n=0}^{n_i} \nu_0^{(\mathrm{H,D})} P_n(E_n^{(\mathrm{H,D})}) \exp(-E_n^{(\mathrm{H,D})}/kT)}{\sum\limits_{n=0}^{n_i} \exp(-E_n^{(\mathrm{H,D})}/kT)}, \tag{11.47}$$

where we have neglected effects due to lattice deformations. $E_n^{(\mathrm{H,D})}$ corresponds to the nth energy level of H or D

$$E_n^{(\mathrm{H,D})} = h\nu_0^{(\mathrm{H,D})}(n + \tfrac{3}{2}), \tag{11.48}$$

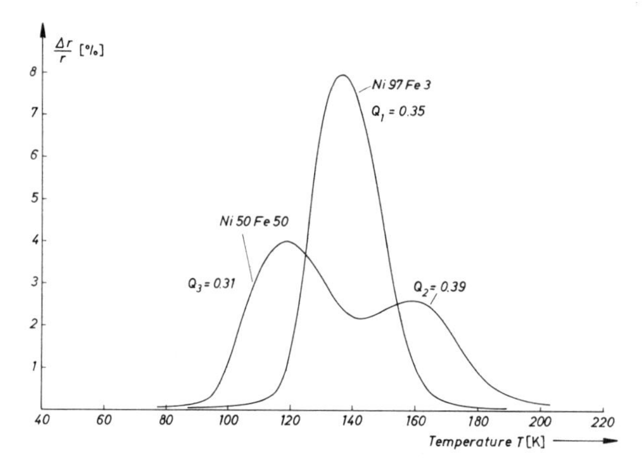

Fig. 11.21. Relaxation spectra of hydrogen-charged **Ni**Fe alloys ($t_1 = 1$ s, $t_2 = 180$ s)

and

$$P_n(E_n^{(\mathrm{H,D})}) = \exp\left\{-\frac{2}{\hbar}\int_{x_1}^{x_2} \sqrt{2m_{\mathrm{H,D}}[V(x)-E_n^{(\mathrm{H,D})}]}\right\} \tag{11.49}$$

represents the tunneling probability from the nth energy level. In order to interpret the measured isotope effects $Q^{(\mathrm{H})} < Q^{(\mathrm{D})}$ and $\tau_0^{(\mathrm{H})} > \tau_0^{(\mathrm{D})}$, we have chosen a potential $V(x)$ where $E_2^{(\mathrm{H})}$ lies just above the metastable configuration T_d and the tunneling of D takes place from the fourth energy level. The theoretical activation energies are $Q_{\mathrm{theor.}}^{(\mathrm{H})} = E_2^{(\mathrm{H})} - E_0^{(\mathrm{H})} = 0.32$ eV and $Q_{\mathrm{theor.}}^{(\mathrm{D})} = E_4^{(\mathrm{D})} - E_0^{(\mathrm{D})} = 0.41$ eV, which gives a satisfactory agreement with the experimental results $Q^{(\mathrm{H})} = 0.34$ eV and $Q^{(\mathrm{D})} = 0.40$ eV. The isotope effects for the diffusion of H in Pd have been explained in a similar way by *Sicking* [11.62].

11.5.5 Hydrogen Relaxation in NiFe Alloys

The magnetic relaxation spectrum of fcc **Ni**Fe alloys after charging electrolytically with hydrogen is shown in Fig. 11.21. Whereas in the diluted alloy Ni-3% Fe, only one relaxation maximum at ~ 135 K is observed, further relaxation maxima appear at 118 K and 160 K in the Ni-50% Fe alloy. The relaxation maxima show the behavior of orientation aftereffects, and the activation parameters may be determined along the lines outlined in Section 11.2. We obtain $\tau_0 = 1.3 \cdot 10^{-12}$ s and $Q_1 = 0.39$ eV for the upper maximum, and the lower maximum is described by $\tau_0 = 3 \cdot 10^{-12}$ s, $Q_2 = 0.35$ eV. Similar values for the activation parameters were determined previously by *Adler* [11.63] and *Adler* and *Radeloff* [11.64–66] in NiFe alloys.

In fcc alloys an orientation aftereffect of hydrogen is only possible if hydrogen forms complexes with the diluted component. Figure 11.22 shows a model for a tetragonal FeH complex. A binding energy of 30 meV was found

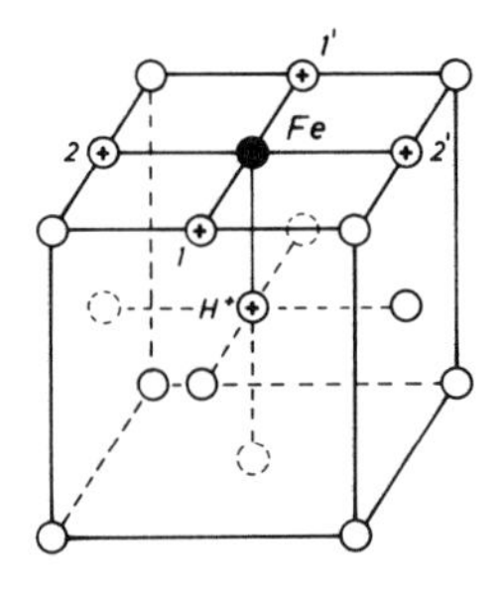

Fig. 11.22. Model of a tetragonal FeH complex in **Ni**Fe

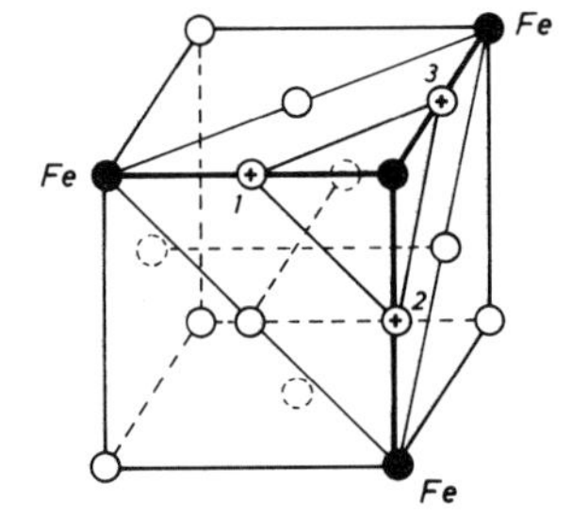

Fig. 11.23. Model of an FeH complex with a trigonal Fe arrangement in **Ni**Fe

for this FeH complex. The H-atom occupies an octahedral interstitial site, and a reorientation of the complex takes place by jumps of the H-atom into the positions 1, 1′, 2, 2′. The relaxation process is described by one relaxation time $\tau = 1/6\nu$. If the Fe-concentration increases, the probability increases that more than one Fe-atom is involved in the complex formation. In Fig. 11.23 such a configuration is shown for a trigonal arrangement of Fe-atoms where the H-atom can occupy three energetically different, but crystallographically equivalent positions. The relaxation time for this complex is given by $\tau = 1/3\nu$. It is suggested that the relaxation peak observed at $-125\,°\mathrm{C}$ is due to such an FeH complex.

11.5.6 Magnetic Relaxation in FePd Alloys

The relaxation spectrum of FePd alloys was measured for 1, 3, 5, 10 at.%Pd. Whereas in the bcc alloy Fe-1%Pd (α phase) only an extended background relaxation in the whole temperature range is observed, we have found a well-developed relaxation maximum in the eutectic alloys (α phase + FePd) at 142 K [11.56]. The relaxation amplitude has a maximum for 10 at.%Pd. According to the results represented in Fig. 11.24, the half-width is relatively large (~ 50 K), thus pointing to a diffusion aftereffect. Figure 11.24 also includes the relaxation curve for deuterium with a maximum at 149 K. An activation energy of $Q^{(\mathrm{H})} = 0.40$ eV was determined for the hydrogen diffusion aftereffect. From the shift of the deuterium maximum to larger temperatures we obtain under the assumption of equal pre-exponential factors for the deuterium $Q^{(\mathrm{D})} = 0.42$ eV. The measured relaxation is due to the diffusion of H or D in the fcc FePd component of the eutectic alloy. No indication for complex formation could be observed.

11.5.7 Other fcc Alloys

From the results of the preceding sections it became evident that hydrogen isotopes may form diatomic complexes with transition metals in diluted alloys.

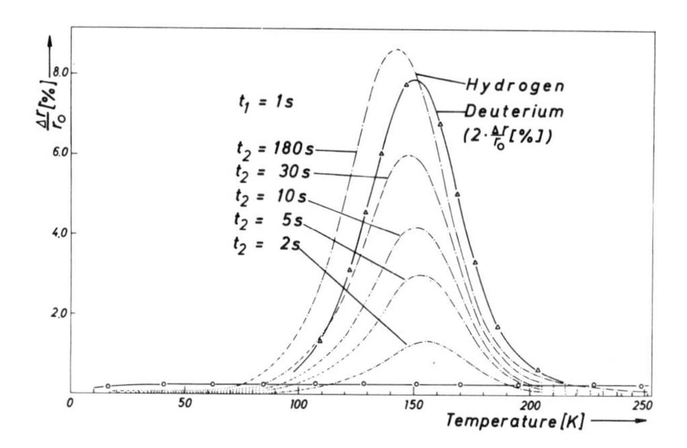

Fig. 11.24. Relaxation spectra of a hydrogen and deuterium charged **Fe**–10% Pd alloy. (–○–○– 100 h–750 °C–H_2 and 2 h–820 °C UHV; –·–·– 1 h doped electrolytically with H, 53 mA cm^{-2}; —△—△— 1 h doped electrolytically with D, 53 mA cm^{-2})

In particular, if the diluted component has a smaller number of 3*d*-electrons than the matrix atoms, such complex formations seem to be favored. Therefore, alloys of the type **Ni**Me seem to be most appropriate for an investigation of MeH complexes. The experimental results on the activation parameters of several MeH complexes are summarized in Table 11.1. It is noteworthy that also in the **Ni**Cu alloys the formation of CuH complexes is observable, giving rise to a relaxation maximum at ~140 K. Obviously in Ni the shielding of neighbouring Cu^+ and H^+ ions is energetically more favorable than the shielding of two separated ions. The activation parameters of all MeH complexes investigated in fcc alloys show similar isotope effects. For example, the difference in the activation energies for the reorientation of the MeD and the MeH complexes is of the order of magnitude 40–50 meV, and the ratio of the pre-exponential factors, $\tau_0^{(H)}/\tau_0^{(D)}$ ranges between 6 and 15. As in the case of the NiFe alloys, we attribute the observed relaxation maxima to the reorientation of tetragonal MeH complexes (see Fig. 11.22). As in the CH complexes in Ni, the transition path of the H-atoms is supposed to lead over a metastable tetrahedral lattice site.

11.5.8 TiH Complexes in Iron

As was discussed in Section 11.5.3, the study of hydrogen in α iron is difficult because of its low solubility. We therefore investigated a ferromagnetic bcc FeTi alloy with larger hydrogen solubility [11.56]. The relaxation spectrum was studied for the Fe-2 at.%Ti and Fe-10 at.%Ti alloy. The first of these alloys forms a bcc solid solution (α phase), and the second one corresponds to a eutectic alloy (α phase $+Fe_2Ti$).

Table 11.1. Activation energies and pre-exponential factors of MeH complexes in fcc crystals [11.67]

Alloy [at. %]	Annealing atmosphere 1120 K	$T_{max}(t_2=180\,s)$ [K]	Q [eV]	τ_0 [s]	$Q^{(D)}-Q^{(H)}$ [eV]	$\tau_0^{(H)}/\tau_0^{(D)}$
Ni−2%Pd	H	130.5	0.325 ± 0.005	$(5\ \pm1)\cdot10^{-12}$	0.05	16.7
	D	137	0.375 ± 0.005	$(3\ \pm1)\cdot10^{-13}$		
Ni−2% Pt	H	130	0.334 ± 0.005	$(3\ \pm1)\cdot10^{-12}$	0.041	7.5
	D	137	0.375 ± 0.005	$(4\ \pm1)\cdot10^{-13}$		
Ni−2% Cu	H	136.1	0.350 ± 0.003	$(2.5\pm0.5)\cdot10^{-12}$	0.045	8.3
	D	142.8	0.395 ± 0.005	$(3\ \pm1)\cdot10^{-13}$		
Ni−2% Ti	H	132.1	0.332 ± 0.005	$(5\ \pm1)\cdot10^{-12}$	0.043	10
	D	138.3	0.375 ± 0.005	$(5\ \pm1)\cdot10^{-13}$		
Ni−2% V	H	136	0.346 ± 0.005	$(3.5\pm1)\cdot10^{-12}$	0.044	7.8
	D	143	0.390 ± 0.005	$(4.5\pm1)\cdot10^{-13}$		
Ni−2% Nb	H	130.8	0.329 ± 0.005	$(4\ \pm1)\cdot10^{-12}$	0.041	8
	D	137.3	0.370 ± 0.005	$(5\ \pm1)\cdot10^{-13}$		
Ni−2% Ta	H	131.5	0.330 ± 0.003	$(4\ \pm1)\cdot10^{-12}$	0.045	10
	D	138.3	0.375 ± 0.003	$(4\ \pm1)\cdot10^{-13}$		
Ni−2% Cr	H	135.5	0.350 ± 0.005	$(2\ \pm1)\cdot10^{-12}$	0.045	6.7
	D	142	0.395 ± 0.005	$(3\ \pm1)\cdot10^{-13}$		
Ni−2% Mo	H	135	0.342 ± 0.003	$(3.5\pm0.5)\cdot10^{-12}$	0.043	8.8
	D	141.8	0.385 ± 0.003	$(4\ \pm0.5)\cdot10^{-13}$		
Ni−2% W	H	136.8	0.341 ± 0.005	$(5.5\pm1)\cdot10^{-12}$	0.045	9.2
	D	143.8	0.386 ± 0.003	$(6\ \pm1)\cdot10^{-13}$		
Ni−2% Mn	H	131.5	0.330 ± 0.005	$(5\ \pm1)\cdot10^{-12}$	0.045	10
	D	138.3	0.375 ± 0.005	$(5\ \pm1)\cdot10^{-13}$		
Ni−2% Co	H	135.5	0.355 ± 0.003	$(1.5\pm0.5)\cdot10^{-12}$	0.04	6
	D	142.3	0.395 ± 0.003	$(2.5\pm0.5)\cdot10^{-13}$		
Ni−2% Fe	H	135	0.345 ± 0.005	$(3\ \pm1)\cdot10^{-12}$	0.045	8.6
	D	141.6	0.390 ± 0.005	$(3.5\pm1)\cdot10^{-13}$		
Ni−2% Os	H	137	0.356 ± 0.005	$(1.5\pm0.5)\cdot10^{12}$	—	—
	D	143.5	—	—		
Co−9% Fe	H	165	0.435 ± 0.01	$(6\ \pm2)\cdot10^{-13}$	0.075	20
	D	173	0.510 ± 0.01	$(2\ \pm2)\cdot10^{-14}$		
Co−15% Pd	H	145	$0.38\ \pm0.01$	$(2\ \pm1)\cdot10^{-12}$	0.06	22
	D	153	$0.44\ \pm0.01$	$(9\ \pm3)\cdot10^{-14}$		

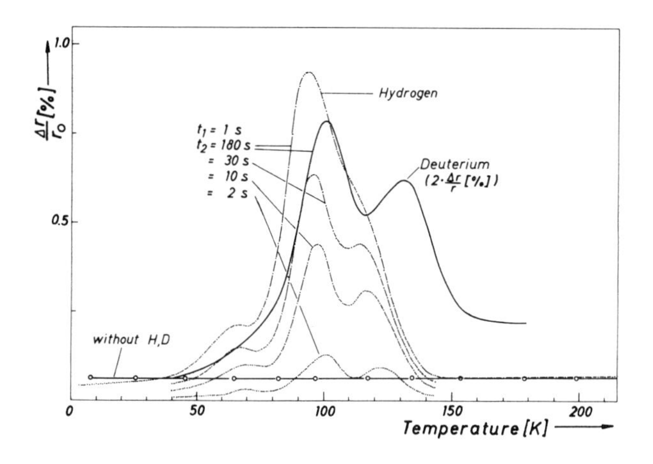

Fig. 11.25. Relaxation spectra of Fe-10 at.%Ti alloys after charging with hydrogen and deuterium

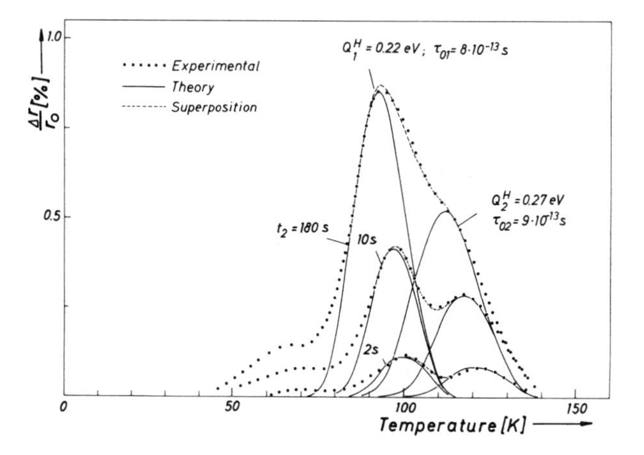

Fig. 11.26. Analysis of relaxation spectra by means of superimposed orientation aftereffects

The relaxation spectrum for the Fe-10 at.%Ti alloy after charging electrolytically with hydrogen or deuterium is shown in Fig. 11.25. In the uncharged specimen, no significant relaxation can be observed. After charging with hydrogen in the temperature range from 40 K–140 K, a relaxation spectrum composed of three maxima appears. The maxima temperatures for H are 65, 95, and 115 K, and for D: ~70, 100, and 130 K. In the Fe-2%Ti alloy the upper two maxima are also observed; the low-temperature peak disappears. The relaxation peaks can be described by orientation aftereffects (Fig. 11.26). The results of an analysis are represented in Table 11.2. For the analysis, a spectrum of activation energies ΔQ was assumed in order to explain the correct half-width and the amplitude of the relaxation maxima.

Two models are proposed for the configuration of the TiH complex. In Model I (Fig. 11.27) the H-atoms occupy octahedral, and in Model II (Fig. 11.28) tetrahedral lattice sites. In Model I the reorientation of the TiH complex

Table 11.2. Activation energies and pre-exponential factors of TiH and TiD complexes in FeTi alloys

	Hydrogen		Deuterium	
T_{max}[K]	95	115	100	130
Q_i [eV]	0.225 ± 0.01	0.27 ± 0.01	0.24 ± 0.01	0.32 ± 0.01
$\tau_{0,i}$ [s]	$(8 \pm 1) \cdot 10^{-13}$	$(9 \pm 1) \cdot 10^{-13}$	$(10 \pm 1) \cdot 10^{-13}$	$(10 \pm 1) \cdot 10^{-13}$
ΔQ_i [eV]	0.037	0.045	0.04	0.045

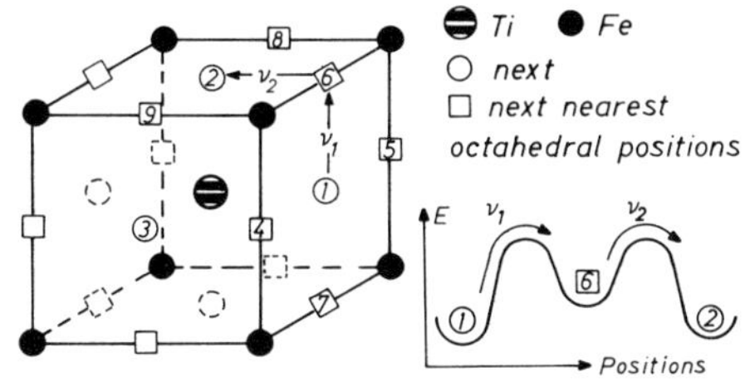

Fig. 11.27. Model of a tetragonal TiH complex in α-Fe with the H-atoms on octahedral lattice sites

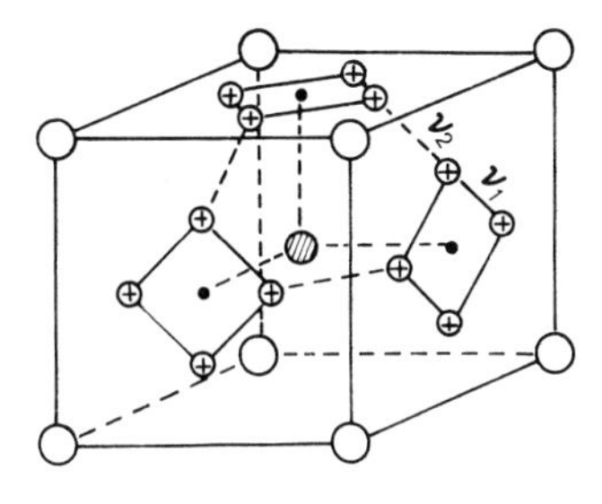

Fig. 11.28. Model of a monoclinic TiH complex in α-Fe with the H-atoms on tetrahedral lattice sites

is characterized by two atomic jump frequencies: ν_1 for the transition from a nearest neighbor (n.n.) position into a next n.n. position, and ν_2 for the reversed process. A solution of the rate equation (11.35) gives for Model I

$$\frac{1}{\chi(t)} = \frac{1}{\chi(0)} + \frac{C_{\mathrm{TiH}}}{kT} \sum_{i=1}^{4} A_i [1 - \exp(-t/\tau_i')], \tag{11.50}$$

where A_i corresponds to the relaxation amplitude of the ith relaxation mode, and where the relaxation times τ_i' are given by

$$\tau_1' = (8\nu_1 + 4\nu_2)^{-1}; \tau_2' = 1/4\nu_2; \tau_{3,4}' = (4\nu_1 + 2\nu_2 \pm 2\sqrt{4\nu_1^2 + \nu_2^2 - 2\nu_1\nu_2})^{-1}. \tag{11.51}$$

Under the assumption $\nu_1 \ll \nu_2$, i.e., if the n.n. position is a stable and the next n.n. position a metastable configuration, (11.50) reduces to

$$\frac{1}{\chi(t)} = \frac{1}{\chi(0)} + \frac{C_{\mathrm{TiH}}}{kT}\{A_4[1-\exp(-t/\tau_1)] + (A_1+A_2+A_3)[1-\exp(-t/\tau_2)]\}. \tag{11.52}$$

The appearance of only two relaxation times, $\tau_1 = 1/6\nu_1$ and $\tau_2 = 1/4\nu_2$, gives a consistent explanation of the measured two-peak spectrum. In our Model I the high-temperature peak is due to transitions from n.n. positions into next n.n. positions, whereas the low-temperature peak is attributed to jumps from next n.n. to n.n. positions. Five relaxation times are obtained for Model II, and a reduction to two relaxation times as for Model I is not possible. This is a strong indication for the validity of Model I. The measured activation energy Q_1 must be attributed to transitions from a n.n. site into a next n.n. site, and the activation energy Q_2 describes the transition from a next n.n. site into a n.n. site.

11.6 Conclusions

1) In fcc (Ni) and hcp (Co) crystals, hydrogen gives rise to a diffusion aftereffect (magnetic Gorsky effect) because hydrogen occupies lattice sites of the same symmetry as the corresponding crystal.

2) In iron neither a diffusion nor an orientation aftereffect of H could be detected.

3) In fcc or bcc crystals where hydrogen forms diatomic complexes with other impurity interstitial atoms (C, N, O), magnetic orientation aftereffects may be observed.

4) In substitutional diluted alloys of the ferromagnetic transition metals (Fe, Co, Ni) with metals on the left of the transitional series, the hydrogen atoms are attracted by the diluted component (Ti, Hf, Zr). The diatomic complexes formed, TiH in **Fe**Ti and FeH in **Ni**Fe, give rise to magnetic orientation aftereffects.

5) Only a diffusion aftereffect is observed in alloys of Fe with transition metals on the right side of the transitional series (**Fe**Ni, **Fe**Pd) because hydrogen is repelled by the diluted components.

6) The isotope effects of τ_0 and Q as measured for H and D in Ni and for CH complexes in NiC alloys are attributed to a quantum mechanical tunneling process from excited energy levels.

7) TiH complexes in FeTi alloys reveal only an isotope effect of the activation energies. The isotope effect is of the order of magnitude of the difference between the zero-point energies (15–60 meV) of the localized H and D oscillators. This peculiar behavior qualitatively may be explained by assuming that the H-atoms are excited from the ground state into states which allow over-barrier motion.

Acknowledgments. The author is very grateful to Prof. Dr. A. Seeger, Prof. Dr. F. Walz, Priv.-Doz. Dr. H. Teichler, Dr. K. Vetter, and Dipl.-Phys. B. Hohler for many stimulating discussions and numerous suggestions.

References

11.1 P.Moser: Dr. Thesis, University of Grenoble (1965)
11.2 A.Seeger, H.Kronmüller, H.Rieger: Z. Angew. Phys. **18**, 377 (1965)
11.3 H.Kronmüller: *Nachwirkung in Ferromagnetika*, Springer Tracts in Natural Philosophy, Vol. 12 (Springer, Berlin, Heidelberg, New York 1968)
11.4 H.Kronmüller: "Studies of Point Defects in Metals by Means of Mechanical and Magnetic Relaxation", in *Vacancies and Interstitials in Metals*, ed. by A.Seeger, D.Schmucker, W.Schilling, and J.Diehl (North-Holland, Amsterdam 1969) pp. 667—728
11.5 R.Cantelli, H.Kronmüller: Z. Metallk. **60**, 384 (1969)
11.6 H.E.Schaefer, G.Lampert: Phys. Stat. Sol. (b) **53**, 113 (1972)
11.7 F.Walz: Phys. Stat. Sol. **29**, 245 (1968)
11.8 G.Schaumann, J.Völkl, G.Alefeld: Phys. Rev. Lett. **21**, 891 (1968)
11.9 G.Alefeld, J.Völkl, G.Schaumann: Phys. Stat. Sol. **37**, 337 (1970)
11.10 C.P.Flynn, A.M.Stoneham: Phys. Rev. B **1**, 3966 (1970)
11.11 L.Néel: J. Phys. Rad. **12**, 339 (1951)
11.12 L.Néel: J. Phys. Rad. **13**, 249 (1952)
11.13 H.Kronmüller, H.R.Hilzinger: Int. J. Magnetism **5**, 27 (1973)
11.14 H.Kronmüller: AIP Conf. Proc. **10**, 1006 (1973)
11.15 G.Rieder: Abhandl. Braunschweig. Wiss. Ges. **11**, 20 (1959)
11.16 E.Kröner: *Kontinuumstheorie der Versetzungen und Eigenspannungen* (Springer, Berlin, Göttingen, Heidelberg 1958)
11.17 J.Friedel: Ber. Bunsenges. Phys. Chem. **76**, 828 (1972)
11.18 W.M.Mueller: In *Metal Hydrides* (Academic Press, New York 1968) p. 389
11.19 T.R.P.Gibb,Jr.: Progr. Inorg. Chem. **3**, 315 (1962)
11.20 R.B.Mc Lellan: Mat. Sci. Eng. **18**, 5 (1975)
11.21 H.Kronmüller: Z. Physik **168**, 478 (1962)
11.22 H.Kronmüller: "Magnetisierungskurve der Ferromagnetika", in *Moderne Probleme der Metallphysik*, Vol. 2, ed. by A.Seeger (Springer, Berlin, Heidelberg, New York 1966) pp. 24—156
11.23 H.Träuble: "Magnetisierungskurve der Ferromagnetika", in *Moderne Probleme der Metallphysik*, Vol. 2, ed. by A.Seeger (Springer, Berlin, Heidelberg, New York 1966) pp. 157—475
11.24 H.D.Dietze: Techn. Mitt. Krupp **17**, 67 (1959)
11.25 W.S.Gorsky: Z. Physik SU **8**, 457 (1935)
11.26 J.Völkl: Ber. Bunsenges. Phys. Chem. **76**, 797 (1972)
11.27 V.A.Somenkov, A.V.Gurskaya, M.G.Bemlyanov, M.E.Kost, N.A.Chernoplekov, A.A.Chertkov: Soviet Phys. Solid State **10**, 1076 (1968)
11.28 W.Gissler, G.Alefeld, T.Springer: J. Phys. Chem. Sol. **31**, 2361 (1969)
11.29 H.D.Carstanjen, R.Sizmann: Ber. Bunsenges. Phys. Chem. **76**, 1223 (1972)
11.30 A.Seeger: Phys. Lett. **58** A, 137 (1976)
11.31 G.Schöck: Acta Met. **11**, 617 (1963)
11.32 W.Köster, L.Bangert, R.Hahn: Arch. Eisenhüttenwes. **25**, 569 (1954)
11.33 A.Kuke, H.Kronmüller, H.Schultz: Phys. Stat. Sol. (a) **36**, K 161 (1969); Phys. Stat. Sol. (a) **25**, 523 (1974)
11.34 H.Kronmüller: Appl. Phys. **4**, 317 (1974)
11.35 H.Kronmüller: Nuovo Cimento **33**, 205 (1976)
11.36 A.Seeger, C.Wüthrich: Nuovo Cimento **33**, 38 (1976)
11.37 C.Wüthrich: Scripta Met. **9**, 641 (1975)

11.38 A. Seeger, P. Schiller: "Kinks in Dislocation Lines and their Effects on the Internal Friction in Crystals", in *Physical Acoustics*, Vol. 3 A, ed. by W. P. Mason (Academic Press, New York, London 1966) Chap. 8, pp. 361—445; Acta Met. **10**, 348 (1962)
11.39 W. Eichenauer, W. Löser, W. Witte: Z. Metallk. **56** (S), 287 (1956)
11.40 L. Katz, M. Guinan, R. J. Borg: Phys. Rev. B **4**, 330 (1971)
11.41 Y. Ebisuzaki, W. J. Kass, M. O'Keeffe: J. Chem. Phys. **49** (8), 3329 (1968)
11.42 W. Dander, H. Kronmüller: Ber. Bunsenges. Phys. Chem. **76**, 826 (1972)
11.43 G. Herbst, H. Kronmüller: To be published
11.44 N. Nishida, R. S. Hayano, K. Nagamine, T. Yamazaki: *Proc. Int. Conf. Magnetism 1976* (North-Holland, Amsterdam 1977) to be published
11.45 W. R. Heller: Acta Met. **9**, 600 (1961)
11.46 R. Gibala: Trans. AIME **239**, 1574 (1967)
11.47 K. Takita, K. Sakamoto: Scripta Met. **4**, 403 (1970)
11.48 L. C. Weiner, M. Gensamer: Acta Met. **5**, 692 (1957)
11.49 C. M. Sturges, A. P. Miodownik: Acta Met. **17**, 1197 (1969)
11.50 M. E. Hermant: Ph.D. Thesis, University of Amsterdam (1966)
11.51 H. Steeb, H. Kronmüller: Phys. Stat. Sol. (a) **16** K, 175 (1973)
11.52 J. J. Au, H. K. Birnbaum: Scripta Met. **7**, 595 (1973)
11.53 J. F. Dufresne, A. Seeger, P. Groh, P. Moser: Phys. Stat. Sol. (a) **36**, 579 (1976)
11.54 R. B. Mc Lellan, D. M. Coldwell: Acta Met. **23**, 57 (1975)
11.55 R. Martinez, F. Walz, H. Kronmüller: Appl. Phys. **7**, 107 (1975)
11.56 K. Vetter: Dr. rer. nat. Thesis, University of Stuttgart (1976)
11.57 W. Eichenauer, H. Künzig, A. Pebler: Z. Metallk. **49**, 220 (1958)
11.58 H. Kronmüller, N. König, F. Walz: Phys. Stat. Sol. (b) **79**, 237 (1977)
11.59 H. Schreyer: Diploma Thesis, University of Stuttgart (1976)
11.60 M. V. Klein, H. Kronmüller: J. Appl. Phys. **33**, 2191 (1962)
11.61 H. E. Schaefer, H. Kronmüller: Phys. Stat. Sol. (b) **67**, 63 (1975)
11.62 G. Sicking: Ber. Bunsenges. Phys. Chem. **76**, 790 (1972)
11.63 E. Adler: Z. Metallk. **56**, 249 (1965)
11.64 E. Adler, C. Radeloff: Z. Angew. Phys. **20**, 346 (1966)
11.65 C. Radeloff, E. Adler: Z. Angew. Phys. **25**, 45 (1968)
11.66 E. Adler, C. Radeloff: J. Appl. Phys. **40**, 1526 (1969)
11.67 H. Kronmöller, B. Hohler, H. Schreyer, K. Vetter: Phil. Mag. (in print)

12. Diffusion of Hydrogen in Metals*

J. Völkl and G. Alefeld

With 22 Figures

12.1 Background

The recent increase of experimental work on diffusion of hydrogen in metals has been motivated from points of view of both basic as well as applied research.

1) The diffusivity of hydrogen in metals is extremely high (e.g., $2 \cdot 10^{12}$ jumps per second at room temperature for H in vanadium [12.1] and exceeds that of heavy interstitials like oxygen and nitrogen at this temperature by 15–20 orders of magnitude. Figure 12.1 shows a comparison for H, O, and N in Nb and for C in Fe [12.1–3]. Phenomenologically this high diffusivity is a consequence of the low activation energy for hydrogen diffusion (e.g., 0.05 eV for hydrogen in vanadium [12.1]).

2) Hydrogen diffusion can be observed at low temperatures so that quantum effects in the diffusion of hydrogen can be expected. In palladium, for example, *Fritz* et al. [12.4] found heat production below 1 K, possibly resulting from a rearrangement of hydrogen atoms. Recent quenching experiments by *Hanada* [12.5, 6] on H in Ta have shown hydrogen mobility at 11 K.

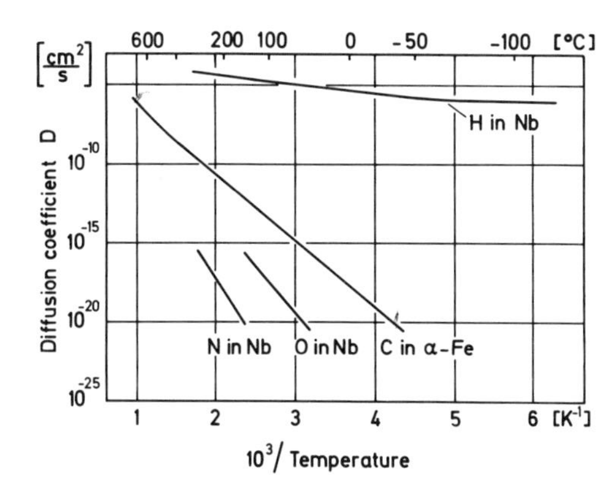

Fig. 12.1. Diffusion coefficients of H [12.1], N, and O [12.2] in Nb, and C in α-Fe [12.3]

* This contribution follows in arrangement an earlier article on "Hydrogen Diffusion in Metals" published in *Diffusion in Solids, Recent Developments*, ed. by A.S. Nowick, J.J. Burton (Academic Press, New York 1975) [12.11a]. The publisher's permission is gratefully acknowledged.

The energy difference between vibrational states of the hydrogen atoms (local modes) can be determined by inelastic neutron scattering. For hydrogen in niobium, one finds 0.11 eV [12.7–9], which corresponds to a temperature of about 1300 K. So for this system, room temperature can already be considered as low.

3) Three isotopes with the largest possible mass ratios are available for hydrogen.

The experimental techniques as well as the results have been reviewed by *Birnbaum* and *Wert* [12.10] and by *Völkl* and *Alefeld* [12.11a]. Therefore, the experimental techniques will be mentioned only briefly in this chapter, whereas in the results, presented in close analogy to [12.11a], the latest measurements available are included. Phase relations and solubility measurements may be found in [Ref. 12.11b, Chapters 2 and 3] as well as in [12.11a].

12.2 Experimental Methods

In general, hydrogen diffusion is being studied by setting up a nonequilibrium distribution and measuring the time to reach equilibrium under the given conditions. Yet there are also three methods in which hydrogen diffusion in equilibrium is being studied, namely, nuclear magnetic resonance (NMR) (see Chap. 9), quasielastic neutron scattering (QNS) (see Chap. 10), and Mössbauer effect [12.12] (see Chap. 6). The first group of experiments has the advantage that, in addition to a time constant, from which the diffusion coefficient is derived, an equilibrium property can be measured as well, namely, the derivative of the chemical potential μ in respect to the hydrogen density ϱ [atoms cm^{-3}]. The same quantity appears as a factor in the measured diffusion coefficient, as can easily be seen as follows:

$$s = -M\varrho \operatorname{grad}\mu(\varrho) = -M\varrho \frac{\partial\mu}{\partial\varrho} \operatorname{grad}\varrho \quad (M = \text{mobility},\ s = \text{current}). \tag{12.1}$$

The diffusion coefficient measured by studying the decay of a concentration gradient is therefore defined as

$$D^* = M\varrho \frac{\partial\mu}{\partial\varrho}. \tag{12.2}$$

For low concentrations, $\partial\mu/\partial\varrho$ is given by

$$\frac{\partial\mu}{\partial\varrho} = \frac{kT}{\varrho}. \tag{12.3}$$

Under this condition (12.2) can be written as

$$D(\varrho \to 0) = M(\varrho \to 0) \cdot kT. \tag{12.4}$$

Equation (12.4) is known as the Einstein relation, connecting the diffusion coefficient with the mobility. The temperature dependence of the diffusion coefficient D^* is (ignoring the trivial factor kT) that of the mobility of the individual particles only in the limit of high dilution. Therefore, in general, the additional measurement of the equilibrium property $\partial\mu/\partial\varrho$ is necessary to eliminate the temperature and concentration dependences not characteristic for the individual jump processes. By this procedure one obtains the diffusion coefficient D, defined by

$$D(\varrho)=M(\varrho)\cdot kT \tag{12.4a}$$

which is called "tracer-diffusion coefficient". The mobility can be concentration dependent, e.g., due to changes in the potentials caused by the lattice expansion or more trivial due to the blocking of sites at higher concentrations. A simple correction for the diffusion coefficient eliminating the concentration and temperature dependence of $\partial\mu/\partial\varrho$ as well as in first order also the concentration dependence of M due to blocking can be achieved as follows: For higher concentrations (12.3) can in mean-field approximation be written as [12.13]

$$\frac{\partial\mu}{\partial\varrho}=\frac{kT}{\varrho}\,\frac{(1-T_s(\varrho)/T)}{(1-\varrho/\varrho_m)}\,. \tag{12.5}$$

ϱ_m = maximum concentration and $T_s(\varrho)$ = temperature of the spinodal, depending on the concentration.

By determining experimentally $T_s(\varrho)$ [12.1, 14] and multiplying the measured diffusion coefficient D^* by $(1-T_s/T)^{-1}$, i.e.,

$$D'=D^*\frac{1}{1-T_s/T}\,, \tag{12.6}$$

one finds the following relation between the tracer-diffusion coefficient D of (12.4a) and D':

$$D'=D\frac{1}{1-\varrho/\varrho_m}\,. \tag{12.7}$$

If in addition one assumes that the concentration dependence of the mobility due to blocking can be written as

$$M(\varrho)=M^0(\varrho)\,(1-\varrho/\varrho_m)\,, \tag{12.8}$$

one gets for (12.7) the relation

$$D'=M^0(\varrho)\cdot kT\,. \tag{12.9}$$

The diffusion coefficient D' is therefore characteristic for the concentration and temperature dependence of the individual jumps, and therefore will be used for comparison with rate theories. The tracer-diffusion coefficient D must be used if comparing NMR (see Chap. 9) and QNS (see Chap. 10) with macroscopic diffusion data. In both of these methods the motion of individual particles which are marked by their spin is being measured. (For QNS this statement holds only for incoherent scattering which is dominant for hydrogen but not for deuterium.) The "motional narrowing" experiments using the Mössbauer line of Ta^{181} have so far only been applied to small concentrations [12.12]. It is expected that at higher concentrations the broadening of the Mössbauer line is determined by a diffusion coefficient according to (12.2), with a thermodynamic factor $\partial\mu/\partial\varrho$ which is characteristic for density fluctuations with short wavelengths.

In this connection, the following further complication must be pointed out: Due to the long-range elastic interaction between the dissolved hydrogen atoms, the quantity $\partial\mu/\partial\varrho$ in (12.2) depends on the geometric details of the macroscopic concentration gradient studied, surprisingly also on the shape of the sample, as predicted theoretically [12.15–18] (see also Chap. 2) and verified experimentally [12.19]. An example for this shape dependence of D^* will be shown in this chapter. Since the thermodynamic factor in D^* is different for different density-fluctuation modes [12.15–18], it is important to use such experimental data for the thermodynamic factor which are compatible with the special diffusion experiment considered. A very careful consideration is required if combining data from different experiments, e.g., D^* from permeation data and $\partial\mu/\partial\varrho$ from solubility data.

The experimental techniques involving concentration gradients can be subdivided either according to the method of producing the nonequilibrium state or according to the method by which the relaxation to equilibrium is measured. Since a description and evaluation of the experimental methods have been performed in [12.11a], we shall give here only a brief list: Permeation methods, electrochemical methods (see [Ref. 12.11b, Chap. 3]), mechanical relaxation methods, magnetic disaccommodation (see Chap. 11), resistivity-relaxation methods, tracer method, gravimetric method, radiographic method, and x-ray method.

12.3 Absolute Values for the Diffusion Coefficients at Small Concentrations (α Phases)

In the subsequent section the absolute values and the temperature dependence of the diffusion coefficients in some representative metals will be discussed.

12.3.1 Palladium

The palladium-hydrogen system together with H in Ni and Fe has of all systems received the greatest amount of attention. Most of the experimental

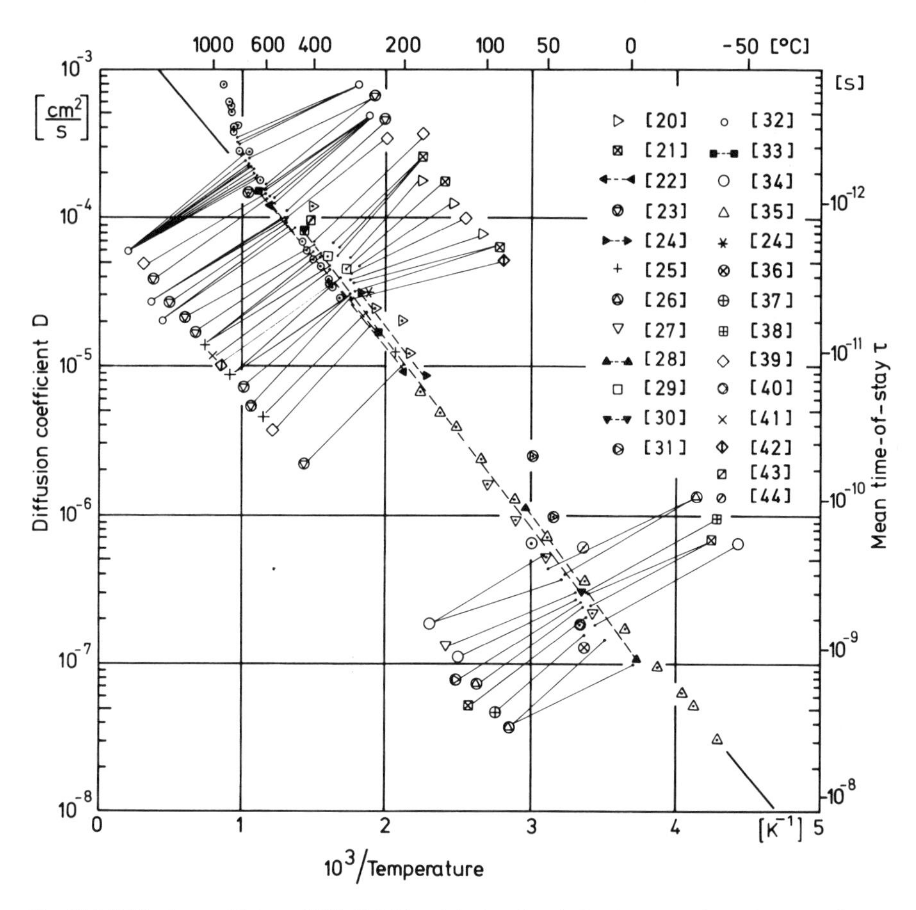

Fig. 12.2. Diffusion coefficient of H in Pd. The right-hand scale for the mean time-of-stay τ refers to octahedral-octahedral jumps. (Numbers in brackets refer to References of this Chapter)

methods have been applied. The data of 25 authors are collected in Fig. 12.2 [12.20–44]. The consistency of the data is remarkably good. Because of relatively well-defined surface conditions, the different permeation techniques give reliable results in this case. The line drawn is, what we think, the best fit. This line yields the following values for D_0 and U:

$$D_0 = 2.90 \cdot 10^{-3}\,\mathrm{cm^2\,s^{-1}} \qquad U = 0.230\,\mathrm{eV}.$$

D_0 is about an order of magnitude larger than those values found for H in V, Nb, or Ta. In contrast to this the vibration frequency of the hydrogen atom (local mode) $\hbar\omega = 0.056$ eV (β phase) [12.45] or $\hbar\omega = 0.066$ (α phase) [12.46] is a factor of two smaller than those of hydrogen in the bcc group mentioned. The local-mode energy 0.066 eV is also small compared to the activation energy

0.23 eV so that the jump model for diffusion appears to be a good approximation. The predictions of the jump model have been verified by quasielastic neutron scattering (see Chap. 10). For the mean jump distance the value $a/\sqrt{2}$ (a = lattice parameter) has been found which holds for jumps between octahedral sites instead of $a/2$ for jumps between tetrahedral sites. The deviation from the Arrhenius behavior at high temperature will be discussed in Section 12.6. The small, although systematic scatter around room temperature towards lower values may be attributed to surface permeation problems [12.47].

12.3.2 Nickel

The consistency of the data on nickel (Fig. 12.3) is as good as in Pd for the region above 100 °C, whereas below, the scatter increases. Since all measurements, with one exception [12.68, 69, 73], have been performed with surface-sensitive methods, this consistency is quite remarkable considering the large scatter for other metals. Apparently, as with palladium, the influence of surface contamination is negligible. Because of the smaller solubility of H in Ni, other experimental methods have limited applicability. At the Curie point of nickel, *Belyakov* and *Ionov* [12.52] have observed a small step in the diffusion coefficient, which still seems to be visible in the compiled data in Fig. 12.3. The best values indicated in Fig. 12.3 can be written as

$$D_0 = 6.9 \cdot 10^{-3}\,\mathrm{cm^2\,s^{-1}} \qquad U = 0.420\,\mathrm{eV} \qquad T > T_c$$

$$D_0 = 4.8 \cdot 10^{-3}\,\mathrm{cm^2\,s^{-1}} \qquad U = 0.408\,\mathrm{eV} \qquad T < T_c\,.$$

The latter value for U has also been found at 200 K using the magnetic disaccommodation technique [12.74]. Compared to Pd, the activation energy is about twice as large, but also D_0 is larger. As a consequence the absolute value of D for nickel at room temperature is three orders of magnitude lower, whereas at 600 °C this ratio has decreased below one order of magnitude.

12.3.3 Iron (α Phase)

In contrast to Figs. 12.2 and 12.3, the measurements for iron in Fig. 12.4 [12.71, 75–116] show quite unsatisfactory scatter in spite of a large number of investigations. Only above 250 °C the scatter is about one order of magnitude and decreases slightly with higher temperature. The absolute value of D for iron in this temperature region is higher than that for Pd.

Several possibilities have been discussed in the literature to explain the large scatter (for detailed references see [12.11a]): Surface effects, trapping of H at lattice imperfections like impurities, dislocations, precipitates, grain boundaries, etc., the formation of immobile diinterstitials, the formation of mole-

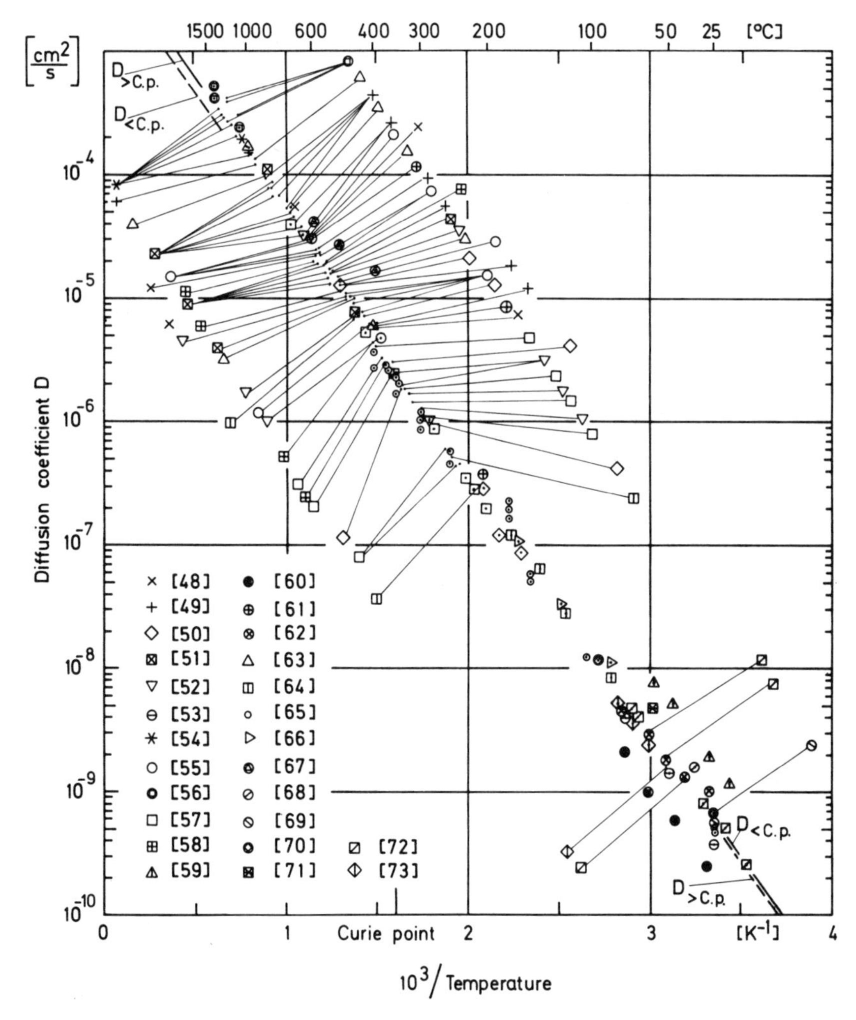

Fig. 12.3. Diffusion coefficient of H in Ni. ($D_{<\mathrm{C.p.}}$ and $D_{>\mathrm{C.p.}}$ are best values below and above the Curie point). (Numbers in brackets refer to References of this Chapter)

cular hydrogen in micro- or macropores, either already existing in the material or produced by excess loading.

Considering the relative significance of the effects mentioned, a comparison of the Fe measurements with those for Ni seems useful. In both metals the solubility is low and for both metals surface-sensitive methods have been applied almost exclusively. If imperfections in the metals play an important role, the Ni data should show comparably large scatter. Furthermore, it has been shown for H in Nb [12.117] that impurities like nitrogen do not significantly change the diffusion constant. Therefore, we think that surface

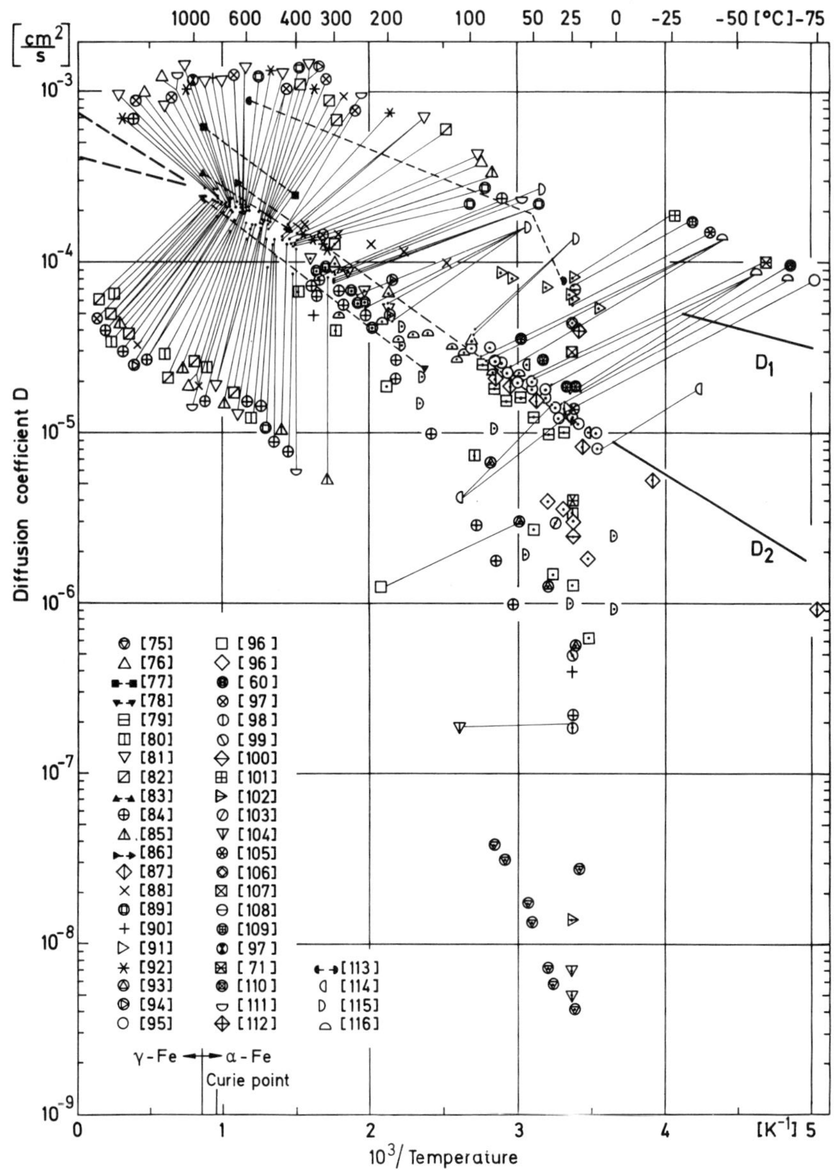

Fig. 12.4. Diffusion coefficient of H in α-Fe. (Numbers in brackets refer to References of this Chapter)

effects are the cause of error. Thus surface-independent measurements on Fe are necessary before the influence of imperfections can be inferred. Assuming that surface effects and possibly trapping are responsible for slowing down of the diffusion, one is inclined to identify the highest values measured with the bulk-diffusion coefficient. Under this condition the measurements of *Nelson* and *Stein* [12.113] on triply zone-refined iron yield

$$D_0 = 2.3 \cdot 10^{-3}\,\mathrm{cm^2\,s^{-1}} \qquad U = 0.069\,\mathrm{eV}.$$

Nevertheless, these measurements show a break at 50 °C like many others. *Heumann* and *Domke* [12.97], combining their own measurements with those of [12.109], suggest an even smaller activation energy (D_1 in Fig. 12.4): $D_0 = 4.15 \cdot 10^{-4}\,\mathrm{cm^2\,s^{-1}}$, $U = 0.044$ eV. On the other hand, if one were to believe the large amount of data at $D \approx 10^{-5}\,\mathrm{cm^2\,s^{-1}}$ for room temperature, one may draw a line as indicated by D_2 with the values

$$D_0 = 7.5 \cdot 10^{-4}\,\mathrm{cm^2\,s^{-1}} \qquad U = 0.105\,\mathrm{eV}.$$

12.3.4 Niobium

The good consistency of the data shown in Fig. 12.5 [12.8, 118–126] only exists if the results of surface-independent methods (Gorsky effect, QNS, resistivity relaxation) are compared. The Nb surface is difficult to control so that even at high temperatures, surface-dependent methods, as applied by *Albrecht* et al. [12.127], *Ryabchikov* [12.128], and *Oğurtani* [12.129], are not very reliable (see [12.11a]), except if the surface of UHV-outgassed Nb is covered by Pd before exposed to air [12.123]. The latter measurements are included in Fig. 12.5. The systematic deviation of the results of *Cantelli* et al. [12.119] below −50 °C is very likely caused by precipitation of hydrogen. These authors used the absolute value of an internal friction peak, measured at one frequency as a function of temperature, to determine D.

The results of applying the Gorsky effect technique all refer to hydrogen concentrations $c \to 0$. Since with the Gorsky effect the quantity $\partial\mu/\partial\varrho$ can be measured independently, possible corrections of D due to this factor are easily made [12.118]. The concentration and isotope dependence as well as the break in temperature dependence will be discussed later.

The absolute values of D at room temperature are larger than for Pd. Suggestions for best values are

Hydrogen in Nb

$$D_0 = 5.0 \cdot 10^{-4}\,\mathrm{cm^2\,s^{-1}} \qquad U = 0.106\,\mathrm{eV} \qquad T > 0\,^\circ\mathrm{C}$$

$$D_0 = 0.9 \cdot 10^{-4}\,\mathrm{cm^2\,s^{-1}} \qquad U = 0.068\,\mathrm{eV} \qquad T < -50\,^\circ\mathrm{C}$$

Deuterium in Nb

$$D_0 = 5.2 \cdot 10^{-4}\,\mathrm{cm^2\,s^{-1}} \qquad U = 0.127\,\mathrm{eV}$$

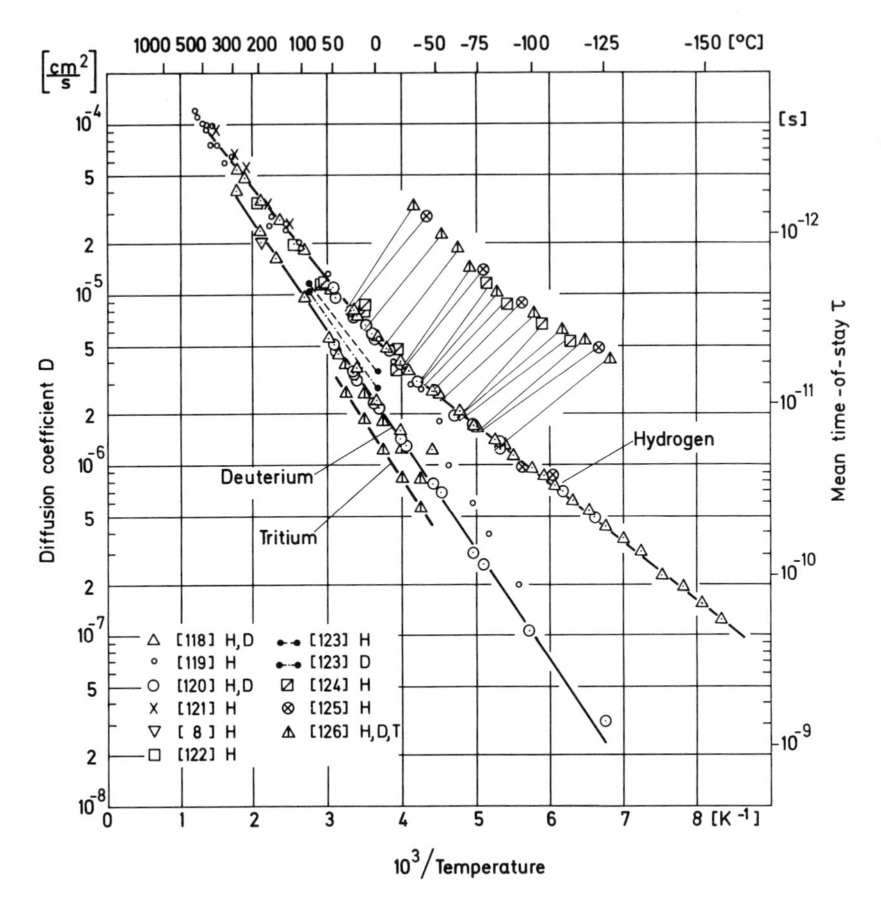

Fig. 12.5. Diffusion coefficient of H, D, and T in Nb. The mean time-of-stay τ is calculated for tetrahedral-tetrahedral jumps. (Numbers in brackets refer to References of this Chapter)

Tritium in Nb

$$D_0 = 4.5 \cdot 10^{-4}\,\mathrm{cm}^2\,\mathrm{s}^{-1} \qquad U = 0.135\,\mathrm{eV}\,.$$

The mean time-of-stay τ on the right-hand scale in Fig. 12.5 is calculated for tetrahedral-tetrahedral jumps; the assumption of this position is justified by the results of *Somenkov* et al. [12.130]. It should be pointed out that the high-temperature activation energy is about equal to the local-mode energy 0.11 eV [12.8, 9], whereas the low-temperature value is considerably lower. It has been shown by *Brand* [12.9], *Conrad* et al. [12.131] and *Stump* et al. [12.132] that the local-mode frequency of D in Nb is about a factor of $1/\sqrt{2}$ lower than for H. Therefore, in this case the activation energy has a value between the first and second excited levels of the deuterium atom.

Internal-friction measurements by *Cannelli* and *Verdini* [12.133] on H and by *Schiller* and *Schneiders* [12.134] on H and D in Nb show, for kilocycle frequencies at about 100 K, internal-friction peaks which originally were discussed as Snoek peaks for H or D. The presence of the peaks depends on the impurity concentration [12.135, 137]. It is therefore unlikely that these processes leading to internal friction are connected with the long-range diffusion process [12.135–137].

12.3.5 Tantalum

Only results of surface-independent methods (Gorsky effect, Mössbauer effect, QNS [12.1. 12. 138–147]) are included in Fig. 12.6. In addition, results with Pd-coated samples [12.123] are shown. Nevertheless, discrepancies in the absolute value (not so much in the slopes) can be noted.

The agreement of the QNS result for $c=2\%$ [12.144] with Gorsky effect measurements is very good. The deviations for $c=15\%$ H/Ta(QNS) [12.143] are easily explained by the concentration dependence of D. The same argument holds for the results of *Zeilinger* and *Pochmann* [12.148], using a neutron-radiographic method. Possible reasons for the systematically lower values of *Cantelli* et al. [12.139, 146] have been discussed in [12.11a].

Similar to H in Nb, a pronounced break in the temperature dependence of the diffusion coefficient has been found recently with the Gorsky effect technique [12.145], using foil-shaped specimens (0.12 mm thick) in order to reduce measuring time. The beginning of this change in activation energy was already detected by *Ströbel* et al. [12.149] between room temperature and 170 K, using nuclear acoustic resonance and can also be supposed from the results of *Cantelli* et al. [12.139]. Preliminary Gorsky effect measurements on Ta–D show a similarly strong break. Quenching experiments by *Hanada* [12.5, 6] and *Engelhard* [12.150] also indicate that H and D stay mobile at 10 K, with an activation energy comparable to the low-temperature value.

The lines shown in Fig. 12.6 correspond to the following values:

Hydrogen in Ta

$$D_0=4.4\cdot10^{-4}\,\mathrm{cm^2\,s^{-1}} \qquad U=0.140\,\mathrm{eV} \qquad T>0\,^\circ\mathrm{C}$$

$$D_0=2.0\cdot10^{-6}\,\mathrm{cm^2\,s^{-1}} \qquad U=0.040\,\mathrm{eV} \qquad T<-75\,^\circ\mathrm{C}$$

Deuterium in Ta

$$D_0=4.6\cdot10^{-4}\,\mathrm{cm^2\,s^{-1}} \qquad U=0.160\,\mathrm{eV}\,.$$

Again the high-temperature activation energy for H in Ta is about equal to the local-mode energy [12.7, 9, 143]. For the internal friction peaks observed at lower temperatures [12.133, 160], the same statements hold as given for Nb in the preceding section.

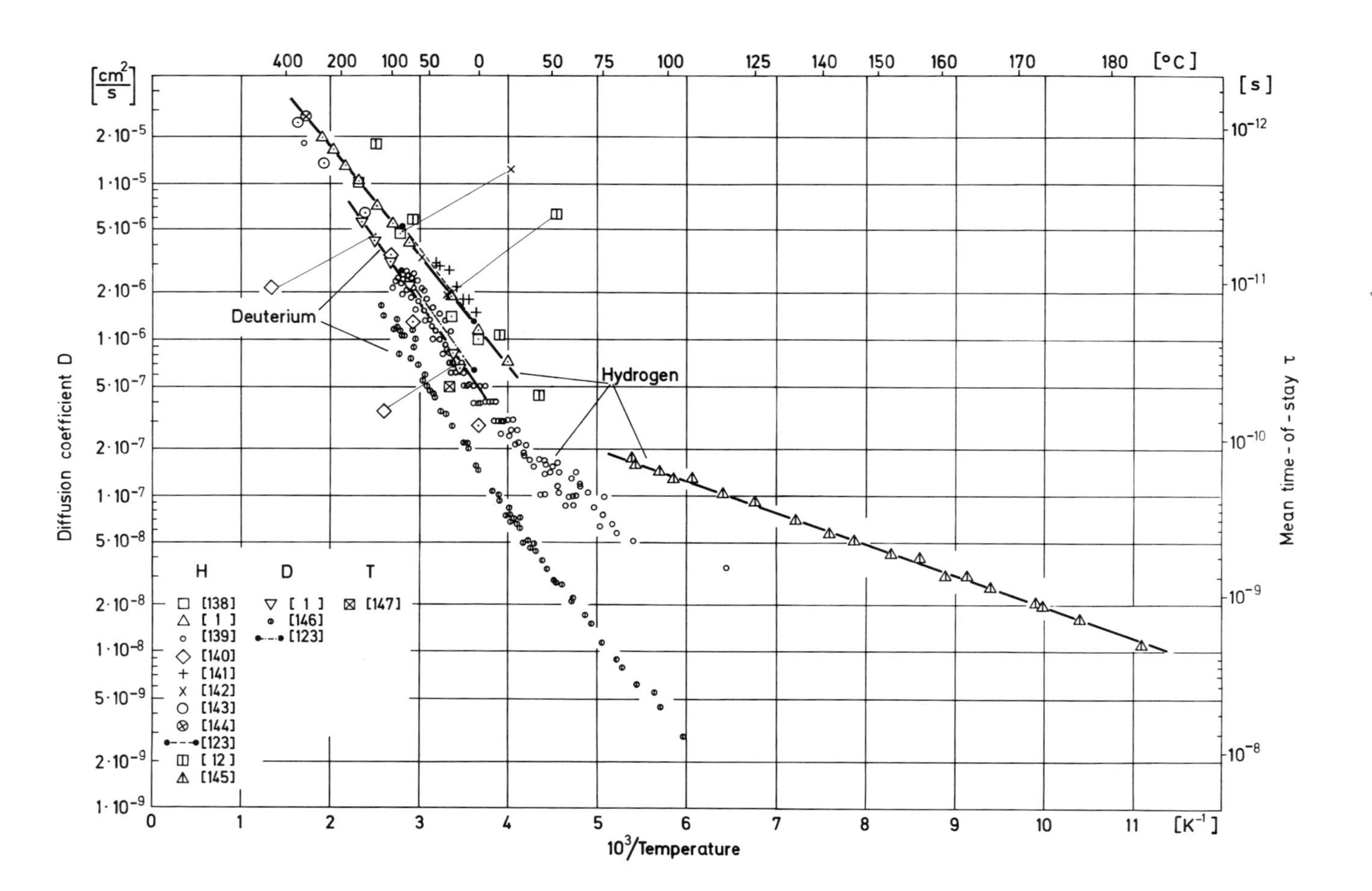

[°C]
400 200 100 50 0 50 75 100 125 140 150 160 170 180
[cm²/s]
Diffusion coefficient D
2·10⁻⁵
1·10⁻⁵
5·10⁻⁶
2·10⁻⁶
1·10⁻⁶
5·10⁻⁷
2·10⁻⁷
1·10⁻⁷
5·10⁻⁸
2·10⁻⁸
1·10⁻⁸
5·10⁻⁹
2·10⁻⁹
1·10⁻⁹
[s]
10⁻¹²
10⁻¹¹
10⁻¹⁰
10⁻⁹
10⁻⁸
Mean time-of-stay τ
Deuterium
Hydrogen
H
D
T
[138]
[1]
[139]
[140]
[141]
[142]
[143]
[144]
[123]
[12]
[145]
[1]
[146]
[123]
[147]
0 1 2 3 4 5 6 7 8 9 10 11
[K⁻¹]
10³/Temperature

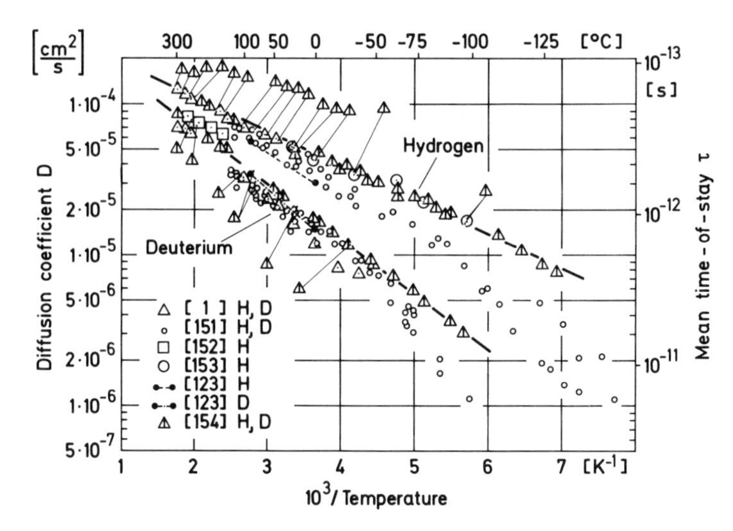

Fig. 12.7. Diffusion coefficient for H and D in V. The mean time-of-stay τ is calculated for tetrahedral-tetrahedral jumps. (Numbers in brackets refer to References of this Chapter)

12.3.6 Vanadium

Fig. 12.7 shows data of surface-independent methods [12.1, 151–154] and results of an electrochemical permeation method on Pd-coated samples [12.123]. The absolute value at room temperature is the largest value so far measured, except maybe for iron (Fig. 12.4) or PdCu-bcc (Fig. 12.21). At $-100\,°C$ the diffusion coefficient for H in V is still as high as for H in Pd at 300 °C.

The scatter of the data by *Cantelli* et al. [12.151] at low temperatures can probably be attributed to the same effects as discussed for Ta in [12.11a]. The quasielastic neutron-scattering data by *Rowe* et al. [12.152] have been performed on a sample with $c=20\%$ H/V. An extrapolation of the 20% H data to $c=0$ would bring them closer to the Gorsky effect data.

The lines drawn in Fig. 12.7 correspond to the following values:

Hydrogen in V

$$D_0=3.1\cdot10^{-4}\,\mathrm{cm^2\,s^{-1}}\qquad U=0.045\,\mathrm{eV}$$

Deuterium in V

$$D_0=3.8\cdot10^{-4}\,\mathrm{cm^2\,s^{-1}}\qquad U=0.073\,\mathrm{eV}.$$

◀ Fig. 12.6. Diffusion coefficient of H, D, and T in Ta. The mean time-of-stay τ is calculated for tetrahedral-tetrahedral jumps. (Numbers in brackets refer to References of this Chapter)

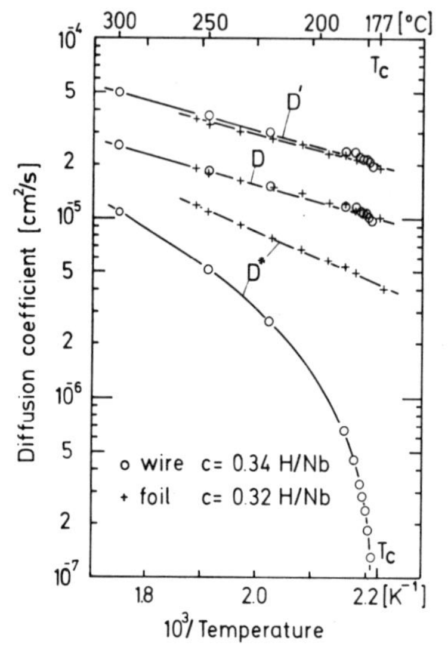

Fig. 12.8. Diffusion coefficients of H in Nb in the region of the critical concentration for wire- and foil-shaped samples. D^*, D, and D' are defined in (12.2), (12.4a), and (12.6) [12.14, 19]

Since the local-mode energy for H in V has been determined as 0.12 eV [12.8, 9, 156], the "activation energy" is far below the local-mode energy, similar to the case for the low-temperature value for H in Nb. *Rowe* et al. [12.152] have shown that the local-mode frequencies for H and D scale as $\sqrt{2}$. Therefore, in this case, the activation energy for deuterium is close to the local-mode frequency. From annealing experiments of quenched-in hydrogen, *Abe* et al. [12.158] concluded that for V, in contrast to Nb and Ta, the extrapolation of the high-temperature diffusion coefficient shown in Fig. 12.7 holds to temperatures of 30–50 K.

12.4 High Hydrogen Concentrations

As shown in Section 12.2, for high concentrations the measured diffusion coefficient depends on the experimental method used. Except for NMR and QNS, the experimental diffusion coefficient is modified by the concentration- and temperature-dependent equilibrium property $\partial\mu/\partial\varrho$. How drastic this influence can be is shown in Fig. 12.8 (see wire!). D^* represents the diffusion coefficient as measured with the Gorsky effect for 34% H/Nb. For $T \to T_c$ it follows that $\partial\mu/\partial\varrho \to 0$. Therefore, D^* approaches zero [12.159], as shown in Fig. 12.8. This effect is known as "critical slowing down". Figure 12.8 also shows D and D' calculated according to (12.4a) and (12.6). The tracer-diffusion coefficient D shows a normal Arrhenius behavior with no indication for the approach of the critical point. The blocking effect amounts to a reduction from D' to D by a factor of two (for 34% H/D), as the ratio of D'/D indicates.

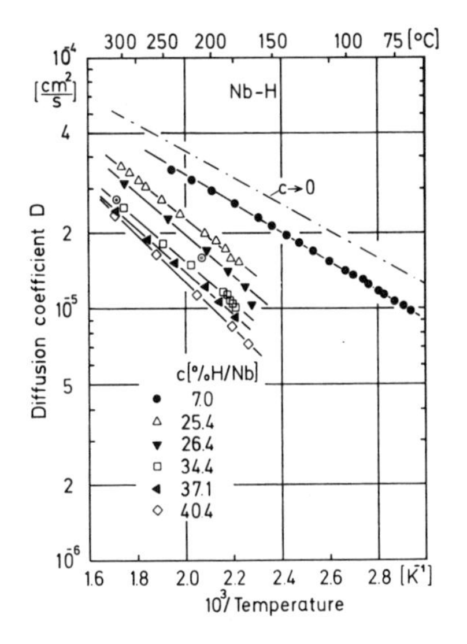

Fig. 12.9. Concentration dependence of the diffusion coefficient D for Nb–H [12.14], ⊙ =QNS results [12.122]

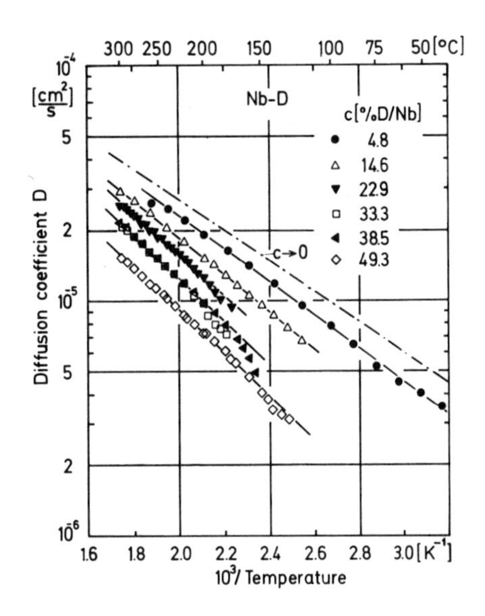

Fig. 12.10. Concentration dependence of the diffusion coefficient D for Nb–D [12.14]

The shape dependence of the diffusion coefficient D^* is also shown in Fig. 12.8 by comparing the results for a foil with those of a wire with about equal concentration (see also Chap. 2). Due to the much lower T_c for the foil [12.19, 155], the effect of critical slowing down is hardly noticeable in this case. Removing the corresponding thermodynamic factors for foil or wire (in these experiments determined by measuring the relaxation strength of the Gorsky effect) yields the diffusion coefficient D and D', characteristic for individual jumps with and without blocking effects. D and D', respectively, are found shape independent as it must be.

Figures 12.9–12.12 show details of the concentration dependence of the diffusion coefficient D of Nb–H, Nb–D, Ta–H, and Ta–D, as measured with the Gorsky effect [12.14]. Analogous values for H in V can be found in [12.155]. Diffusion measurements for high concentrations of H in these metals have also been performed, for example, by *Gissler* et al. [12.122] for $NbH_{0.33}$ (QNS), by *Birchall* and *Ross* [12.161] for $NbH_{0.14}$ (QNS), by *Zogał* and *Cotts* [12.162] for $NbH_{0.6}$ (NMR), by *Lütgemeier* et al. [12.175] for $NbH_{0.58}$ and $NbH_{0.78}$ (NMR), by *Rush* et al. [12.143] for Ta_2H (QNS), by *Pedersen* et al. [12.186] for $TaH_{0.1-0.66}$ (NMR), by *Wicke* and *Obermann* [12.141] for $TaH_{0.03-0.14}$ (gravimetric method), by *Züchner* [12.142] for $TaH_{0.02-0.15}$ (x-ray technique), by *de Graaf* et al. [12.163] for $VH_{0.198}$ and $VH_{0.57}$ (QNS), and by *Krüger* and *Weiß* [12.164] for $VH_{0.3}$ (NMR). In Fig. 12.9 the measurements of *Gissler* et al. [12.122] are included. The agreement of the results of both methods is good.

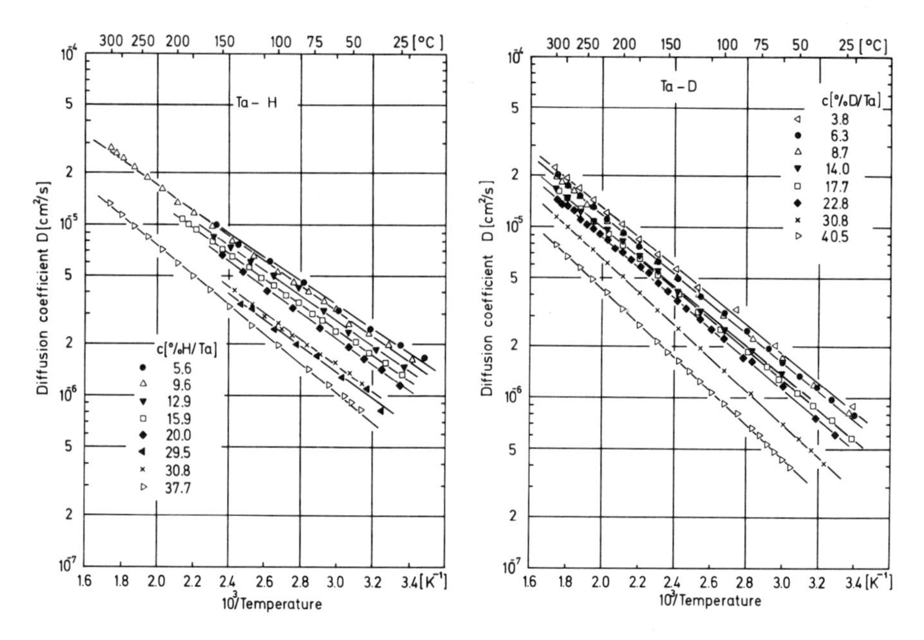

Fig. 12.11. Concentration dependence of the diffusion coefficient D for Ta–H [12.14]

Fig. 12.12. Concentration dependence of the diffusion coefficient D for Ta–D [12.14]

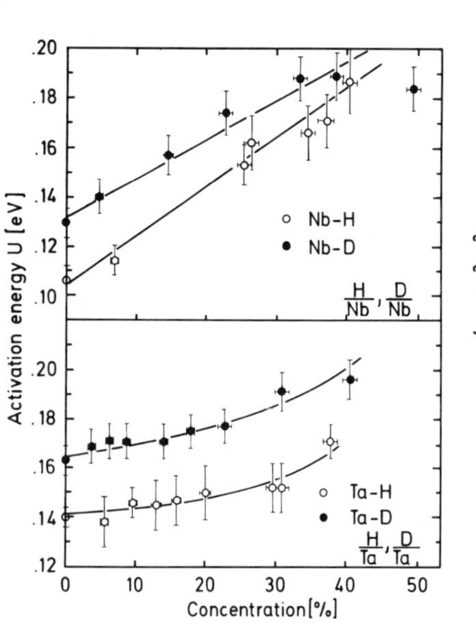

Fig. 12.13. Concentration dependence of the activation energies for Nb–H, Nb–D, Ta–H, and Ta–D [12.14]

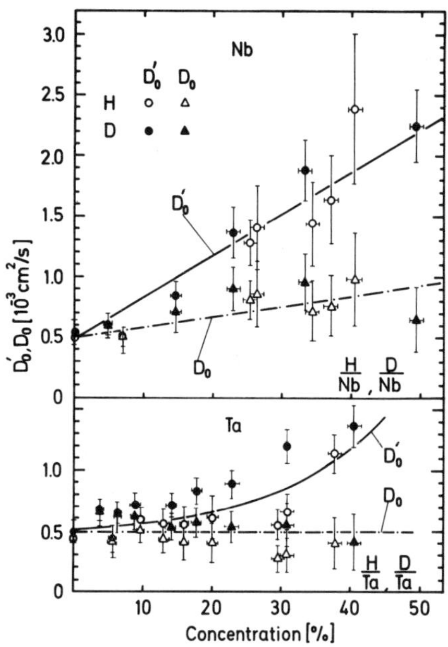

Fig. 12.14. Concentration dependence of the pre-exponential factors D_0 and D_0' for Nb–H, Nb–D, Ta–H, and Ta–D [12.14]

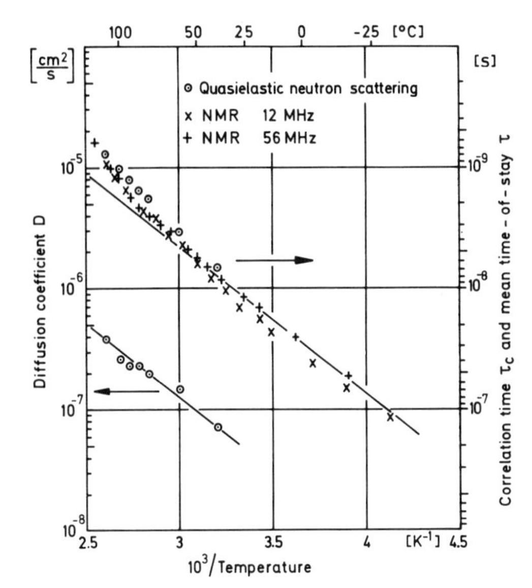

Fig. 12.15. Diffusion coefficient and mean time-of-stay of H in the β phase of Nb (0.9 H/Nb) [12.169]

In Figs. 12.13 and 12.14 the concentration dependence of the activation energies and the D_0 and D'_0 values, respectively, are plotted [12.14]. Whereas D_0 shows nearly no concentration dependence, the activation energies increase (in spite of the expansion of the lattice) by nearly a factor of two. As a total, the change in the diffusion coefficients D (not for D^*) as a function of concentration is remarkably small.

In the transition from the α or α' to the β phases of the bcc metals, more drastic changes in D have been reported (see [12.11a] and, in addition, e.g., [12.165, 166] for Ta and [12.167, 168] for V). The absolute values decrease by one or two orders of magnitude. As an example, Fig. 12.15 shows the diffusion coefficient for H in Nb for the β phase (90% H/Nb) [12.169]. Comparing with Fig. 12.5, we find a reduction by two orders of magnitude. The upper curves in Fig. 12.15 represent measurements of the mean time-of-stay with NMR and QNS [12.169]. The agreement is remarkably good. From these data and the measured diffusion coefficient the mean jump distance can be calculated, with the result $\Delta l = 4.8 \pm 0.8$ Å [12.169]. This value is large compared with 1.17 Å for tetrahedral-tetrahedral jumps and 1.65 Å for octahedral-octahedral jumps, yet it is consistent with the distance of occupied sites in the ordered β structure as suggested by *Somenkov* et al. [12.130].

12.5 Isotope Dependence

Deviations from the predictions of classical rate theory [12.170, 171] $D_1/D_2 = \sqrt{m_2/m_1}$ have been observed for all metal-hydrogen systems. This is not too

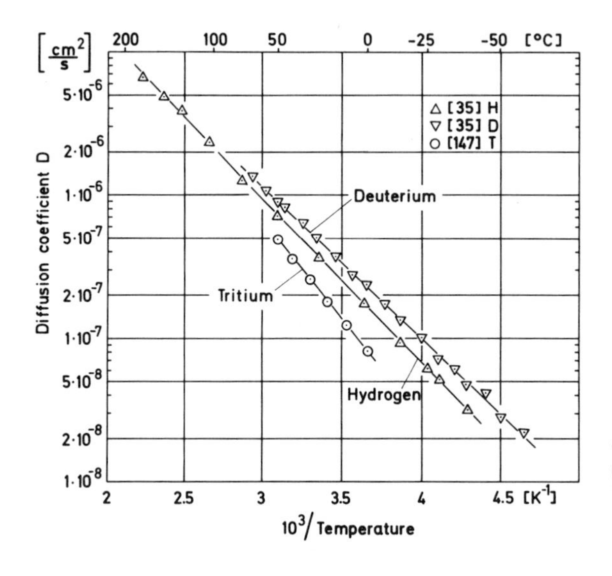

Fig. 12.16. Diffusion coefficient of isotopes of hydrogen in Pd (Numbers in brackets refer to References of this Chapter)

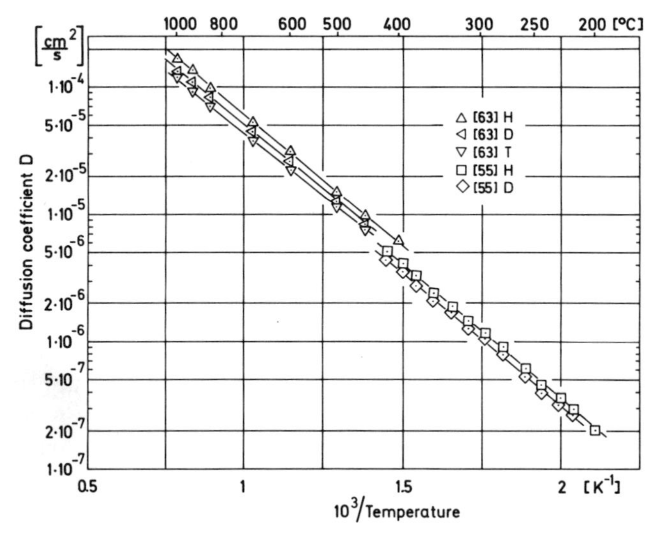

Fig. 12.17. Diffusion coefficient of isotopes of hydrogen in Ni. (Numbers in brackets refer to References of this Chapter)

unexpected in the light of the small masses and the high Einstein temperatures of the hydrogen isotopes. The characteristic features of the isotope dependence seem to be correlated with the structure of the host metal, as demonstrated for the bcc metals V (Fig. 12.7), Nb (Fig. 12.5), and Ta (Fig. 12.6) and for the fcc metals Pd [12.35, 147] (Fig. 12.16), Ni [12.55, 63] (Fig. 12.17), and Cu [12.55, 63] (Fig. 12.18).

For the three bcc metals (Figs. 12.5–7), hydrogen diffuses faster than deuterium in the complete temperature region investigated. The pre-

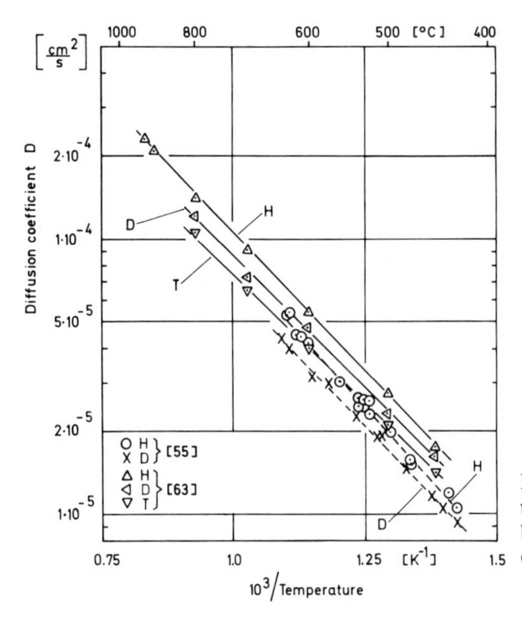

Fig. 12.18. Diffusion coefficient of isotopes of hydrogen in Cu. (Numbers in brackets refer to References of this Chapter)

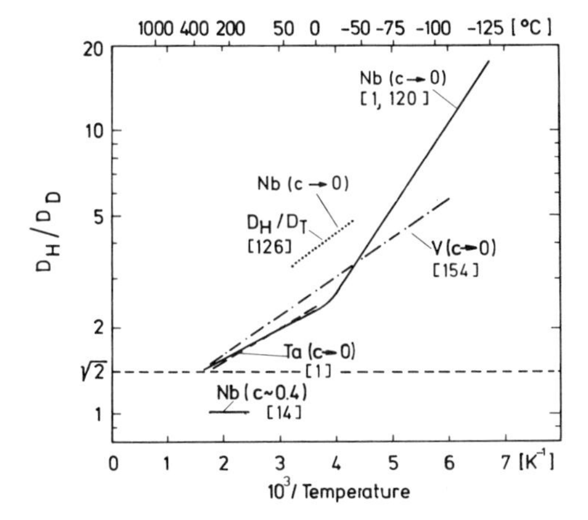

Fig. 12.19. Ratios of diffusion coefficients of H and D in V, Nb, and Ta. (Numbers in brackets refer to References of this Chapter)

exponentials are almost isotope independent (for Nb and Ta above 300 K), whereas for the activation energies the relation $U_H < U_D$ holds. As a consequence, the ratio D_H/D_D shown in Fig. 12.19 for the three metals is temperature dependent. Only at a definite temperature which is about the same for all three metals is this ratio $\sqrt{2}$, whereas at lower temperatures, ratios up to about 20 have been observed [12.120]. For tritium diffusion the temperature dependence has been measured for Nb–T by *Matusiewicz* and *Birnbaum* [12.126], whereas for Ta only one value (at room temperature, without error bars) has been published

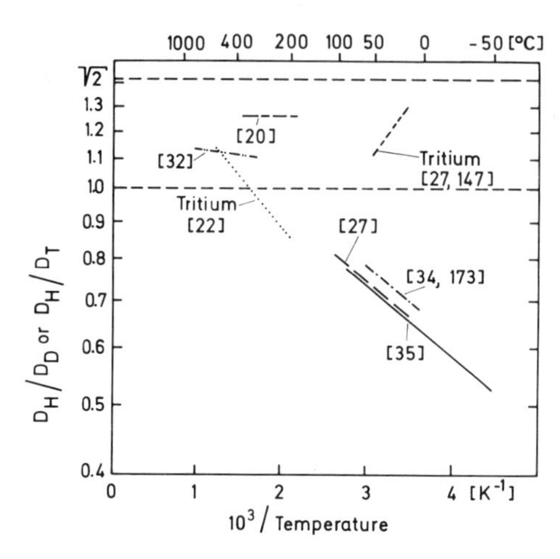

Fig. 12.20. Ratios of diffusion coefficients of isotopes of hydrogen in Pd. (Numbers in brackets refer to References of this Chapter)

[12.147]. The ratio between the diffusion coefficient of D to T in Nb is closer to the classical value than between H and D [12.126].

For the fcc metals the pre-exponential factors are mass dependent and scale as $\sqrt{m_2/m_1}$ for H and D in all three metals within error bars. The activation energies are mass dependent as well, but, in contrast to the three bcc metals, one finds $U_H > U_D$. This fact leads to a reversed isotope dependence below a certain temperature ($\approx 500\,^\circ$C for Pd), i.e., deuterium diffuses faster than hydrogen. This was first observed experimentally by *Bohmholdt* and *Wicke* [12.27] for Pd and is shown in Fig. 12.16 for data obtained with the Gorsky effect [12.35]. An extrapolation of high-temperature diffusion data by *Katz* et al. [12.63] for Ni and Cu would yield similar results. In Fig. 12.20 the ratio D_H/D_D for diffusion in Pd is shown. In contrast to bcc metals, all values are below $\sqrt{2}$ and furthermore for $T < 500\,^\circ$C below 1, indicating the reversed isotope dependence. For tritium the experimental situation is as follows: *Katz* et al. [12.63] found that for Ni and Cu the tritium-diffusion coefficient follows the same mass dependence as observed for H and D, i.e., $D_{0H}:D_{0D}:D_{0T} \approx 1:(\sqrt{2})^{-1}:(\sqrt{3})^{-1}$ and $U_H > U_D > U_T$. Therefore, extrapolating to low temperatures, tritium would diffuse fastest. The same result has been found by *Salmon* and *Randall* [12.22] for H and T in Pd using the permeation technique. Since solubility data for T in Pd were not available, the authors calculated the diffusion coefficient of T by using solubility data of hydrogen [12.172] and correcting for the higher binding energy of the tritium molecule. In contrast to this result obtained in the temperature region 200–600 °C, *Sicking* and *Buchold* [12.147] found a normal isotope dependence (see Figs. 12.16 and 12.20) for tritium in Pd in the room-temperature region.

A comparison of the isotope dependence measured at high hydrogen concentrations is in general more difficult, since samples with identical concentrations are hard to prepare. Yet this is a necessary requirement due to the observed dependence of the diffusion coefficient upon concentration. Comparing Figs. 12.11 and 12.12, one finds that in Ta the nonclassical isotope dependence of the diffusion coefficient continues to exist at high interstitial concentrations. In contrast to this, for Nb (Figs. 12.9 and 12.10) the ratio of diffusion coefficients for comparable concentrations decreases with increasing concentration and surprisingly approaches 1 at about 40% [12.14] (see also Fig. 12.19).

As an example for the β phase of Nb (78%H or D), the following parameters for the mean time-of-stay are reported [12.175]: Hydrogen: $\tau_0=1\cdot 10^{-12}$ s; $U_H=0.22$ eV. Deuterium: $\tau_0=2\cdot 10^{-13}$ s; $U_D=0.31$ eV. More data on the α' phase in Pd are listed in Ref. [12.11a], Table 3.

12.6 Deviations of the Diffusion Coefficient from the Arrhenius Relation

Because of the nonclassical diffusion behavior of H and the large temperature region over which the diffusion coefficient has been measured, it is by no means evident that the diffusion coefficient of H in metals should obey an Arrhenius relation over the complete temperature region. The break in the temperature dependence of the diffusion coefficient D for H in Nb, shown in Fig. 12.5, was first observed with the Gorsky effect method [12.1], and was confirmed both by *Wipf* [12.120] using resistivity measurements and recently by the QNS technique using a high resolution backscattering spectrometer [12.125]. Also NMR experiments [12.176] and quenching experiments by *Faber* and *Schultz* [12.177], *Hanada* [12.6], and *Engelhard* [12.150] are consistent with this change of slope. It should be mentioned that for impure samples the process causing this enhanced diffusion at low temperatures disappears [12.117], as has been confirmed recently by QNS [12.178].

An even more pronounced change in the temperature dependence of D was found by *Kokkinidis* [12.145] for H in Ta (Fig. 12.6). Whereas in Nb for deuterium no change in slope has been observed so far, preliminary experiments [12.145] show this effect also for D in Ta. It should be mentioned that *Guil* et al. [12.165] observed similar deviations from the Arrhenius relation for both isotopes at low temperatures in the β phase of Ta by measuring the kinetics of absorption. Also for the β phases of V, an inconsistency of the temperature dependence of the longitudinal relaxation time T_1 with an Arrhenius relation is reported [12.168].

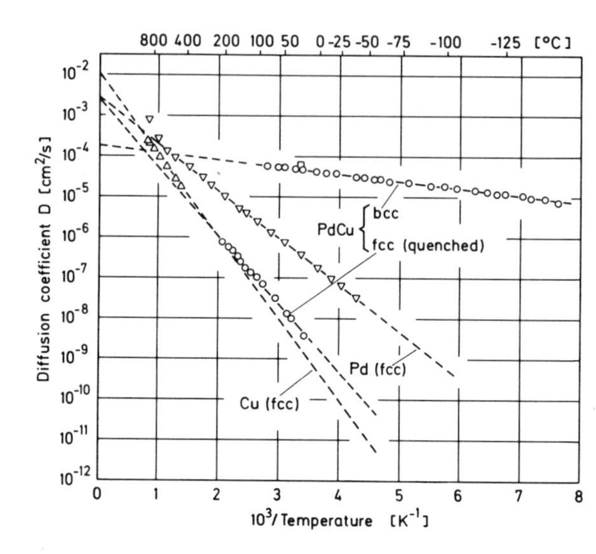

Fig. 12.21. Diffusion coefficient D for Pd (from Fig. 12.2), Cu [12.63] and $Pd_{0.47}Cu_{0.53}$ (bcc) [12.179] and $Pd_{0.47}Cu_{0.53}$ (fcc) [12.180]; □ = *Piper* [12.174]

12.7 Influence of Structure

Structure-dependent properties of H diffusion have been discussed in connection with isotope effects. First it is worth noting that the smallest values for the activation energy have been found for the bcc metals V, Nb, Ta, and α-Fe. There are several examples in which the transition from a close-packed to a bcc structure leads to a drastic increase of the diffusion coefficient, caused by a decrease in activation energy.

In Fig. 12.21 the diffusion coefficient of Pd (fcc) (from Fig. 12.2), Cu (fcc) [12.63], $Pd_{0.47}Cu_{0.53}$ (fcc, stabilized by quenching), and $Pd_{0.47}Cu_{0.53}$ (bcc, equilibrium phase) are plotted. The measurements for the alloy have been performed with the Gorsky effect by *Lang* [12.179] and *Steinhauser* [12.180]. The previous measurement by *Piper* [12.174] is also included. Whereas D for the fcc phase of the alloy falls close to that of copper, the bcc phase has at room temperature a D which is four orders of magnitude larger. The activation energy of $Pd_{0.47}Cu_{0.53}$ (bcc), U (bcc) = 0.035 eV is about a factor of ten smaller than U (fcc), whereas D_0 (fcc) is a factor ten larger than D_0 (bcc). The absolute values of D (bcc) are as high as for V (Fig. 12.7). Similar observations have been made for FeNi [12.71, 181], when changing from fcc to bcc. For Ti, *Wasilewski* and *Kehl* [12.182] found: U (bcc) = 0.29 eV, U (hcp) = 0.54 eV.

A further example for structure dependence of D is shown in Fig. 12.22 for the $\alpha-\gamma$ transition of iron; data of those authors who have used the same measuring technique in both phases have been included. For the γ phase an activation energy of $U(\gamma) \approx 0.5$ eV was reported rather consistently (see summary by *Bester* and *Lange* [12.183]) compared to $U(\alpha)$ between 0.05 and 0.1 eV, i.e., again a change of an order of magnitude for the transition fcc-bcc.

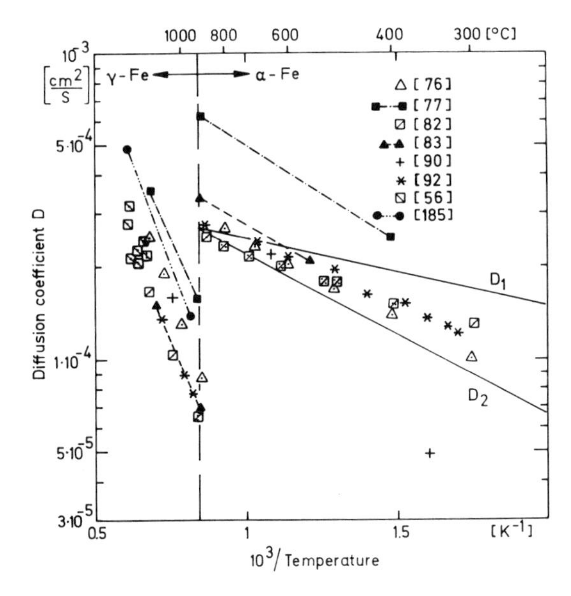

Fig. 12.22. Diffusion coefficients of H in α- and γ-Fe, D_1 and D_2 represent the lines of Fig. 12.4. (Numbers in brackets refer to References of this Chapter)

One would be inclined to attribute the higher diffusivity to the less close-packed structure of the bcc metals. Yet, as *Piper* [12.174] has pointed out, for Ti as well as for PdCu the close-packed structures are less dense than the bcc phases. The following consideration may be helpful in understanding this structure dependence [12.184]: In fcc metals there is one octahedral site per host atom, in bcc metals three octahedral or six tetrahedral sites per host atom. With d = nearest neighbor distance, the average distance of the interstitial sites is d (fcc), $d/\sqrt{3}$, or $d/\sqrt{6}$ (bcc), respectively. The smaller distance for bcc suggests that the activation barrier is smaller or can easier be tunneled through.

Finally, it should be mentioned that correlations between the H-diffusion coefficient and the degree of order have also been noted (for discussion see [12.11a]), which may be superimposed on the effects shown in Fig. 12.21.

12.8 Conclusion

In contrast to the accumulation of a large amount of experimental data, the theoretical interpretation of the diffusion of H in metals is still quite unsatisfactory (see Chap. 8). For example, the change in temperature dependence, as observed for H in Nb or Ta, is completely unexplained. From the experimental point of view the situation is unsatisfactory for the metal Fe and its alloys. Also, more measurements on tritium diffusion are necessary. In general, the experimental interest will continue to focus on low-temperature diffusion measurements, the structure dependence, and the structure dependence of the isotope effect.

References

12.1 G.Schaumann, J.Völkl, G.Alefeld: Phys. Stat. Sol. **42**, 401 (1970)
12.2 R.W.Powers, M.V.Doyle: J. Appl. Phys. **30**, 514 (1959)
12.3 A.E.Lord, Jr., D.N.Beshers: Acta Met. **14**, 1659 (1966)
12.4 J.J.Fritz, H.J.Maria, J.G.Aston: J. Chem. Phys. **34**, 2185 (1961)
12.5 R.Hanada: Scripta Met. **7**, 681 (1973)
12.6 R.Hanada: In *Effect of Hydrogen on Behavior of Materials*, ed. by A.W.Thompson and I.M.Bernstein (AIME 1976) p. 676
12.7 M.Sakamoto: J. Phys. Soc. (Japan) **19**, 1862 (1964)
12.8 G.Verdan, R.Rubin, W.Kley: *Proc. IAEA Symp. Inelastic Scattering of Neutrons*, Kopenhagen (1968) Vol. 1, p. 223
12.9 K.Brand: Atomkernenergie **17**, 113 (1971)
12.10 H.K.Birnbaum, C.A.Wert: Ber. Bunsenges. Physik. Chem. **76**, 806 (1972)
12.11a J.Völkl, G.Alefeld: In *Diffusion in Solids, Recent Developments*, ed. by A.S.Nowick and J.J.Burton (Academic Press, New York 1975) p. 231
12.11b G.Alefeld, J.Völkl (eds.): *Hydrogen in Metals II, Application-Oriented Properties*, Topics in Applied Physics, Vol. 29 (Springer, Berlin, Heidelberg, New York 1978) in preparation
12.12 A.Heidemann, G.Kaindl, D.Salomon, H.Wipf, G.Wortmann: Phys. Rev. Lett. **36**, 213 (1976)
12.13 J.R.Lacher: Proc. Roy. Soc. Ser. A **161**, 525 (1937)
12.14 H.C.Bauer, J.Völkl, J.Tretkowski, G.Alefeld: Z. Physik B **29**, 17 (1978)
12.15 H.Wagner, H.Horner: Advan. Phys. **23**, 587 (1974)
12.16 H.Horner, H.Wagner: J. Phys. C: Solid State Phys. **7**, 3305 (1974)
12.17 R.Bausch, H.Horner, H.Wagner: J. Phys. C: Solid State Phys. **8**, 2559 (1975)
12.18 H.-K.Janssen: Z. Physik B **23**, 245 (1976)
12.19 J.Tretkowski, J.Völkl, G.Alefeld: Z. Physik B **28**, 259 (1977)
12.20 W.Jost, A.Widmann: Z. Phys. Chem. B **45**, 285 (1940)
12.21 R.M.Barrer: Trans. Faraday Soc. **36**, 1235 (1940)
12.22 O.N.Salmon, D.Randall: USAEC Report KAPL-984 (1954)
12.23 W.D.Davis: USAEC Report KAPL-1227 (1954)
12.24 G.Toda: Hokkaido Univ. Res. Inst. Catalysis J. **6**, 13 (1958)
12.25 O.M.Katz, E.A.Gulbransen: Rev. Sci. Instr. **31**, 615 (1960)
12.26 J.W.Simons, T.B.Flanagan: J. Phys. Chem. **69**, 3581 (1965)
12.27 G.Bohmholdt, E.Wicke: Z. Phys. Chem. N.F. **56**, 133 (1967)
12.28 G.Holleck, E.Wicke: Z. Phys. Chem. N.F. **56**, 155 (1967)
12.29 K.Sköld, G.Nelin: J. Phys. Chem. Sol. **28**, 2369 (1967)
12.30 S.A.Koffler, J.B.Hudson, G.S.Ansell: Trans. Met. Soc. AIME **245**, 1735 (1969)
12.31 J.O'M.Bockris, M.A.Genshaw, M.Fullenwider: Electrochim. Acta **15**, 47 (1970)
12.32 V.A.Gol'tsov, V.B.Demin, V.B.Vykhodets, G.Ye.Kagan, P.V.Gel'd: Phys. Metals Metallog. **29**, 195 (1970)
V.A.Goltsov, G.E.Kagan: *Proc. Congrès L'Hydrogène dans les Métaux*, Paris (1972) p. 249
12.33 G.L.Holleck: J. Phys. Chem. **74**, 503 (1970)
12.34 H.Züchner: Z. Naturforsch. **25**a, 1490 (1970)
12.35 J.Völkl, G.Wollenweber, K.-H.Klatt, G.Alefeld: Z. Naturforsch. **26**a, 922 (1971)
12.36 M.A.V.Devanathan, Z.Stachurski: Proc. Roy. Soc. Ser. A **270**, 90 (1962)
12.37 M.von Stackelberg, P.Ludwig: Z. Naturforsch. **19**a, 93 (1964)
12.38 A.C.Makrides, D.N.Jewett: Engelhard Ind. Techn. Bull. **7**, 51 (1966)
12.39 J.Knaak, W.Eichenauer: Z. Naturforsch. **23**a, 1783 (1968)
12.40 V.Breger, E.Gileadi: Electrochim. Acta **16**, 177 (1971)
12.41 J.M.Rowe, J.J.Rush, L.A.de Graaf, G.A.Ferguson: Phys. Rev. Lett. **29**, 1250 (1972)
12.42 C.J.Carlile, D.K.Ross: Solid State Commun. **15**, 1923 (1974)
12.43 W.Kley, W.Drexel: Commission Europ. Comm. Eur. 5466e (1976)

12.44 N. Boes, H. Züchner: J. Less-Common Metals **49**, 223 (1976)
12.45 J. Bergsma, J. A. Goedkoop: Physica **26**, 744 (1960)
12.46 W. Drexel, A. Murani, D. Tocchetti, W. Kley, I. Sosnowska, D. K. Ross: J. Phys. Chem. Sol. **37**, 1135 (1976)
12.47 E. Wicke, K. Meyer: Z. Physik. Chem. N.F. **64**, 225 (1969)
12.48 C. E. Ransley, D. E. J. Talbot: Z. Metallk. **46**, 328 (1955)
12.49 M. L. Hill, E. W. Johnson: Acta Met. **3**, 566 (1955)
12.50 A. G. Edwards: Brit. J. Appl. Phys. **8**, 406 (1957)
12.51 H. H. Grimes: Acta Met. **7**, 783 (1959)
12.52 Yu. I. Belyakov, N. I. Ionov: Soviet Phys.—Techn. Phys. **6**, 146 (1961)
12.53 K. M. Olsen, C. F. Larkin: J. Electrochem. Soc. **110**, 86 (1963)
12.54 L. N. Ryabchikov: Ukrains'kyi Fiz. Zhu. **9**, 303 (1964)
12.55 W. Eichenauer, W. Löser, H. Witte: Z. Metallk. **56**, 287 (1965)
12.56 H. Schenck, K. W. Lange: Arch. Eisenhüttenw. **37**, 809 (1966)
12.57 Y. Ebisuzaki, W. J. Kass, M. O'Keeffe: J. Chem. Phys. **46**, 1378 (1967)
12.58 W. Fischer: Z. Naturforsch. **22**a, 1581 (1967)
12.59 S. Scherrer, G. Lozes, B. Deviot: C.R. Acad. Sci. Sèr. B **264**, 1499 (1967)
12.60 J. O'M. Bockris, M. A. Genshaw, M. Fullenwider: Electrochim. Acta **15**, 47 (1970)
12.61 P. Combette, P. Azou: Mem. Sci. Rev. Met. **67**, 17 (1970)
12.62 W. Beck, J. O'M. Bockris, M. A. Genshaw, P. K. Subramanyan: Met. Trans. **2**, 883 (1971)
12.63 L. Katz, M. Guinan, R. J. Borg: Phys. Rev. B **4**, 330 (1971)
12.64 M. R. Louthan, Jr., R. G. Derrick, A. H. Dexter: Report DPST (NASA)-71-4 (1971)
M. R. Louthan, Jr., J. A. Donovan, G. R. Caskey, Jr.: Scripta Met. **8**, 643 (1974)
12.65 W. M. Robertson: Jül-Bericht Jül-Conf-6, Vol. II (KFA-Jülich 1972) p. 449; Z. Metallk. **64**, 436 (1973)
12.66 G. Euringer: Z. Physik **96**, 37 (1935)
12.67 K. H. Lieser, H. Witte: Z. Phys. Chem. **202**, 321 (1954)
12.68 J. Čermák, A. Kufudakis: Mem. Sci. Rev. Met. **63**, 767 (1966)
12.69 J. Čermák, A. Kufudakis: Mem. Sci. Rev. Met. **65**, 375 (1968)
12.70 Y. Ebisuzaki, M. O'Keeffe: J. Chem. Phys. **48**, 1867 (1968)
12.71 W. Dresler: Thesis, Techn. Univ. Berlin (D83) Germany (1971)
12.72 K. Yamakawa, M. Tada, F. E. Fujita: Scripta Met. **10**, 405 (1976); Japan J. Appl. Phys. **15**, 769 (1976)
12.73 J. Čermák, A. Kufudakis: J. Less-Common Metals **49**, 309 (1976)
12.74 H. Kronmüller, H. Steeb, N. König: Nuovo Cimento **33**B, 205 (1976)
12.75 R. M. Barrer: Trans. Faraday Soc. **36**, 1235 (1940)
12.76 C. Sykes, H. H. Burton, C. C. Gegg: J. Iron Steel Inst. **156**, 155 (1947)
12.77 W. Geller, T.-H. Sun: Arch. Eisenhüttenw. **21**, 423 (1950)
12.78 T. M. Stross, F. C. Tompkins: J. Chem. Soc. **159**, 230 (1956)
12.79 W. Raczyński: Arch. Hutnictwa **3**, 59 (1958); see also
M. Smialowski: Neue Hütte **2**, 611 (1957)
12.80 H. Zitter, H. Krainer: Arch. Eisenhüttenw. **29**, 401 (1958)
12.81 W. Eichenauer, H. Künzig, A. Pebler: Z. Metallk. **49**, 220 (1958)
12.82 H. Maas: Thesis, Univ. Münster, Germany (1958)
12.83 H. Schenck, H. Taxhet: Arch. Eisenhüttenw. **30**, 661 (1959)
12.84 E. W. Johnson, M. L. Hill: Trans. Met. Soc. AIME **218**, 1104 (1960)
12.85 D. C. Carmichael, J. R. Hornaday, A. E. Morris, N. A. Parlee: Trans. Met. Soc. AIME **218**, 826 (1960)
12.86 R. W. Lee, D. E. Swets, R. C. Frank: Mem. Sci. Rev. Met. **58**, 36 (1961)
12.87 W. Raczyński, S. Stelmach: Bull. Acad. Pol. Sci., Sér. Sci. Chim. **9**, 633 (1961)
12.88 W. L. Bryan, B. F. Dodge: Am. Inst. Chem. Eng. (A.I.Ch.E) J. **9**, 223 (1963)
12.89 R. Wagner, R. Sizmann: Z. Angew. Phys. **18**, 193 (1964)
12.90 W. Schwarz, H. Zitter: Arch. Eisenhüttenw. **36**, 343 (1965)
12.91 W. Beck, J. O'M. Bockris, J. McBreen, L. Nanis: Proc. Roy. Soc. Ser. A **290**, 220 (1966)

12.92 Th. Heumann, D. Primas: Z. Naturforsch. **21**a, 260 (1966)
12.93 G.M. Evans, E.C. Rollason: J. Iron Steel Inst. **207**, 1484 (1969)
12.94 F. Erdmann-Jesnitzer, H. Hieber: Arch. Eisenhüttenw. **40**, 73 (1969)
12.95 J.Y. Choi: Met. Trans. **1**, 911 (1970)
12.96 B.K. Reiermann: Thesis, Techn. Univ. Berlin (D83) Germany (1970)
12.97 E. Domke: Thesis, Univ. Münster, Germany (1971)
Th. Heumann, E. Domke: Jül-Bericht Jül-Conf-6, Vol. II (KFA-Jülich 1972) p. 492
12.98 A.E. Schuetz, W.D. Robertson: Corrosion **13**, 437t (1957)
12.99 B. Baranowski, M. Śmiałowski, Z. Szkłarska-Śmiałowska: Bull. Acad. Pol. Sci., Sér. Sci. Chim. **5**, 191 (1957)
12.100 P. Bastien, P. Amiot: Rev. Met. **55**, 24 (1958)
12.101 M.A.V. Devanathan, Z. Stachurski, W. Beck: J. Electrochem. Soc. **110**, 886 (1960) and M.A.V. Devanathan, Z. Stachurski: J. Electrochem. Soc. **111**, 619 (1964)
12.102 V.P. Alikin: Isv. Estestvennonauchn. Inst. pri Permsk. Univ. **14**, 19 (1960) [Abstract in: Chem. Abstr. **57**, Col. 4468c (1962)]
12.103 W. Palczewska, I. Ratajczyk: Bull. Acad. Pol. Sci., Sér. Sci. Chim. **9**, 267 (1961)
12.104 V.V. Kuznietsov, N.I. Subbotina: Elektrokhimija **1**, 1096 (1965)
12.105 T.P. Radhakrishnan, L.L. Shreir: Electrochim. Acta **12**, 889 (1967)
12.106 S. Wach. A.P. Miodownik: Corrosion Sci. **8**, 271 (1968)
12.107 K.-D. Joppien: Thesis, Techn. Univ. Clausthal, Germany (1968)
12.108 W. Raczyński, S. Talbot-Besnard: C.R. Acad. Sci. Sér. C **269**, 1253 (1969)
12.109 J.-L. Dillard: C.R. Acad. Sci. Sér. C **270**, 669 (1970)
J.-L. Dillard, S. Talbot-Besnard: *Proc. Congrès L'Hydrogène dans les Metaux*, Paris 1972, p. 159
12.110 H.-G. Ellerbrock, G. Vibrans, H.-P. Stüwe: Acta Met. **20**, 53 (1972)
12.111 P.V. Geld, R.A. Ryabov, V.I. Saly: *Proc. Congrès L'Hydrogène dans les Metaux*, Paris 1972, p. 167
12.112 T.K. Govindan Namboodhiri, L. Nanis: Acta Met. **21**, 663 (1973)
12.113 H.G. Nelson, J.E. Stein: NASA-Report TN D-7265 (1973)
12.114 A.J. Kumnick, H.H. Johnson: Met. Trans. **5**, 1199 (1974)
12.115 V. Safonov, J. Chêne, J. Galland, P. Azou, P. Bastien: C.R. Acad. Sci. Sér. C **278**, 445 (1974)
12.116 H.-J. König, K.W. Lange: Arch. Eisenhüttenw. **46**, 669 (1975)
K.W. Lange, H.-J. König: Forschungsber. d. Landes NRW Nr. 2566 (1976)
12.117 W. Münzing, J. Völkl, H. Wipf, G. Alefeld: Scripta Met. **8**, 1327 (1974)
12.118 G. Schaumann, J. Völkl, G. Alefeld: Phys. Rev. Lett. **21**, 891 (1968)
G. Schaumann: Thesis, Techn. Hochschule Aachen, Germany (1969); Jül-Bericht Jül-606-FN (KFA-Jülich 1969) and Ref. 12.1
12.119 R. Cantelli, F.M. Mazzolai, M. Nuovo: Phys. Stat. Sol. **34**, 597 (1969)
12.120 H. Wipf: Thesis, Techn. Univ. München, Germany (1972);
Jül-Bericht Jül-876-FF (KFA-Jülich 1972)
H. Wipf, G. Alefeld: Phys. Stat. Sol. (a) **23**, 175 (1974)
12.121 J.H.L. Birchall, D.K. Ross: Jül-Bericht Jül-Conf-6, Vol. I (KFA-Jülich 1972) p. 313
12.122 W. Gissler, G. Alefeld, T. Springer: J. Phys. Chem. Sol. **31**, 2361 (1970)
12.123 N. Boes, H. Züchner: Z. Naturforsch. **31**a, 760 (1976)
12.124 D.G. Westlake, S.T. Ockers, D.W. Regan: J. Less-Common Metals **49**, 341 (1976)
12.125 D. Richter, B. Alefeld, A. Heidemann, N. Wakabayashi: J. Phys. F: Metal Phys. **7**, 569 (1977)
12.126 G. Matusiewicz, H.K. Birnbaum: J. Phys. F: Metal. Phys. **7**, 2285 (1977)
12.127 W.M. Albrecht, W.D. Goode, M.W. Mallett: J. Electrochem. Soc. **106**, 981 (1959)
12.128 L.N. Ryabchikov: Ukrains'kyj Fiz. Zhu. **9**, 293 (1964)
12.129 T.Ö. Oğurtani: Met. Trans. **2**, 3035 (1971)
12.130 V.A. Somenkov, A.V. Gurskaya, M.G. Zemlyanov, M.E. Kost, N.A. Chernoplekov, A.A. Chertkov: Soviet Phys.—Solid State **10**, 1076 (1968)
12.131 H. Conrad, G. Bauer, G. Alefeld, T. Springer, W. Schmatz: Z. Physik **266**, 239 (1974)

12.132 N.Stump, G.Alefeld, D.Tocchetti: Solid State Commun. **19**, 805 (1976)
12.133 G.Cannelli, L.Verdini: Ricerca Sci. **36**, 98 (1966)
12.134 P.Schiller, A.Schneiders: Jül-Bericht Jül-Conf-2, Vol. II (KFA-Jülich 1968) p. 871
12.135 G.Cannelli, F.M.Mazzolai: Appl. Phys. **1**, 111 (1973)
12.136 C.Baker, H.K.Birnbaum: Acta Met. **21**, 865 (1973)
12.137 P.Schiller: Nuovo Cimento **33**B, 226 (1976)
12.138 B.A.Merisov, V.I.Khotkevich, A.I.Karnus: Phys. Metals Metallog. **22**, 163 (1966)
12.139 R.Cantelli, F.M.Mazzolai, M.Nuovo: J. Phys. (Paris) **32**, C 2–59 (1971)
12.140 B.A.Merisov, A.D.Serdyuk, I.I.Fal'ko, G.Ya. Khadzhay, V.I.Khotkevich: Phys. Metals Metallog. **32**, 154 (1971)
12.141 E.Wicke, A.Obermann: Z. Phys. Chem. N.F. **77**, 163 (1972)
12.142 H.Züchner: Z. Phys. Chem. N.F. **82**, 240 (1972)
12.143 J.J.Rush, R.C.Livingston, L.A.de Graaf, H.E.Flotow, J.M.Rowe: J. Chem. Phys. **59**, 6570 (1973); see also:
L.A.de Graaf, J.J.Rush, R.C.Livingston, H.E.Flotow, J.M.Rowe: Jül-Bericht Jül-Conf-6, Vol. I (KFA-Jülich 1972) p. 301
12.144 J.M.Rowe, J.J.Rush, H.E.Flotow: Phys. Rev. B**9**, 5039 (1974)
12.145 M.Kokkinidis: Diploma Thesis, Techn. Univ. München, Germany (1977)
12.146 R.Cantelli, F.M.Mazzolai, M.Nuovo: Appl. Phys. **1**, 27 (1973); Jül-Bericht Jül-Conf-6, Vol. II (KFA-Jülich 1972) p. 770
12.147 G.Sicking, H.Buchold: Z. Naturforsch. **26**a, 1973 (1971)
12.148 A.Zeilinger, W.A.Pochmann: J. Phys. F: Metal Phys. **7**, 575 (1977)
12.149 B.Ströbel, K.Läuger, H.E.Bömmel: Appl. Phys. **9**, 39 (1976)
12.150 J.Engelhard: Verhandl. DPG (VI) **12**, 296 (1977)
12.151 R.Cantelli, F.M.Mazzolai, M.Nuovo: J. Phys. Chem. Sol. **31**, 1811 (1970)
12.152 J.M.Rowe, K.Sköld, H.E.Flotow, J.J.Rush: J. Phys. Chem. Sol. **32**, 41 (1971)
12.153 R.Heller, H.Wipf: Phys. Stat. Sol. (a) **33**, 525 (1976)
R.Heller: Diploma Thesis. Techn. Univ. München, Germany (1973)
12.154 U.Freudenberg: Diploma Thesis, Techn. Univ. München, Germany (1976)
12.155 J.Völkl, G.Alefeld: Nuovo Cimento **33**B, 190 (1976)
12.156 J.J.Rush, H.E.Flotow: J. Chem. Phys. **48**, 3795 (1968)
12.157 J.M.Rowe: Solid State Commun. **11**, 1299 (1972)
12.158 F.Abe, R.Hanada, H.Kimura: Scripta Met. **8**, 955 (1974)
12.159 G.Alefeld, G.Schaumann, J.Tretkowski, J.Völkl: Phys. Rev. Lett. **22**, 697 (1969)
12.160 G.Cannelli, L.Verdini: Ricerca Sci. **36**, 246 (1966)
12.161 J.H.L.Birchall, D.K.Ross: Jül-Bericht Jül-Conf-6, Vol. I (KFA-Jülich 1972) p. 313
12.162 O.J.Zogał, R.M.Cotts: Phys. Rev. B **11**, 2443 (1975)
12.163 L.A.de Graaf, J.J.Rush, H.E.Flotow, J.M.Rowe: J. Chem. Phys. **56**, 4574 (1972)
12.164 G.J.Krüger, R.Weiß: *Proc. 18th Ampere Congr.*, Nottingham, 1974, p. 339
12.165 J.M.Guil, D.O.Hayward, N.Taylor: Proc. Roy. Soc. (London) A **335**, 141 (1973)
12.166 K.Tanaka, T.Hashimoto: J. Phys. Soc. Japan **34**, 379 (1973)
12.167 R.R.Arons, H.G.Bohn, H.Lütgemeier: *Proc. 18th Ampere Congr.*, Nottingham 1974, p. 343
12.168 Y.Fukai, S.Kazama: Acta Met. **25**, 59 (1977)
12.169 B.Alefeld, H.G.Bohn, N.Stump: Jül-Bericht Jül-Conf-6, Vol. I (KFA-Jülich 1972) p. 286
12.170 C.Wert, C.Zener: Phys. Rev. **76**, 1169 (1949)
12.171 C.H.Vineyard: J. Phys. Chem. Sol. **3**, 121 (1957)
12.172 R.L.Favreau, R.E.Patterson, D.Randall, O.N.Salmon: USAEC Rep. KAPL-1036 (1954)
12.173 N.Boes: Diploma Thesis, Univ. Münster, Germany (1971); see also
H.Züchner, N.Boes: Ber. Bunsenges. Physik. Chem. **76**, 783 (1972)
12.174 J.Piper: J. Appl. Phys. **37**, 715 (1966)
12.175 H.Lütgemeier, H.G.Bohn, R.R.Arons: J. Magn. Resonance **8**, 74 and 80 (1972)
H.G.Bohn: Jül-Bericht Jül-853-FF (KFA Jülich 1972)
12.176 A.Seeger: Private communication
12.177 K.Faber, H.Schultz: Scripta Met. **6**, 1065 (1972)

12.178 D. Richter, J. Töpler, T. Springer: J. Phys. F: Metal Phys. **6**, L93 (1976)
12.179 G. Lang: Diploma Thesis, Techn. Univ. München, Germany (1976)
12.180 K. A. Steinhauser: Diploma Thesis, Techn. Univ. München, Germany (1977)
12.181 W. Dresler, M. G. Frohberg: Jül-Bericht Jül-Conf-6, Vol. II (KFA-Jülich 1972) p. 516
12.182 R. J. Wasilewski, G. L. Kehl: Metallurgica **50**, 225 (1954)
12.183 H. Bester, K. W. Lange: Arch. Eisenhüttenw. **43**, 207 (1972)
12.184 K. Kehr: Private communication
12.185 H. Bester, K. W. Lange: Arch. Eisenhüttenw. **47**, 333 (1976)
12.186 B. Pedersen, T. Krogdahl, O. E. Stokkeland: J. Chem. Phys. **42**, 72 (1965)

13. Positive Muons as Light Isotopes of Hydrogen

A. Seeger

With 11 Figures

13.1 Background

Muons were discovered in cloud-chamber experiments by *Anderson* and *Neddermeyer* [13.1–3] and by *Street* and *Stevenson* [13.4] in a systematic attempt to resolve a number of apparent paradoxes in the properties of cosmic radiation. The newly discovered particles were named "mesotrons", since their mass was found to be intermediate between that of the proton and that of the electron–positron pair, the only charged elementary particles known at the time. The "mesotrons" were thought to constitute the penetrating components of the cosmic radiation[1].

Shortly before the discovery of the muon, *Yukawa* [13.8] had proposed his theory of nuclear forces in terms of a field associated with a particle of about the same mass as that of the muon. Although *Yukawa*'s prediction of a "meson" had not played a rôle in the discovery of the muon, in the following decade muons were generally identified with the particles of the Yukawa theory. Only gradually was the contradiction realized that as Yukawa particles, the mesotrons should possess a very strong interaction with nuclei and thus become rapidly absorbed in the upper atmosphere, whereas as the penetrating component of the cosmic radiation, they would have to have very little interaction with matter. A particularly significant experiment was performed by *Conversi* et al. [13.9], who demonstrated that the lifetimes of positive or negative muons stopped in light elements were about same. Since negative muons are electrostatically attracted by the nuclei whereas positive muons are repelled, this observation meant that the interaction between muons and nuclei must be extremely weak.

The paradox just described was resolved in 1947 by *Powell* and co-workers [13.10, 11], who showed that the mesotrons were *not* the original particles of cosmic radiation but that they were generated by the decay of another type of particle of intermediate mass. In order to indicate that these newly discovered mesons were the *primary* mesons of the cosmic radiation they were called π-mesons or *pions*. The mesotrons were renamed μ-mesons and later *muons*, by which name they will be referred to in the remainder of this chapter. It is now

---

[1] A brief account of the discovery of muons (and positrons) has been given by *Anderson* [13.5]. For more background information and historical references the reader is referred to the books by *Weissenberg* [13.6] and by *Marshak* [13.7].

generally accepted that the pion is the particle predicted by *Yukawa* [13.8]. The name μ-meson has fallen into disuse, since muons are no longer classified as mesons but as *leptons*, together with positive and negative electrons, with neutrinos and antineutrinos [13.12], and the recently discovered heavier $\tau^{\pm}$ particles [13.13].

Muons are Dirac particles of spin 1/2 and possess, apart from their larger mass and their finite lifetimes, the same properties as electrons (= negatons) and positrons [13.14, 15][2]. Like the negaton–positron particle–antiparticle pairs, they exist in a positive or negative charge state, forming the $(\mu^- - \mu^+)$ particle–antiparticle pair.

In the present chapter we are concerned only with the positively charged muon, i.e., the muon antiparticle. From the point of view of the solid-state physicist, the positive muon is in fact more closely related to the proton (or the other hydrogen nuclei) than to the positron. This results from the fact that the muon mass is by only a factor of nine smaller than the proton mass, whereas the mass ratio of muons and electrons is 207. Like the hydrogen nuclei, the muons may be considered as very heavy compared with the electrons but nevertheless light compared with most metal atoms. With protons and tritons the positive muons have further in common that they possess spin 1/2 and hence no electrical quadrupole moment. It is true that their interactions with nuclei are completely different from those of the hydrogen nuclei, but this can be taken into account easily in solid-state experiments. Through the positive muon, nature has thus provided us with a 'light isotope' of hydrogen which, as we shall see, has extremely interesting and useful properties. It is the aim of this chapter to explore the use of positive muons as light hydrogen isotopes in metals and to summarize the results so far obtained.

13.2 Basic Properties and Interactions of Muons

13.2.1 The Muon Decay

Table 13.1 summarizes the properties of positive muons together with those of positrons and of the three hydrogen nuclei proton, deuteron, and triton[3]. In addition to mass m, spin I, magnetic moment μ, and gyromagnetic ratio γ defined by

$$\boldsymbol{\mu} = \gamma \boldsymbol{I}, \tag{13.1}$$

---

[2] This statement is now becoming invalid, since at extremely high energies differences between muons and electron–positrons are being unravelled that relate presumably to the existence of an intermediate vector boson. However, for the physics of the condensed matter, for nuclear physics, and even for elementary particle physics at intermediate energies, these differences are without practical consequences.

[3] A fourth hydrogen isotope, ^{4}H, may be formed by nuclear reactions, but plays no rôle in solid-state physics.

Table 13.1. Properties of particles with positive elementary charge

Particle Symbol	Positron e^+	Positive muon μ^+	Proton p	Deuteron d	Triton t
Rest mass m	$m_e = 0.91095 \cdot 10^{-30}$ kg $= 0.51100$ MeV/c^2	$m_\mu = 206.769\, m_e$ $= 0.012610\, m_p$ $= 105.659$ MeV/c^2	$m_p = 1836.152\, m_e$ $= 1.6726 \cdot 10^{-27}$ kg $= 938.28$ MeV/c^2	$m_d = 1.998\, m_p$	$m_t = 2.993\, m_p$
Spin I	1/2	1/2	1/2	1	1/2
Magnetic moment μ	$\mu_e = 1.001160\, \mu_B$ $= 658.2107\, \mu_p$	$\mu_\mu = 1.001166 \frac{m_e}{m_\mu} \mu_B$ $= 3.18334\, \mu_p$ $= 4.4905 \cdot 10^{-26}$ JT^{-1}	$\mu_p = 2.7928\, \mu_N$ $= 1.41062 \cdot 10^{-26}$ JT^{-1}	$\mu_d = 0.8574\, \mu_N$	$\mu_t = 2.9788\, \mu_N$
Gyromagnetic ratio γ[rad s^{-1}T^{-1}]	$\gamma_e = 1.7608 \cdot 10^{11}$	$\gamma_\mu = 8.5161 \cdot 10^8$	$\gamma_p = 2.6752 \cdot 10^8$	$\gamma_d = 0.4106 \cdot 10^8$	$\gamma_t = 2.853 \cdot 10^8$
Quadrupole moment Q[e m^2]	0	0	0	$2.77 \cdot 10^{-31}$	0
Lifetime $\tau(= T_{1/2}/\ln 2)$	stable, annihilating with e^- in 10^{-10} to 10^{-7} s	$\tau_\mu = 2.197 \cdot 10^{-6}$ s	stable	stable	$\tau_t = 17.81$ a
Atom	Positronium $e^+ + e^- = \mathrm{Ps}$	Muonium $\mu^+ + e^- = \mathrm{Mu}$	Hydrogen $p + e^- = \mathrm{H} \equiv {}^1\mathrm{H}$	Deuterium $d + e^- = \mathrm{D} \equiv {}^2\mathrm{H}$	Tritium $t + e^- = \mathrm{T} \equiv {}^3\mathrm{H}$

Bohr magneton $\mu_B = \frac{e\hbar}{2m_e} = 9.2741 \cdot 10^{-24}$ JT^{-1}; nuclear magneton $\mu_N = \frac{e\hbar}{2m_p} = 5.0508 \cdot 10^{-27}$ JT^{-1}; elementary charge $e = 1.60219 \cdot 10^{-19}$ C.

Table 13.1 lists the lifetimes $\tau(=T_{1/2}/\ln 2$, where $T_{1/2}$ denotes the half-life) and the decay or principal annihilation modes of the particles. It will be noted that of the five particles considered, three are stable. Both the triton and the muon decay through weak interaction (β-decay). The lifetime of the muon is by a factor $3.9 \cdot 10^{-15}$ shorter than that of the triton. We may thus say that hydrogen possesses two radioactive isotopes with extremely different lifetimes. Naturally, the enormous difference in lifetimes requires quite different detection methods for the two 'isotopes'. As we shall see below, *parity non-conservation* in both the production and the decay of muons provides the essential tool for studying muons.

The lifetime of positrons is even shorter than that of muons, but in contrast to muon and triton lifetimes it is not governed by radioactive decay but by annihilation with the omnipresent antiparticle e^-. (As the positively charged particle with the smallest mass the positron cannot decay.) Positron lifetimes depend on the precise circumstances under which the positrons annihilate (in particular on the electron density at the location of the positron) and may be as large as $1.4 \cdot 10^{-7}$ s when ortho-positronium is formed. In metals, typical positron lifetimes are about $2 \cdot 10^{-10}$ s. By contrast to this and to the behaviour of negative muons in matter of intermediate or large atomic numbers, the lifetimes of *positive muons* and of *tritons* are practically independent of the environment in which they decay.

The information on the interaction of muons with their environment is obtained from the positrons emitted in the muon decay reaction

$$\mu^+ \to e^+ + \nu_e + \bar{\nu}_\mu . \tag{13.2}$$

In (13.2), ν_e denotes the neutrino associated with the electron–positron pair and $\bar{\nu}_\mu$ the antineutrino associated with the muons. The maximum kinetic energy of the positrons in (13.2) is $E_{\max}(e^+) = 52.83$ MeV (in the μ^+ rest-system); the average value is $\overline{E(e^+)} = 36$ MeV.

The preceding values are in striking contrast to the maximum electron energy in the β^--decay of tritons, $E_{\max}(e^-) = 17.95$ keV. Whereas the small kinetic energy of the β^--decay electrons of tritium leads to considerable experimental difficulties, the high positron energies in (13.2) guarantee us that we can easily get the information on the decaying muons out of massive specimens. In this respect the situation is similar to the case of positron annihilation, in which the information on the interaction of the positrons with their environment is contained in the γ quanta resulting from the annihilation reaction. In metals the dominant positron annihilation mode results in the emission of two γ-rays with energies $E_\gamma \simeq 0.511$ MeV; these are practically not absorbed or scattered in metal specimens.

A very important feature of the muon decay reaction (13.2) is that *it does not conserve parity*. Within experimental error the observations are in agreement with the two-component Dirac theory of the neutrinos (based on the *Weyl* equation [13.16]) [13.17–21] and with a weak-interaction Hamiltonian with equal and opposite vector and axial vector coupling constants (so-called V–A theory) [13.6, 12, 14, 15, 22–25].

If the rest energy of the emitted positrons may be neglected compared with their kinetic energy $E(e^+)$, the V–A theory predicts for the probability p that per

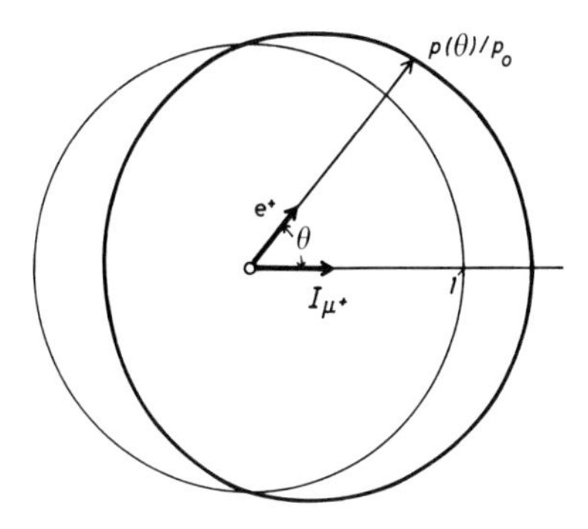

Fig. 13.1. Polar diagram of normalized probability $p(\theta)/p_0$ for positron emission at an angle θ from the muon spin direction ($\bar{a}=1/3$) (full curve). The thin curve represents the unit circle

unit time a positron is emitted into a solid angle $d\Omega$ in a direction forming the angle θ with the direction of the μ^+-spin

$$dp(x,\theta)=\frac{3-2x}{2\tau_\mu}[1+a(x)\cos\theta]\,x^2\,dx\,d\Omega\,. \tag{13.3}$$

Here

$$x\equiv\frac{E(e^+)}{E_{\max}(e^+)} \tag{13.4}$$

is the kinetic energy of the emitted positrons measured in units of the maximum energy and

$$a(x)\equiv\frac{2x-1}{3-2x} \tag{13.5}$$

the asymmetry coefficient.

From (13.3) one obtains for the average positron energy

$$\overline{E(e^+)}=4\pi\tau_\mu E_{\max}(e^+)\int\limits_{x=0}^{x=1}p(x)x\,dx=\tfrac{7}{10}E_{\max}(e^+)=36\,\mathrm{MeV}\,. \tag{13.6}$$

The asymmetry coefficient $a(x)$ is negative for $E(e^+)<E_{\max}(e^+)/2$ and positive for $E(e^+)>E_{\max}(e^+)/2$. At $E(e^+)=E_{\max}(e^+)$ the asymmetry coefficient assumes its maximum value of unity. If we integrate (13.3) over all positron energies from 0 to $E_{\max}(e^+)$, we find an average asymmetry coefficient

$$\bar{a}=\tfrac{1}{3}\,. \tag{13.7}$$

This means that positrons are preferentially emitted in the direction of the μ^+-spin (Fig. 13.1). When this asymmetry is used to detect the polarization of muons in solids, the low-energy end of the positron spectrum [in which according to (13.5) the asymmetry has the opposite sign] is usually cut off. Thus the *effective* asymmetry coefficient $\bar{a}_{\mathrm{eff}}$ will be slightly larger than (13.7)[4].

13.2.2 Muon Production

The principal source of muons, both in the cosmic radiation and in the laboratory, is the decay of charged pions according to the semileptonic reactions

$$\begin{aligned}\pi^+ &\rightarrow \mu^+ + \nu_\mu \\ \pi^- &\rightarrow \mu^- + \bar{\nu}_\mu\end{aligned}. \qquad (13.8)$$

The lifetime of the charged pions associated with the reactions (13.8) is

$$\tau_{\pi^\pm} = 2.60 \times 10^{-8}\,\mathrm{s}, \qquad (13.8a)$$

corresponding to a pion "mean free path" $c\tau_{\pi^\pm} = 7.80\,\mathrm{m}$, where c is the velocity of light. In the pion rest-system the muons resulting from (13.8) possess a unique energy of 4.12 MeV, corresponding to a muon momentum $p_\mu = 29.79\,\mathrm{MeV}/c$ and a muon velocity of about one-quarter of the velocity of light.

It is easily seen that in the pion rest-system the muons resulting from (13.8) must be longitudinally polarized: Pions possess zero spin. By their very nature neutrinos are always longitudinally polarized, spin and momentum forming a left-handed screw [13.12]. Hence balance of momentum and angular momentum requires that the *muons* produced by (13.8) are also *longitudinally polarized.* The spin of positive muons is directed opposite to their momentum in the pion rest-system.

In the muon production facilities[5], the pions decay in flight in so-called muon channels. This means that the kinetic energy of the muons in the laboratory system is much higher than that in the pion rest-system. The separation of a muon beam by momentum selection in the laboratory system leads to 'kinematic depolarization' [13.6]. *Kinematic depolarization* arises from the necessity to extract the muon beam under an angle to the pion beam that differs from 0° or 180°. This means that the collimated muons come from a

[4] In the (p, θ) polar diagram (Fig. 13.1), in which the radius is identified with the probability p of positron emission under an angle θ to the spin direction, the locus of $p(\theta) = p_0(1 + \bar{a}_{\mathrm{eff}} \cos\theta)$ is the inverse of an ellipse with numerical excentricity $\varepsilon = \bar{a}_{\mathrm{eff}}$ and one of the foci located at the origin.

[5] As of 1977, the six most important facilities for the production of muon beams are at the meson factories of SIN (Villigen, Switzerland), LAMPF (Los Alamos, New Mexico), TRIUMF (Vancouver, British Columbia), and at the synchrocyclotrons at the Laboratory for Nuclear Problems at the Joint Institute for Nuclear Research (Dubna, Soviet Union), at SREL (Newport News, Virginia), and at CERN (Geneva, Switzerland).

finite angular range in the pion rest-system. Since muons emitted in opposite directions in the pion rest-system possess opposite polarizations in the laboratory system, it is clear that the muon beam emerging from the collimator cannot be completely polarized. The degree of polarization which may be achieved in practice is in the range 80–85%. For the study of muon interactions in solids it is most important that not only can the muon polarization be detected through the preferential emission of positrons in the direction of polarization [(13.3)] but also that the muon channels are capable of supplying muon beams of fairly high polarization.

Muon experiments on metals require the stopping of the muon beam in the sample. In order to be able to keep the size of target specimens reasonable, one usually reduces the energy of the muon beam by letting it pass through light materials (e.g., water) of the order of magnitude of $150\,\mathrm{kg\,m^{-2}}$. In the metal the muons lose their kinetic energy very rapidly by the excitation of electrons; similarly to positrons they are thermalized (i.e., their kinetic energy is reduced to $\approx 3k_BT/2$) within about 10^{-13} s to 10^{-12} s, i.e., in times that are very short compared with the muon lifetime. The depolarization during the slowing-down period in metals is negligible.

13.2.3 Interactions of Thermalized Positive Muons

The fate of positive muons after thermalization in matter depends critically on whether they do or do not form 'muonium'

$$\mathrm{Mu} = \mu^+ + e^-, \tag{13.9}$$

i.e., the hydrogenlike 'atoms' with μ^+ nuclei. Muonium may participate in chemical bonding in much the same way as atomic hydrogen. If muonium is *not* formed (as appears to be the case in metals, see Sect. 13.4.5), the dominant interaction of the muons is that with the *electrostatic fields* in the sample. This results—as in the case of the hydrogen nuclei—in the preferential location of positive muons on *interstitial* sites of metals and alloys.

The interaction of the magnetic moments of the muons with *magnetic fields*, though energetically much less important than the electrostatic interaction of the muon charge, provides the principal information on the behaviour of thermalized muons in matter. In a time-independent magnetic field $\boldsymbol{B}_\mu$ at the muon location the magnetic moment and the spin of a muon rotate with the angular velocity

$$\omega_\mu = \gamma_\mu \boldsymbol{B}_\mu . \tag{13.10}$$

This has given rise to the name μSR = 'muon spin rotation' for the study of muons in matter by means of their magnetic interactions.

The interaction between the magnetic moments of the muons and magnetic fields may lead to a partial or complete loss of the polarization of the thermalized muons. As will be discussed in Section 13.4.5, the depolarization associated with *muonium formation* occurs on a time scale that is long compared with the thermalization time (see Sect. 13.2.2) but still short compared with the muon lifetime. The 'fast' depolarization observed in many non-metals is generally attributed to this process; it takes place so rapidly that its time dependence cannot be studied.

In metals one usually does not find such a fast depolarization.[6] This offers the possibility of investigating the decay of the muon polarization in detail and of deriving information on the fate of the thermalized muons until they disintegrate by the radioactive decay (13.2). The general ideas and the theoretical concepts involved are the same as in the relaxation of nuclear spins in nuclear magnetic resonance (NMR) experiments (see Chap. 9). The experimental techniques, however, are quite different and will be described in Section 13.3.

As in the case of proton resonance, we have to distinguish between the decay of the *longitudinal spin polarization* and that of the *transverse spin polarization.* The first involves the reversal of the muon spins in a field B_l parallel (or antiparallel) to the muon polarization. The accompanying change of the muon energy in the magnetic field must be transferred from the spin system to another system, which is generally referred to as "the lattice". The decay of the longitudinal polarization is therefore also called "spin–lattice relaxation".

The decay of the *transverse spin polarization* may be observed if the magnetic field $\boldsymbol{B}_\mu$ at the muon sites has a component perpendicular to the original polarization direction. If this field component is spatially inhomogeneous, the rate of precession of the muon spins around the magnetic field direction according to (13.10) will depend on the muon location. The phase coherence of the precession and hence the 'transverse polarization' will get gradually lost. The decay of the transverse polarization is therefore also referred to as "spin-phase relaxation".

Let us now consider the various magnetic interactions that may or may not lead to the relaxation of the muon spin polarization. Examples for the calculations of relaxation rates due to some of these interactions will be treated in Section 13.4.

1) We discuss first local fields B_μ that are *time independent* and are the *same for all muon locations.* Since the interaction energy does not vary with time, transitions between the energy levels in a longitudinal field do *not* take place. If there is a non-vanishing transverse component, all muon spins precess at the same rate. Hence such magnetic fields lead neither to longitudinal nor transverse relaxation. Examples are provided by homogeneous static external

[6] Exceptions may occur in ferromagnetic materials, whose internal magnetic fields may be so strong that under certain conditions the relaxation rate can no longer be measured (see Sect. 13.5.2).

fields, by the demagnetizing fields in ferromagnetic samples of ellipsoidal shapes, and by the contact field $\boldsymbol{B}_{\text{Fermi}}$, the dipolar field $\boldsymbol{B}_{\text{dip}}$, and the Lorentz field $\boldsymbol{B}_{\text{Lorentz}}$ in ferromagnets in which all muon locations are crystallographically equivalent (see Sect. 13.5.2).

2) Next we consider the dipole–dipole interaction between the magnetic moment of the muon $\boldsymbol{\mu}_\mu$ and the *nuclear magnetic moments* $\boldsymbol{\mu}_j$, located at vectorial distances $\boldsymbol{r}_j$ from the muon. The interaction Hamiltonian is given by

$$\mathscr{H}_{\text{dip}} = \frac{\mu_0}{4\pi} \sum_j \left[\frac{\boldsymbol{\mu}_\mu \cdot \boldsymbol{\mu}_j}{r_j^3} - \frac{3(\boldsymbol{\mu}_\mu \cdot \boldsymbol{r}_j)(\boldsymbol{\mu}_j \cdot \boldsymbol{r}_j)}{r_j^5} \right], \tag{13.11}$$

where the summation extends over all nuclei of the crystal. In metals with a high abundance of isotopes with fairly large nuclear gyromagnetic ratios this interaction gives the most important contribution to μ^+-spin relaxation, with the possible exceptions of metals containing high densities of ordered electron spins (i.e., ferromagnetic, ferrimagnetic, or antiferromagnetic metals) or of paramagnetic impurities.

3) In treating the interaction between muons and *electron magnetic moments* we have to make a distinction whether the electron moments may be considered concentrated on atomic sites, or whether they are associated with itinerant electrons. In the first case the interaction Hamiltonian is again given by (13.11), where now $\boldsymbol{\mu}_j$ denotes the magnetic moment at an atom located at $\boldsymbol{r}_j$ from the muon spin. Since the electrostatic repulsion by the ion cores causes the thermalized positive muons to avoid the sites of atoms or ions, (13.11) does not contain a term with $\boldsymbol{r}_j = 0$ when applied to solids or liquids in which the electron magnetic moments are located at paramagnetic ions, or to the magnetic moments at the ion cores of ferromagnetic, ferrimagnetic, or antiferromagnetic substances. Because of the term with $\boldsymbol{r}_j = 0$ Eq. (13.11) is not directly applicable, however, to *itinerant electrons*, say conduction electrons in metals. In general these will give rise to a finite electron spin density $\varrho_{\text{spin}}(0)$ and hence to a finite magnetization *at the location of the muons*. Subjecting (13.11) to the appropriate limiting process [13.26] results in the interaction Hamiltonian

$$\mathscr{H}_{\text{Fermi}} = -\frac{4\mu_0}{3} \mu_\mu \mu_e \varrho_{\text{spin}}(0), \tag{13.12}$$

where μ_e is the electron magnetic moment, hence $-2\mu_e \varrho_{\text{spin}}(0)$ the electron magnetization at the muon location.

The interaction (13.12) is often referred to as the *Fermi interaction* [13.27] or the *contact interaction*. From the preceding remarks it should be clear that the physical origin of the contact interaction is the same as that of (13.11), namely the dipolar interaction of the magnetic moment of the muon with the magnetic moments of other particles. Equation (13.12) may also be used to treat the hyperfine interaction in muonium (see Sect. 13.4.5).

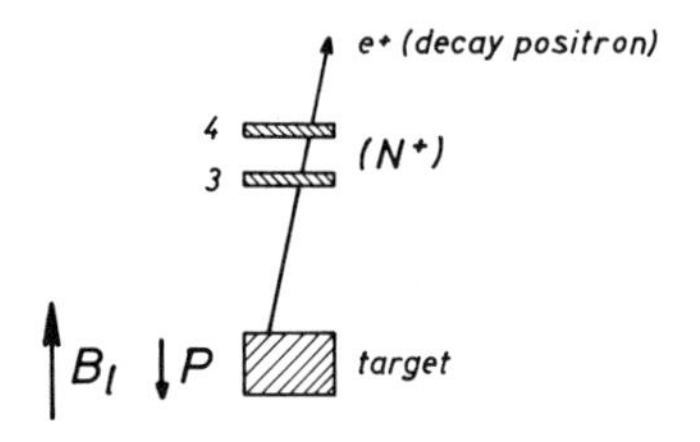

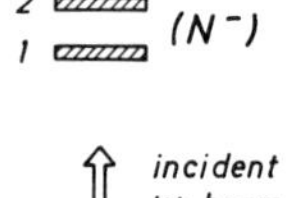

Fig. 13.2. Sketch of experimental arrangement for measuring the residual polarization of a muon beam stopped in a target and the relaxation of the polarization $\boldsymbol{P}$ in a longitudinal applied field $\boldsymbol{B}_l$(N^+, N^- = positron counting rates in forward or backward counters)

4) The interaction between magnetic moments of muons is always negligible (in contrast to the proton–proton interaction in hydrides), since in muon experiments the samples contain at best a few muons simultaneously.

13.3 Experimental Techniques[7]

On a polarized muon beam stopped in a metal, essentially four different quantities may be measured: 1) the residual polarization, 2) the longitudinal relaxation, 3) frequency (and sense) of the spin precession, and 4) the transverse relaxation. We shall see that in order to perform these measurements, one needs essentially two different types of experimental arrangements, namely one for the measurements 1) and 2), and another one for 3) and 4). All measurements depend on the asymmetry of the positron emission in (13.2) with respect to the direction of polarization $\boldsymbol{P}$. With the exception of 1), they make use of the time-differential technique (see below).

1) *Residual polarization.* The residual polarization P_{res} is defined as the modulus P of the polarization at the time of the muon decay. The principle of the measurement is indicated in Fig. 13.2. The target in which the muons are stopped is placed between two "positron telescopes", each consisting of two scintillation counters. The forward and backward counting rates, N^+ and N^-, of positrons passing through the telescopes are related to the residual polarization P_{res}, the effective asymmetry $\bar{a}_{\text{eff}}$ (see Sect. 13.2.1)[8], and a geometry factor g somewhat smaller than unity (allowing for the finite apertures of the telescopes and the finite size of target and beam) by

$$\frac{N^- - N^+}{N^- + N^+} = P_{\text{res}} \bar{a}_{\text{eff}} g\,. \tag{13.13}$$

[7] For further details see [13.28].

[8] The effects of kinematic depolarization may be included in $\bar{a}_{\text{eff}}$.

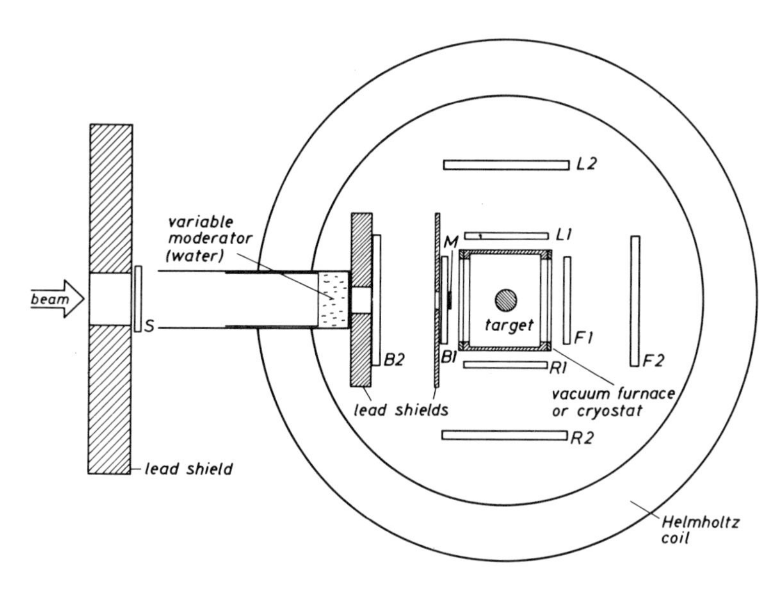

Fig. 13.3. Spectrometer for measuring muon spin rotation and transverse polarization. Counter S monitors the incoming muon beam, counter M the muons entering the target. The counters B, L, F, R form the backward, left-hand, foreward, and right-hand positron telescopes. Positive muons are polarized antiparallel to the beam direction. A transverse magnetic field (homogeneous over the target) $\boldsymbol{B}_t$ is applied perpendicular to the plane of the drawing

For determining absolute values of P_{res}, $\bar{a}_{\text{eff}}$ g must be calculated from the geometries of beam, target, and the counters, or be determined by measurements on a test substance with negligible depolarization.

2) *The longitudinal relaxation.* The experimental arrangement sketched in Fig. 13.2 may be used to determine the time dependence $P_l(t)$ of the longitudinal polarization by applying a static magnetic field in the direction of the muon beam. The beam intensity and the size of the target must be chosen in such a way that with a probability close to unity at any given time only one muon is present in the target. The passage of a muon through counters 1 and 2 is used to trigger an electronic clock, which is stopped when the emitted positron passes through either the forward or the backward telescope. By anticoincidence with counters 3 and 4 it is assured that the muon actually has been stopped in the target. By this *time-differential technique* the individual muon lifetimes are determined for each registered decay event, and the function $P_l(t)$ may be constructed with the help of a multichannel analyzer.

3) *Frequency and sense of the muon precession.* Figure 13.3 shows a μSR spectrometer as built up at SIN by a Heidelberg–Stuttgart collaboration. The muon spins precess in a magnetic field $\boldsymbol{B}_t$ perpendicular to the plane of the drawing. The spin precession may be made visible by the time-differential technique. As indicated in Fig. 13.4, the probability of positron emission rotates

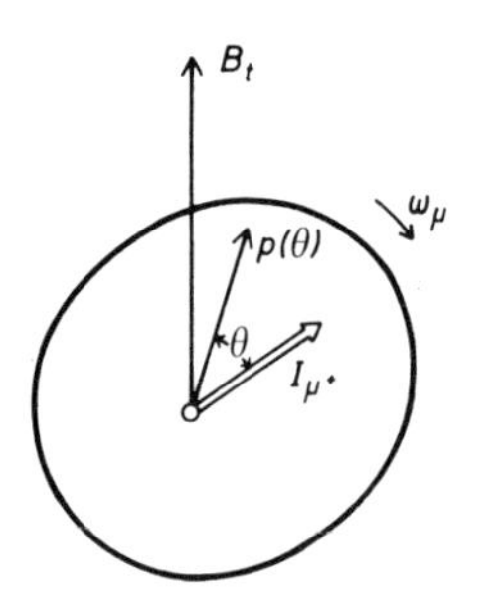

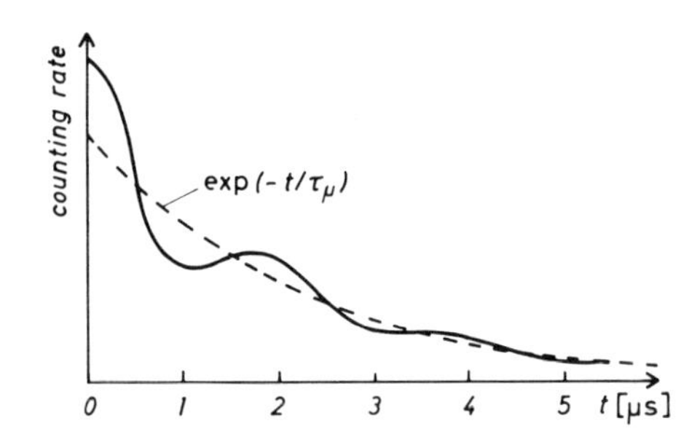

Fig. 13.4. Rotation of muon spin and positron emission probability $p(\theta)$ around a transverse magnetic field B_t. Counting rate in positron telescope located in the plane of rotation as a function of time t after entering of the target by the muons

with the muon spin. The counting rate in one of the positron telescopes varies as a function of the time t after the muons have entered the target as shown on the right-hand side of Fig. 13.4. The dashed exponential curve represents the β-decay of the muons with lifetime τ_μ, the superimposed wiggles are associated with the periodicity of the muon precession. The counting rate in any of the telescopes may be written as

$$\dot{N}_j = N_j^0 \tau_\mu^{-1} \exp(-t/\tau_\mu)[1 - P_t(t)\bar{a}_{\text{eff}} g_j \cos(\omega_\mu t - \phi_j)]. \tag{13.14}$$

Here N_j^0 and g_j are normalization constants and geometrical factors for the jth telescope. $P_t(t)$ is the transverse relaxation function, ω_μ the spin precession frequency, and ϕ_j the phase constant pertaining to the jth telescope. By comparing the ϕ_j with each other, one may determine the sense of rotation of the spins and hence whether the field $\boldsymbol{B}_\mu$ at the muon sites has a component parallel or antiparallel to the applied field $\boldsymbol{B}_{\text{appl}}$ (for ferromagnetic materials this question is non-trivial, see Sect. 13.5.2).

In general, muons stopped at different locations will have different spin precession frequencies ω_μ. As discussed in Section 13.2.3, spreads in the ω_μ distribution give rise to a transverse depolarization $P_t(t)$. In this case one may use (13.10) to determine the mean value of the field B_μ acting at the muon sites. If the ω_μ distribution has several peaks, i.e., if there exist several distinct ω_μ values (e.g., due to non-equivalent muon sites, see Sect. 13.4.4), the spin precession frequencies must be obtained by Fourier transforming $\dot{N}_j(t)$.

4) *The transverse relaxation.* If the spin precession frequencies are centred around one mean value, the transverse relaxation function $P_t(t)$ may be found by fitting (13.14) to the measurements. The elimination of the muon decay from the data may be achieved by combining the $\dot{N}_j(t)$ curves of different telescopes. When a Fourier analysis is performed in order to determine peaks in the ω_μ distribution, the relaxation functions associated with the individual ω_μ values may be found from the shape and the widths of the ω_μ peaks.

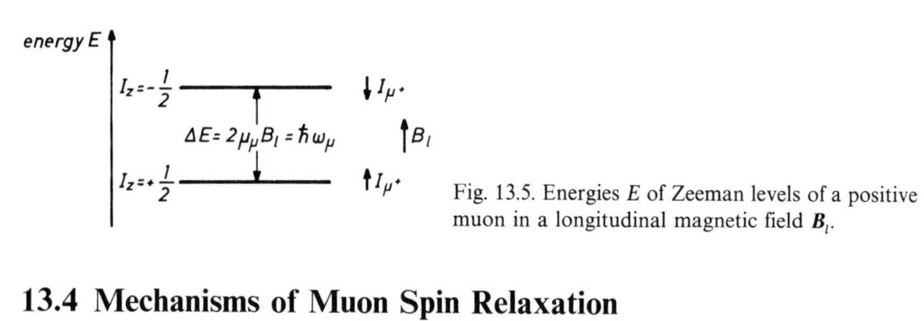

Fig. 13.5. Energies E of Zeeman levels of a positive muon in a longitudinal magnetic field $\boldsymbol{B}_l$.

13.4 Mechanisms of Muon Spin Relaxation

13.4.1 Relaxation Functions and Relaxation Times

We have seen in Section 13.3 that by means of the time-differential technique the time dependences $P_l(t)$ and $P_t(t)$ of the longitudinal and the transverse polarization of the muons may be measured. The information on the physical nature of the relaxation processes contained in these time dependences (which we shall henceforth call relaxation functions) is usually extracted by comparison with calculations based on models. In simple cases these calculations give relaxation functions that may be characterized by a single parameter [apart from the normalization constants $P_i(0)$]. Examples are provided by the *exponential decay*

$$P_i(t)=P_i(0)\exp(-\Gamma_i t) \quad (i=l,t) \tag{13.15}$$

or the *Gaussian decay*

$$P_t(t)=P_t(0)\exp(-\sigma^2 t^2). \tag{13.16}$$

The exponential relaxation function (13.15) shows up in a number of problems. The reciprocals of the longitudinal (Γ_l) and transverse (Γ_t) relaxation rates, $T_1\equiv 1/\Gamma_l$ and $T_2\equiv 1/\Gamma_t$, are known as relaxation times (see Chap. 9) but will not be used in the present chapter in order to avoid confusion with the absolute temperature T. Equation (13.16) will be derived in Section 13.4.2 as an approximation to the transverse relaxation function in a "rigid lattice"; an explicit expression for the parameter σ characterizing the relaxation rate will be given in (13.25).

In the case of longitudinal relaxation, (13.15) follows from a very simple but nevertheless fairly general model (Fig. 13.5) [13.26, 29]. In a longitudinal field B_l the states with muon spin up ($I_z=+1/2$) and muon spin down ($I_z=-1/2$) differ by the Zeeman energy

$$\Delta E=2\mu_\mu B_l=\hbar\omega_\mu, \tag{13.17}$$

where ω_μ is the spin precession frequency. We characterize the transitions between the two levels by downward and upward transition probabilities, $W_\downarrow$

and $W_\uparrow$. Then the populations of the two levels, n_+ and n_-, are governed by the rate equations

$$\frac{dn_+}{dt} = W_\downarrow n_- - W_\uparrow n_+ , \tag{13.18a}$$

$$\frac{dn_-}{dt} = -W_\downarrow n_- + W_\uparrow n_+ . \tag{13.18b}$$

Subtraction of (13.18b) from (13.18a) gives us

$$\frac{d(n_+ - n_-)}{dt} = -(W_\downarrow + W_\uparrow)(n_+ - n_-) + (W_\downarrow - W_\uparrow)(n_+ + n_-) . \tag{13.19}$$

From (13.19) it follows that the longitudinal polarization

$$P_l(t) \equiv \frac{n_+ - n_-}{n_+ + n_-} \tag{13.20}$$

obeys to a good approximation (13.15) with

$$\Gamma_l = W_\downarrow + W_\uparrow . \tag{13.21}$$

In thermal equilibrium we have on the one hand $dn_+/dt = dn_-/dt = 0$; on the other hand the population ratio is given by a Boltzmann factor. From this follow the relationships

$$\left(\frac{n_-}{n_+}\right)_{\text{eq}} = \frac{W_\uparrow}{W_\downarrow} = \exp(-\Delta E/k_{\mathrm{B}}T) . \tag{13.22}$$

13.4.2 Interaction with Nuclear Spins

The calculation of the muon spin relaxation $P_t(t)$ caused by the dipolar interaction with nuclear spins is based on the Hamiltonian (13.11) and follows closely the treatment of the relaxation in a system with two different spins S_i and I_j in a *strong* applied magnetic field $\boldsymbol{B}_{\text{appl}}$ [13.30]. A number of simplifications apply: 1) The muon spin is $S = 1/2$. 2) We may neglect interactions between the magnetic moments of the muons (item 4 of Sect. 13.2.3). 3) The gyromagnetic ratio of the muons is large compared to those of the nuclei, $\gamma_\mu \gg \gamma_I$. From this it follows that in non-ferromagnetic materials, where essentially the same field acts on the muons and on the nuclei, the muon spin precession frequency $\omega_\mu = \gamma_\mu B_{\text{appl}}$ is much larger than the nuclear spin precession frequency $\omega_I = \gamma_I B_{\text{appl}}$.[9]

If we confine ourselves to 'slow' motion of the muons in the sense that the muon jump frequency is less than ω_I, we may apply the so-called adiabatic approximation [13.30, 32]. This gives us [13.33]

$$P_t(t) = P_t(0) \prod_{j=1}^{N} \frac{\sin[(2I+1)\frac{1}{2}a_j(t)]}{(2I+1)\sin\frac{1}{2}a_j(t)} \tag{13.23}$$

with

$$[a_j(t)]^2 = \gamma_\mu^2 \gamma_I^2 \hbar^2 \left(\frac{\mu_0}{4\pi}\right)^2 2\int_0^t (t-t')g_j(t')dt' . \tag{13.23a}$$

Here $g_j(t')$ denotes the time average (= correlation function)

$$g_j(t') = \langle b_j(t) b_j(t-t') \rangle_t , \tag{13.23b}$$

where

$$b_j(t) \equiv \frac{1 - 3\cos^2\theta_j(t)}{r_j^3(t)} . \tag{13.23c}$$

In (13.23c), $r_j = |\boldsymbol{r}_j|$ is the distance between the jth nuclear spin and the muon, and θ_j the angle between the direction of the applied field and the vector $\boldsymbol{r}_j$ from the site of the muon at time t to the jth spin.

Expansion of (13.23) up to second powers in a_j, valid for small t, gives

$$P_t(t)/P_t(0) = 1 - \tfrac{1}{6}I(I+1)\sum_{j=1}^{N}[a_j(t)]^2 \approx \exp\left\{-\tfrac{1}{6}I(I+1)\sum_{j=1}^{N}[a_j(t)]^2\right\} . \tag{13.24}$$

Special cases already known in the literature [13.34] are the following:

1) The muons are immobile. Then the b_j are time independent, $g_j(t') = g_j(0) = b_j^2$, and $a_j = \hbar\gamma_\mu\gamma_I b_j(\mu_0|4\pi)t$. With the approximation (13.24) this gives a Gaussian depolarization function (13.16) with [13.35]

$$\sigma^2 \equiv \tfrac{1}{6}I(I+1)\hbar^2\gamma_\mu^2\gamma_I^2 \left(\frac{\mu_0}{4\pi}\right)^2 \sum_{j=1}^{N} \frac{(1-3\cos^2\theta_j)^2}{r_j^6} . \tag{13.25}$$

[9] This statement is in general not true for ferromagnets, since the hyperfine fields acting on the nuclear magnetic moments may be much larger than the magnetic fields B_μ at the muon sites. An important example is cobalt, which has a 100% abundancy of the isotope ^{59}Co with $I = 7/2$ and $\gamma_I = 63.5 \cdot 10^6$ (rad s^{-1} T^{-1}). The low-temperature hyperfine field at the Co nuclei is -23.3 T [13.31]; hence the precession frequency of the nuclei in the hyperfine field $\omega_I = 1.5 \cdot 10^9$ Hz, which is indeed larger than ω_μ in the usual applied fields. The relaxation effects due to the interactions between muons and the dipolar fields of Co nuclei are therefore not covered by the present theory and require separate treatment.

2) If

$$\sum_{j=1}^{N} g_j(t) = \exp(-t/\tau_c) \sum_{j=1}^{N} [b_j(0)]^2 , \tag{13.26}$$

then

$$2\int_0^t (t-t') \sum_{j=1}^{N} g_j(t')dt' = -2\tau_c^2 \left[\exp\left(-\frac{t}{\tau_c}\right) - 1 + \frac{t}{\tau_c}\right] \sum_{j=1}^{N} [b_j(0)]^2 , \tag{13.27}$$

so that the approximation (13.24) gives us [13.30, 32]

$$P_t(t) = P_t(0) \exp\left\{-2\sigma^2\tau_c^2 \left[\exp\left(-\frac{t}{\tau_c}\right) - 1 + \frac{t}{\tau_c}\right]\right\}. \tag{13.28}$$

In (13.26–28) τ_c is the so-called correlation time. It is related to the mean time of residence of the muons on a given interstitial site and hence to the diffusion coefficient D^{μ^+} of the muons. As τ_c decreases, i.e., with increasing diffusion coefficient, the decay of the polarization according to (13.28) takes place over larger times. At the same time (13.28) undergoes a transition from the Gaussian (13.16) for $t \ll \tau_c$ to an exponential (13.15) with

$$\Gamma_t = 2\sigma^2\tau_c . \tag{13.29}$$

The physical origin of this important phenomenon, which may be termed 'motional averaging', is the following. Muons residing at different interstices feel different magnetic fields; their spins precess, therefore, at different angular velocities. As discussed in Sections 13.2.3 and 13.3, this gives rise to transverse spin relaxation. If, however, the muons visit many sites before an appreciable decay of the transverse polarization has taken place, the spin precession is governed by the average field seen by the muons and is therefore more uniform. For times $t > \tau_c$ this results in a slower decay of the transverse polarization. In magnetic resonance experiments this is associated with a narrowing of the resonance line. The phenomenon is therefore known as 'motional narrowing' [13.30, 36].

When applying (13.28) to the determination of muon diffusion coefficients, it must be kept in mind that (13.26) represents only a rough approximation to the correct correlation functions. For future quantitative work the $g_j(t')$ have to be determined by calculations based on detailed diffusion models [13.37, 38].

The requirement of a *strong* applied field mentioned above means that the following conditions have to be fulfilled for the preceding treatment to be valid:

Table 13.2. Numerical values of $\sum_{j=1}^{n}(1-3\cos^2\theta_j)^2$ for n neighbouring atoms in various directions from muon sites in a cubic crystal and for two different crystallographic directions of the magnetic field

	$\boldsymbol{B}$ in $\langle 100\rangle$	$\boldsymbol{B}$ in $\langle 111\rangle$
6 neighbours in $\langle 100\rangle$	12	0
12 neighbours in $\langle 110\rangle$	6	12
8 neighbours in $\langle 111\rangle$	0	$10\frac{2}{3}$

I) The applied field B_{appl} must exceed the local field $B_{\mathrm{L}}^{(\mu)}$ at the muon sites generated by the nuclear magnetic moments. From an expression given by *Wolf* [13.38] we deduce

$$(B_{\mathrm{L}}^{(\mu)})^2 = \tfrac{1}{8}\left(\frac{\mu_0}{4\pi}\right)^2 \gamma_\mu^2 I(I+1)\hbar^2 \sum_j |b_j|^2 = \tfrac{3}{4}\sigma^2/\gamma_I^2 . \tag{13.30}$$

It follows from (13.30) that the condition

$$B_{\mathrm{L}}^{(\mu)} \ll B_{\mathrm{appl}} \tag{13.31a}$$

is equivalent to

$$\sigma \ll \omega_I = \frac{\gamma_I}{\gamma_\mu}\omega_\mu . \tag{13.31b}$$

II) B_{appl} must exceed the local fields $B_{\mathrm{L}}^{(\mathrm{N})}$ at the sites of the nuclei. This condition is always satisfied if condition I) is fulfilled.

III) The interaction energy of the electrical field gradient generated by the muons at the nuclear sites with the nuclear quadrupole moments must be small compared with the Zeeman energy of the nuclei in the applied field [13.39].

For a given muon site in a *rigid* crystal, the expression (13.25) for σ^2 contains only well-known quantities. It is furthermore strongly dependent on the crystallographic direction of the magnetic field. This may be seen from Table 13.2, which gives the contribution $\sum_j(1-3\cos^2\theta_j)^2$ for various configurations that occur for muons on interstitial or lattice sites in cubic lattices (muons on *lattice* sites may be realized by the trapping of muons by monovacancies). For comparison we mention that the average of $(1-3\cos^2\theta_j)^2$ over all spatial directions is 4/5.

The comparison of calculated σ values with the measured magnitude and dependence on the magnetic field direction allows us to distinguish between different possibilities for the location of the muons. Small differences between experiment and theory may be attributed to two effects: The finite extension of the muon wave function gives a (usually very small) reduction of σ^2 relative to the value predicted by (13.25) (see *McMullen* and *Zaremba* in the list of "Additional References"). The main reduction of σ^2 is expected to come from the outward displacements of neighbouring ions due to the elastostatic repulsion by the positive muons. The comparison between experimental and theoretical σ^2 values thus permits to obtain an upper limit for the outward displacement of first- and possibly also of second-nearest neighbours.

If condition III) is not satisfied, the dependence of σ on the crystallographic direction of $\boldsymbol{B}_{\text{appl}}$ is reduced, since the quantization axis of the nuclear spins is now determined by both the applied field and the electric field gradient generated by the muons [13.39]. In a simple model[10], σ becomes a function of ω_Q/ω_I, where

$$\omega_Q = \frac{Q}{\hbar} \frac{V_{zz}(1+\gamma)}{4I(2I+1)}. \tag{13.32}$$

In (13.32), Q denotes the electrical quadrupole moment and I the spin of the nuclei, V_{zz} the electrical field gradient at the nearest-neighbour nuclei, and $\gamma > 0$ an effective Bloch enhancement factor. *Hartmann* [13.39] gives explicit results for octahedral and tetrahedral interstices in both the face-centred cubic and body-centred cubic lattices for $I = 3/2$ and $\boldsymbol{B}$ parallel to $\langle 100 \rangle$, $\langle 110 \rangle$, or $\langle 111 \rangle$. In addition, the cases $I = 5/2$ or $7/2$ have been treated for octahedral interstices in the face-centred cubic lattice and $\boldsymbol{B}$ parallel $\langle 111 \rangle$. An application of Hartmann's theory to Cu ($I = 3/2$) will be given in Section 13.5.1.

13.4.3 Interaction with Itinerant Electrons

In an applied magnetic field, the Pauli paramagnetism of conduction electrons gives rise to a finite density ϱ_{spin} of electron spins at interstitial sites. The Fermi interaction (13.12) of the magnetic moments of the electrons with the magnetic moments of positive muons leads to a "Knight shift" of the muon spin precession frequency and to a contribution to the muon spin relaxation rates resulting from the simultaneous flipping of electron and muon spins. The calculation of this contribution is analogous to that of the electronic relaxation in NMR [13.29] and gives

$$\Gamma_l = \Gamma_t = \frac{64\pi^3}{9} \left(\frac{\mu_0}{4\pi}\right)^2 \gamma_e^2 \gamma_\mu^2 \hbar^3 k_B T [\omega_{\varepsilon_F}(0)]^2 \tag{13.33}$$

[10] The tensor ellipsoids of the electric field gradients are assumed to be spheroids with axes of rotation parallel to the directions from the muon interstitial site to the nearest-neighbour sites. The electric field gradients at more distant sites are neglected.

with

$$\omega_{\varepsilon_F}(\boldsymbol{r}) = \iint_{\varepsilon_F} |\psi_{\boldsymbol{k}}(\boldsymbol{r})|^2 \frac{dS_{\boldsymbol{k}}}{|\nabla\varepsilon(\boldsymbol{k})|} . \tag{13.33a}$$

Here k_B denotes Boltzmann's constant, T the absolute temperature, $\omega_{\boldsymbol{k}}(\boldsymbol{r})$ the electron wave functions, $\nabla\varepsilon(\boldsymbol{k})$ the gradient of the electron energy in the wave number space, and $dS_{\boldsymbol{k}}$ a surface element of the surfaces of constant energy $\varepsilon(\boldsymbol{k})$ in wave number ($\boldsymbol{k}$-)space. The integration in (13.33a) extends over the Fermi surface. The muon location has been chosen as $\boldsymbol{r}=0$.

Numerical estimates of (13.33) are difficult, even for metals with well-known band structures, since it is essential to take into account the modification of the electron wave functions by the presence of the muon. Electrons are attracted towards the positive muons, and therefore the relaxation rates (and the Knight shifts) are enhanced compared to the values calculated if for $\psi_{\boldsymbol{k}}(\boldsymbol{r})$ the wave functions of the perfect metals are used. The theoretical estimates of this enhancement [13.40, 41] are strongly model dependent but indicate nevertheless that the electronic contributions to the relaxation rate are much smaller than τ_μ^{-1} and therefore difficult to measure.

The experiments on the Knight shift both in ferromagnetic and non-ferromagnetic metals have been reviewed recently by *Schenck* [13.42]. Pd is the only case in which experimental data are available for both muons and dilute hydrogen. The agreement between the two measurements is good.

13.4.4 Interaction with the Dipolar Fields in Ferromagnets

a) Calculation of Dipolar Fields

A prerequisite for a calculation of the muon relaxation rate in ferromagnets is the determination of the magnetic fields $\boldsymbol{B}_\mu$ at the sites of the muons. In addition to the applied field $\boldsymbol{B}_{\text{appl}}$ and the fields of the nuclear magnetic moments, one has, in general, large contributions from the electronic magnetization. It is convenient to decompose the magnetic field resulting from the electronic magnetization into a contribution of magnetic dipole moments $\boldsymbol{\mu}_{\text{ion}}$ located at the sites of the ion cores, and into a contribution associated with the magnetization of the conduction electrons. The latter may be handled by means of (13.12).

The cumbersome task of having to carry out the summation over the long-ranging fields of all the dipole moments in a ferromagnet may be reduced to manageable proportions by the following procedure:

1) The shape of the sample is taken into account by introducing the demagnetizing field

$$\boldsymbol{B}_{\text{demag}} = -\mu_0 \underset{\sim}{\boldsymbol{D}} \cdot \boldsymbol{M}_{\text{sample}} . \tag{13.34}$$

Here $\boldsymbol{M}_{\text{sample}}$ denotes the (average) magnetization of the sample and $\underset{\sim}{\boldsymbol{D}}$ the so-called demagnetizing tensor. We consider here only specimens of ellipsoidal shape, for which $\underset{\sim}{\boldsymbol{D}}$ is a constant ($\underset{\sim}{\boldsymbol{D}}=\underset{\sim}{\boldsymbol{I}}/3$ for spheres; $\underset{\sim}{\boldsymbol{I}}$ = unit tensor).

2) Taking the positions of the muon as the centre, we circumscribe a sphere of a radius that is large compared with the interionic distance but small compared with the size of the ferromagnetic domains. The (fictitious) magnetic charges on the surface of this so-called Lorentz sphere give rise to the Lorentz field

$$\boldsymbol{B}_{\text{Lorentz}}=\frac{\mu_0}{3}\boldsymbol{M}_{\text{domain}}. \tag{13.35}$$

In (13.35) $\boldsymbol{M}_{\text{domain}}$ denotes the magnetization inside the ferromagnetic domains. Its magnitude M_{domain} varies slightly between the spontaneous magnetization M_{sp} in zero applied field and the saturation magnetization M_{sat} in very large applied fields.

3) The contribution $\boldsymbol{B}_{\text{dip}}$ from the dipole moments *inside* the Lorentz sphere is obtained by summing over the magnetic fields of these dipoles. With increasing radius of the Lorentz sphere, the sum $\boldsymbol{B}_{\text{Lorentz}}+\boldsymbol{B}_{\text{dip}}$ converges towards the magnetic field coming from *all* dipole moments within the ferromagnetic domain, provided this is large enough to perform the limiting process in a physically meaningful way.

Lorentz [13.43] showed that $\boldsymbol{B}_{\text{dip}}$ is zero at positions surrounded by an arrangement of cubic symmetry of magnetic dipole moments of equal magnitude and direction[11]. An example for this theorem is an octahedral interstitial site (cube center) in a face-centred cubic (fcc) ferromagnetic crystal, where $\boldsymbol{B}_{\text{dip}}=0$. This remains true even when a muon located at such a position displaces the surrounding ions, as long as the displacement field maintains the cubic symmetry.

By contrast, in body-centred (bcc) lattices there are no interstitial sites of cubic symmetry. Both the octahedral and the tetrahedral interstitial positions possess tetragonal symmetry. Hence we expect that in general $\boldsymbol{B}_{\text{dip}}$ does not vanish at these sites. Nevertheless, Lorentz's theorem may be used to demonstrate the following for bcc ferromagnets: Superposition of the lattices surrounding three interstices of the same type but with different directions of their tetragonal axes results in a cubic arrangement of equal dipole moments, hence a vanishing magnetic field at the interstices of the structure obtained by this superposition. This means that the *sum* of the dipolar fields at three differently orientated interstitial sites must be zero, irrespective of the direction of the magnetization.

---

[11] For detailed discussions of Lorentz's theory the reader is referred to *Brown* [13.44] or *Hellwege* [13.45].

The preceding result will be illustrated by two examples.

1) Bcc *crystal with magnetization parallel* [100] (e.g., α-Fe in the absence of an applied or demagnetizing field). We denote the dipole field at an octahedral or tetrahedral interstice by $\boldsymbol{B}_{\mathrm{dip}}^{\|}$ if the tetragonal axis of the site is parallel to the magnetization direction, and by $\boldsymbol{B}_{\mathrm{dip}}^{\perp}$ if it is perpendicular to it. By the argument of the preceding paragraph we have

$$\boldsymbol{B}_{\mathrm{dip}}^{\|}+2\boldsymbol{B}_{\mathrm{dip}}^{\perp}=0\,. \tag{13.36}$$

By symmetry, $\boldsymbol{B}_{\mathrm{dip}}$ must be parallel or antiparallel to the magnetization direction, i.e., in the present example parallel or antiparallel to [100].

2) Bcc *crystal with magnetization parallel* [111] (e.g., α-Fe saturated by a strong applied field in the [111] direction). We denote the dipole fields at the interstitial sites with differently orientated tetragonal axes by $\boldsymbol{B}_{\mathrm{dip}}^{(i)}$ $(i=x,y,z)$. The argument given above tells us that

$$\boldsymbol{B}_{\mathrm{dip}}^{(x)}+\boldsymbol{B}_{\mathrm{dip}}^{(y)}+\boldsymbol{B}_{\mathrm{dip}}^{(z)}=0\,. \tag{13.37}$$

Equation (13.37) admits two different types of solutions. Either we must have $\boldsymbol{B}_{\mathrm{dip}}^{(i)}\equiv 0$ for $i=x,y,z$, or the $\boldsymbol{B}_{\mathrm{dip}}^{(i)}$ must be coplanar. In the latter case, for reasons of symmetry, the common plane must be (111). Again for reasons of symmetry $\boldsymbol{B}_{\mathrm{dip}}^{(x)}$ must then lie in the plane containing both [100] and [111], i.e., in the $(01\bar{1})$ plane. This means that $\boldsymbol{B}_{\mathrm{dip}}^{(x)}$ points parallel or antiparallel to $[\bar{1}\bar{1}2]$. The three vectors must have equal magnitudes, i.e.,

$$B_{\mathrm{dip}}^{(x)}=B_{\mathrm{dip}}^{(y)}=B_{\mathrm{dip}}^{(z)}\,. \tag{13.37a}$$

For a given site of tetragonal symmetry, the preceding relationships may be written as a tensor equation, e.g., for sites with the x-axis as tetragonal axis

$$\boldsymbol{B}_{\mathrm{dip}}=\pm B_{\mathrm{dip}}^{\|}\begin{pmatrix}1 & 0 & 0\\ 0 & -\frac{1}{2} & 0\\ 0 & 0 & -\frac{1}{2}\end{pmatrix}\cdot\boldsymbol{\mu}_{\mathrm{ion}}^{0}\,, \tag{13.38}$$

where $\boldsymbol{\mu}_{\mathrm{ion}}^{0}$ denotes the unit vector in the direction of the magnetic moments.

The application of (13.38) to a bcc crystal with magnetic moments parallel to [111] verifies the results found in example 2) by symmetry arguments. Considering again the specific case of an interstitial site with tetragonal axis parallel to the x-direction, we obtain from (13.38)

$$\boldsymbol{B}_{\mathrm{dip}}^{(x)}=\pm\frac{B_{\mathrm{dip}}^{\|}}{2\sqrt{3}}[2\bar{1}\bar{1}]\,. \tag{13.39}$$

Equation (13.39) demonstrates that in general $\boldsymbol{B}_{\mathrm{dip}}$ is *not* parallel to $\boldsymbol{\mu}_{\mathrm{ion}}$ and that $\boldsymbol{B}_{\mathrm{dip}}$ may even be *perpendicular* to the magnetization direction.

The preceding examples show that considerable information on the dipole fields in ferromagnets may be obtained without detailed computation. In order to find the absolute magnitude of the dipolar fields as well as the appropriate signs in (13.38, 39), one has to evaluate the lattice sums

$$\boldsymbol{B}_{\mathrm{dip}}=\frac{\mu_0}{4\pi}\sum_j\frac{3\boldsymbol{r}_j(\boldsymbol{r}_j\cdot\boldsymbol{\mu}_{\mathrm{ion}})-(\boldsymbol{r}_j\cdot\boldsymbol{r}_j)\,\boldsymbol{\mu}_{\mathrm{ion}}}{r^5}\,. \tag{13.40}$$

In (13.40) $\boldsymbol{r}_j$ denotes the vectors from an interstitial site to the positions of the magnetic moments. The summation extends over the interior of the Lorentz sphere.

For the octahedral sites in an undistorted bcc crystal with $\boldsymbol{\mu}_{\text{ion}}$ parallel to one of the cubic axes (a_0 = edge length of the elementary cube), (13.40) gives

$$B_{\text{dip}}^{\|} = -2B_{\text{dip}}^{\perp} = \frac{\mu_0}{4\pi} 21.33 \frac{\mu_{\text{ion}}}{a_0^3}. \tag{13.41a}$$

The analogous result for the tetrahedral sites reads

$$B_{\text{dip}}^{\|} = -2B_{\text{dip}}^{\perp} = -\frac{\mu_0}{4\pi} 6 \frac{\mu_{\text{ion}}}{a_0^3}. \tag{13.41b}$$

This means that in (13.38, 39) the + sign applies to the octahedral sites and the − sign to the tetrahedral sites.

If we assume for simplicity that the domain magnetization $\boldsymbol{M}_{\text{domain}}$ may be entirely attributed to the magnetic dipoles (i.e., we neglect for the time being the contribution of the conduction electrons to $\boldsymbol{M}_{\text{domain}}$), we have for bcc crystals $\boldsymbol{M}_{\text{domain}} = 2\boldsymbol{\mu}_{\text{ion}}/a_0^3$. We may then rewrite (13.38) as a linear tensor equation connecting $\boldsymbol{B}_{\text{dip}}$ and $\boldsymbol{M}_{\text{domain}}$

$$\boldsymbol{B}_{\text{dip}} = \underset{\sim}{\Xi} \cdot \boldsymbol{M}_{\text{domain}}. \tag{13.42}$$

The second-rank tensor $\underset{\sim}{\Xi}$ reads for octahedral interstices

$$\underset{\sim}{\Xi}^{\text{oct},x} = \begin{pmatrix} 10.706 & & \\ & -5.353 & \\ & & -5.353 \end{pmatrix}, \tag{13.42a}$$

and for tetrahedral interstices

$$\underset{\sim}{\Xi}^{\text{tet},x} = \begin{pmatrix} -3.00 & & \\ & 1.50 & \\ & & 1.50 \end{pmatrix}. \tag{13.42b}$$

Equations (13.42a) and (13.42b) refer to sites with tetragonal axes parallel to the x-direction; the $\underset{\sim}{\Xi}$ tensors for the other sites may be obtained by cyclic permutation. Equation (13.42b) shows that in a bcc crystal magnetized along one of the cube axes, the dipole field $\boldsymbol{B}_{\text{dip}}$ at the tetrahedral interstices with tetragonal axis *parallel* to the magnetization points in the direction *opposite* to the magnetization. This means that $\boldsymbol{B}_{\text{dip}}$ opposes the Lorentz field.

Numerical results for α-Fe ($M_{\text{sp}} = 1.07 \cdot 10^6\ \text{A m}^{-1}$) are as follows:[12] For *octahedral* interstices $B_{\text{dip}}^{\|} = 1.86\ \text{T}$ (in agreement with an earlier result of *Stearns* [13.47]) and $B_{\text{dip}}^{(i)} = 1.32\ \text{T}$; for tetrahedral interstices $B_{\text{dip}}^{\|} = 0.522\ \text{T}$ and $B_{\text{dip}}^{(i)} = 0.369\ \text{T}$.

---

[12] They are taken from a paper by *Kronmüller* et al. [13.46], which contains also a more general formulation of the theory, covering hexagonal close-packed metals, too.

b) Calculation of Relaxation Rates

As an example for the muon spin relaxation associated with the dipolar fields of ferromagnets, we treat muons in a ferromagnetic bcc metal (e.g., α-Fe) that are located either on octahedral or tetrahedral interstitial sites. The frequency of muon jumps from an interstitial site to a neighbouring site is denoted by ν. (Only octahedral–octahedral or tetrahedral–tetrahedral jumps are considered; the generalization to mixed jumps is straightforward.)

As discussed in Section 13.4.4a, the dipolar fields depend on the orientation of the tetragonal axes of the interstitial sites. For a muon of given spin direction this means that jumps between adjacent sites lead to transitions between energy levels as discussed in Section 13.4.1. In simple cases (see below), in which the two-level model of Section 13.4.1 is applicable, (13.15) predicts an exponential decay law for the longitudinal relaxation. An exponential law is also predicted for the transverse relaxation if the correlation time τ_c is short enough (see [13.26]).

A detailed treatment [13.48] shows that the relaxation rates Γ_i depend on the direction of the magnetization. In the two limiting cases of magnetization along $\langle 100 \rangle$ or $\langle 111 \rangle$ directions, one finds

$$\Gamma_l^{\langle 100 \rangle} = 0, \qquad (13.43a)$$

$$\Gamma_t^{\langle 100 \rangle} = \tfrac{1}{2}(\gamma_\mu B_{\text{dip}}^{\|})^2 \tau_c \qquad (13.43b)$$

or

$$\Gamma_l^{\langle 111 \rangle} = \tfrac{1}{2}(\gamma_\mu B_{\text{dip}}^{\|})^2 \frac{\tau_c}{1+\omega_\mu^2 \tau_c^2}, \qquad (13.44a)$$

$$\Gamma_t^{\langle 111 \rangle} = \tfrac{1}{2}\Gamma_l^{\langle 111 \rangle} = \tfrac{1}{4}(\gamma_\mu B_{\text{dip}}^{\|})^2 \frac{\tau_c}{1+\omega_\mu^2 \tau_c^2}. \qquad (13.44b)$$

In the specific examples of octahedral–octahedral or tetrahedral–tetrahedral jumps, the correlation time τ_c is given by

$$\tau_c = \frac{1}{6\nu} = \frac{2}{3}\bar{\tau}, \qquad (13.45)$$

where $\bar{\tau}$ is the mean time of residence of a muon at an interstitial site. It is related to the muon diffusion coefficient D^{μ^+} by

$$D^{\mu^+} = \frac{a_0^2}{36\tau_c} \qquad (13.46a)$$

for octahedral occupancy and by

$$D^{\mu^+} = \frac{a_0^2}{72\tau_c} \qquad (13.46b)$$

for tetrahedral occupancy.

Since $\langle 100 \rangle$ is the axis of easy magnetization in α-Fe, experiments in zero or low applied fields correspond to (13.43). In this case the information obtainable on the interstitial sites (contained in $B_{\mathrm{dip}}^{\|}$, see Sect. 13.4.4a) and on the diffusion coefficient (contained in τ_c) are coupled in such a way that they cannot be separated on the basis of relaxation measurements alone. It is therefore very interesting to consider the case of $\langle 111 \rangle$ magnetization, which may be realized by applying a sufficiently large magnetic field in a $\langle 111 \rangle$ direction.

According to (13.44), both $\Gamma_l^{\langle 111 \rangle}$ and $\Gamma_t^{\langle 111 \rangle}$ show a maximum as a function of τ_c, hence of the temperature, at $\omega_\mu \tau_c = 1$. The maximum values are

$$\hat{\Gamma}_l^{\langle 111 \rangle} = \tfrac{1}{4}(\gamma_\mu B_{\mathrm{dip}}^{\|})^2 \tau_c = 2\hat{\Gamma}_t^{\langle 111 \rangle} . \tag{13.47}$$

Since at the maximum $\tau_c = (\gamma_\mu B_\mu)^{-1}$, where B_μ is given by the sum of the applied field, the Lorentz field, and the demagnetizing field, we may determine $B_{\mathrm{dip}}^{\|}$ and, by comparison with the theoretical values of Section 13.4.4a, the type of interstitial site occupied by the muons. The numerical values of $B_{\mathrm{dip}}^{\|}$ are sufficiently different for octahedral and tetrahedral sites, so that it should be easy to distinguish between these two sites. Beyond that it should be possible to obtain information on the displacement of the neighbouring ions by the muons from the deviation of $B_{\mathrm{dip}}^{\|}$ from the value for an ideal lattice. The fact that both $\Gamma_t^{\langle 111 \rangle}$ and $\Gamma_l^{\langle 111 \rangle}$ are different from zero will be particularly valuable for studying the trapping of muons by impurities or lattice defects. Whereas trapping suppresses the motional averaging discussed in Section 13.4.2 and hence enhances the transverse relaxation, it does not affect the longitudinal relaxation. Simultaneous measurements of $\Gamma_l^{\langle 111 \rangle}$ and $\Gamma_t^{\langle 111 \rangle}$ will thus permit the measurement of muon diffusion coefficients in Fe even in the presence of defects capable of trapping muons.

13.4.5 Muonium Formation

Positive muons may capture electrons from nearby atoms (or from the electron gas of metals) and form muonium (Mu) according to (13.9). The cross section for Mu formation depends on the speed v of the muons and is expected to pass through a maximum at

$$v \approx a \Delta E_{\mathrm{ion}} / h , \tag{13.48}$$

where a is of the order of magnitude of the atomic radii and ΔE_{ion} denotes the difference in the ionization energies of Mu and of the atoms from which the electrons are captured[13]. The maximum cross-section is of the order of magnitude of the geometrical cross-section of these atoms. The muon kinetic energies corresponding to the maximum capture rate are of the order of magnitude 10^2 eV.

If we consider the direction of motion of the muons as axis of quantization, either triplet muonium (electron and muon spin parallel) or singlet muonium (electron and muon spin antiparallel) may be formed (in non-ferromagnetic materials with equal probabilities). Whereas in the absence of external disturbances muons in the triplet state will maintain their polarization, in the singlet state muons and electrons exchange their spin directions with the muonium

[13] The ionization energy of muonium is 13.539 eV; its Bohr radius $a_\mu = 0.532 \cdot 10^{-10}$ m.

hyperfine splitting frequency[14]

$$\omega_{\mathrm{HF}}=\frac{16\mu_0}{3}\frac{\mu_\mu\mu_e}{ha_\mu^3}=2.804\cdot 10^{10}\,\mathrm{s}^{-1}\,. \tag{13.49}$$

On the one hand, the reciprocal of this frequency is large compared with the slowing-down and thermalization time of muons, so that the depolarization before thermalization is negligible irrespective of whether muonium is formed or not (see Sect. 13.2.2). On the other hand, polarization measurements are carried out on time scales large compared with 10^{-10} s. This means that muons in the singlet Mu state should appear always as fully depolarized.

By the above arguments one expects that in non-ferromagnetic materials the formation of thermalized muonium leads to a reduction of the original muon polarization by one-half (or more, if the 'muonium electrons' exchange their spins with electrons of opposite spin in the surrounding 'matrix' on a time scale between 10^{-10} s and the measuring time). In metals, such a reduction of the initial polarization by at least one-half has *not* been found. However, the conclusion that muonium atoms (and, hence, presumably also hydrogen atoms) are *not* formed in metals is premature, since the exchange scattering of the electrons may lead to electron spin flips at a rate large compared with ω_{HF}, thus making ineffective the relaxation mechanism for singlet muonium discussed above.

The question of the detection of muonium in metals has been considered in more detail by *Gorelkin* and *Smilga* [13.49] and by *Grebinnik* et al. [13.50, 51]. Since only conduction electrons close to the Fermi surface can participate in the spin-flip mechanism, it is clearly necessary to work at low temperatures. Qualitative arguments indicate that in normal-conducting metals the spin-flip frequency resulting from the exchange interaction between 'muonium electrons' and conduction electrons is of the order of magnitude $10^{12}(T/\mathrm{K})\,\mathrm{s}^{-1}$, i.e., at all accessible temperatures large or comparable with the hyperfine splitting frequency. *Gorelkin* and *Smilga* [13.49] proposed to measure the muon polarization in superconductors well below the transition temperature, where the number of electrons with unpaired spins decreases exponentially with decreasing temperature. At sufficiently low temperatures the reduction of the polarization by a factor of one-half due to muonium formation should become observable. In order to avoid complications due to the dipole–dipole interactions between muons and nuclei (see Sect. 13.4.2), the experiments should be performed on superconductors in which the abundant isotopes have zero nuclear spin (e.g., Pb or Hg). Such experiments have not been reported to date.

Grebinnik et al. [13.50] proposed the following method to search for muonium in metals. Muonium may be considered as a paramagnetic impurity, which may be polarized according to Curie's law. The resulting electron spin

[14] This vacuum value of ω_{HF} may be calculated from (13.12) by taking the electron spin density at the muon, $\varrho_{\mathrm{spin}}(0)$, from the theory of the hydrogen atom ground state.

polarization at the muon site gives rise, through the Fermi interaction (13.12), to an increase of the muon precession frequency somewhat similar to the Knight shift (which, however, in contrast to the present situation is caused by the polarization of the *conduction* electrons). This results in a temperature-dependent muon spin precession frequency

$$\omega_\mu = \gamma_\mu B_t \left(1 + \frac{\hbar\omega_{\mathrm{HF}}}{4k_{\mathrm{B}}T} \frac{\mu_e}{\mu_\mu}\right). \tag{13.50}$$

In (13.50), ω_{HF} denotes the hyperfine splitting frequency of Mu in the metal. Due to the modification of the muonium wave function by the metallic matrix, it will differ from the 'vacuum' value (13.49) as calculated from the theory of the free hydrogen atom. Measurements of the muonium precession frequency with an accuracy of $\sim 3 \cdot 10^{-4}$ on Al, Cu, Zn, and C (graphite) down to 4 K [13.50] did not show a temperature dependence as predicted by (13.50), indicating that ω_{HF} is smaller than the vacuum value by at least a factor of 10^{-4}. *Grebinnik* et al. [13.50] took this as evidence that in the above-mentioned metals and in graphite, muonium is *not* formed.

13.5 Experiments on Muon Diffusion and Location

13.5.1 Nonferromagnetic Metals

a) Copper

The temperature dependence of muon diffusion was first investigated in copper [13.34]. At the present time Cu is the non-ferromagnetic metal studied in most detail.

Figure 13.6 shows the precession signal of polarized muons in a copper single crystal at three different temperatures [13.52]. The β-decay of the muons has been corrected for in Fig. 13.6. The damping of the precession wiggles begins to become detectable at about room temperature and increases with decreasing temperature (motional-averaging effect, Sect. 13.4.2). It becomes practically temperature independent below about 90 K. As one expects from the simple theory leading to (13.28), at low temperatures the depolarization curve may be described by a *Gaussian* [(13.16)], whereas at higher temperatures the simple exponential law (13.15) provides a better description of the experimental results.

The full lines in Fig. 13.6 correspond to (13.14) and (13.28) with N^0, $\bar{a}_{\mathrm{eff}}g$, ω_μ, ϕ, τ_c, and σ^2 chosen by a maximum likelihood method. These fits were used to obtain the times t_e at which $P(t)$ had decayed to e^{-1}. The reciprocal of these times, which may be taken as a global measure of the rate of transverse

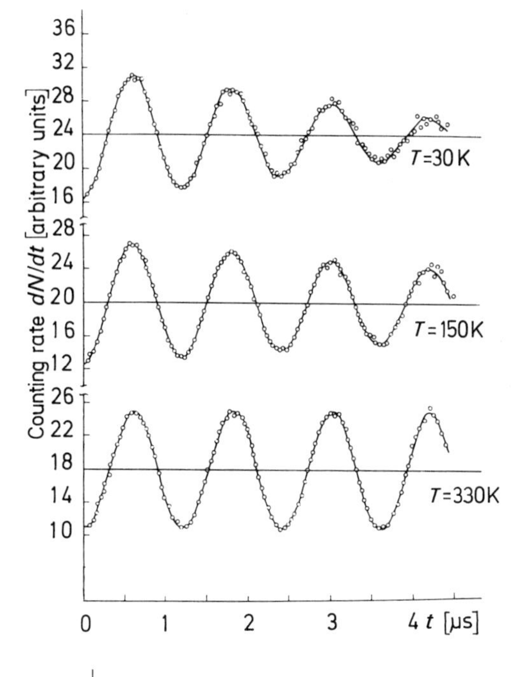

Fig. 13.6. Time dependence of positron counting rate due to μ^+ precession in copper single crystal in transverse magnetic field $B_t = 0.0062$ T at three different temperatures [13.52]. The muon decay has been corrected for

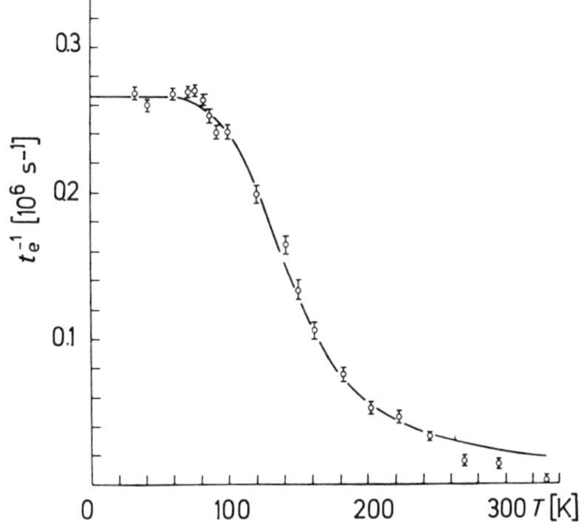

Fig. 13.7. Reciprocal of time t_e in which precession signal in copper single crystal decays by a factor e^{-1} as a function of temperature [13.52]

relaxation, are plotted in Fig. 13.7 as a function of temperature. If (13.28) is assumed to hold, t_e and τ_c are related by

$$(t_e/\tau_c) + \exp[-(t_e/\tau_c)] = 1 + (2\sigma^2\tau_c^2)^{-1}. \tag{13.51}$$

The temperature dependence of τ_c following from (13.51) (with $\sigma = 0.266 \cdot 10^6\,\mathrm{s}^{-1}$, see Fig. 13.7) has been fitted by *Grebinnik* et al. [13.52] to

$$\tau_c = \tau_c^\infty \exp(Q^{\mu^+}/k_B T) \tag{13.52}$$

with

$$Q^{\mu^+} = (0.048 \pm 0.002)\,\mathrm{eV}\,, \tag{13.52a}$$

$$\tau_c^\infty = 10^{-(7.61 \pm 0.04)}\,\mathrm{s}\,. \tag{13.52b}$$

The 'activation energy' (13.52a) is very much smaller than the migration energy of hydrogen in Cu measured at high temperatures (about 0.40 eV [13.54–56], see also Chap. 12). The small value of Q^{μ^+} and the accompanying large value of τ_c^∞ have been interpretated by *Grebinnik* et al. [13.52] as evidence for the tunnelling of muons ("below-barrier diffusion").

From the experiments of *Camani* et al. [13.53], to be discussed below, it follows that at low temperatures (20–80 K), positive muons in Cu reside in octahedral interstices. If we make the simple assumption that the muons diffuse by jumps between adjacent octahedral interstices, the relationship between the muon diffusion coefficient D^{μ^+} and the time of residence on an octahedral site, $\bar{\tau}$, is given by

$$D^{\mu^+} = a_0^2/12\bar{\tau}\,. \tag{13.53}$$

With the additional assumption $\bar{\tau} = \tau_c$, the fit of *Grebinnik* et al. [13.52] leads to

$$D^{\mu^+} = 4.4 \cdot 10^{-13} \exp(-0.048\,\mathrm{eV}/k_\mathrm{B}T)\,\mathrm{m}^2\,\mathrm{s}^{-1}\,. \tag{13.54}$$

In Fig. 13.8 we have plotted the muon diffusion coefficients calculated in this way from the data of Fig. 13.7, together with the hydrogen diffusion coefficients measured at high temperatures by *Katz* et al. [13.54], and *Perkins* and *Begeal* [13.55]. Polycrystalline specimens [13.34, 51] gave similar values. The increased error bars at about 300 K are due to the fact that at these temperatures the muons move so rapidly that the loss of transverse polarization within a few muon lifetimes is extremely small. At about 90 K the time of residence becomes about 5 times the muon lifetime, so that the motional averaging effect, from which τ_c is derived, becomes very small, again resulting in a large error of τ_c.

Teichler [13.57] has analyzed the data of *Grebinnik* et al. [13.52] within the framework of the quantum theory of diffusion (for the general background see [13.58, 59] and Chap. 8) and has concluded that they may be understood in terms of tunnelling transitions between the ground-state energy levels of adjacent octahedral sites. Making use of a simple picture for the phonon spectrum and the phonon–muon interaction, he derives

$$D(T) = \frac{a_0^2}{\hbar}|J|^2 \left[\frac{\pi}{4E_\mathrm{a} k_\mathrm{B} T_0 h_1(T/T_0)}\right]^{1/2} \exp\left[-\frac{E_\mathrm{a}}{k_\mathrm{B} T_0 h_2(T/T_0)}\right]. \tag{13.55}$$

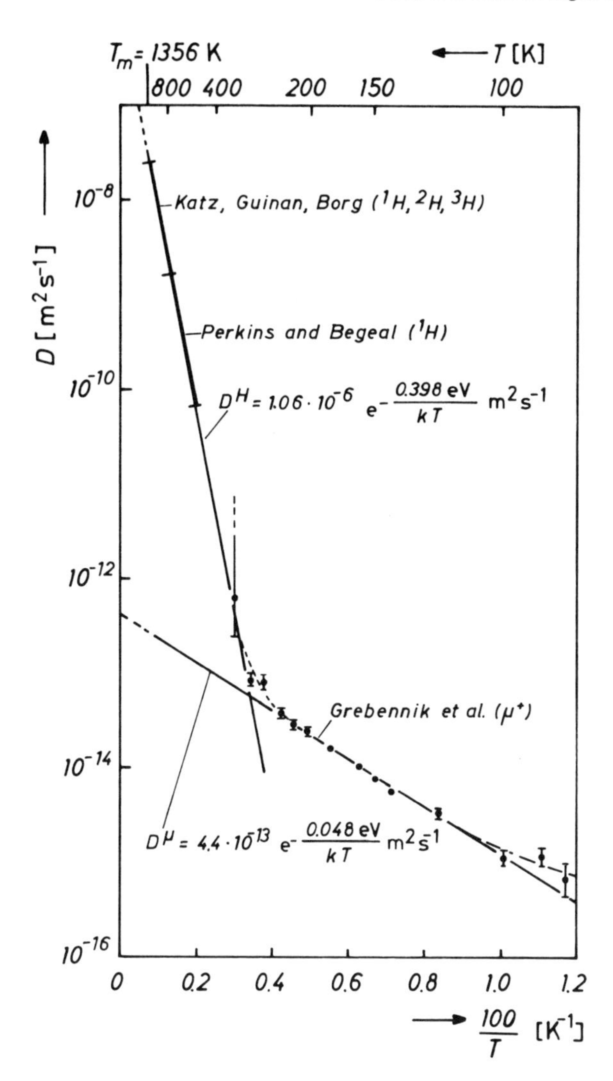

Fig. 13.8. Copper: Temperature dependence of the diffusion coefficients of hydrogen and positive muons. The thin straight lines are Arrhenius-law fits to the hydrogen or muon data, respectively. The dashed curve represents a fit of (13.55) to the muon diffusion data, the dotted curve a plausible interpolation between the hydrogen and the muon measurements

Here J is the matrix element for muon transfer between adjacent interstitial sites and E_a the 'shifting energy' necessary to shift the local deformation of the lattice from one muon site to an adjacent one. The temperature T_0 is of the order of magnitude of one-half the Debye temperature. The functions $h_1(T/T_0)$ and $h_2(T/T_0)$ have been calculated by *Teichler* [13.57]; both approach T/T_0 at $T \gg T_0$. Fitting (13.55) to the muon diffusion coefficient gives $E_a = 0.075$ eV and $J = 18 \cdot 10^{-6}$ eV.

Equation (13.55) accounts for the deviations of the data from (13.52) at low temperatures (dashed curve in Fig. 13.8) but not at high temperatures. At these temperatures, other transition modes are likely to play a rôle (e.g., jumps over the potential energy barriers or tunnelling between excited states) and to increase the muon diffusion coefficient over and above the value given by (13.55). Since, according to *Katz* et al. [13.54] and *Eichenauer* et al. [13.60], the diffusion coefficients of hydrogen, deuterium, and tritium differ only insignificantly on the scale of Fig. 13.8, we may expect that D^{μ^+} is of the same order of magnitude at high temperatures. A possible interpolation between the hydrogen data and the muon data is indicated in Fig. 13.8, but it must be emphasized that at the present time the precise relationship between the high-temperature diffusion coefficients of the hydrogen isotopes and of positive muons is unknown because of the lack of an adequate theory of the isotope effect. Nevertheless, it is remarkable that the muon diffusion coefficient at 330 K deduced from the last data point of *Grebinnik* et al. [13.52] is compatible with the extrapolation of the hydrogen diffusion data to low temperatures.

In the low-temperature regime, in which motional averaging is negligible, *Camani* et al. [13.53] investigated the dependence of the transverse relaxation function on the magnitude and the crystallographic orientation of the magnetic field. Their main results may be summarized as follows:

1) The decay of the transverse polarization is well described by a *Gaussian* (damping parameter σ).

2) The decay curves do not show significant differences between 20 K and 80 K.

3) The damping parameter σ depends on both the magnitude and the crystallographic direction of the applied magnetic field $\boldsymbol{B}_{\text{appl}}$.

4) The dependence of σ on the crystallographic orientation of $\boldsymbol{B}_{\text{appl}}$ is weak at small field strengths and quite marked at large fields.

Items 1) and 2) agree with the work of *Gurevich* et al. [13.34] and also with more recent results of *Hartmann* et al. [13.61] on Cu single crystals at $B_{\text{appl}} = 0.035$ T. Items 3) and 4) are explained in terms of the quadrupole effects discussed in Section 13.4.2. Figure 13.9 shows the fit of Hartmann's theory to the data of *Camani* et al. [13.53].

From the transition of the low-field to the high-field region the ratio ω_Q/ω_I and from this

$$(1+\gamma)\,V_{zz} \approx 0.32 \cdot 10^{30}\,e\,\text{m}^{-3} \tag{13.56}$$

has been obtained (e = elementary electrical charge; for Cu: $I = 3/2$, quadrupole moment $Q \simeq -0.16 \cdot 10^{-28}\,e\,\text{m}^2$). According to *Camani* et al. [13.53], a model for the screening of positive muons due to *Meier* [13.40] (using a Hulthen potential) predicts $V_{zz} = 0.068 \cdot 10^{30}\,e\,\text{m}^{-3}$, a calculation by *Jena* and *Singwi* [13.62] based on the density functional formalism gives $V_{zz} = 0.05 \cdot 10^{30}\,e\,\text{m}^{-3}$. In view of the uncertainty in the knowledge of the enhancement factor γ, these values are certainly not incompatible with (13.56). Nevertheless, it is clear that a quantitative theoretical understanding of the quadrupole effects is possible only in terms of an in-depth treatment of the screening and enhancement effects.

At high magnetic fields (13.25) should be applicable. The observed variation of the high-field σ with the crystallographic orientation of $\boldsymbol{B}_{\text{appl}}$ is compatible

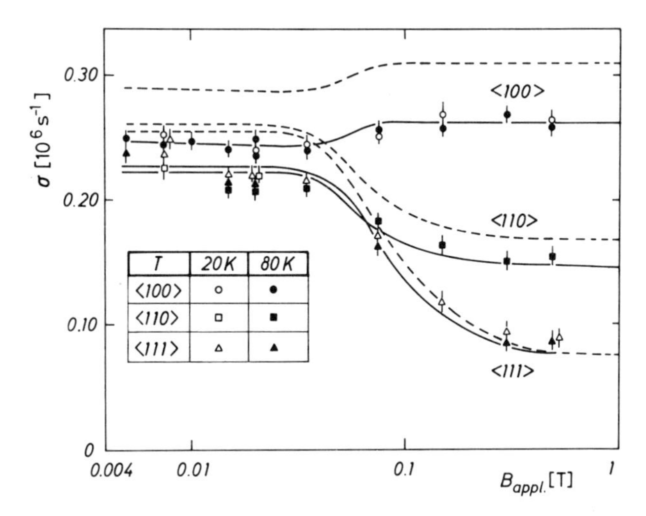

Fig. 13.9. Comparison of Hartmann's theory [13.31] of quadrupole effects with the measurements of *Camani* et al. [13.53] on copper single crystals at 20 K and 80 K for three different crystallographic orientations of the applied magnetic field $\boldsymbol{B}_{\rm appl}$. The full lines represent the fit to muon location on octahedral sites, the dashed lines represent the predictions for tetrahedral sites (after [13.53])

with the location of the positive muons on octahedral interstitial sites and incompatible with location on tetrahedral interstitial sites. Small systematic deviations from the expression (13.25) as calculated for a rigid lattice indicate that the Coulomb repulsion has increased the separation between muons and their six nearest neighbours (see Sect. 13.4.2). If the finite extension of the muon wave function is disregarded, this increase comes out to about 5%. Such an increase is compatible with a theoretical estimate of about 4% based on the phonon–muon interaction following from the fit of (13.55) to the experimental data [13.63].

In summary it may be said that the muon experiments on Cu are in excellent agreement with μ^+ location on octahedral interstitial sites, that the accord with the hydrogen data is good, and that they permit us to deduce quantitative information on the coupling between a positively charged particle and phonons.

b) Group-V Transition Metals

From the point of view of their nuclear properties, the Group-V transition metals are about as suited as Cu for experiments based on the interaction between the magnetic moments of positive muons and nuclei. Each of these metals consists to virtually 100% of one isotope with large spin and large gyromagnetic ratio (^{51}V: $I=7/2$, $\gamma_I=7.033\cdot10^7\,\mathrm{rad\,T^{-1}\,s^{-1}}$, $Q=0.2\cdot10^{-28}\,e\,\mathrm{m}^2$; ^{93}Nb: $I=9/2$, $\gamma_I=6.549\cdot10^7\,\mathrm{rad\,T^{-1}\,s^{-1}}$, $Q=-0.16\cdot10^{-28}\,e\,\mathrm{m}^2$;

^{181}Ta: $I=7/2$, $\gamma_I=3.198\cdot10^7\,\mathrm{rad\,T^{-1}\,s^{-1}}$, $Q=4.0\cdot10^{-28}\,e\,\mathrm{m}^2$). The scientific and technological interest in the properties of hydrogen in the Group-V transition metals (see Chap. 1, [13.108]) makes μ^+SR studies in these metals very attractive. Several groups have recently reported such investigations (V: [13.64–66]; Nb: [13.60, 65, 67, 68]; Ta: [13.65, 66, 69, 70]). However, the experimental information available is not nearly as complete as in the case of Cu; thus at the present time a number of questions that could be answered in principle have to remain open.

From the work on hydrogen in transition metals, one expects that trapping of positive muons by residual impurities will be very important, the more so since μ^+SR experiments are usually carried out at lower temperatures than the experiments on H. Trapping of positive muons leads to an increased transverse relaxation [13.71], hence to an apparent lowering of D^{μ^+} as calculated by the theory of Section 13.4.2. It is very helpful to compare the transverse relaxation of the muons with the hydrogen diffusion coefficients extrapolated from higher temperatures in order to get a feeling for possible trapping effects.

A further important step in the analysis of the muon data is the comparison of the measured σ values with those calculated from (13.25) for specific interstitial sites. The most likely candidates for the locations of positive muons in bcc metals are the tetrahedral and the octahedral interstices. Both types of interstices possess tetragonal symmetry. In single-crystal experiments, well-defined σ values are therefore expected only for $\langle 111\rangle$ orientations of $\boldsymbol{B}_{\mathrm{appl}}$. For $\boldsymbol{B}_{\mathrm{appl}}$ parallel to $\langle 100\rangle$, theory predicts two different σ values, but in none of the published papers has this been taken fully into account.

All three metals show motional averaging, possibly modified by the trapping effects mentioned above. The most detailed experiments on V are those of *Fiory* et al. [13.64] on a polycrystalline specimen ($B_{\mathrm{appl}}=0.0058\,\mathrm{T}$). The relaxation rate as characterized by t_e^{-1} [defined by (13.28)] drops continuously from 10 K to 300 K (the extremal temperatures investigated). The extrapolation to $T=0\,\mathrm{K}$ is consistent with the value $\sigma=0.399\cdot10^6\,\mathrm{s}^{-1}$ calculated from (13.25) for tetrahedral interstices in a rigid polycrystalline sample, but less so with the corresponding value $\sigma=0.429\cdot10^6\,\mathrm{s}^{-1}$ for octahedral interstices. It is incompatible with *Birnbaum* and *Flynn's* [13.72] suggestion (originally made for H in Nb), according to which at low temperatures a light interstitial in a bcc metal might tunnel between an array of four neighbouring tetrahedral sites, since this would lead to partial motional averaging. *Fiory* et al. [13.64] conclude that at the lowest temperatures positive muons in V are localized on single interstitial sites; it appears likely that these are tetrahedral sites. The results of *Heffner* et al. [13.66] but not those of *Hartmann* et al. [13.65] are in agreement with this conclusion; the latter authors found at low temperatures a temperature-independent value $\sigma=0.21\cdot10^6\,\mathrm{s}^{-1}$.

On a Nb single crystal with $B_{\mathrm{appl}}=0.0103\,\mathrm{T}$ parallel to $\langle 100\rangle$, *Lankford* et al. [13.68] observed a plateau at $\sigma=0.324\cdot10^6\,\mathrm{s}^{-1}$, extending from the lowest temperature investigated (5.3 K) to about 55 K. It is interrupted by a dip to a σ value as low as $0.24\cdot10^6\,\mathrm{s}^{-1}$ at 20 K. Data on a Nb polycrystal [13.65] show more scatter but are otherwise similar to the results just mentioned. Here the low-temperature plateau corresponds to $\sigma = 0.29 \cdot 10^6 \, s^{-1}$.

The polycrystalline average of (13.25) for rigid Nb crystals is $\sigma = 0.341 \cdot 10^6 \, s^{-1}$ for tetrahedral interstices and $0.367 \cdot 10^6 \, s^{-1}$ for octahedral interstices. For Nb single crystals with $\boldsymbol{B}_{appl}$ parallel to $\langle 100 \rangle$, *Lankford* et al. [13.68] quote $\sigma = 0.407 \cdot 10^6 \, s^{-1}$ and $\sigma = 0.506 \cdot 10^6 \, s^{-1}$ for tetrahedral or octahedral interstices, respectively[15].

We see that the straightforward agreement between the experimental low-temperature σ and the calculations for the tetrahedral sites that was found in V does not exist for Nb. Let us examine whether *quadrupole effects* (see Sect. 13.5.1a) play a rôle by considering higher magnetic fields. *Hartmann* et al. [13.65] report a weak field dependence between 0.005 T and 0.2 T for magnetic fields in $\langle 100 \rangle$ or $\langle 111 \rangle$ directions. Even at the highest fields employed, the absolute values of σ do not agree with the theoretical predictions for tetrahedral or octahedral sites. For the ratios $\sigma^{\langle 100 \rangle}/\sigma^{\langle 111 \rangle}$, the calculations of *Lankford* et al. [13.68] predict 1.24 for tetrahedral sites and 2.79 for octahedral sites. Experimental values for this ratio are 1.2 ([13.68], $B_{appl} = 0.176$ T at 10 K) and 1.35 ([13.65], $B_{appl} = 0.20$ T). These results favour the muon location in tetrahedral rather than octahedral interstices. Nevertheless, the situation remains unclear because the absolute values of σ come out too low even at the highest magnetic fields employed. (Further uncertainty exists about the nature of the above-mentioned 'dip'; for a discussion see below.) The *Birnbaum–Flynn* proposal [13.72] may be excluded for positive muons in Nb, since it would require substantially *lower* σ values than those measured. A further result of the Nb work is that quadrupole effects do not appear to be very important. This lends additional credibility to the conclusions drawn on V, whose quadrupole moment has a magnitude similar to that of Nb.

Dorenburg et al. [13.69] measured the transverse relaxation rate of a high-purity Ta crystal between 18 K and 252 K with a magnetic field $B_{appl} = 0.0202$ T in a $\langle 111 \rangle$ direction. This field direction has the advantage that interstitial sites with differently oriented crystallographic axes are equivalent, so that an analysis in terms of a single σ value is appropriate. Furthermore, for this crystallographic orientation of the transverse field, the difference in the theoretical predictions of (13.25) for octahedral and tetrahedral interstitial sites in the bcc lattice is particularly large.

The relaxation rate is rather low at low temperatures $[\sigma(18\,\mathrm{K}) = (0.090 \pm 0.01) \cdot 10^6 \, s^{-1}]$, passes through a maximum $[\sigma(40\,\mathrm{K}) = (0.112 \pm 0.003) \cdot 10^6 \, s^{-1}]$, and shows motional averaging up to 100 K. Recent measurements by *Camani* et al. [13.70] confirmed and extended these results. They show clearly that a plateau exists between 25 K and 45 K. It is independent of the orientation and the strength (up to 0.6 T) of $\boldsymbol{B}_{appl}$ [13.70]. The measurements

[15] The meaning of these values is not completely clear. They represent presumably mean values for the three possible orientations of $\boldsymbol{B}_{appl}$ with respect to the tetragonal axes of the interstices.

on a polycrystal in a field $B_{appl}=0.04$ T by *Hartmann* et al. [13.65] give good agreement with those of *Dorenburg* et al. [13.69] at 40 K and in the motional averaging region, but do not show an indication of the low-temperature drop.

The σ(40 K) value is compatible with the prediction $\sigma=0.128\cdot10^6\,s^{-1}$ of (13.25) for a tetrahedral site in a rigid Ta crystal. The corresponding theoretical value for an octahedral site is $\sigma=0.078\cdot10^6\,s^{-1}$, so that in Ta the experimental evidence appears to favour muon location on tetrahedral sites over that on octahedral sites. The field independence of the σ value of the plateau may mean either that the electric field gradients at the neighbouring nuclei happen to be extremely small, or that the quadrupole effect is so strong that even at the highest fields the quadrupolar interaction and not the Zeeman energy determines the spin orientation of the adjacent nuclei. Since the magnitude of Q of the Ta nuclei is about 25 times that of the Cu, V, or Nb nuclei, the second possibility is more likely to apply. In contrast to the first possibility, it accounts for the independence of σ of the crystallographic orientation of $\boldsymbol{B}_{appl}$ without any further assumptions.

Let us now turn to the information that may be derived from the *temperature dependence* of the transverse muon spin relaxation rate in the Group-VB metals. The analysis is based on fits to (13.28) or, in the case of exponential relaxation functions, to (13.29). Correlation times τ_c are obtained by using $\sigma=0.40\cdot10^6\,s^{-1}$ (V), $\sigma=0.324\cdot10^6\,s^{-1}$ (Nb), and $\sigma=0.125\cdot10^6\,s^{-1}$ (Ta). From these values muon diffusion coefficients D^{μ^+} have been deduced by means of (13.46b), thus assuming jumps between adjacent tetrahedral interstices.

In Fig. 13.10 the results are compared with extrapolations of hydrogen diffusion coefficients D^H to low temperatures. In V, D^{μ^+} shows a much weaker temperature dependence than D^H, whose activation energy is already low compared with the values observed on other metals. $D^{\mu^+}(T)$ and $D^H(T)$ intersect at about 30 K at a level of about $10^{-15}\,m^2\,s^{-1}$. This suggests the following interpretation: Above 30 K the muon results are strongly affected by impurity trapping; the true muon diffusion coefficients are comparable to or larger than the extrapolated hydrogen values. Since $D^{\mu^+}=10^{-15}\,m^2\,s^{-1}$ corresponds to a mean diffusion distance within the muon lifetime of about 0.1 nm, i.e., less than the muon jump distance, the muon diffusion coefficient at 30 K must be several powers of ten larger in order to permit muon trapping by impurities. By contrast, the muon diffusion coefficient obtained for 10 K lies by almost ten powers of ten above the extrapolated D^H value and is thus likely to represent the true D^{μ^+}. The conclusion is that, similar to Cu, the Arrhenius plot of $D^{\mu^+}(T)$ in V shows a low-temperature branch due to tunnelling with an effective activation energy much smaller than that measured for hydrogen.

In Nb the comparison is made with the low-temperature branch of the $\ln D^H$-vs-T^{-1} diagram (comp. Chap. 12). The diffusion coefficients obtained from the μ^+SR experiments lie below the extrapolated D^H values. This suggests that they are all affected by trapping. In order to permit trapping within the muon lifetime, the true D^{μ^+} at the lowest temperature at which the correlation

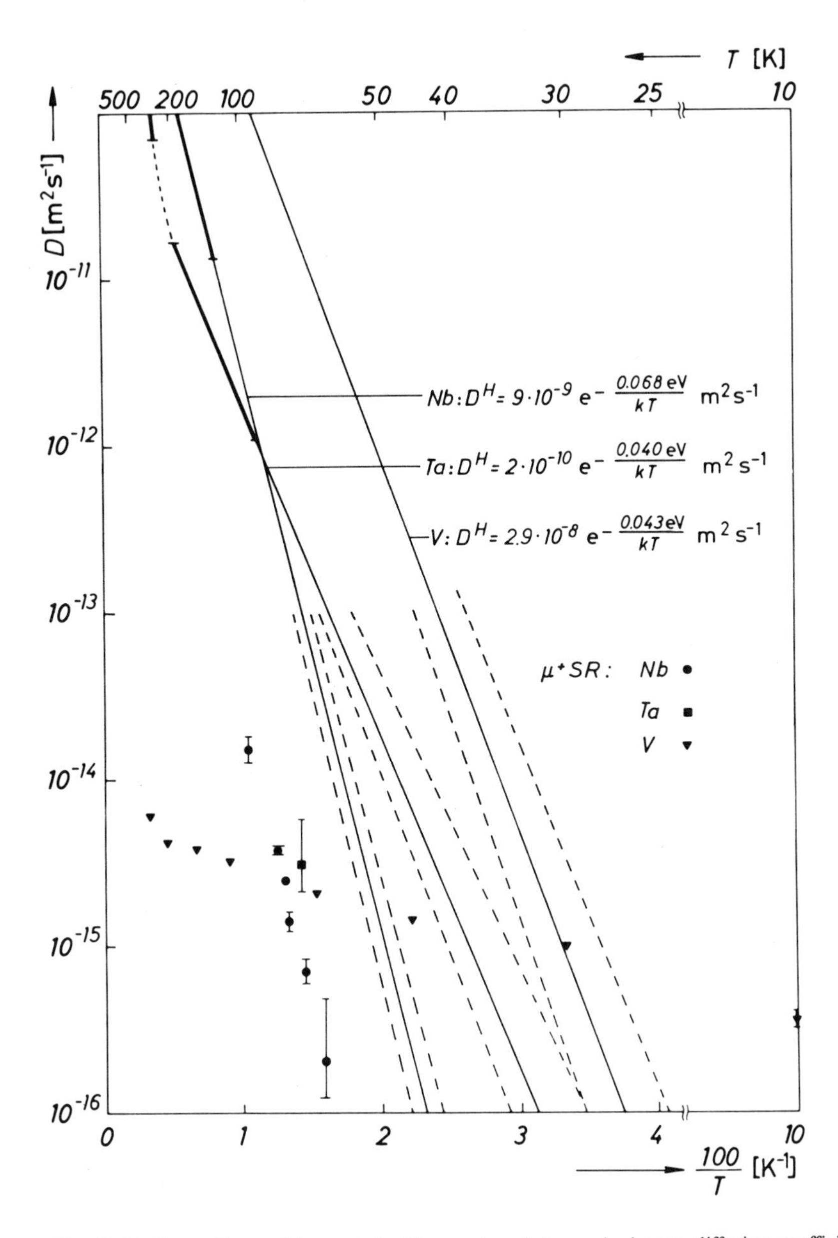

Fig. 13.10. Group-V transition metals: Comparison between hydrogen diffusion coefficients D^{H} (extrapolations to low temperatures; estimated uncertainty limits indicated by dashed lines) and muon diffusion coefficients D^{μ^+} calculated from (13.46b). Hydrogen measurements: full lines (comp. Chap. 12); μ^+ SR measurements: vanadium = ▼, niobium = ●, tantalum = ■

time τ_c has been determined (63 K) must be considerably larger than the extrapolated D^H. This is in marked contrast to V. It appears feasible that in Nb even the σ values in the low-temperature plateau mentioned above are affected by trapping.

In Ta the muon diffusion coefficient calculated at 70 K lies below the extrapolated D^H values, too. It is therefore likely that this measurement is affected by trapping and that the true D^{μ^+} is considerably higher.

The situation in the Group-VB metals may be summarized as follows: Of all the possible locations of positive muons discussed, that in tetrahedral interstices of the bcc lattice is most compatible with the available evidence. However, the agreement between experiment and theory is not nearly as good as in the case of Cu.—The majority of the 'motional averaging' measurements on Group-VB metals is presumably strongly affected by trapping at impurities. With the exception of the V point at 10 K, the muon diffusion coefficients plotted in Fig. 13.10 must be regarded as lower limits to the true values. The nature of the traps and the magnitude of the muon binding energy to the traps is not known at present. Future measurements of the transverse muon spin relaxation rate should be carried out on carefully characterized high-purity specimens. They should be supplemented by measurements of the longitudinal relaxation rate, since these provide information on D^{μ^+} that is less affected by trapping than that obtainable from transverse measurements.

Finally, a few remarks will be made on the increase of the relaxation rate with decreasing temperature observed in some experiments in the liquid-hydrogen—liquid-helium temperature range. Such a low-temperature increase is predicted by theory [due to so-called coherent transitions, in which tunnelling between adjacent interstices takes place without change in the phonon occupation numbers (comp. Chap. 8)]. The problem is to understand why at even lower temperatures in Bi (see the following subsection) and Nb (but according to the available evidence [13.70], not in Ta) the transverse relaxation rate returns to approximately the same level as above the low-temperature drop. It is true that at very low temperatures the muons may be immobilized again by trapping at dislocations or impurities, but then the relaxation rate should depend on the nature of these imperfections and differ from that of muons immobilized at higher temperatures in an otherwise unperturbed environment.

c) Other Metals

The experimental evidence on the behaviour of positive muons in nonmagnetic metals other than those discussed in the two preceding subsections is quite limited. We discuss here a few results that are of interest either from a theoretical point of view or for the understanding of the behaviour of hydrogen in metals.

In transverse spin relaxation experiments on Al, very little damping of the 'wiggles' in muon spin rotation experiments has been found over the temperature range 1–700 K [13.61, 66]. This is most surprising, since the interaction between the magnetic moments of muons and nuclear spins and the diffusion properties of hydrogen at high temperatures (see Chap. 12) are similar to those in Cu. The crystal structures of the two elements are the same, so one might have expected low-temperature diffusion coefficients of muons in Al of the same order of magnitude as those found in Cu. The experiments, however, show that over the entire temperature range investigated the muon diffusion coefficients exceed $D^{\mu^+} = 10^{-13}\,\mathrm{m^2\,s^{-1}}$, with the possible exception of temperatures at about 100 K. Here the experiments of *Heffner* et al. [13.66] are compatible with a muon diffusion coefficient of about $0.5 \cdot 10^{-13}\,\mathrm{m^2\,s^{-1}}$.

Another interesting case is that of Bi, recently investigated by *Grebinnik* et al. [13.73]. The experiments extend from 250 K to liquid-helium temperatures and have been analyzed by means of (13.16) in terms of a temperature-dependent σ. Between 150 K and 100 K, σ increases very rapidly with falling temperature until a plateau at $\sigma = 0.15 \cdot 10^6\,\mathrm{s^{-1}}$ is attained between 100 K and 75 K. This plateau is followed by a drop with a sharp minimum of $\sigma = 0.08 \cdot 10^6\,\mathrm{s^{-1}}$ at 25 K. Then σ rises rapidly again to $\sigma \approx 0.15 \cdot 10^6\,\mathrm{s^{-1}}$ at liquid-helium temperatures, i.e., to the level of the plateau between 75 K and 100 K.

Grebinnik et al. [13.73] interprete their results on Bi in the manner outlined in the last paragraph of Section 13.5.1b. Why the trapping of positive muons at dislocations or other imperfections should lead to the same σ as the immobilization of the muons on interstitial sites in an undisturbed environment remains unexplained. Although *Grebinnik* et al. [13.73] stated that their experiments are in good agreement with the estimate of *Andreev* and *Lifshitz* [13.74] for the onset of coherent diffusion, an element of doubt about the correctness of this interpretation remains.

The discussion of muon diffusion in Cr, which has been studied by *Kossler* et al. [13.75], will be postponed until Section 13.5.2e, since the principal interaction of the muon magnetic moments is not with the nuclear spins [the abundance of ^{53}Cr, the only stable isotope with a (rather small) magnetic moment, is only about 10%] but with electron spins and thus similar to that prevailing in ferromagnets.

For first measurements on β-Pd hydride (Pd–$H_{0.97}$) the reader is referred to the literature [13.76].

13.5.2 Ferromagnetic and Antiferromagnetic Metals

a) α-Iron

As in the other bcc metals, positive muons in α-iron are likely to be located either on tetrahedral or on octahedral interstitial sites. In ferromagnetic iron in zero or small applied fields, in which the magnetization within the ferromagnetic domains points in one of the $\langle 100 \rangle$ directions, these sites fall into two classes with very different local magnetic fields, depending on whether their tetragonal axes are parallel or perpendicular to the direction of magnetization (see Sect. 13.4.4a). However, in none of the numerous experiments on α-Fe [13.77–84] has more than one spin precession frequency been detected, indicating a strong motional averaging effect [13.47, 80].

In a high-purity α-Fe single crystal in zero external field the spin precession of muons, whose original spin polarization was parallel to [110], was observed by *Nishida* et al. [13.83] down to temperatures as low as 22 K. This means that down to this temperature the muons are sufficiently mobile to average over the different dipolar fields. Under these conditions the muon spin precession frequency is given by

$$\omega_\mu = \gamma_\mu B_\mu(B_{\mathrm{appl}}) \tag{13.57}$$

with

$$\boldsymbol{B}_\mu(\boldsymbol{B}_{\mathrm{appl}}) = \boldsymbol{B}_{\mathrm{appl}} + \boldsymbol{B}_{\mathrm{demag}} + \boldsymbol{B}_{\mathrm{Lorentz}} + \boldsymbol{B}_{\mathrm{Fermi}} \,. \tag{13.58}$$

From the Hamiltonian (13.11) of the Fermi interaction follows the magnetic field due to the conduction electrons

$$B_{\mathrm{Fermi}} = -\tfrac{4}{3}\mu_0 \mu_e \varrho_{\mathrm{spin}}(0)\,. \tag{13.59}$$

In ferromagnets, $\boldsymbol{B}_{\mathrm{Fermi}}$ contains two contributions, a paramagnetic one of the 'Knight-shift type' (see Sect. 13.4.3) and a contribution resulting from the 'ferromagnetic' polarization of the conduction electrons at the muon site, which is thought to be parallel (or antiparallel) to $\boldsymbol{M}_{\mathrm{domain}}$[16]. In a demagnetized specimen in zero applied field, the second contribution is the only one that needs to be taken into account in addition to $\boldsymbol{B}_{\mathrm{Lorentz}}$.

In several experiments [13.77, 81, 82] $B_\mu(0)$ has been shown to be smaller than the Lorentz field $B_{\mathrm{Lorentz}} = 0.715$ T as calculated from (13.35), the most accurate determination being that of *Gurevich* et al. [13.82], giving $B_\mu(0) = (0.3509 \pm 0.0004)$ T. This means that at the muon sites the conduction electron magnetization must have a direction opposite to $\boldsymbol{M}_{\mathrm{domain}}$. *Seeger* [13.85] and independently *Nishida* et al. [13.83] pointed out that this result constitutes experimental evidence for the occupancy of tetrahedral sites rather than octahedral sites, since neutron diffraction has shown [13.86] the conduction electron magnetization to be negative (i.e., opposite to $\boldsymbol{M}_{\mathrm{domain}}$) at tetrahedral interstices and positive at octahedral interstices.

[16] We use the following sign convention: A plus sign indicates that $\boldsymbol{B}_{\mathrm{Fermi}}$ is parallel to $\boldsymbol{M}_{\mathrm{domain}}$. ϱ_{spin} is given a positive sign if there is an excess of electrons with spins parallel to $\boldsymbol{M}_{\mathrm{domain}}$.

From the functional dependence $B_\mu(B_{\text{appl}})$, *Gurevich* et al. [13.80, 82] deduced that in α-Fe $\boldsymbol{B}_\mu(0)$ has the opposite sign of $\boldsymbol{M}_{\text{domain}}$[17]. In the sign convention introduced above, this means that the Fermi field acting on the muons in Fe is the negative sum of $B_\mu(0)$ and the Lorentz field, hence

$$B_{\text{Fermi}}^{(\mu)} = -(B_\mu(0) + B_{\text{Lorentz}}) = -1.061\,\text{T}\,. \tag{13.60}$$

Comparison with the Fermi field at tetrahedral sites in an unperturbed Fe crystal [13.86]

$$B_{\text{Fermi}}^{(\text{tet})} = -(0.11 \pm 0.03_5)\,\text{T} \tag{13.61}$$

indicates that the presence of the muons enhances the conduction electron spin density $\varrho_{\text{spin}}(0)$ at the muon site by about a factor of ten. This factor is surprisingly large and not yet satisfactorily accounted for by theory [13.42, 86–89], but nevertheless easier to understand than would be the reversal of the sign of the spin density required if the muons were assumed to occupy octahedral interstices.

The relationship between the transverse relaxation rate in a $\langle 100 \rangle$ magnetization, $\Gamma_t^{\langle 100 \rangle}$, and the muon diffusion coefficient may be found from (13.43b) and (13.46). It reads for muons jumping between adjacent tetrahedral sites

$$D^{\mu^+} = \frac{a_0^2}{144} \frac{(\gamma_\mu B_{\text{dip}}^{||})^2}{\Gamma_t^{\langle 100 \rangle}}\,. \tag{13.62}$$

Figure 13.11 shows the muon diffusion coefficients D^{μ^+} that may be deduced from the transverse relaxation measurements [13.79, 83, 84] by means of (13.62), and, for purposes of comparison, several temperature dependences of the diffusion coefficient of hydrogen, D^{H}. Whereas the majority of the high-temperature determinations of D^{H} in α-Fe are reasonably consistent with each other, the room-temperature values given in the literature scatter by several powers of ten (see [13.89]). This scatter is thought to be mainly due to the trapping of hydrogen by impurities and other crystal defects and to surface barriers. The highest values are presumably least affected by trapping and therefore most likely to represent true hydrogen diffusion coefficients. As representative of these data, the recent measurement of *Dillard* and *Talbot-Besnard* [13.91] (*D.–T.B.*) has been included in Fig. 13.11. The three $D^{\text{H}}(T)$ curves

[17] In the other ferromagnetic metals, Ni, Co, and Gd, $\boldsymbol{B}_\mu(0)$ has the same direction as M_{domain} [13.82] with the exception of Co below 450 K (for details see Sect. 13.5.2c). By a less transparent argument based on the sense of rotation of the muon spins, both *Foy* et al. [13.77] and *Gurevich* et al. [13.81] had deduced that in α-Fe $\boldsymbol{B}_\mu(0)$ and $\boldsymbol{M}_{\text{domain}}$ were parallel. Since this erroneous result was used in [13.85], the numerical result obtained there is not in agreement with (13.61).

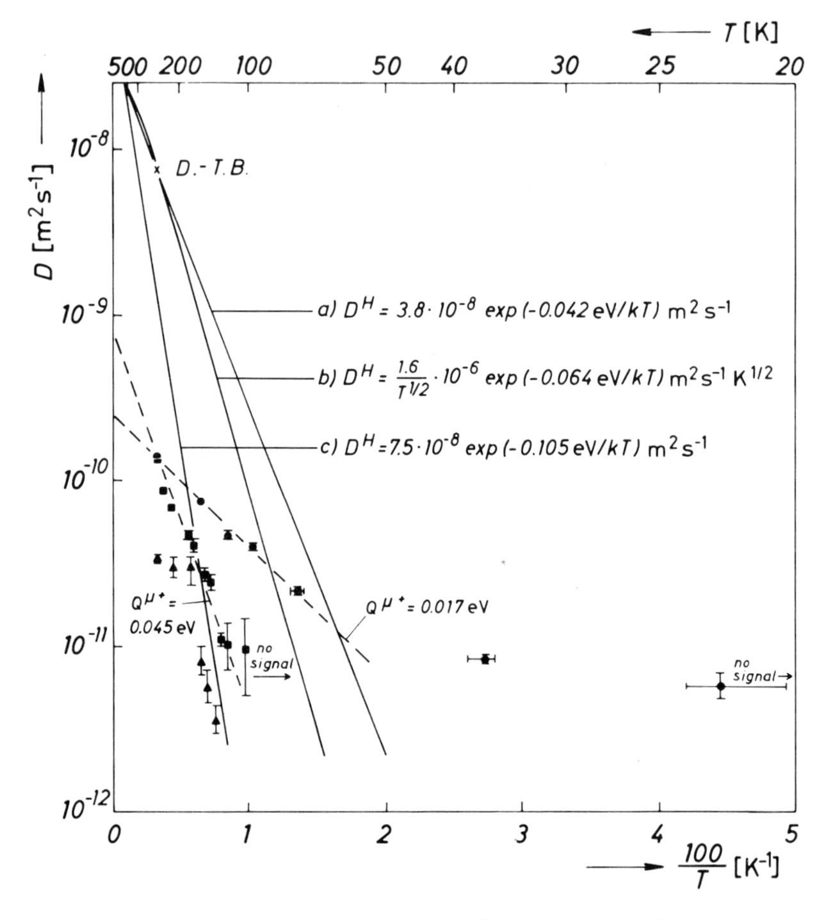

Fig. 13.11. α-iron: Comparison between hydrogen diffusion coefficients D^{H} (extrapolated to low temperatures according to various procedures, see text; *D.–T.B.* = room-temperature measurement by *Dillard* and *Talbot-Besnard* [13.91]) and muon diffusion coefficient D^{μ^+} as determined by means of (13.62). μ^+ SR measurements: *Gurevich* et al. [13.79]: ▲; *Nishida* et al. [13.83]: ●; *Graf* et al. [13.84]: ■

shown have been obtained in the following ways: 1) The high-temperature data and the *D.–T.B.* result were fitted by an Arrhenius law [13.92]. 2) The same data were fitted [13.92] by the high-temperature limit of the quantum tunnelling theory (see [13.57] and Chap. 8)

$$D^{\mathrm{H}} = A\,T^{-1/2} \exp(-Q^{\mathrm{H}}/kT)\,. \tag{13.63}$$

3) The high-temperature data and the most frequently obtained room-temperature values were fitted by an Arrhenius law [13.90].

The highest muon diffusion coefficients have so far been observed by *Nishida* et al. [13.83]. These authors adjusted the high-temperature end of their data to an Arrhenius law with an activation energy of $Q^{\mu^+}=(17\pm0.3)$ meV, which is very much lower than any of the values compatible with the hydrogen measurements. Let us suppose for the sake of the argument that the muon measurements have to be interpreted, as has been done in deriving the D^{μ^+} values of Fig. 13.11, in terms of muon diffusion by jumps between adjacent tetrahedral sites. Then the small value of Q^{μ^+} means that these jumps must occur by tunnelling. In the tunnelling regime, however, one expects the muon diffusion coefficient to be larger than the hydrogen diffusion coefficient (see Sect. 13.5.1a). According to Fig. 13.11 this is not the case, and we must hence conclude that the D^{μ^+} values deduced from the $\Gamma_t^{\langle 100\rangle}$ measurements do not represent intrinsic diffusion coefficients. Three alternative interpretations of the data appear possible.

1) According to (13.62), D^{μ^+} is inversely proportional to $\Gamma_t^{\langle 100\rangle}$. Any additional contributions to the transverse relaxation rate, e.g., interference with the motional averaging over the dipolar fields because of trapping by impurities, reduce the calculated diffusion coefficients. Such an interpretation is supported by the fact that the measured relaxation rates depend on the quality and the pre-history of the samples [13.79, 83, 84].

2) If it is assumed that the muons jump between adjacent octahedral rather than tetrahedral sites, the calculated diffusion coefficients must be multiplied by $2(1.86/0.522)^2\approx 25$ [see Sect. 13.4.4a, and Eqs. (13.45) and (13.46)]. Then the room-temperature values of D^{μ^+} become comparable with D^{H}. The assumption is in conflict, however, with the sign of $B_\mu(0)$ discussed above and does not, by itself, explain the sample dependence of $\Gamma_t^{\langle 100\rangle}$. We consider it therefore unlikely that it provides a complete explanation of the available measurements.

3) The true explanation may be a mixture of 1) and 2). The muons may spend a fraction of their time on octahedral sites, small enough to maintain the negative sign of $B_\mu(0)$, but long enough to lead to an appreciable increase of the relaxation rate. In addition, trapping at imperfections as discussed in 1) plays a rôle.

In view of the technological and scientific interest in the diffusion of hydrogen in α-Fe and the existing uncertainties in the intermediate- and low-temperature values of D^{H}, it is clearly highly desirable to find the correct interpretation of the muon spin relaxation measurements on α-Fe and to obtain reliable muon diffusion coefficients. The way to attempt this are measurements on $\langle 111\rangle$-magnetized single crystals over a wide range of $\boldsymbol{B}_{\mathrm{appl}}$. It follows from the theory of Section 13.4.4b that such measurements should permit not only an unambiguous determination of the muon interstitial sites and a separation of diffusion and trapping effects, but also an extension of the temperature range over which muon diffusion in α-Fe may be investigated experimentally.

b) Nickel

The spin precession of positive muons in Ni was first measured on polycrystalline samples by *Foy* et al. [13.77] and by *Gurevich* et al. [13.79–81], and was later also studied in single crystals [13.93–95]. A unique precession frequency ω_μ was found, and from (13.57) B_μ was determined between 0.12 K and 800 K [13.77, 93, 94, 96].

The transverse relaxation rate Γ_t depends only weakly on temperature [13.77, 93] but appears to be affected by the pre-history of the samples. At 77 K and 300 K, *Patterson* et al. [13.93] report $\Gamma_t < 5 \cdot 10^5 \, \mathrm{s}^{-1}$ on a Ni single crystal, whereas on a polycrystalline sample, *Foy* et al. [13.77] found $\Gamma_t \approx 5 \cdot 10^6 \, \mathrm{s}^{-1}$ near room temperature and no precession signal at lower temperatures. This indicates that the observed transverse relaxation is not caused by the diffusion of the muons but is due to spatial variations of the magnetic fields at the muon sites.

The experimental results suggest strongly that in Ni the dipolar fields vanish at the muon locations, i.e., that these are interstices with cubic symmetry (see Sect. 13.4.4). In the fcc structure there are two such sites, the octahedral and tetrahedral interstices. At both sites (13.58) is applicable. In the absence of applied and demagnetizing fields, one has (taking into account Footnote 16)

$$B_{\mathrm{Fermi}} = -[B_{\mathrm{Lorentz}} - B_\mu(0)] \,. \tag{13.64}$$

Inserting the low-temperature value $B_\mu(0) = (0.1496 \pm 0.0021)\,\mathrm{T}$ [13.94] and the Lorentz field $B_{\mathrm{Lorentz}} = 0.2136\,\mathrm{T}$ calculated by means of (13.64) from $M_{\mathrm{sp}} = 510 \cdot 10^3 \,\mathrm{A\,m^{-1}}$ gives us

$$B_{\mathrm{Fermi}} = -0.064\,\mathrm{T} \,. \tag{13.65}$$

This value agrees in sign and in magnitude with the magnetic field due to the conduction electron magnetization at octahedral sites in unperturbed Ni crystals as deduced from neutron diffraction data [13.86]. Since the sign and the order of magnitude of the conduction electron magnetization at the tetrahedral interstices are the same [13.86], it is not possible to use the arguments of Section 13.5.2a to distinguish between muon location in octahedral and tetrahedral interstices. The analogy to Cu, however, suggests that in Ni the octahedral sites are preferred.

The agreement in magnitude between B_{Fermi} and neutron diffraction data is surprising in view of the large spin polarization found for Fe (see Sect. 13.5.2a) and has given rise to various comments in the literature [13.87, 89, 97, 98]. Another surprising feature is the fact that in Ni up to 300 K, B_{Fermi} varies much less with temperature than either M_{sp} or B_μ [13.94]. This appears also to be true for the hyperfine field at implanted ^{12}B nuclei residing at octahedral sites in Ni [13.99]. Various attempts to understand these observations have been made [13.89, 100].

c) Cobalt

Because of its hexagonal crystal structure in the α phase and the possibility of changing the crystallographic directions of the magnetization easily by varying the temperature and the magnitude of the applied field [13.101–103], muon spin rotation in cobalt (and also in gadolinium; see Sect. 13.5.2d) shows a number of features that go beyond those found in the cubic ferromagnets iron and nickel. After a number of exploratory experiments [13.81, 82], a detailed study was carried out by *Graf* et al. [13.104], covering the temperature range between 4 K and 1100 K (i.e., both the hcp and the fcc phases of Co) and applied fields B_{appl} up to 0.2 T [13.104].

In the temperature regime up to 450 K, in which the directions of easy magnetization are parallel to the hexagonal axis, $\boldsymbol{B}_\mu$ has a direction opposite to $\boldsymbol{M}_{\text{domain}}$ [13.104]. Since in this temperature regime $|B_\mu(0)|$ is quite small ($\lesssim 0.03$ T), this conclusion is mainly based on considerations of the general consistency of the experimental results. It refutes earlier statements [13.82, 84] based on room-temperature observations only but is supported by recent measurements by *Nishida* et al. [13.105]. In the temperature regimes of "conical" distributions of the directions of easy magnetization and of magnetizations lying in the basal plane, $\boldsymbol{B}_\mu(0)$ possesses the direction of $\boldsymbol{M}_{\text{domain}}$ and increases with increasing temperature [13.104, 105]. At the hcp–fcc transition, $B_\mu(0)$ jumps from 0.040 T in the hcp phase to 0.073 T in the fcc phase, where it remains almost temperature independent up to the highest temperatures investigated [13.104][18].

The facts that in the hcp phase of Co only one muon spin precession frequency is observed and that a measurable transverse relaxation rate is found at virtually all temperatures suggest strongly that, in contrast to α-Fe, the muons in hcp Co are localized at only one kind of interstitial site with the full point symmetry of the hexagonal structure. The two most likely possibilities for such sites in the hcp structure are octahedral or tetrahedral interstices. The observed dependence of B_μ on the temperature and on the magnitude and direction of $\boldsymbol{B}_{\text{appl}}$ is very strongly in favour of the first possibility and rules out occupancy of the tetrahedral interstices [13.104, 105]. In the fcc phase the experimental results are analogous to those in nickel, so that here the muons are presumably located on octahedral sites, too.

After correcting for $\boldsymbol{B}_{\text{dip}}$ as calculated for the octahedral sites, one finds at 4 K $B_{\text{Fermi}} = -0.61$ T, which has the same sign as and differs by about a factor of three from the magnetization due to conduction electrons as determined by neutron diffraction [13.104, 105].

The rate of transverse depolarization increases rapidly around 520 K, i.e., in the temperature range in which the axes of easy magnetization begin to leave the hexagonal axis [13.104, 105]. *Gurevich* et al. [13.82] report that at room

[18] *Graf* et al. [13.104] remark that the rather large room-temperature value $B_\mu(0) = (0.086 \pm 0.008)$ T reported by *Gurevich* et al. [13.81] (but not in a later paper [13.82]) was presumably due to retained fcc phase.

temperature Γ_t increases rapidly with increasing B_{appl}. From the observations on Γ_t it should be possible to derive detailed information on the diffusion and/or trapping of positive muons in Co, but this has not yet been done.

d) Gadolinium

Muon spin rotation in gadolinium was first studied by *Gurevich* et al. [13.82, 106], with subsequent work by other groups [13.105, 107]. On a single crystal, muon spin precession has been observed between 4.2 K and 300 K [13.105]. $B_\mu(0)$ has the direction of $\boldsymbol{M}_{\text{domain}}$ in the entire temperature range and shows a maximum of about 0.15 T at about 150 K [13.105–107]. From a comparison of the temperature dependence of B_μ with that calculated for $\boldsymbol{B}_{\text{dip}}$ it was concluded that in Gd positive muons do *not* occupy tetrahedral interstices and that they are likely to stay mainly at octahedral interstices [13.105, 107]. The diffusion of the muons among these sites has not yet been studied.

e) Chromium

Chromium is antiferromagnetic up to its Néel temperature $T_N = 312$ K. In this temperature regime the electronic magnetic moments give rise to dipolar fields $\boldsymbol{B}_{\text{dip}}$ at the octahedral and tetrahedral interstices that are smaller than those in ferromagnetic α-Fe but much larger than those due to the nuclear magnetic moments. They lead to a much stronger muon spin relaxation than that occurring in V, Nb, and Ta, and permit the study of much higher muon diffusivities than in other non-ferromagnetic bcc metals.

Kossler et al. [13.75] measured the transverse muon spin depolarization in a polycrystalline specimen of high-purity Cr between 77 K and 312 K. Their experiments do not allow them to differentiate between octahedral and tetrahedral occupancy, although such a distinction should be possible by future experiments on single crystals.

The correlation times τ_c deduced by *Kossler* et al. [13.75] from their measurements by theoretical considerations similar to those of Section 13.4.4b have been fitted between 130 K and 250 K to an Arrhenius law

$$\tau_c = (4.4 \pm 1.5) \exp[(0.046 \pm 0.05)\,\text{eV}/kT]\, 10^{-12}\,\text{s}\,. \tag{13.66}$$

According to (13.46b) this corresponds to a pre-exponential factor of the muon diffusion coefficient

$$D_0^{\mu^+} = 2.6 \cdot 10^{-10}\,\text{m}^2\,\text{s}^{-1}\,. \tag{13.67}$$

This value is comparable to the pre-exponential factor found in the low-temperature branch of the hydrogen and deuterium diffusion coefficients of Ta;

hence (13.66) may well represent the effect of motional averaging due to muon diffusion. Between 250 K and 312 K, τ_c is found to increase with increasing temperature. *Kossler* et al. [13.75] attribute this to the effects of muon trapping by imperfections [13.71]. This explanation is quite plausible since, according to (13.67), at 300 K the mean distance travelled by a muon during its lifetime is about 20 nm.

13.6 Summary of the Results Obtained So Far

The study of positive muons in metals is only in its beginning and constitutes, in the opinion of the writer, a technique of considerable future potential. In this section we summarize the main results obtained so far (indicating in each case the section in which details or further information may be found) and their possible relevance for the understanding of the behaviour of hydrogen in metals.

1) Positive muons may be considered as short-lived (lifetime $\tau_\mu = 2.2 \cdot 10^{-6}$ s) light isotopes of hydrogen (mass $m_\mu \approx 1/9$ of the proton mass) (Sect. 13.1). The behaviour of positive muons in metals is studied through the interaction of their magnetic moments with local magnetic fields, in particular those resulting from nuclear magnetic moments or (in the case of ferromagnetic or antiferromagnetic metals) ordered electronic magnetic moments (Sect. 13.2.3). These interactions affect the precession rate of the muon spins or the loss of the muon spin polarization in transverse or longitudinal magnetic fields (Sect. 13.4) and are detected by means of the non-conservation of parity in the muon decay (Sects. 13.2.1 and 13.3).

2) In spite of extensive search, no evidence has been found for formation of muonium atoms in metals (Sect. 13.4.5). This suggests that hydrogen is dissolved in metals not in the form of atoms but as protons, deuterons, or tritons.

3) In copper, positive muons are located on octahedral interstitial sites, in agreement with the present knowledge on hydrogen in fcc metals. The distance of the nearest neighbour atoms from this site is increased by about 5% by the presence of the muons (Sect. 13.5.1a).

4) In nickel and the fcc phase of cobalt, positive muons must be located on sites of cubic symmetry, i.e., in either octahedral or tetrahedral interstices (Sects. 13.5.2b and c), in accord with what is known on hydrogen in fcc metals.

5) In the bcc metals V, Nb, Ta, and α-Fe, the available evidence is in favour of muon location in tetrahedral interstices, but the experimental data are not yet complete and a—possibly part-time—location on octahedral interstices cannot be completely ruled out at present (Sects. 13.5.1b and 13.5.2a). These results are in agreement with the experiments on hydrogen in these metals. It is to be expected that in the near future definitive results on the muon location in the above-mentioned group of bcc metals will be obtained.

6) In the hexagonal metals Co and Gd, location of positive muons in tetrahedral interstices can be ruled out with a high degree of certainty. There is fairly strong evidence that in both metals positive muons are located in octahedral interstices (Sects. 13.5.2c and d). In view of the good accord between muon and hydrogen results on fcc metals [comp. 3), 4)], these conclusions are likely to apply to hydrogen in hcp metals as well.

7) The enhancement of the conduction electron density at the muon location in non-ferromagnetic metals is rather large. A quantitative comparison between positive muons and hydrogen has been possible by means of the Knight shift measurements on Pd and gives reasonable agreement (Sect. 13.4.3). The effect of the muons on the conduction electron spin density at the muon locations in ferromagnetic metals varies strongly from metal to metal. It is small in Ni (Sect. 13.5.2b), intermediate in Co (Sect. 13.5.2c), and very large in α-Fe (Sect. 15.5.2a).

8) Muon diffusion in Cu between 85 K and 250 K can be well described by 'incoherent' tunnelling between the ground states in neighbouring octahedral interstices. Above this temperature range the available data appear to join smoothly to the extrapolation of the hydrogen diffusion coefficients measured at high temperatures (Sect. 13.5.1a).

9) The muon diffusivity in Al is much higher than that in Cu under comparable conditions (Sect. 13.5.1c). Muons in Al are very mobile down to the lowest temperatures investigated (1.2 K). The relationship to hydrogen diffusion, which at higher temperatures is similar to that in Cu, is not clear.

10) Except for liquid-helium–liquid-hydrogen temperatures, the muon diffusivities D^{μ^+} in V, Nb, Ta, and α-Fe deduced from measurements of the transverse relaxation rate are *lower* than the extrapolations of hydrogen diffusivities from high-temperature measurements (Sects. 13.5.1b and 13.5.2a). This indicates that in the temperature range between room temperature and liquid-hydrogen temperature, muon trapping by impurities and other imperfections is important. The same is likely to be true for hydrogen in these metals. At lower temperatures, muon trapping may become unimportant because D^{μ^+} may be too small for a significant fraction of the muons to reach trapping sites before they decay. Muon studies under non-trapping conditions should be a useful guide to the low-temperature diffusion behaviour of hydrogen in these metals.

11) The muon diffusivity in Cr is similar to the hydrogen diffusion coefficient of Ta in a comparable temperature range (Sect. 13.5.2e). This may serve as a first indication as to the order of magnitude of the hydrogen diffusivity in Cr.

12) In a number of metals (Bi, Group-V transition metals), indications have been found that at low temperatures 'coherent' tunnelling (i.e., tunnelling without accompanying change of the phonon-states occupation) may take place, but clarifying experiments are still required (Sects. 13.5.1b and c).

References

13.1 C.D.Anderson, S.H.Neddermeyer: Phys. Rev. **50**, 263 (1936)
13.2 C.D.Anderson, S.H.Neddermeyer: Phys. Rev. **51**, 884 (1937)
13.3 C.D.Anderson, S.H.Neddermeyer: Phys. Rev. **54**, 88 (1938)
13.4 J.C.Street, E.C.Stevenson: Phys. Rev. **51**, 1005 (1937)
13.5 C.D.Anderson: Am. J. Phys. **29**, 825 (1961)
13.6 A.O.Weissenberg: *Muons* (North-Holland, Amsterdam 1967)
13.7 R.E.Marshak: *Meson Physics* (McGraw-Hill, New York 1952)
13.8 H.Yukawa: Proc. Phys.-Math. Soc. Japan **17**, 48 (1935)
13.9 M.Conversi, E.Pancini, O.Piccioni: Phys. Rev. **71**, 209 (1947)
13.10 C.M.G.Lattes, G.P.S.Occhialini, C.F.Powell: Nature (London) **160**, 435 (1947)
13.11 C.F.Powell: Rep. Progr. Phys. **13**, 350 (1950)
13.12 E.Segré: *Nuclei and Particles* (W.A.Benjamin, New York, Amsterdam 1964)
13.13 M.L.Perl and other authors: Phys. Rev. Lett. **35**, 1489 (1975)
13.14 J.J.Sakurai: Progr. Nucl. Phys. **7**, 234 (1959)
13.15 G.Feinberg, L.M.Lederman: Ann. Rev. Nucl. Sci. **13**, 431 (1963)
13.16 H.Weyl: Z. Physik **56**, 330 (1929)
13.17 W.Pauli: In *Handbuch der Physik*, Bd. 24/1, 2. Aufl., ed. by H.Geiger and K.Scheel (Springer, Berlin 1933)
13.18 T.D.Lee, C.N.Yang: Phys. Rev. **105**, 1671 (1957)
13.19 L.D.Landau: Zh. Eksp. Teor. Fiz. **32**, 407 (1957)
13.20 L.D.Landau: Nucl. Phys. **3**, 127 (1957)
13.21 A.Salam: Nuovo Cimento **5**, 299 (1957)
13.22 C.S.Wu: "The Neutrino" in *Theoretical Physics in the Twentieth Century*, ed. by M.Fierz and V.F.Weisskopf (Interscience, New York 1960)
13.23 J.C.Taylor: Rept. Progr. Phys. **27**, 407 (1964)
13.24 T.D.Lee, C.S.Wu: Ann. Rev. Nucl. Sci. **15**, 381 (1965)
13.25 C.Rubbia: In *High Energy Physics*, Vol. III, ed. by E.H.S.Burhop (Academic Press, New York, London 1969) p. 238
13.26 C.P.Slichter: *Principles of Magnetic Resonance* (Springer, Berlin, Heidelberg, New York 1978)
13.27 E.Fermi: Z. Physik **60**, 320 (1930)
13.28 J.H.Brewer, K.M.Crowe, F.N.Gygax, A.Schenk: In *Muon Physics*, Vol. III, ed. by V.W.Hughes and C.S.Wu (Academic Press, New York 1975)
13.29 J.Winter: *Magnetic Resonance in Metals* (Oxford University Press, London 1971)
13.30 A.Abragam: *The Principles of Nuclear Magnetism* (Oxford University Press, London 1961)
13.31 M.Kawakami, T.Hirota, T.Koi, T.Wakiyama: J. Phys. Soc. Japan **33**, 1591 (1972)
13.32 R.Kubo, K.Tomita: J. Phys. Soc. Japan **9**, 888 (1954)
13.33 P.Jung: Unpublished
13.34 I.I.Gurevich, E.A.Mel'eshko, I.A.Muratova, B.A.Nikol'skiĭ, V.S.Roganov, V.I.Selivanov, B.V.Sokolov: Phys. Lett. **40** A, 143 (1972)
13.35 J.H.van Vleck: Phys. Rev. **74**, 1168 (1948)
13.36 N.Bloembergen, E.M.Purcell, R.V.Pound: Phys. Rev. **73**, 679 (1948)
13.37 H.C.Torrey: Phys. Rev. **92**, 962 (1953)
13.38 D.Wolf: *Spin Temperature and Nuclear Spin Relaxation in Matter: Basic Principles and Applications* (Oxford University Press, London) in press
13.39 O.Hartmann: Phys. Rev. Lett. **39**, 832 (1977)
13.40 P.F.Meier: Helv. Phys. Acta **48**, 228 (1975)
13.41 E.Mann: Unpublished
13.42 A.Schenck: Hyperfine Interactions **4**, 282 (1978)
13.43 H.A.Lorentz: *The Theory of Electrons* (B.G.Teubner, Leipzig 1909)
13.44 W.F.Brown,Jr.: In *Encyclopedia of Physics*, Vol. XVII, ed. by S.Flügge (Springer, Berlin, Göttingen, Heidelberg 1957) p. 1

13.45 K.-H. Hellwege: *Einführung in die Festkörperphysik* (Springer, Berlin, Heidelberg, New York 1976) Sect. 24
13.46 H. Kronmüller, H.-R. Hilzinger, P. Monachesi, A. Seeger: To be published
13.47 M.B. Stearns: Phys. Lett. **47** A, 397 (1974)
13.48 A. Seeger: To be published
13.49 V.N. Gorelkin, V.P. Smilga: Soviet Phys.—JETP **42**, 482 (1976)
13.50 V.G. Grebinnik, I.I. Gurevich, V.A. Zhukov, I.G. Ivanter, A.P. Manych, B.A. Nikol'skiĭ, V.I. Selivanov, V.A. Suetin: JETP Lett. **22**, 16 (1975)
13.51 V.G. Grebinnik, I.I. Gurevich, V.A. Zhukov, I.G. Ivanter, A.P. Manych, B.A. Nikol'skiĭ, V.I. Selivanov, V.A. Suetin: JETP Lett. **23**, 8 (1976)
13.52 V.G. Grebinnik, I.I. Gurevich, V.A. Zhukov, A.P. Manych, E.A. Meleshko, I.A. Muratova, B.A. Nikol'skiĭ, V.I. Selivanov, V.A. Suetin: Soviet Phys.—JETP **41**, 777 (1976)
13.53 M. Camani, F.N. Gygax, W. Rüegg, A. Schenck, H. Schilling: Phys. Rev. Lett. **39**, 836 (1977)
13.54 L. Katz, M. Guinan, R.J. Borg: Phys. Rev. B **4**, 330 (1971)
13.55 G. Perkins, D.R. Begeal: Ber. Bunsenges. Phys. Chem. **76**, 863 (1972)
13.56 D.B. Butrymowicz, J.R. Manning, M.E. Read: J. Phys. Chem. Ref. Data **4**, 201 (1975)
13.57 H. Teichler: Phys. Lett. **64**A, 78 (1977)
13.58 C.P. Flynn, A.M. Stoneham: Phys. Rev. B **1**, 3966 (1970)
13.59 A. Seeger, H. Teichler: In *La diffusion dans les milieux condensés — théories et applications*, Vol. I (Centre d'Études Nucleaires de Saclay, I.N.S.T.N. 1976) p. 217
13.60 W. Eichenauer, W. Loser, H. Witte: Z. Metallk. **48**, 373 (1957)
13.61 O. Hartmann, E. Karlsson, K. Pernestål, M. Borghini, T.O. Niinikoski, L.-O. Norlin: Phys. Lett. **61** A, 141 (1977)
13.62 P. Jena, K.S. Singwi: Phys. Rev. B **17**, 3518 (1978)
13.63 H. Teichler: Phys. Lett. (in press)
13.64 A.T. Fiory, K.G. Lynn, D.M. Parkin, W.J. Kossler, W.F. Langford, C.E. Stronach: Phys. Rev. Lett. **40**, 968 (1978)
13.65 O. Hartmann, E. Karlsson, L.O. Norlin, K. Pernestål, M. Borghini, T. Niinikoski, E. Walker: Hyperfine Interactions **4**, 824 (1978)
13.66 R.H. Heffner, W.B. Gauster, D.M. Parkin, C.Y. Huang, R.L. Hutson, M. Leon, M.E. Schillaci, M.L. Simmons, W. Triftshäuser: Hyperfine Interactions **4**, 838 (1978)
13.67 A.T. Fiory, D.E. Murnick, M. Leventhal, W.J. Kossler: Phys. Rev. Lett. **33**, 969 (1974)
13.68 W.F. Lankford, H.K. Birnbaum, A.T. Fiory, R.P. Minnich, K.G. Lynn, C.E. Stronach, L.H. Bieman, W.J. Kossler, J. Lindemuth: Hyperfine Interactions **4**, 833 (1978)
13.69 K. Dorenburg, M. Gladisch, D. Herlach, M. Krenke, H. Metz, H. Orth, G. zu Putlitz, A. Seeger, H. Teichler, M. Wigand: Schweizerisches Institut für Nuklearforschung, Jahresbericht 1977, E 62
13.70 M. Camani, F.N. Gygax, W. Rüegg, A. Schenck, H. Schilling: Schweizerisches Institut für Nuklearforschung, Jahresbericht 1977, E 58
13.71 A. Seeger: Phys. Lett. **53** A, 324 (1975)
13.72 H.K. Birnbaum, C.P. Flynn: Phys. Rev. Lett. **37**, 25 (1976)
13.73 V.G. Grebinnik, I.I. Gurevich, V.A. Zhukov, A.I. Klimov, V.N. Maĭorov, A.P. Manych, E.V. Mel'nikov, B.A. Nikol'skii, A.V. Pirogov, A.N. Ponomarev, V.I. Selivanov, V.A. Suetin: JETP Lett. **25**, 298 (1977)
13.74 A.F. Andreev, I.M. Lifshitz: Soviet Phys.—JETP **29**, 1107 (1969)
13.75 W.J. Kossler, A.T. Fiory, D.E. Murnick, C.E. Stronach, W.F. Lankford: Hyperfine Interactions **3**, 287 (1977)
13.76 A.T. Fiory, D.E. Murnick, R.P. Minnich, W.J. Kossler: Solid State Commun. **21**, 747 (1977)
13.77 M.L. Foy, N. Heiman, W.J. Kossler, C.E. Stronach: Phys. Rev. Lett. **30**, 1064 (1973)
13.78 N. Heiman, M.L.G. Foy, W.J. Kossler, C.E. Stronach: "Magnetism and Magnetic Materials—1973", in *AIP Conference Proceedings*, No. 18 (American Institute of Physics, New York 1974) p. 525
13.79 I.I. Gurevich, A.N. Klimov, V.N. Maĭorov, E.A. Meleshko, B.A. Nikol'skiĭ, V.S. Roganov, V.I. Selivanov, V.A. Suetin: JETP Lett. **18**, 332 (1973)

13.80 I.I.Gurevich, A.I.Klimov, V.N.Maĭorov, E.A.Meleshko, B.A.Nikol'skiĭ, A.V.Pirogov, V.I.Selivanov, V.A.Suetin: JETP Lett. **20**, 254 (1974)
13.81 I.I.Gurevich, A.I.Klimov, V.N.Maĭorov, E.A.Meleshko, I.A.Muratova, B.A.Nikol'skiĭ, V.S.Roganov, V.I.Selivanov, V.A.Suetin: Soviet Phys.—JETP **39**, 178 (1974)
13.82 I.I.Gurevich, A.I.Klimov, V.N.Maĭorov, E.A.Meleshko, B.A.Nikol'skiĭ, V.I.Selivanov, V.A.Suetin: Soviet Phys.—JETP **42**, 222 (1976)
13.83 N.Nishida, R.S.Hayano, K.Nagamine, T.Yamazaki, J.H.Brewer, D.M.Garner, D.G.Fleming, T.Takeuchi, Y.Ishikawa: Solid State Commun. **22**, 235 (1977)
13.84 H.Graf, W.Kündig, B.D.Patterson, W.Reichart, P.Roggwiller, M.Camani, F.N.Gygax, W.Rüegg, A.Schenck, H.Schilling: Helv. Phys. Acta **49**, 730 (1976)
13.85 A.Seeger: Phys. Lett. **58** A, 137 (1976)
13.86 C.G.Shull: In *Magnetic and Inelastic Scattering of Neutrons by Metals*, ed. by T.J.Rowland and P.A.Beck (Gordon and Breach, New York 1968) p. 15
13.87 P.Jena: Solid State Commun. **19**, 45 (1976)
13.88 P.Jena, K.S.Singwi, R.M.Nieminen: Phys. Rev. B **17**, 301 (1978)
13.89 T.Yamazaki: *Proc. Intern. Conf. on Magnetism* 1976, ed by P.F. de Châtel and J.J.M. Franse, pt. III (North-Holland, Amsterdam 1977) p. 1053
13.90 J.Völkl, G.Alefeld: "Hydrogen Diffusion"; in *Diffusion in Solids, Recent Developments*, ed. by A.S.Nowick and J.J.Burton (Academic Press, London 1975)
13.91 J.L.Dillard, S.Talbot-Besnard: *L'hydrogène dans les métaux*, Vol. 1 (Editions Science et Industrie, Paris 1972) p. 159
13.92 J.F.Dufresne, A.Seeger, P.Groh, P.Moser: Phys. Stat. Sol. (a) **36**, 579 (1976)
13.93 B.D.Patterson, K.M.Crowe, F.N.Gygax, R.F.Johnson, A.M.Portis, J.H.Brewer: Phys. Lett. **46** A, 453 (1974)
13.94 K.Nagamine, S.Nagamiya, O.Hashimoto, N.Nishida, T.Yamazaki, B.D.Patterson: Hyperfine Interactions **1**, 517 (1976)
13.95 M.Camani, F.N.Gygax, W.Rüegg, A.Schenck, H.Schilling: Phys. Lett. **60** A, 439 (1977)
13.96 W.J.Kossler: In *Magnetism and Magnetic Materials*—1974, ed. by C.D.Graham,Jr., G.H.Lander and J.J.Rhyne, AIP Conf. Proceedings No. 24 (American Institute of Physics, New York 1975)
13.97 B.D.Patterson, L.M.Falicov: Solid State Commun. **15**, 1509 (1974)
13.98 K.G.Petzinger, R.Munjal: Phys. Rev. B **15**, 1560 (1977)
13.99 H.Hamagaki, K.Nakai, Y.Nojiri, I.Tanihata, K.Sugimoto: Hyperfine Interactions **2**, 187 (1976)
13.100 K.G.Petzinger: Hyperfine Interactions **4**, 307 (1978)
13.101 Y.Barnier, R.Pauthenet, G.Rimet: Compt. Rend. **252**, 3024 (1961)
13.102 H.Träuble, O.Boser, H.Kronmüller, A.Seeger: Phys. Stat. Sol. **10**, 283 (1965)
13.103 H.Kronmüller, H.Träuble, A.Seeger, O.Boser: Mat. Sci. Eng. **1**, 91 (1966)
13.104 H.Graf, W.Kündig, B.D.Patterson, W.Reichart, P.Roggwiller, M.Camani, F.N.Gygax, W.Rüegg, A.Schenck, H.Schilling, P.F.Meier: Phys. Rev. Lett. **37**, 1644 (1976)
13.105 N.Nishida, K.Nagamine, R.S.Hayano, T.Yamazaki, D.G.Fleming, R.A.Duncan, J.H.Brewer, A.Ahktar, H.Yasuoka: Hyperfine Interactions **4**, 318 (1978)
13.106 I.I.Gurevich, A.I.Klimov, V.N.Maĭorov, E.A.Meleshko, B.A.Nikol'skiĭ, A.V.Purogov, V.S.Roganov, V.I.Selivanov, V.A.Suetin: Soviet Phys.—JETP **42**, 741 (1976)
13.107 H.Graf, W.Hofmann, W.Kündig, P.F.Meier, B.D.Patterson, W.Reichart: Solid State Commun. **23**, 653 (1977)
13.108 G.Alefeld, J.Völkl (eds.): *Hydrogen in Metals II, Application-Oriented Properties*, Topics in Applied Physics, Vol. 29 (Springer, Berlin, Heidelberg, New York 1978)

Contents of **Hydrogen in Metals II**

Application-Oriented Properties (Topics in Applied Physics, Vol. 29)

Additional References with Titles

Chapter 5

D.A. Papaconstantopoulos, B.M. Klein, E.N. Economou, L.L. Boyer: Band structure and superconductivity of PdD_x and PdH_x. Phys. Rev. B **17**, 141 (1978)

M. Gupta, A.J. Freeman: Electronic structure and proton spin-lattice structure of PdH. Phys. Rev. B **17**, 3029 (1978)

N. Kulikov, A. Zvonkov, O. Mayluchkov: "The Calculation of the Band Structure and Some Electronic Properties of Titanium and Nickel Hydride", Second International Conference on Hydrogen in Metals **2A1**, Paris June 6–11, 1977

H. Wenzl, J.-M. Welter: "Properties and Preparation of Nb–H Interstitial Alloys", in *Current Topics in Materials Science*, ed. by E. Kaldis, Vol. 1 (North-Holland, Amsterdam 1978)

T. Schober, H. Wenzl: "The Systems NbH(D), TaH(D), VH(D): Structures, Phase Diagrams, Morphologies, Methods of Preparation", in *Hydrogen in Metals II, Application-Oriented Properties*, ed. by G. Alefeld and J. Völkl, Topics in Applied Physics, Vol. 29 (Springer, Berlin, Heidelberg, New York 1978) Chap. 2

J.H. Weaver, D.T. Peterson: The influence of hydrogen on the bulk band structure of Nb and Ta: An optical study of $NbH_{0.453}$ and $TaH_{0.257}$. Phys. Lett. **62**A, 433 (1977)

J.H. Weaver, J.A. Knapp, D.E. Eastman, D.T. Peterson, C.B. Satterthwaite: Electronic structure of thorium hydrides ThH_2 and Th_4H_{15}. Phys. Rev. Lett. **39**, 639 (1977)

Chapter 8

J. Völkl, H.C. Bauer, U. Freudenberg, M. Kokkinidis, G. Lang, K.-A. Steinhauser, G. Alefeld: "Isotope and Structure Dependence of the Diffusion Coefficient of Hydrogen in Metals"; in *Internal Friction and Ultrasonic Attenuation in Solids*, Proceedings of the VIth International Conference on Internal Friction and Ultrasonic Attenuation in Solids July 4–7, 1977, Tokyo, ed. by R.R. Hasiguti and N. Mikoshiba (University of Tokyo Press, Tokyo 1977) p. 485

A. Madhukar, W. Post: Exact solution for the diffusion of a particle in a medium with site diagonal and off-diagonal dynamic disorder. Phys. Rev. Lett. **39**, 1424 (1977)

H. Teichler: On the theory of muon diffusion in metals. Phys. Lett. **64**A, 78 (1977)

T. Holstein: Quantal occurrence-probability treatment of smallpolaron hopping. Phil. Mag. B **37**, 49 (1978)

L.B. Schein, C.B. Duke, A.R. McGhie: Observation of the band-hopping transition for electrons in naphtalene. Phys. Rev. Lett. **40**, 197 (1978)

C. Morkel, K. Neumaier, H. Wipf: Nitrogen-hydrogen interstitial pair in niobium as a new system showing atomic tunneling. Phys. Rev. Lett. **40**, 947 (1978)

Chapter 12

J.J. Au, H.K. Birnbaum: Magnetic relaxation studies of the motion of hydrogen and deuterium in iron. Acta Met. **26**, 1105 (1978)

V. Heintze: Anwendung der Kernreaktionsmethode zur Untersuchung der Diffusion von Deuteronen in mono- und polykristallinen Nickel-Absorbern. Z. Phys. B **27**, 133 (1977)

B. Huber, G. Sicking: Tritium diffusion in palladium alloys with Pt and Cu. Phys. Stat. Sol. (a) **47**, K85 (1978)

Y.S. Hwang, D.R. Torgeson, A.S. Khan, R.G. Barnes: Hydrogen locations and motions in the metallic-hydride system Ta_6SH_x. I. Temperature dependence of the proton NMR linewidth and second moment. Phys. Rev. B **15**, 4564 (1977)

A. J. Kumnick, H. H. Johnson: Hydrogen and deuterium in iron, 9–73 °C. Acta Met. **25**, 891 (1977)
K. F. Lau, R. W. Vaughan, C. B. Satterthwaite: NMR study of thorium hydride (Th_4H_{15}). Phys. Rev. B **15**, 2449 (1977)
K. Papp, E. Kovács-Csetényi: Diffusion of hydrogen in solid aluminum. Scr. Met. **11**, 921 (1977)
N. R. Quick, H. H. Johnson: Hydrogen and deuterium in iron, 49–506 °C. Acta Met. **26**, 903 (1978)
S. Tanaka, J. D. Clewly, T. B. Flanagan: Kinetics of hydrogen absorption by $LaNi_5$. J. Phys. Chem. **81**, 1684 (1977)
J. Töpfler, E. Lebsanft, R. Schätzler: Determination of the hydrogen diffusion coefficient in Ti_2Ni by means of quasielastic neutron scattering. J. Phys. F: Metal Phys. **8**, L25 (1978)
J. Völkl, H. C. Bauer, U. Freudenberg, M. Kokkinidis, G. Lang, K.-A. Steinhauser, G. Alefeld: "Isotope and Structure Dependence of the Diffusion Coefficient of Hydrogen in Metals", in *Internal Friction and Ultrasonic Attenuation in Solids*, Proceedings of the VIth International Conference on Internal Friction and Ultrasonic Attenuation in Solids, July 4–7, 1977, Tokyo, ed. by R. R. Hasiguti and N. Mikoshiba (University of Tokyo Press, Tokyo 1977) p. 485
M. Yamaguchi, I. Yamamoto, T. Ohta: The motion of hydrogen in Mg_2NiH_x studied by proton magnetic resonance. Phys. Lett. **66**A, 147 (1978)

Chapter 13

V. G. Baryshevskii, S. A. Kuten': Theory of μ^+ relaxation in metals. Sov. Phys.–Solid State **18**, 1677 (1976)
H. K. Birnbaum, M. Camani, A. T. Fiory, F. N. Gygax, W. J. Kossler, W. Rüegg, A. Schenck, H. Schilling: Anomalous temperature dependence in the depolarization rate of positive muons in pure niobium. Phys. Rev. B**17**, 4143 (1978)
M. Borghini, O. Hartmann, E. Karlsson, K. W. Kehr, T. O. Niinikoski, L. O. Norlin, K. Pernestål, D. Richter, J. C. Soulé, E. Walker: Muon diffusion in niobium in the presence of traps. Phys. Rev. Lett. **40**, 1723 (1978)
D. K. Brice: Lattice atom displacements produced near the end of implanted μ^+ tracks. Phys. Lett. **66**A, 53 (1978)
K. Dorenburg, M. Gladisch, D. Herlach, M. Mansel, H. Metz, H. Orth, G. zu Putlitz, A. Seeger, W. Wahl, M. Wigand: Trapping by vacancies and mobility of positive muons in neutron-irradiated aluminum. Z. Physik B **31** (2) (1978)
A. T. Fiory, K. G. Lynn, D. M. Parkin, W. J. Kossler, W. F. Lankford, C. E. Stronach: Diffusion of positive muons in vanadium. Phys. Rev. Lett. **40**, 1214 (1978) (Erratum or Ref. 13.64)
S. Fujii, Y. Uemura: Theory of spin depolarization rate of μ^+SR in pure Fe Solid State Commun. (in press)
W. B. Gauster, A. T. Fiory, K. G. Lynn, W. J. Kossler, D. M. Parkin, C. E. Stronach, W. F. Lankford: On the study of defects in metals with positive muons. J. Nucl. Mater. **69**, **70**, 147 (1978)
W. B. Gauster, R. H. Heffner, C. Y. Huang, R. L. Hutson, M. Leon, D. M. Parkin, M. E. Schillaci, W. Triftshäuser, W. R. Wampler: Measurement of the depolarization rate of positive muons in copper and aluminum. Solid State Commun. **24**, 619 (1977)
P. Jena, Sh. G. Das, K. S. Singwi: Electric-field gradient at Cu nuclei due to an interstitial positive muon. Phys. Rev. Lett. **40**, 264 (1978)
T. McMullen, E. Zaremba: Positive muon spin depolarization in solids. Phys. Rev. B (in press)
K. Nagamine, N. Nishida, S. Nagamiya, O. Hashimoto, T. Yamazaki: Conduction-electron polarization in very dilute **Pd**Fe alloys studied by positive muons. Phys. Rev. Lett. **38**, 99 (1977)
N. Nishida, K. Nagamine, R. S. Hayano, T. Yamazaki, D. G. Fleming, R. A. Duncan, J. H. Brewer, A. Ahktar, H. Yasuoka: The local magnetic fields probed by μ^+ in hcp ferromagnets: Co and Gd. J. Phys. Soc. Jpn. **44**, 1131 (1978)
A. Schenck: "On the Application of Polarized Positive Muons in Solid State Physics", in *Nuclear and Particle Physics at Intermediate Energies (1976)*, ed. by J. B. Warren (Plenum Press, New York 1976)
H. Schilling, M. Camani, F. N. Gygax, W. Rüegg, A. Schenck: Diffusion of positive muons in single crystals of tantalum. Submitted to Phys. Lett.

Subject Index

Dynamics of Solids and Liquids by Neutron Scattering

Editors: S.W. Lovesey, T. Springer

1977. 156 figures, 15 tables. XI, 379 pages
(Topics in Current Physics, Volume 3)
ISBN 3-540-08156-9

Contents:
S.W. Lovesey: Introduction. – *H.G. Smith, N. Wakabayashi:* Phonons. – *B. Dorner, R. Comès:* Phonons and Structural Phase Transformations. – *J.W. White:* Dynamics of Molecular Crystals, Polymers, and Adsorbed Species. – *T. Springer:* Molecular Rotations and Diffusion in Solids, in Particular Hydrogen in Metals. – *R.D. Mountain:* Collective Modes in Classical Monoatomic Liquids. – *S.W. Lovesey, J.M. Loveluck:* Magnetic Scattering.

Neutron Diffraction

Editor: H. Dachs

1978. 111 figures, 32 tables. Approx. 370 pages
(Topics in Current Physics, Volume 6)
ISBN 3-540-08710-9

Contents:
H. Dachs: Principles of Neutron Diffraction. – *J.B. Hayter:* Polarized Neutrons. – *P. Coppens:* Combining X-Ray and Neutron Diffraction: The Study of Charge Density Distributions in Solids. – *W. Prandl:* The Determination of Magnetic Structures. – *W. Schmatz:* Disordered Structures. – *P.-A. Lindgård:* Phase Transitions and Critical Phenomena. – *G. Zaccai:* Application of Neutron Diffraction to Biological Problems. – *P. Chieux:* Liquid Structure Investigation by Neutron Scattering. – *H. Rauch, D. Petrascheck:* Dynamical Diffraction on Perfect Crystals and its Application in Neutron Physics.

Neutron Physics

1977. 40 figures, 11 tables. VII, 135 pages
(Springer Tracts in Modern Physics, Volume 80)
ISBN 3-540-08022-8

Contents:
I. Koester: Neutron Scattering Lengths and Fundamental Neutron Interactions. – *A. Steyerl:* Very Low Energy Neutrons.

X-Ray Optics

Applications to Solids

Editor: H.-J. Queisser

1977. 133 figures, 14 tables. XI, 227 pages
(Topics in Applied Physics, Volume 22)
ISBN 3-540-08462-2

Contents:
H.-J. Queisser: Introduction. – Structure and Structuring of Solids. – *M. Yoshimatsu, S. Kozaki:* High Brilliance X-Ray Sources. – *E. Spiller, R. Feder:* X-Ray Lithography. – *U. Bonse, W. Graeff:* X-Ray and Neutron Interferometry. *A. Authier:* Section Topography. *W. Hartmann:* Live Topography.

Springer-Verlag
Berlin
Heidelberg
New York